Northern Sustainable Development Forum 2020

SHS Web of Conferences Volume 112 (2021)

Yakutsk, Russia
27-30 September 2020

Editor:

Daryana Maksimova

ISBN: 978-1-7138-3105-1

Printed from e-media with permission by:

Curran Associates, Inc.
57 Morehouse Lane
Red Hook, NY 12571

Some format issues inherent in the e-media version may also appear in this print version.

No changes have been made to the content of these proceedings. There may be changes to pagination and minor adjustments for aesthetics.

Printed with permission by Curran Associates, Inc. (2021)

For additional information, please contact EDP Sciences – Web of Conferences at the address below.

EDP Sciences – Web of Conferences
17, Avenue du Hoggar
Parc d'Activité de Courtabœuf
BP 112
F-91944 Les Ulis Cedex A
France

Phone: +33 (0) 1 69 18 75 75
Fax: +33 (0) 1 69 28 84 91

contact-edps@webofconferences.org

Additional copies of this publication are available from:

Curran Associates, Inc.
57 Morehouse Lane
Red Hook, NY 12571 USA
Phone: 845-758-0400
Fax: 845-758-2633
Email: curran@proceedings.com
Web: www.proceedings.com

TABLE OF CONTENTS

Northern Sustainable Development Forum

September 27-30, 2020, Yakutsk, Republic of Sakha (Yakutia), Russia

From 27 to 30 September, the 2nd Northern Sustainable Development Forum was held. The event had brought to Yakutsk more than 100 guests and many more participated via videoconferencing tools. Dozens of international participants from circumpolar countries such as Iceland, Canada, Finland, and Norway were registered at the events and enjoy workshops, discussions, and virtual exhibitions.

This volume comprised of the selected reviewed papers dedicated to the social, cultural, and economic wellbeing of the inhabitants of the North. Keynote papers come from renowned ethnographers, policymakers and dedicated to the culture and modern lifestyle of indigenous people of Northern Russia.

Organizing Institutions

- Northern Forum (https://www.northernforum.org/en/)
- Government of Yakutia (https://www.sakha.gov.ru/)
- North-Eastern Federal University (https://www.s-vfu.ru/en/)

In cooperation with local institutions

- Academy of Sciences of Yakutia
- Yakutsk Scientific Centre of Siberian Branch of Russian Academy of Science
- Arctic State Institute of Culture and Arts
- Association of Russian Social Geographers (Yakutsk Regional Branch)
- Yakutsk State Agricultural Academy
- Arctic College of Northern Peoples
- Philharmonic of Yakutia

In cooperation with Russian institutions

- Arctic Development Project Office
- Russian International Affairs Council
- Saint-Petersburg State University
- Surgut State Pedagogical University
- Siberian Federal University

In international cooperation with

- International Arctic Scientific Committee
- University of the Arctic
- Association of Reindeer Herders of the World
- International Freezing Association Education Committee
- Association of Mayors of World Winter Cities
- Oceanological University of China
- Korea Maritime Institute
- Hokkaido University
- Hokkaido Think Tank Association
- Northern University of Norway
- University of Alaska Fairbanks
- Linking Tourism & Conservation

About the forum

Goal

Organizing a comprehensive discussion of the challenges and prospects of implementing the principles and goals of sustainable development in the regions of the Arctic and the North.

Objectives

- discussing the opportunities and prospects of the interaction of the Arctic and northern regions in solving the problems of sustainable development;
- discussing the role, functions, and tasks of the regional authorities, academia and education, business and public organizations in supporting the sustainable development of the Arctic and northern regions;
- assisting with specifying the priority areas for the socio-economic development of the Arctic and northern regions;
- analyzing the ways for improving the living standards of the northern residents, including the indigenous peoples of the Arctic.

Official Languages

The official working languages are Russian and English. Simultaneous Russian-English and English-Russian interpretation will be provided at the Forum's plenary, business sessions, and for keynotes reports.

Expected Outcomes

- adopting a general resolution on the Forum's main theme;
- signing the cooperation agreements;
- reviewing the projects proposed for the implementation within the Northern Forum;
- publishing the materials of some research and practical events of the Forum in the indexed journal;
- posting the Forum's materials (papers, presentations, recommendations, final reports) on the Forum's official website www.nsdf.ru;
- establishing international interregional cooperation and exchange of information, best practices, and experiences in the sustainable development of the Arctic and the North, including the Northern Forum participants.

Expected Participants

- employees and representatives of public authorities in local and federal levels;
- employees and representatives of international organizations;
- employees and representatives of non-commercial and public organizations;
- representatives of diplomatic missions and trade representatives of foreign countries;
- academics and educators;
- professionals, employees of national and international companies.

Programme Committee

Global Partnership for Northern Development

Chairman

Anatoly Nikolaev	*Dr. of Biolology, Rector of North-Eastern Federal University*

Regular Members

Alexander Vorotnikov	*Cand. of Chemistry, Coordinator of Project Office for the Development of the Arctic*
Lars Kullerud	*PhD., President of the University of the Arctic*
Guo Peiqing	*PhD., Oceanological University of China*
Yuri Grigoriev	*Dr. of Physcis, Vice-rector for Internationalization of North-Eastern Federal University*
Yulia Zaika	*Head of the International Department, Moscow State University*
Jong Di Kim	*PhD., Head of Strategic Development of Korea Maritime Institute*
Vladimir Krever	*Academic Supervisor of the WWF Russia Biodiversity Conservation Program*
Vladlen Kugunurov	*Cand. of Pedagogy, Vice-Rector for International Cooperation of North-Eastern Federal University*
Lassi Heininen	*PhD., Coordinator of Arctic Yearbook Project, University of Helsinky*
Daryana Maksimova	*Cand. of Politics, Deputy Executive Director of Northern Forum*
Alexander Pilyasov	*Dr. of Geology, Chairman of the Russian Section of the European Regional Science Association*
Sugimoto Atsuko	*PhD., Professor of Arctic Research Center in Hokkaido University*
Nadezhda Kharlampieva	*Cand. of History, Senior Researcher of Arctic and Antarctic Research Institute in Saint Petersburg State University*
Mikhail Sworn	*Dr. of Geology, First Deputy Minister of Education and Science of Yakutia*
Trofim Maksimov	*Head of laboratory in the Institute of Biological Problems of the Permafrost Zone of Siberian Branch of Russian Academy of Science*

Transport Infrastructure in the North and in the Arctic

Chairman

Anatoly Nikolaev	*Dr. of Biolology, Rector of North-Eastern Federal University*

Regular Members

Guo Peiqing	*PhD., Oceanological University of China*
Yuri Danilov	*Cand. of Geography, Vice-Rector for Science and Innovation of North-Eastern Federal University*
Jong Di Kim	*PhD., Head of Strategic Development of Korea Maritime Institute*
Alexander Kugaevsky	*Cand. of Economics, Vice-Rector for Analytics and Expertise of North-Eastern Federal University*
Daryana Maksimova	*Cand. of Politics, Deputy Executive Director of Northern Forum*
Frédéric Lasserre	*PhD., Director of the Quebec Council for Geopolitical Research*
Victor Razbegin	*Director of the Russian Branch of the Transcontinental Consortium Ltd.*
Rodion Stepanov	*Head of the Directorate for Arctic programs of the Moscow State Technical University*
Dmitry Filippov	*PhD., Dean of the Road Faculty of North-Eastern Federal University*

Management of the Arctic territories

Chairman

Anatoly Nikolaev	*Dr. of Biolology, Rector of North-Eastern Federal University*

Regular Members

Alexander Vorotnikov	*Cand. of Chemistry, Coordinator of Project Office for the Development of the Arctic*
Evgeny Gololobov	*Dr. of History, Vice Rector for Research of Surgut State Pedagogical University*
Yuri Zakharinsky	*PhD., Head of the Department for Comprehensive Arctic Studies of the Siberian Federal University*
Vladimir Krever	*Academic Supervisor of the WWF Russia Biodiversity Conservation Program*
Daryana Maksimova	*Cand. of Politics, Deputy Executive Director of Northern Forum*
Alexander Pilyasov	*Dr. of Geology, Chairman of the Russian Section of the European Regional Science Association*
Frédéric Lasserre	*PhD., Director of the Quebec Council for Geopolitical Research*
Petr Gogolev	*Dr. of Law, Chairman of the State Assembly of Yakutia*
Natalia Vyakhireva	*Cand. of Politics, Program Manager of Russian International Affairs Council*
Alexander Kugaevsky	*Cand. of Economics, Vice-Rector for Analytics and Expertise of North-Eastern Federal University*

Man in the Arctic

Chairman

Anatoly Nikolaev	*Dr. of Biolology, Rector of North-Eastern Federal University*

Regular Members

Gail Fondahl	*PhD., Professor of University of Northern British Columbia*
Kondratyeva Valentina	*Advisor of North-Eastern Federal University*
Daryana Maksimova	*Cand. of Politics, Deputy Executive Director of Northern Forum*
Vasily Savvinov	*Cand. of Pedagogics, Vice-Rector for Strategic Development of North-Eastern Federal University*
Svein D. Mathiesen	*PhD., professor at the Uarctic EALAT Institute , International Center for Circumpolar Reindeer Herding*
Florian Stammler	*PhD., Professor of Arctic Center of the University of Lapland*
Anatoly Zhozhikov	*Dr. of Pedagogics, Chair of the UNESCO Department of North-Eastern Federal University*
Ulyana Vinokurova	*Dr. of Sociology, Head of the Center for Circumpolar Civilization of Arctic State Institute of Culture and Arts*
Feodosia Gabysheva	*Dr. of Pedagogics, Head of the Standing Committee on Science, Education, Culture, Mass Media and Public Organizations within the State Assembly of Yakutia*
Elena Golomareva	*Head of the Standing Committee on Affairs of Indigenous Peoples of the North within the State Assembly of Yakutia*
Mikhail Pogodaev	*Deputy Minister for the Development of the Arctic and Affairs of the Indigenous Northern Peoples of Sakha (Yakutia)*
Alexander Vorotnikov	*Cand. of Chemistry, Coordinator of Project Office for the Development of the Arctic*

edp sciences

SHS Web of Conferences

Statement of Peer review

In submitting conference proceedings to SHS Web of Conferences, I certify to the Publisher that I adhere to the **Policy on Publishing Integrity** of the journal in order to safeguard good scientific practice in publishing.

1. All articles have been subjected to peer review administered by the proceedings editors.
2. Reviews have been conducted by expert referees, who have been requested to provide unbiased and constructive comments aimed, whenever possible, at improving the work.
3. Proceedings editors have taken all reasonable steps to ensure the quality of the materials they publish and their decision to accept or reject a paper for publication has been based only on the merits of the work and the relevance to the journal.

Title, Date, Location of the Conference

Northern Sustainable Development Forum (NSDF), September 27-30 2020, Yakutsk, Yakutia, Russia

Proceedings editor(s):

Daryana Maksimova

Date and editor's signature

11. 06. 2021.

17, avenue du Hoggar - PA de Courtabœuf – BP 112 - 91944 Les Ulis Cedex A (France)
Tél. : 33 (0)1 69 18 75 75 - Fax : 33(0)1 69 07 45 17 – www.edpsciences.org

Development of a quality management system for a machine-building enterprise

Inna P. Bandurina [1,*], *Mikhail A.* Bandurin[1], and *Alexander P.* Bandurin[2]

[1] Kuban State Agrarian University named after I.T. Trubilin, Kalinina st, 13/20-161, Krasnodar, Russia
[2] Don State Agrarian University, Dubovskogo st. 46-56, Novocherkassk, Russia

Abstract. The article presents proposals for the development of the quality management system of a machine-building enterprise based on a risk-oriented process approach. The theoretical and methodological foundations of the risk management subsystem formation as the basis of continuous development of the quality management system based on ISO 9000 standards are considered. Methodological recommendations of the development of existing enterprise methods for managing the processes of the quality management system are proposed. A methodology of risk management of quality management system processes has been developed. The practical application of the proposals formulated in the paper is shown by the example of an enterprise engaged in the production of electric traction machines for railway rolling stock. At the present stage of development of machine-building production, insufficient attention is paid to the issue of methodological support for risk management of quality management system processes developed on the basis of ISO 9000 series standards, despite the fact that this direction is very relevant and practically in demand in all industries. It is revealed that the process monitoring system, which is based on monitoring process indicators, information about internal audits, self-assessment of the quality management system, allows you to present a real picture of the state of processes and develop management decisions to improve the effectiveness and improvement of processes and the quality management system. And all the information obtained during process monitoring can be used as input data for managing risks in processes.

Keywords: risk-oriented approach, management, enterprise, production

1 Introduction

Modern science-intensive production at an industrial enterprise includes the production of products related to the need for more theoretical calculations, scientific research and experiments. Usually, a high-tech production is considered to be one in which research and development work accounts for at least 60 % of all costs associated with preparation and production of products [1]. Therefore, special attention should be paid to product quality, which is one of the three main components that affect the competitiveness of manufactured products, which in turn directly affects the coverage of all R & d costs and the planned profit from the project [2].

It is recognized that the ISO 9000 series standards [3] are a good tool with a minimal foundation for achieving the main goal of enterprises in the field of designing an effective quality system in order to meet customer requirements. When producing a particular product, obtaining an ISO 9000 series certificate by organizations contributes not only to improving the competitiveness of the product on the market, but also positively forms the business image and reputation of the company [4].

The consumer is increasingly demanding that next to the product certificate, which is considered in this situation only as evidence of the company's engineering capabilities, there is a certificate for the quality system. This confirms that the company has a system that guarantees stable output of products of the quality level that is confirmed by the certificate for the products themselves.

To achieve a stable and competitive position, a machine-building enterprise needs to make timely management decisions that are adequate to changes in the external and internal environment. Therefore, one of the main factors in the development of an organization is the continuous improvement of its quality management system [5].

In connection with the need to switch to a new version of the ISO 9000 series of standards, organizations in various fields of activity, including industrial enterprises, faced the question of combining the already proven "process approach" with a new concept of risk-based thinking [6]. The inherent advantage of this integration is the continuous improvement of the quality management system processes by identifying in advance the "bottlenecks" of processes aimed at preventing the release of non-conforming products and the provision of non-conforming services, which will ultimately increase the satisfaction of consumers and other stakeholders.

* Corresponding author: Chepuraib@gmail.com

Currently, insufficient attention is paid to the issue of methodological support for risk management of quality management system processes developed on the basis of ISO 9000 series standards, despite the fact that this area is very relevant and practically in demand in all industries.

2 Figures and tables

To achieve a stable and competitive position, a machine-building enterprise needs to make timely management decisions that are adequate to changes in the external and internal environment. Therefore, one of the main factors in the development of an organization is the continuous improvement of its quality management system. At the present stage of development of machine-building production, insufficient attention is paid to the issue of methodological support for risk management of quality management system processes developed on the basis of ISO 9000 series standards, despite the fact that this direction is very relevant and practically in demand in all industries.

3 Equations and mathematics

The object of research is the quality management system of a machine-building enterprise that implements and develops a quality management system based on ISO 9000 series standards [9].

The subject of the research is the organizational and economic relations that arise in the development, development and implementation of methodological support of the quality management system of a machine-building enterprise within the framework of a risk-oriented process approach.

The scientific novelty of the research consists of the development of methodological recommendations for improving the quality management system of an enterprise based on a risk-oriented process approach by integrating methods for monitoring the processes of the quality management system and including risk management in the processes of the quality management system [10].

4 Purpose of the Study

The aim of the work is the monitoring system, which is based on monitoring process indicators, information about internal audits, self-assessment of the quality management system, allows you to present a real picture of the state of processes and develop management decisions to improve the effectiveness and improvement of processes and the quality management system. And all the information obtained during process monitoring can be used as input data for managing risks in processes.

5 Research Methods

The development of theoretical and methodological foundations of the development of the quality management system of a machine-building enterprise based on a risk-oriented process approach has set the corresponding research objectives [7]:

- to analyze modern scientific approaches to the development of the quality management system of a machine-building enterprise based on ISO 9000 standards;
- to suggest the direction of development of the methodology for determining and documenting the processes of the quality management system of a machine-building enterprise in the framework of a risk-based approach [8];
- to substantiate the direction of development of methods for monitoring, measuring, analyzing and influencing the processes of the quality management system of a machine-building enterprise in the framework of a risk-oriented approach;
- to develop a methodology of risk management of quality management system processes in the framework of a risk-oriented process approach;
- to conduct testing of the proposals formulated in the process of solving the tasks set in the branch "NEVZ-Electrotech" LLC "TMH Electrotech".

6 Findings

In the course of the study, the following results were obtained, characterized by elements of scientific novelty:

1 It was revealed that the development of the process approach to quality management is reflected in the improvement of ISO 9000 series standards, which require enterprises to constantly improve their quality management system and, accordingly, develop its methodological support.

2 It is proved that the procedure for determining and documenting the processes of the quality management system in the framework of a risk-based process approach required adding the "Process risks" block to the process map and developing an algorithm for describing process risks, including a list of process risks, their level and measures to prevent, reduce or transfer risks [11].

3 It was revealed that in the framework of the risk-based approach, the process manager should include information in the report on the functioning of the process not only about the results of the processes, about the reasons of deviations, developed and implemented corrective actions and corrections, but also information about the effectiveness of actions in relation to risks.

4 The method of risk management of processes of the quality management system of a machine-building enterprise was developed, which included: the algorithm of the risk management process, the methodology of assessing the level of risks and ways to reduce risks [12].

The theoretical significance of the study is that the solutions of conceptual and methodological issues presented in it, develop the scope of application of theoretical aspects of quality management in the direction of developing the organization's quality management system, based on the introduction of a risk-based process monitoring system.

The practical significance of the research results is that the proposed guidelines can be used by machine-building enterprises to improve processes and the quality management system as a whole.

Management development is based on the following scientific approaches:

- system approach;
- process approach;
- risk-based approach.

These scientific approaches are the basis of the ISO 9000 series standards, which are used in the world practice of quality management. ISO 9000 series standards are developed by ISO / TS 176 "Quality management and quality assurance" and are one of the most popular regulatory documents in the world.

Changing the approach to management inevitably led to a revision of the ISO 9000 series standards. According to the report the ISO Survey-2018, in 2018, more than 850 thousand organizations were registered that were certified according to the ISO 9001 standard [13].

The system approach is based on the general theory of systems, founded by Ludwig von Bertalanfi. Such scientists as J. Lorsch, P. Lawrence, and E. G. Yudin made a huge contribution to this scientific direction. For the first time, a theory of systems was applied in the exact sciences, in the late 1950s in management (contribution of the school of management science). The main message of the theory is in proper understanding of the functioning of an organization, it must be considered as a system [14].

The key concept in this approach is the concept of a system, which is a certain integrity consisting of interrelated parts, each of which contributes to the characteristics of the whole. The main idea of the system theory is that any decision (action) has consequences for the entire system [15]. A systematic approach to management allows you to avoid a situation where a decision made in one area turns into a problem for another.

The main advantages of such approach consist in merging processes that best help achieve the expected results, the ability to focus on key processes and providing confidence to all parties involved in the efficiency, effectiveness and consistency of enterprise activity.

Risk-based thinking provides confidence that risks are identified, considered, and managed during the design and application of a quality management system. Risk-based thinking is something that we all do automatically in everyday life [16], it implies the implementation a set of agreed measures and methods for managing and controlling numerous risks (positive and negative) by an organization that affect its ability to achieve planned goals [17].

In addition, the release of the fifth edition of ISO 9000 series was also caused by the following reasons. First, over the past period, there have been significant changes in the world management practice, the business environment; new knowledge has been accumulated. Second, new standards for management systems have appeared, which caused the need to create a unified approach. Third, there are some requirements that need to be, for example, such as "mandatory documented procedures" and "documented quality guidelines». Fourth, vertical integration of the organization's management is also needed [18].

The changes made to the new version of the ISO 9001:2015 standard will create a stable set of requirements for the next ten years or more. This set of requirements can remain unchanged in an ever-changing external economic and business environment.

The key changes that have been introduced to the new version of the standard compared to the version of ISO 9001: 2008 are as follows [19]:

1. The structure of the ISO 9001:2015 standard has been changed. The number of sections in the new version has been increased to ten. This extension was made in order to ensure the compatibility of the various standards for management systems. All standards for management systems will have the same structure with the same section names. This structure is defined by the so-called "SL application model" (Annex SL). The goal of creating a unified framework of standards for management systems is to simplify the application of integrated systems (for example, such as ISO 9001, ISO 14001, ISO 27001 and ISO 22301) [20].

2. Departure from classic corrective and preventive actions. Instead, the new version of ISO 9001: 2015 suggests applying a risk management model. This model is more general than the rigid set of actions specified in ISO 9001:2008 (in the sections corrective and preventive actions). Organizations can use the ISO 31000:2018 standard for risk management.

3. Application of the concept of "organization context". The introduction of this concept implies a broader scope of the quality management system. The new version of the standard requires organizations to take into account many factors that can affect the system and its sustainability (external – energy use, materials, procurement, environment, etc., internal-corporate culture, organizational discipline, etc.). All these factors must be taken into account when planning, creating and operating a quality system.

4. Transition from the concepts of "document" (ISO 9001:2008 p.4.2.3) and "records" (ISO 9001:2008 p. 4.2.4) to the concept of "documented information". This transition eliminated the use of documented procedures and quality guidelines. In the new version of ISO 9001: 2015, these documents are not required. Instead, the organization can use various types of documentation (for example, it can be both paper and electronic documents, as well as video and audio recordings).

A machine-building enterprise must develop, implement, maintain and continuously improve a quality management system, including the necessary processes and their interactions. The processes of the quality management system include management processes, resource provision, product lifecycle processes, measurement, analysis, and improvement.

In practice, guidelines of the definition, design and management of quality management system processes are set in the form of process maps. The procedure of the process actions can also be issued in the form of an organization's standard, instructions, etc.

The process is controlled using a PDCA cycle:

a) planning (P) – setting □go□; pl□nning □ctivities necessary to achieve the goal; planning work with the calculation of the allocation and allocation of resources necessary to achieve the goal, identifying and considering risks and opportunities for improvement;

b) execution (D) – (execution of the process in accordance with the action plan;

c) verification (C) – monitoring and measurement of the process (where applicable), analysis of the information received;

d) impact of (A) – action on the basis of the conclusions drawn in the previous step, aimed at improvement.

The process is managed by the process manager. At the same time, its responsibilities include:

- documentation of the process in accordance with the requirements of regulatory documents established at the enterprise;
- implementation of process planning based on the company's quality goals and business plan;
- defining the output and input data, criteria, and methods required for effective execution and management of the process;
- planning of the availability of resources needed to complete the process and its set criteria with maximum efficiency;
- implementation of process management with the organization of clear interaction between its stages and between departments;
- evaluating the effectiveness of the process at regular intervals;
- conducting a systematic analysis of the actual values of the criteria and the root causes of process variability at regular intervals;
- analysis of the impact of process variability on the output of non-conforming products;
- identification of the need for corrective actions to improve the performance of the process and increase its effectiveness.

At the same time, the process Manager has the following rights [21]:

- submitting the proposals of improving the quality management system processes to the top management of the company or a representative of the management;
- requirement from departments that ensure the functioning of the process (including those that are not functionally subordinate to the process manager) to provide data for accounting/monitoring and analysis of information on the relevant process;
- giving instructions aimed at improving the process to the heads of departments that ensure its functioning.

The process manager plans the goals, criteria, outputs and inputs of the process, as well as the availability of resources necessary for its implementation, evaluates the process and makes changes necessary to achieve the intended results and improve the process.

When planning processes should be taken into account:

- needs and expectations of stakeholders;
- goals to be achieved;
- criteria necessary to ensure the effective functioning of processes and their management;
- legal and regulatory requirements;
- interactions with other processes;
- required inputs and expected outputs;
- resources and information;
- duties, responsibilities and powers in relation to the process;
- potential risks;
- necessary records;
- monitoring, measurement, analysis, assessment and improvement.

The process manager develops process performance criteria that are used for operational process management, staff motivation, and continuous improvement of the quality management system.

The process is documented by the process manager in the form of a process map, which consists of:

- graphical representation of the process as a diagram describing the process elements;
- tables of process performance criteria (Table 1).

Table 1. Rules for filling out the process map

No. of the element of a process	Content
1	Title of a process
2	A list of the main activities with the designation of the document regulating the process, including monitoring, measurement, analysis and evaluation of the process, as well as corrective actions, or a link to the flowchart of the process. A list of processes that are part of a higher-level process, indicating the designations and names of process maps, or a link to the drawing with the decomposition of the process (if necessary).
3	Designation and name of the higher-level process (if available).
4	Statement of the process goal.
5	The position of the head in the process.
6	Products or information obtained as a result of the process. Designations and names of processes that process outputs are directed to.
7	Products or information to be transformed during the process. Designations and names of incoming processes.
8	Names of documents of external origin that regulate the performance of relevant activities and reflect quality requirements.

9	Names of documents of internal origin that regulate the performance of relevant activities and reflect quality requirements.
10	List of required resources.
11	Names of performance criteria and (or) a link to the table of performance criteria.
12	Names and designations of required records and (or) a link to the table of records.
13	Knowledge required for the operation of the process.

In accordance with the proposed risk management procedure, measures were identified, analyzed, evaluated and developed to prevent, reduce or transfer risks of the selected management system processes. Risks were embedded in the processes, so risk maps were not drawn up and risks were included in the process maps [22].

In order to monitor and analyze the functioning of the quality management system processes in the "NEVZ-Electrotech" branch, performance indicators are set for each process, which are calculated and analyzed with the established process map at regular intervals. If the calculated values do not exceed the set base values, the process is considered effective. In this case, the process manager determines the reasons that led to the failure of the process, develops corrections and corrective actions to eliminate the discrepancy, and also reviews the risks of the process in order to identify previously unaccounted risks and assess the effectiveness of previously developed measures to prevent, reduce or transfer the risk.

The risks of the "Procurement" process, their sources, and measures to prevent, reduce, or transfer the risks are shown in Table 2.

Table 2. Rules for filling out the process map

Source of risk	Measures to prevent, reduce or transfer risk
The volume of planned purchases was incorrectly determined due to the variability of information about the need	Approval of the possibility of providing goods and materials for the conclusion of contracts for the supply of products and execution of additional agreements.
	Approval of adjustments of the plan of the release of commercial products, taking into account the timing of the formation of the purchase plan for the planned month and the time to make adjustments.
	Approval of the maintenance work plan, taking into account the terms of forming the purchase plan for the planned month and the time to make adjustments.
The volume of planned purchases was incorrectly determined due to insufficient staff qualifications	Improving the level of competence of employees of the commercial directorate.
Non-fulfillment of the declared volume by the supplier of goods and services due to the delivery of non-conforming products.	Prompt replacement of non-conforming products or delivery from an alternative supplier of goods and materials.
Delivery of non-declared products by the supplier of goods and materials	Sending the task to the supplier of goods and materials for shipment and control of the range of products to be shipped at the time of shipment.
Stop of production at the supplier of goods and materials	Monitoring the activities of suppliers of goods and materials.
Late delivery of materials and purchased components	Organization of additional delivery in the shortest possible time (making an additional order, conducting negotiations, going on a business trip, searching for alternative suppliers, making a replacement).
Disruption of production due to the termination of the activity of the supplier of goods and materials	Approval of the possibility of changing the product design, modernization.

7 Conclusion

The research touched the approaches to management based on ISO 9000 series, covered the basics of the formation of the subsystem of risk management of processes of the quality management system and process of risk management, the methods of risk assessment under the risk-oriented process approach the quality management system were considered.

It is revealed that the process monitoring system, which is based on monitoring process indicators, information on internal audits, self-assessment of the quality management system, allows you to present a real picture of the state of processes and develop management decisions to improve the effectiveness and improvement of processes and the quality management system. And all the information obtained during process monitoring can be used as input data for process risk management. Methods of risk management processes of the quality management system of the organization needs to provide flexible management of internal and external risk processes inherent to modern conditions of functioning of the organization. They allow one to move to a continuous risk management processes in conjunction with the operation of the monitoring system processes of the quality management system of the organization. Analysis of risk assessment methods in the framework of the risk-oriented process approach of the quality management system showed that the methodology of managing process risks should be developed and implemented based on the use of such quality management methods as FMECA (for general risk assessment and development of control actions), and Ishikawa diagrams (for hazard identification). This will provide a comprehensive scientific approach to risk management

References

1. W. Mellis. Software quality management in turbulent times – are there alternatives to process oriented software quality management. Softw. Qual. J., **7(3-4)**, 277–295 (1998)
2. P.B. Petersen. Total quality management and the deming approach to quality management. J. of Manag. Hist., **5(8)**, 468–488 (1999)
3. I. Shuli. Earnings management and the quality of the financial reporting. Perspect. of Innov., Econ. and Busin., **8(2)**, 45–48 (2011)
4. S.S. Skobkin, Ya.A. Belavina, G.V. Kalita. Methodological approaches to the formation of the quality management system for hotels in Russian Federation. Int. Rev. of Manag. and Mark., **7(1)**, 128–136 (2017)
5. F.K. Abdrazakov, S.S. Orlova, T.A. Pankova, E.N. Mirkina, O.V. Mikheeva. Risk assessment and the prediction of breakthrough wave during a dam accident. J. of Interdiscipl. Res., **8(1)**, 154–161 (2018)
6. L.V. Kireicheva, O.B. Khokhlova. Elemental composition of different fractions from the sapropel organic matter. Euras. Soil Sci., **33(9)**, 947–949 (2000)
7. N. Fen, D.B. Kozlov, I.S. Rumyantsev. Hydraulic studies of stepped spillways of various design. Pow. Technol. and Eng., **49(5)**, 337–344 (2016)
8. L.V. Kireicheva, O.A. Zakharova. The effect of cyclic irrigation with wastewater on the properties of gray forest soils. Euras. Soil Sci., **35(9)**, 990–995 (2002)
9. V.I. Olgarenko, G.V. Olgarenko, I.V. Olgarenko. A method of integral efficiency evaluation of water use on irrigation systems. Int. Multidiscipl. Sci., GeoConf. SGEM, **18(3.1)**, 3–9 (2018)
10. N.F. Ryzhko, F.K. Abdrazakov, S.N. Ryzhko, S.V. Botov. The increase of qualitative indicators during watering with multiple support sprinklers. J. of Fund. and Appl. Sci., **10(6S)**, 1482–1497 (2018)
11. D.V. Kozlov. A linear dynamic model for calculating the transverse oscillation of free-floating ice cover. Water Res., **28(2)**, 215–219 (2001)
12. V.I. Olgarenko, I.V. Olgarenko, V.I. Selyukov. Computer technology of planning water use in irrigation systems. Bull. of the Rus. Acad. of Agricult. Sci., **4**, 12–15 (2012)
13. D.V. Kozlov, A.N. Nasonov, I.M. Zhogin, I.V. Tsvetkov. Multifractal principles of aquatic ecosystem development control by algacenosis correction. Water Res., **44(2)**, 259–266 (2017)
14. M.A. Bandurin, V.A Volosukhin,. A.V. Mikheev, Y.V. Volosukhin, V.V. Vanzha. Finite-element simulation of possible natural disasters on landfall dams with changes in climate and seismic conditions taken into account. J. of Phys. Conf. Ser., **1015**, 032011 (2018)
15. B.P. Chesnokov, O.V. Naumova, V.A. Strelnikov, F.K. Abdrazakov, B.A. Tronin. Polyethylene production from granules using high voltage. Int. J. of Appl. Eng. Res., **11**, 2140–2144 (2016)
16. V.I. Olgarenko, G.V. Olgarenko, I.V. Olgarenko. Integrated assessment of the technical level of the irrigation and drainage systems. Meliorat. and water manag., **6**, 8–11 (2013)
17. D.V. Kozlov, A.N. Nasonov, I.M. Zhogin, I.V. Tsvetkov. Multifractal principles of aquatic ecosystem development control by algacenosis correction. Water Res., **44(2)**, 259–266 (2017)
18. L.V. Kireicheva, N.P. Karpenko. Evaluation of the efficiency of irrigation in a zonal soil sequence. Euras. soil sci., **48(5)**, 524–532 (2015)
19. L.V. Kirejcheva et al. *New technologies for designing, substantiating the construction, operation and management of land reclamation systems* (Publ. House of GNU VNIIA Rus. Agricult. Acad., Moscow, 2010), 240 p
20. M.A. Bandurin, V.A. Volosukhin, A.V. Mikheev, Ya.V. Volosukhin, I.P. Bandurina. Finite element simulation of cracks formation in parabolic flume above fixed service live. IOP Conf. Ser. Mat. Sci. and Eng., **327**, 022010 (2018)
21. L.V. Kireicheva, F. Yurchenko, V.M. Yashin. *Scientific principles of creation and management of reclamation systems in Russia* (All-Rus. Res. Inst. of Agrochem., Moscow, 2017), 295 p
22. L.V. Kirejcheva et al. *New technologies for designing, substantiating the construction, operation and management of land reclamation systems* (Publ. House of GNU VNIIA Rus. Agricult. Acad., Moscow, 2010), 240 p

Improving the assessment of the environmental and economic efficiency of measures

Inna P. Bandurina [1,*], *Mikhail A.* Bandurin[1], and *Alexander P.* Bandurin[2]

[1] Kuban State Agrarian University named after I.T. Trubilin, 13/20-161, Kalinina str., Krasnodar, Russia
[2] Don State Agrarian University, 46-56, Dubovskogo str. Novocherkassk, Russia

Abstract. The purpose of the study is to improve methods for improving the regulatory and methodological framework for assessing the environmental and economic effectiveness of land reclamation investment projects. The research methodology is based on the Monte Carlo method, which corresponds to international standards, the theory and algorithms of innovative methods for determining risks, as well as the ability to provide the user with information about the content of risk assessment operations and ensure that their preferences are taken into account in the calculation process. Currently, many tools have been developed for automating risk analysis procedures, including those performed by the Monte Carlo method, which are described in various studies with varying degrees of completeness. The analysis of priority methods of assessing the risks of achieving the projected economic indicators of project solutions is performed and the prospects of the simulation method for practical use in the field of land reclamation are shown. The results of the risk assessment of environmental and economic efficiency of anti-filtration coatings of hydraulic structures of irrigation systems performed by the Monte Carlo method in the environment of the Crystal Ball software product are presented, and the need to improve the reliability of the predicted results of the effectiveness of the designed measures is established. Future research is risk assessment of profitability of the designed activities promote the development of existing and formation of new theories to justify the feasibility of implementation of reclamation investment projects.

Keywords: environmental problems, monitoring, risk, projects, management

1 Introduction

The absence in the economic theory and practice of reclamation investment projects of a unified method of the effectiveness of project solutions that uniquely meets the user's multi-faceted requirements for analyzing the risk of investments has led to the priorities of heuristic models that use the skills of managers. Quantitative (analytical), qualitative and probabilistic methods of calculation have been developed to the maximum, each with its own advantages and disadvantages [1], without an explicit right of preference. The methods used for analyzing and accounting for investment project risks and comparative analysis of their functional capabilities based on research [2] are divided into: qualitative, assessing risk on a certain scale (high, medium, low risk level); quantitative, assessing risks by the numerical value of the indicator, and mixed.

The success of reducing the risk of reclamation investment projects is largely due to the methods used to analyze the effectiveness of the designed solutions, which significantly differ taking into account the variability (in space and time) of the project performance parameters that form quantitative risk indicators [3].

The need to analyze possible risks from the introduction of innovative production technologies already at the design stage is due to the need to exclude incompatibility of the introduced innovative technology with the conditions of use. Risks of investment in projects of the development of technological processes of production and management of activities are classified in theory according to many characteristics, among which the following are most often distinguished: possible result; source and period of risk occurrence [4].

Priority risk factors of reclamation investment projects include [5–7]:

- uncertainty of accounting for natural and economic conditions of the design object: farming systems; the intensity of agricultural production; completeness and accuracy of knowledge about the results of reclamation measures, capital investment standards in reclamation measures, conditions for the formation of fixed assets and working capital of production on reclaimed land, crop yields, costs of agricultural production, etc.;
- production and technological conditions associated with the possibility of non-fulfillment by the contractor implementing reclamation investment projects of its obligations to the customer; with the quality of reclamation work and requirements for technological processes of creation and operation, properties of building materials, etc. project characteristics that depend on the study of the

* Corresponding author: Chepuraib@gmail.com

technical and technological environment for the implementation and functioning of the reclamation object;

• availability of financing, which characterizes the probability of the customer fulfilling financial obligations, the ability to attract borrowed funds, and the liquidity of agricultural products;

• investment climate that takes into account the depreciation of financial investments due to unforeseen changes in the political situation, legislation, etc. reasons caused by the procedures for institutionalizing the economy; unsuccessful investment of money in their own real investments;

• market conditions that determine the instability of the exchange rate, discount rates, and inflation indices.

The principal factor of profitable investment of capital is knowledge of the specifics of the implementation of specific investment projects and the use of effective innovative methods for calculating the effectiveness of planned activities [8]. The most general relationship between the probability of risk and the predicted return of an investor is reflected by a pattern: the higher the return on investment, the higher the risk of not achieving it; reducing the return increases the probability of no risk of investment [9].

2 Figures and tables

Assessment of the effectiveness of reclamation investment projects is characterized by a high level of risk of projected environmental and economic indicators due to the long, more than fifty-year period of implementation of project solutions, a large temporary benefit from investment to return. The success of reducing the risk of reclamation investment projects is largely due to the methods used to analyze the effectiveness of the designed solutions, which significantly differ taking into account the variability (in space and time) of the project performance parameters that form quantitative risk indicators.

3 Equations and mathematics

The object of the research is the Monte Carlo method, which complies with international standards, the theory and algorithms of innovative methods for determining risks, as well as the ability to provide the user with information about the content of risk assessment operations and ensure that his preferences are taken into account in the process of performing calculations.

The subject of the research is the analysis and calculations of the risk of economic efficiency of specific measures in the practice of the irrigation water management complex, to ensure high quality of strategic decisions and, as a result, the competitiveness and viability of the land reclamation sector.

The scientific novelty of the research lies in the development of the reliability of the predicted socio - ecological and economic efficiency of reclamation investment projects. Simulation modeling or the Monte Carlo method is proposed as a promising method for monitoring the effectiveness of reclamation investment projects.

4 Purpose of the Study

The purpose of the study is to improve methods for improving the regulatory and methodological framework for assessing the environmental and economic effectiveness of land reclamation investment projects. The research methodology is based on the Monte Carlo method, which corresponds to international standards, the theory and algorithms of innovative methods for determining risks, as well as the ability to provide the user with information about the content of risk assessment operations and ensure that their preferences are taken into account in the calculation process.

5 Research Methods

The research methodology is based on the Monte Carlo method, which corresponds to international standards, the theory and algorithms of innovative methods for determining risks, as well as the ability to provide the user with information about the content of risk assessment operations and ensure that their preferences are taken into account in the calculation process. To date, many tools have been developed for automating risk analysis procedures, including those performed by the Monte Carlo method, which are described with varying degrees of completeness in various studies. Studying the degree of automation, the level of financial security, availability of experts etc, indicators of the system of evaluating the effectiveness of reclamation activities allowed establishing an initial period of the formation of the practice of risk assessment yields, the promise of using the tools of Crystal Ball, now owned by Oracle Corporation [10].

6 Findings

Qualitative analysis is performed to detect and characterize the likely risks of investment decisions, reveal the causes and conditions of their impact, determine the damage from expected risks and the cost of measures to reduce and /or eliminate it [11]). This approach usually precedes quantitative monitoring of the risk characteristics of investment decisions.

The generally recognized priority of quantitative accounting of risks and uncertainty of investment decisions is explained by the use of methods that increase the accuracy of calculating possible losses and minimize the subjectivity of the situation perception by the decision-maker [12].

The availability of many ways to control, account, and forecast of the risks of reclamation investment projects did not guarantee ease of evaluating the effectiveness of the latter in practice [9].

It connects with:

• wide range of requirements for taking into account the effectiveness of specific reclamation investment projects;

• insufficient attention to the specifics of reclamation investment projects when choosing the method of risk assessment, usually due to the preferences of the developer and the business entity;

- lack of a differentiated approach to assess the risk of reclamation investment projects depending on the stage of its life cycle;
- replacing the risk assessment with an assessment of the uncertainty of random events;
- lack of proper unification of certain methods of risk assessment;
- complexity of calculating and using project-specific parameters in theoretical models.

At the same time, the priority of sensitivity, scenario and simulation modeling are traced.

The relative leadership of the statistical method of monitoring the risk of investment projects is explained by the accessibility of the applied theoretical approaches to the understanding of managers who form the decision and the simplicity of calculations [13]. The criteria of risk accounting and analysis are the average values of randomly changing performance indicators of investment decisions and their variation, which characterizes the degree of deviation of the predicted value from the average. The best solutions correspond to a smaller variation of the estimated parameter. The advantages of statistical risk analysis are often reduced by the lack of actual information about the probability distribution law of the estimated indicator, and replacing it, in most cases, with a normal one, which does not contribute to the accuracy of the result. Monitoring requires data from long-term series of observations of the parameters of the effectiveness of investment decisions, which is also not always achievable.

Project risks are divided into systematic or general risks that affect all forms of investment activity and all investment participants, and specific, internal risks that reflect the characteristics of a particular project. The priority of the specific risks of investments in the projects belongs to [14]:

- defects in design and estimate documentation, etc.;
- non-fulfillment of obligations by the contractor implementing reclamation investment projects;
- changing the goals and priorities of project participants in the development of activities;
- unreliability of knowledge about the financial situation and business reputation of project participants;
- attracting borrowed funds due to the customer's failure to fulfill its obligations to investors.

General risks include changes caused by changes in the external and internal economic environment, which determine a decrease in stock market rates, exchange rates, purchase prices for agricultural products, an increase in the discount rate, and other negative factors. Methods of monitoring the risks of investment decisions that are common in the economy and a comparative analysis of their functionality are given in the next section.

The simulation method or Monte Carlo method, successfully combining and improving the advantages of sensitivity and scenario analysis methods for assessing the risks of investment decisions, determines the impact of uncertainty on investment decisions based on simulation data, which provides the selection of key variables for the effectiveness of investment projects randomly by generating random numbers [15]. Investment performance indicators are recalculated using a given mathematical model that represents their dependence on random and deterministic components of project activities. The risk of investment decisions is calculated based on the probability of distribution of the performance indicator values.

The advantages of the method include compliance with the requirements of international standards, methodology and algorithms for determining risks, the ability of the user to understand the content of the risk procedures, which provides a sufficiently detailed description of the processes and required actions, and ensures that their preferences are taken into account in the calculation process.

The complexity and labour capacity of performing the necessary calculations, which become obsolete with the development of information technologies, initially served as a deterrent to the introduction of the method of risk assessment of investment decisions in practice.

The advantages of the method of parameter variation (sensitivity analysis) include high theoretical validity, simplicity of calculations, the ability to visualize the result, and so, on properties. The disadvantages of the method are that it is impossible to account for the correlation between risk factors and obtain a probability distribution of risk estimates. The profitability of reclamation investment projects is most dependent on the discount rate. This is due to the exponential dependence of the projected cost of money on this factor, with a fixed discount rate; priority is given to purchase prices and agricultural production costs [16].

Scenario analysis, being a logical development of sensitivity analysis, increases the accuracy of calculations and, consequently, the resulting data, and provides for accounting for the variability of events for the estimated performance indicators of the project. The assessment is based on the optimistic, pessimistic, and most likely scenarios of simultaneously changing factors, under which the estimated performance indicators of the project are calculated. The risk is assessed by the deviation of the weighted average of the values of the optimistic, pessimistic and optimal options for the profitability of reclamation investment projects from the value of the base option. Based on the difference between the planned and forecast values of investment performance indicators in the project, conclusions are drawn about the risk of investment decisions and/or the presence of additional risk factors in the project to be taken into account. The disadvantages of the method include the lack of formal methods for forming the boundaries of scenarios and estimating the values of variables for them [17].

The advantages of the Monte Carlo method are compliance with international standards, the theory and algorithms of innovative methods for determining risks, providing the user with information about the content of risk assessment operations and ensuring that their preferences are taken into account in the calculation process [18].

The widespread use of the Monte Carlo method in the practice of monitoring the risk of project decisions was initially hindered by the complexity and complexity of calculations, which, with the development of information technologies, lose their significance.

The risk assessment of investments in the planned measures of anti filtration protection of water supply structures and irrigation systems is carried out as part of the following procedures and their sequence:

• development of an analytical expression that characterizes the relationship between the desired indicator of project profitability and the source data presented by factors that affect the change in the result;

• setting the distribution function for the model's input indicators;

• computer simulation that involves generating random values of source factors;

• formation of the model distribution function, and calculation of project profitability risk parameters;

• analysis of the final results.

To date, many tools have been developed for automating risk analysis procedures, including those performed by the Monte Carlo method, which are described with varying degrees of completeness in studies [19].

Based on the results of the analysis of existing approaches to the development of a system for evaluating the effectiveness of reclamation investment projects, the Crystal Ball toolkit developed by Oracle Corporation is proposed to develop the practice of analyzing and accounting for the risks of their profitability [20].

In this publication, the procedures of analyzing and accounting of the risk of environmental and economic efficiency of reclamation investment projects using the simulation method (Monte Carlo method) are presented by the example of an existing project aimed at improving the sustainability and reducing unproductive irrigation water waste, developed and implemented by employees of the JSC "Institute of safety of hydraulic structures".

As the effect (result) of reclamation investment projects, the prevented damage from the elimination of unauthorized irrigation water costs was taken, and the cost of creating and operating a protective device was taken as the cost (table 1).

The extension of the period of effective operation of irrigation system structures is provided by the device of anti-filtration two-layer coating of irrigation channels made of geotextile material with soil filler [21].

The discounted balance of the incremental flow calculated for the standard period of operation of the protective complex of a hydraulic structure is positive and amounts to 210.6 thousand rubles, the payback period of total capital investments based on the increase in discounted net income does not exceed one and a half years.

The risk of reclamation investment projects was calculated based on the probability of not achieving the planned return if the deviation from the predicted values of the prevented damage from reducing water losses, the cost of installing and operating a protective device, and the discount rate. These factors are significant when calculating the profitability of reclamation investment projects [22]. The calculations were performed using the commercial software product Crystal Ball in the Excel program.

Table 1. Data for calculating the effectiveness of anti-filtration protection

Title	Unit of measurement	Data	
		Before repair	After repair
Reduction of water losses *	th. m3/km per year	-	92.5
Unit costs of water supply*	rub./ m3	0.12	0.12
Prevented damage from reducing irrigation water losses	th. rub./ per year	-	11.1
Cost of the protective device	th. rub./km	-	50
Operating costs of anti-filtration protection	th. rub./km per year	-	1.0*
Service life of the protective device	year	30	30
Discount rate	%	8	8

Risk assessment of reclamation investment projects required:

• development of analytical dependence of the project's economic efficiency on the factors that determine it;

• formation of distribution functions for the dependence of the economic efficiency of reclamation investment projects and its determining factors;

• computer generated random values of effective analytical dependence factors;

• calculation of risk parameters of profitability of reclamation investment projects [23];

• analysis of the final results.

The probability of the predicted values of parameters and the dependence of the profitability of reclamation investment projects is characterized by normal distribution functions, which are described by knowing the average value and the standard deviation from the average.

Thus, the analysis and calculations of the risk of economic efficiency of specific measures in the practice of irrigation water management clearly showed the need for a comprehensive and in-depth assessment of the effectiveness of planned measures. This will benefit the resurgent innovation activity, ensuring high quality of strategic decisions and, as a result, the competitiveness and viability of the land reclamation sector.

The results of calculating the parameters of distribution functions for key factors of dependence on the profitability of reclamation investment projects are presented in Table 2.

Table 2. Parameters of distribution functions

Factor Reduction of water losses, th. m^3 / km per year	Limits of 90 % confidential interval from 70 to 115	Average value	
		92.5	13.68
Specific costs to water supply, rub./ m^3	from 0,12 to 0.20	0.16	0.024
Costs for protective unit, th. rub./km	from 65 to 45	55	6.069
Operating costs, th.rubles / km per year	from 1.3 to 0.9	1.1	0.122
Coefficient of discount %	from 0.29 to 0.51	0.40	0.092

This requires increased attention of the operations service to compliance with the rules for implementing production processes and technologies regulated by the project, as well as to the market environment, which determines the variability of the result in the first place. Account of these factors is also necessary to determine the scope of implementation of the proposed technological improvements. The low risk of the effectiveness of reclamation investment projects is indicated by the absolute probability of a positive value of the effectiveness of the investment project.

But at the same time, it is obvious that the probability of profitability in the amount of 210.6 thousand rubles, declared in the project, does not reach 10 %, which is lower than the average value. Therefore, it is necessary to pay more attention to the implementation of production processes and technologies regulated by the project, as well as to market conditions as priority factors that determine the variability of the result.

7 Conclusion

Improving the reliability of the predicted socio-ecological and economic performance of reclamation investment projects requires improving the regulatory framework for assessing the risk of not achieving the predicted result. As a promising method for monitoring the effectiveness of reclamation investment projects, we propose simulation modeling or the Monte Carlo method, which has been successfully tested as part of these studies.

References

1. L.V. Kireicheva, O.B. Khokhlova. Elemental composition of different fractions from the sapropel organic matter. Euras. Soil Sci., **33(9)**, 947–949 (2000)
2. L.V. Kireicheva, O.A. Zakharova. The effect of cyclic irrigation with wastewater on the properties of gray forest soils. Euras. Soil Sci., **35(9)**, 990–995 (2002)
3. N. Fen, D.B. Kozlov, I.S. Rumyantsev. Hydraulic studies of stepped spillways of various design. Pow. Technol. and Eng., **49(5)**, 337–344 (2016)
4. V.I. Olgarenko, G.V. Olgarenko, I.V. Olgarenko. A method of integral efficiency evaluation of water use on irrigation systems. Int. Multidiscipl. Sci., GeoConf. SGEM, **18(3.1)**, 3–9 (2018)
5. N.F. Ryzhko, F.K. Abdrazakov, S.N. Ryzhko, S.V. Botov. The increase of qualitative indicators during watering with multiple support sprinklers. J. of Fund. and Appl. Sci., **10(6S)**, 1482–1497 (2018)
6. D.V. Kozlov. A linear dynamic model for calculating the transverse oscillation of free-floating ice cover. Water Res., **28(2)**, 215–219 (2001)
7. V.I. Olgarenko, I.V. Olgarenko, V.I. Selyukov. Computer technology of planning water use in irrigation systems. Bull. of the Rus. Acad. of Agricult. Sci., **4**, 12–15 (2012)
8. I.P. Kruzhilin, M.A. Ganiev, V.V. Melikhov, K.A. Rodin, N.N. Dubenok, A.S. Ovchinnikov, S.D. Fomin, N.M. Abdou. Mode of rice drip irrigation. ARPN J. of Eng. and Appl. Sci., **12(24)**, 7118–7123 (2017)
9. D.V. Kozlov, A.N. Nasonov, I.M. Zhogin, I.V. Tsvetkov. Multifractal principles of aquatic ecosystem development control by algacenosis correction. Water Res., **44(2)**, 259–266 (2017)
10. M.A. Bandurin, V.A Volosukhin,. A.V. Mikheev, Y.V. Volosukhin, V.V. Vanzha. Finite-element simulation of possible natural disasters on landfall dams with changes in climate and seismic conditions taken into account. J. of Phys. Conf. Ser., **1015**, 032011 (2018)
11. B.P. Chesnokov, O.V. Naumova, V.A. Strelnikov, F.K. Abdrazakov, B.A. Tronin. Polyethylene production from granules using high voltage. Int. J. of Appl. Eng. Res., **11**, 2140–2144 (2016)
12. V.I. Olgarenko, G.V. Olgarenko, I.V. Olgarenko. Integrated assessment of the technical level of the irrigation and drainage systems. Meliorat. and water manag., **6**, 8–11 (2013)
13. F.K. Abdrazakov, N.F. Ryzhko, S.N. Ryzhko, S.A. Horin, S.V. Botov. Electricity consumption decrease at pump stations during watering by multi-support sprinkling units. J. of Fund. and Appl. Sci., **10**, 1464–1481 (2018)
14. O.G. Degtyareva, D.A. Dac'o, G.V. Degtyarev, A.D. Gumbarov. Design in cae system of low-head weir tiled foundation sinking. Proc. of the Kuban State Agrar. Univ., **64**, 221–226 (2017)
15. L.V. Kireicheva, N.P. Karpenko. Evaluation of the efficiency of irrigation in a zonal soil sequence. Euras. soil sci., **48(5)**, 524–532 (2015)
16. L.V. Kirejcheva et al. *New technologies for designing, substantiating the construction, operation and management of land reclamation systems* (Publ. House of GNU VNIIA Rus. Agricult. Acad., Moscow, 2010), 240 p
17. O.G. Degtyareva, G.V. Degtyarev, N.L. Lavrov, D.U. Aliev. Constructive-technological decisions in

regulating the flow of atmospheric precipitation. Magaz. of Civil Eng., **82(6)**, 32–48 (2018)

18. M.A. Bandurin, V.A. Volosukhin, A.V. Mikheev, Ya.V. Volosukhin, I.P. Bandurina. Finite element simulation of cracks formation in parabolic flume above fixed service live. IOP Conf. Ser. Mat. Sci. and Eng., **327**, 022010 (2018)
19. L.V. Kireicheva, F. Yurchenko, V.M. Yashin. *Scientific principles of creation and management of reclamation systems in Russia* (All-Rus. Res. Inst. of Agrochem., Moscow, 2017), 295 p
20. O.G. Degtyareva, T.I. Safronova, I.I. Rudchenko, I.A. Prikhodko. Nonlinearity account in the foundation soils when calculating the piled rafts of buildings and constructions. IOP Conf. Ser. Mat. Sci. and Eng. **698(2)** (2019)
21. O. Degtyareva, G. Degtyarev, I. Togo, V. Terleev, A. Nikonorov, Yu. Volkova. Analysis of stress-strain state rainfall runoff control system – buttress dam. Proc. Eng., **165**, 1619–1628 (2016)
22. V.G. Degtyarev, N.V. Kozhenko. Consolidation processes in clay soils, taking into account the multistory building construction staging. IOP Conf. Ser. Mat. Sci. and Eng., **698(2)**, 022016 (2019)

Approaches and features of economic efficiency assessment of the infrastructure transport projects

Dmitry Filippov*

NEFU Road Construction Faculty, 13, Krasilnikov street, Yakutsk, 677007, Russia

Abstract. This article contains an example of the applied application of the DCR 21-83 guidelines. This article contains a description of the grouped components and assessment criteria, in particular, a description of the calculated indicators in the context of internal transport, non-transport and socio-economic effects arising from capital investments in transport infrastructure facilities. A specific example shows the procedure for assessing and calculating summary economic indicators for groups and for the project as a whole. This shows the possibility of using a rather complicated methodology and its effectiveness as a scientific tool for a comprehensive assessment for solving applied problems in the construction and rehabilitation of transport infrastructure.

1 Introduction

Capital investments efficiency assessment in highways infrastructure objects is carried out using departmental building regulations DCR 21-83 (VSN 21-83) "Guidelines for the Assessment of Return on Investment in Construction and Rehabilitation of Roads and Highways" [1]. The existing methodology based on these rules makes it possible to define economic consequences of the realization of investments in transport infrastructure projects. This makes the rules an excellent tool for justifying the economic efficiency of large-scale capital-intensive transport infrastructure projects involving government funding. It allows us to take into account all possible measurable effects that occur not only in transport, but also in the economy and public life of the territory as a whole. At the same time, the application of rules is associated with the need to collect and process a significant amount of statistical information, and to conduct a large number of calculations. As a result, for reasons of high labor intensity, substantial cost, which raises the cost of work at the design stage of the facility, as well as the lack of economic agents in the regions that have the competence and information base to conduct large-scale applied scientific research, the use of such an assessment has not been widely adopted. Limitations are also the sufficiency, reliability and quality of the available basic data.

Thus, the initial goal of testing the methodology at the regional level showed not only the feasibility of its use, but also its universal and complex nature, taking into account the effects that arise and are measurable, which is important for assessing the investment qualities of large-scale infrastructure projects with implicit commercial profitability. The study demonstrated the possibility of using a rather complicated method and its effectiveness as a scientific tool for integrated assessment for solving applied problems in road construction. And its use in combination with dynamic assessment methods based on cash flow analysis can significantly increase the scientific validity of management decisions and the quality of investment projects in the field of transport infrastructure development.

2 Materials and methods

Practice shows that the use of this technique is advisable in conjunction with dynamic financial modeling. Since the model allows you to evaluate the aggregated integral indicators based on the analysis of cash flows during the settled period [2]. What is stipulated and regulated by the provisions of the methodological recommendations of VK 477 [3].

The use of the state (national economic) approach is an advantage of the assessment. This approach ensures that all direct and indirect receipts and costs are taken into account not only directly on the assessed road and transport, but in other sectors of the economy of the studied object territory of gravity.

Cost effectiveness is determined by comparing the net cumulative difference between inflows and outflows with the volume of capital investment. Changes resulting from the construction and reconstruction of transport infrastructure facilities are expressed in improving the quality of the road network and, as a result, in reducing transportation costs, reducing economic losses in industry, agriculture and other sectors. That is, improving the satisfaction of the need for the transportation of

* Corresponding author: fil_dv@mail.ru

goods and passengers has a positive effect on the value of the national income.

The economic effect consists of the following elements: an increase in net product in industries and profits of organizations in the non-production sphere. Also, the overall economic effect arising from investments in linear objects can be supplemented by accompanying monetized social results. The social result is associated with a reduction in the economic losses of society from road traffic accidents (RTA) and decrease in the loss of passengers' travel time.

3 Results of research

The components of the economic effect are internal transport, non-transport and social effects.

1. Consider the internal transport economic effect. It represents the difference between the costs in the reference and design conditions, and also takes into account the profit of enterprises from the implementation of additionally arising transportation. This type of effect is generated during the operation of roads, on vehicles, and during cargo handling.

2. The economic non-transport effect, expressed in the increase in net production, is formed from the reduction of costs and losses in conditions of growth in industries not directly related to the road. This performance category takes into account the incentive effect of highways on industry, agriculture, tourism and other industries.

3. The economic effect in the social sphere. This is a measurable social effect. It includes reducing the travel time of passengers, reducing material losses from road accidents, and environmental damage.

Let us consider the use of DCR 21-83 on the example of the construction of the Nyurba - Nyurbachan - Bysytakh - Chukar - Kirov - Khatyn-Ssyy - Zharkhan highway. Under the existing (reference) conditions at the time of the assessment, the road had the following characteristics [4]: the length of 177 km, the length of the all-year-round use section of 27 km, 150 km in the spring and autumn thaw, as well as impassable most of the summer and can only be used as a winter road.

1. Internal transport effect

The project involves the improvement of the technical category of the road. The road surface improves and it becomes year-round. Achievement of design values contributes to the generation of an internal transport effect, in particular:

1. Improving the performance of road maintenance enterprises.
2. Increase in the volume of cargo handling.
3. Reducing the costs of transport organizations.
4. Growth in the volume of cargo transportation.
5. Growth in income from freight road transport.

Table 1. Summary calculation of the internal transport effect for the period.

Statement	**Effect, thousand rub.**
Increase in income in road economy	366138,2
Growth of volumes of loading and unloading works	1066500
Reduction of expenses of the transportation enterprises	616796
Total:	2049434,2

Total internal transport effect that can be estimated for the entire estimated period of the project will amount to 2,049.4 million rubles.

2. Non-transport effect

It is accepted that the projected road will cause additional cash flow. New receipts will be the result of general economic growth in the branches of material production of the Nyurba and adjacent regions, that is, in the zone of gravity of the new highway.

The calculation of the effect arising outside the direct organizations of the transport complex was carried out in two directions.

1. Since the road becomes year-round, the need to create seasonal stocks of various products (food, fuel, building materials, technical products, consumer goods) is eliminated.

2. The new road stimulates the development of branches of material production. The key industries for our example - the Nyurba region - are: diamond mining, coal mining and agriculture.

The effect of reducing the cost of creating seasonal reserves for the entire calculation horizon will reach 10,607.4 million rubles. For the promising period of the project, the cumulative positive effect in agriculture will amount to 114 million rubles. The effect in the coal industry will be expressed by a decrease in the cost of transporting coal by improving the quality of the road and increasing its technical characteristics. The expected amount of the effect is 1,995 million rubles. for the period of the project.

Table 2. Summary calculation of non-transport effect for the period.

Statement	**Effect, mil rub.**
Reduction of expenses for seasonal stocks	10607,4
Effect in diamond mining	218,0
Effect in agriculture	114,0
Effect in coal mining	1995,0
Total:	12934,4

3. Social effect

The new road will help meet the pent-up demand for passenger traffic. Simultaneously with an increase in the comfort of the speed of transportation. This is very important, since it raises the issue of improving safety, taking into account the harsh natural and climatic conditions of the North and the large distances between settlements. Let us recall that Yakutia is characterized by the isolation of territories and the discrete planning nature of the location of productive forces, energy facilities and settlements.

Therefore, transport infrastructure projects in the northern regions of Russia are distinguished by a more significant impact of social efficiency compared to commercial. That is, the social and environmental effects generated by the infrastructure project become an important and significant factor in assessing the overall economic efficiency of the project and can influence the decision to invest in the construction or reconstruction of the road.

Local or private effects are expressed as follows.

1. Reducing the loss of time of the population during travel.
2. Positive impact on human health due to the reduction of accidents and acceleration of the provision of emergency medical care.
3. Improving the comfort of passenger transportation through the use of high-comfort vehicles (compared to all-terrain vehicles and off-road vehicles), reducing the negative impact of cold air in winter and heat in summer (the territory is characterized by a serious seasonal temperature difference from +35 degrees Celsius to -55 degrees Celsius).
4. Reduction of car accidents on the projected road. The cost estimate of the economic effect of reducing accidents was made based on the maximum amount of insurance compensation for car insurance.

Note that bad and low-quality roads can cause negative social consequences.

1. Mortality due to non-provision or untimely provision of medical care.
2. In areas with low transport accessibility, the development of education, sports and culture is hampered.
3. Reduction of free time of the employed population due to the forced stay on the road.
4. Growth of unemployment.
5. Decrease in transport mobility of the population.
6. Mortality and injuries, including children, due to road traffic accidents.
7. Increased crime due to reduced police mobility.
8. Material losses of organizations and the population from fires and natural disasters due to transport inaccessibility or poor operational quality of highways.

The listed and other consequences, due to the low level of development of the road network, become the reasons for the lack of positive prospects, the outflow of the permanent local population and the possible degradation of the territory in the future.

Table 3. Summary calculation of social effect for the period.

Statement	Effect, mil rub.
Effect of reduction of trip time	924
Comfort ability	Improvement
Reduction of accident rate	3,2
Reduction of accident rate, %	20-25%

Table 4. Summary calculation of economic effect.

Statement	Effect, mil rub.
Internal transport effect	2049,4
Non-transport effect	12934,4
Social effect	927,2
Total effect	15911,0

Insufficient development of highways and their unsatisfactory condition is a limiting factor in the socio-economic development of any territory.

Table 5. Aggregate economic indicators and calculation of efficiency of capital investments.

Statement	Unit	Value	
		Current	Project
Extent of year round roads, including:	km		157,2
- year-round roads with hard coating	km	27,3	157,2
- winter	km	150	
Volume of cargo transportation	thousand tons	593	830
Volume of passenger transportation	thousand pass	1427	2922
Passenger turnover of the enterprises of all kinds of activity	thousand pass.-km	16293	57515
Capital investments	mil. rub	-	5247,8
Full current expenses and effects considered for the period	thousand rub		15911,0
Coefficient of capital investments overall effectiveness			3,03
E_{ci} = 15911/5247,8 = 3,03			

The cumulative effect of investments in the construction of the road, which is amenable to cost estimation, in three areas of assessment for the entire period will amount to 15,911.0 million rubles. Table 5 shows the generalized results of calculations and estimates of the efficiency ratio [5].

The calculated value of the overall efficiency indicator (3.03) indicates the positive efficiency and feasibility of investments in the construction of the highway.

As already noted, the performance assessment according to DCR 21-83 can be supplemented by the implementation of a dynamic financial model of the project for VK 477 with the calculation of final integral indicators [6]. At the same time, the example considered by us in this article shows that methodological recommendations for assessing the effectiveness of investment projects well complement the Guidelines for determining the economic efficiency of capital investments in the construction and rehabilitation of highways and proves the feasibility of joint applied use of both methods.

The combination of methodological approaches to the implementation of calculations allows you to calculate the potential effects that can be estimated by value, helps to maximize the cumulative effect by taking into account the positive social consequences. An integrated approach can improve the quality of project development, scientifically substantiate management decisions, and positively influence government decision-making. We believe that such an assessment is necessary when justifying the economic conditions for the implementation of the project to build a bridge across the Lena River near the city of Yakutsk.

4 Conclusion

This article contains the results of a study on the applied application of «Guidelines for the Assessment of Return on Investment in Construction and Rehabilitation of Roads and Highways». A concrete example presents the assessment procedure, local and summary calculations of economic indicators by types of efficiency and for the project as a whole. Assessment of internal transport, non-transport and social effects resulting from capital investments in a linear road infrastructure is executed. The results showed the feasibility of applying the guidelines, their universal and complex nature, taking into account the effects that arise and are evaluable, which is important for evaluating the investment qualities of large-scale infrastructure projects with implicit commercial profitability. The study demonstrated the possibility of using a rather complicated method and its effectiveness as a scientific tool for integrated assessment for solving applied problems in road construction [7]. And its use in combination with dynamic valuation methods based on cash flow analysis can greatly enhance the scientific soundness of management decisions and the quality of investment projects in the field of transport infrastructure development.

References

1. Departmental construction norms of VSN 21-83 *«Guidelines for the Assessment of Return on Investment in Construction and Rehabilitation of Roads and Highways»*. Library of normative documentation, Retrieved from: www.files.stroyinf.ru
2. Erik W. Larson, Clifford F. Gray. *Project Management: The Managerial Process* (McGraw-Hill, 2011)
3. Department of Economic Affairs of the Russian Federation «Methodical recommendations about assessment of efficiency of investment projects and their selection for financing» (the second edition), No. VK 477. Electronic fund of legal and normative-technical documents, www.docs.cntd.ru (1999)
4. The Federal Law No. 257-FZ «About highways and road activity in the Russian Federation and introduction of amendments to separate acts of the Russian Federation» (2007)
5. Feasibility Study of investments on construction of the highway «Nyurba – Nyurbachan – Bysyttakh – Chukar – Kirov – Hatyn-Sysyy – Zharkhan» with bridges. (Yakutsk: JSC RPII Yakutproekt, NCP Sakhainfocentre, 2008)
6. N.A. Loginova. *Economic assessment of investments in transport* (St.Petersburg, 2007)
7. D.V. Filippov, A.E. Ivanova. Amazonia Investiga: Revista Amazonia Investiga Universidad de la Amazonia, **7 (17)** 109-120 (2018)

SHS Web of Conferences **112**, 00004 (2021) https://doi.org/10.1051/shsconf/202111200004
Northern Sustainable Development Forum 2020

Differentiation of normative standards of budgetary appropriations for the maintenance of winter roads in the North of Russia

Elena Eduardovna Grigoryeva[1,*] and *Evgeniy Petrovich* Grigoryev[1]

[1]North Eastern Federal University, 58, Belinsky street, Yakutsk, 677000, Russia

Abstract. The article discusses the results of a study of the issues of maintaining highways of regional and intermunicipal importance in the northern regions of Russia. A special role in the article is given to the category of highways as land and ice winter roads. In the hard-to-reach areas of the Arctic and North of Russia, winter roads are the most important element of the transport system. The maintenance and service of winter roads are the responsibility of the regional authorities. The existing issue of insufficient budgetary support for the maintenance of winter roads is, inter alia, associated with the imperfection of the methodology for calculating target norms of budgetary appropriations. The article presents well-reasoned facts and substantiation of the need to update the norms of budgetary appropriations for the maintenance of winter roads, taking into account difficult territorial natural and climatic conditions of work and transport accessibility of goods delivery. Recommendations are given for changing the methodology for calculating monetary costs for the maintenance of winter roads of regional and intermunicipal importance when determining the amount of appropriations from the budget of the constituent entities of the Russian Federation.

1 Introduction

When developing regional regulations and rules for calculating monetary costs for the maintenance, repair and thorough overhaul of highways of regional and intermunicipal importance, for determining the amount of appropriations from the budget of the constituent entities of the Russian Federation, the Government of the Russian Federation recommends relying on Decree No. 658of the Government of the Russian Federation "On the Standards of Financial Costs and the Rules for Calculating the Amount of Federal Budgetary Appropriations for Thorough Overhaul, Repairs and Maintenance of Federal Highways" dated May 30, 2017. The standards specified in the Decree are calculated taking into account the increase to 12 years of the turnaround time for the operation of federal highways with an improved type of coverage, subject to the requirements of technical regulations in terms of standard design loads. The general concept of developing standards allows for 5% optimization of the standards for monetary costs for the maintenance and repair of highways [1].

This paper presents the results of a study of the regional characteristics of the northern regions of Russia affecting the maintenance of winter roads and provides a quantitative assessment and justification of the need to update the normative standards of budgetary appropriations, taking into account difficult territorial natural and climatic conditions of work and transport accessibility of goods delivery.

2 Research methods

The analysis of information on the ongoing work on thorough overhaul, current repairs and maintenance of highways of regional and intermunicipal importance of the Republic of Sakha (Yakutia) was carried out based on the provided source materials and as a result of processing information from other reliable sources.

Winter roads include seasonal highways with snow, ice and frozen ground and road surfaces. According to the classification, winter roads are subdivided into the following types: according to the duration of operation, according to their location on the ground (land, ice), as well as according to the duration of the season. In turn, regular and temporary winter roads, depending on the estimated annual traffic density or the estimated traffic intensity, are divided into three categories.

When studying the methodological basis of the research issue, a variety of scientific works [2-5] related to various methods and methodological approaches to the calculation of standards for monetary costs for the maintenance of highways was considered. For example, I.A. Kurbatov [6] proposed a methodology, where the calculation of the standards of compensation from the main expenditure components is carried out on the basis of determining the cost of carrying out transportation and the cost of maintaining and developing the road network.

* Corresponding author: elena.grigoreva80@mail.ru

The author attributes the costs of purchasing a car and operating and repair costs to the components of transport costs. The use of compensation standards makes it possible to finance highways. Egorova T.P. made efforts to adapt standard models of vehicles for operation under Arctic conditions and to introduce cross-country transport machines have unraveled a number of challenging issues in the complex cycle of developing a vehicle and delivering it to consumer [7-9].

3 Studying application of territorial coefficient in calculation rules

In the study, a sample of the most costly regions of the Russian Federation [10-12] was made, including the Far Eastern Federal District, using the territorial coefficient of 1.59 for maintenance, 1.68 for repairs, and 1.68 for thorough overhaul.

The results of a comparative analysis of the current regulatory acts of the constituent entities of the Russian Federation that are part of the Far Eastern Federal District show that some constituent entities of the Russian Federation continue using Decree of the Government of the Russian Federation approved in 2008-2010 when calculating the standards for financial costs and the size of budgetary appropriations for thorough overhaul, repair and maintenance of regional and intermunicipal highways.

The results of a comparative analysis of the normative financial costs of the current normative acts of the constituent entities of the Russian Federation included in the Far Eastern Federal District are presented in Table 1.

Analyzing the range of asymmetry of the differentiation of costs of maintaining highways of regional and intermunicipal importance in the constituent entities of the Russian Federation in the Far Eastern Federal District in comparison with the Russian Federation as a whole, it can be argued that the values of the standards of financial costs for maintenance differ from 0.28 to 1.9 times with the standard coefficient of differentiation of 1.68 for the Far Eastern Federal District (according to the Decree of the Government of the Russian Federation No. 658). At the same time, low indicators (below 1) do not indicate that the cost of work in the region is lower than the average in Russia, but indicate the unreliability of this standard of the real cost of work in the region, which leads to underestimated insufficient values of budgetary appropriations.

Insufficient funding for repair and thorough overhaul of the regional road network of the Republic of Sakha (Yakutia) causes the following negative consequences (Table 2):

– It does not allow maintaining the proper technical condition of highways and leads to a decrease in the level of road maintenance.

– A low level of road maintenance leads to an increase in the number of accidents, including due to shortcomings in the transport and operational state of the road network sections.

– The limited financial security of work leads to non-observance of turnaround times with their required increases, taking into account Rules No. 658.

Table 1. Information on the financial costs of maintenance of regional and intermunicipal roads as of 01.01.2020.

No	Subjects of the Russian Federation	Financial cost standards per 1 km of category V, (RUB thous.)				Differentiation *
		Year of adoption of standards	The standard is approved in prices for X year	Content		
				in approval prices	The same in 2019 prices	
1	Republic of Buryatia	2018	2016	547	611	0.67
2	Republic of Sakha (Yakutia)	2009	2008	319	527	0.58
3	Zabaykalsky Krai	2018	2017	641	683	0.75
4	Kamchatka Krai	2011	2010	900	1595	1.76
5	Primorsky Krai	2012	2014	336	469	0.52
6	Khabarovsk region	2009	2009	526	1078	1.19
7	Amur region	2018	2020	321	313	0.34
8	Magadan Region	2018	2018	466	488	0.54
9	Sakhalin Region	2011	2007	709	1734	1.91
10	Jewish Autonomous Region	2008	2013	181	251	0.28
11	Chukotka Autonomous District	2010	2011	709	1100	1.21
12	Russian Federation (RF)	2017	2016	806	908	1.00
13	Far Eastern Federal District	2017	2016	1282	1432	1.58

Explanation: the index-deflator of consumer prices and investments in fixed assets corresponds to the data of Rosstat posted on the official website *Differentiation of maintenance costs in comparison with the RF (prices in 2019)

Let's consider one of such regions, the Republic of Sakha (Yakutia). A regional features of the structure of length of regional and intermunicipal roads in Yakutia are as follows: low transport accessibility, significant distances between 36 municipal districts (especially between the Arctic regions) and high remoteness from federal highways, which justifies the need to maintain land and ice seasonal (winter) roads. A characteristic feature of the Republic of Sakha (Yakutia) is that 73% of the regional road network is occupied by seasonal roads (35% ice, 38% land). Consequently, the presence of a significant share of regional and intermunicipal winter roads justifies the need to include separate coefficients in the rules for calculating the standards of financial costs, taking into account the differentiation of the cost of work on winter roads [13-15].

Table 2. Information on the financial costs of repair and overhaul of regional and intermunicipal roads as of 01.01.2020.

No	Subjects of the Russian Federation	Financial cost standards per 1 km of category V, (RUB thous.)			
		Repairs		Overhaul	
		in approval prices	The same in 2019 prices	in approval prices	The same in 2019 prices
1	Republic of Buryatia	4738	5216	12008	13221
2	Republic of Sakha (Yakutia)	3798	6270	7150	11804
3	Zabaykalsky Krai	4704	4414	10396	9754
4	Kamchatka Krai	4651	3996	14 699	12629
5	Primorsky Krai	2578	2278	11627	10274
6	Khabarovsk region	3539	2858	11985	9677
7	Amur region	2634	2571	8242	8044
8	Magadan Region	3122	3569	9867	11278
9	Sakhalin Region	3665	2118	11583	6692
10	Jewish Autonomous Region	2806	1495	9090	4841
11	Chukotka Autonomous District	3198	6130	9506	18221
12	Russian Federation (RF)	4738	5169	12008	13099
13	Far Eastern Federal District	7960	8940	20173	22658

Explanation: the index-deflator of consumer prices and investments in fixed assets corresponds to the data of Rosstat posted on the official website.

In the Republic of Sakha (Yakutia), the length of the regional road network in 2019 was 13,079.09 km with a decrease of 0.02% or 23.45 km as compared to 2018. The base normative standard is applied in prices of 2008, taking into account the investment deflator index by planning years to the base one in accordance with calculation rules No. 352. It should be borne in mind that the investment deflator index for years does not fully take into account the dynamics of the growth in the cost of goods and services (work) in the region; therefore it is proposed using the consumer price deflator index for the region in the calculation rules for road maintenance, as well as it is used in similar calculation rules by other constituent entities of the Russian Federation.

As a result of data statistical processing, an average of 24.7% of the normative need was allocated for the maintenance of regional and intermunicipal highways in the Republic of Sakha (Yakutia) in 2019, which is lower than the level for the Russian Federation as a whole (29.6%). This does not allow maintaining the proper technical condition of the roads. Consequently, the full scope of work on the maintenance of the regional road network is not being carried out. In fact, in 2019, 61% of the regional road network of the Republic of Sakha (Yakutia) financed by their road fund was covered, where maintenance work was carried out with an average cost of RUB 157,700 per 1 km, which is 45% lower than the national level (RUB 288,100).

In the Arctic zone of the Republic of Sakha (Yakutia), regional and intermunicipal highways are assigned on the basis of the right of operational management to the Arctic Roads Public Enterprise of the Republic of Sakha (Yakutia). In 2019, 5,708.61 km of regional and intermunicipal highways were assigned to the Arctic Roads Public Enterprise of the Republic of Sakha (Yakutia), whereof 92% are seasonal roads, including: 44% land roads and 48% ice roads. The actual cost of maintaining of highways of regional and intermunicipal importance in the Arctic zone of the Republic of Sakha (Yakutia) assigned to the Arctic Roads Public Enterprise of the Republic of Sakha (Yakutia) was RUB 102.38 per 1 km, covering 93% of the total fixed length of highways. Consequently, the real cost of maintaining of 1 km of highway of the regional road network of the Republic of Sakha (Yakutia) is three times lower than the normative one.

4 Approaches to winter roads coefficient development

In order to determine a methodological approach to calculating the amount of budgetary appropriations for the maintenance of winter roads of regional and intermunicipal importance of the Republic of Sakha (Yakutia), an analysis of the practical experience of the northern regions in the development of rules for calculating financial costs for the maintenance of seasonal (winter) highways, including Chukotka Autonomous District, Yamalo-Nenets Autonomous Area, Sakhalin Region and Krasnoyarsk Territory was carried out.

In Sakhalin Region, in Decree No. 1042 of the Administration of Yuzhno-Sakhalinsk City dated June 3, 2010, when establishing standards for financial costs for the maintenance of common use highways of local significance, winter roads are allocated to the group of non-category roads which are equated to the 5th category.

In Yamalo-Nenets Autonomous Area, all winter roads of regional or intermunicipal importance are equated to the 3rd category. The normative standard in the amount of RUB 392,700 (at the price level of 2017) is established by separate Resolution No. 1092-п of the Government of Yamalo-Nenets Autonomous Area dated October 18, 2017. When determining the amount of budgetary appropriations for the maintenance of winter highways in Yamalo-Nenets Autonomous Area, the level of budgetary provision of the regional budget expenditures for the maintenance of winter highways in the corresponding year is taken into account as a percentage (Kbp) and is calculated according to formula 1.

$$A_{maintaince} = H * \square L_j * (K_{bp}/100) \quad (1)$$

where $A_{maintaince}$ is the amount of appropriations from the district budget for the maintenance of winter highways (RUB thous.);

L is the length of winter highways (km);

K_{bp} is the level of budgetary provision of the regional budget expenditures for the maintenance of winter highways in the corresponding year (%).

In Chukotka Autonomous District, the distribution of winter roads categories is applied taking into account the arrangement of ice crossings along the river or sea, as well as the costs of preparatory work for winter roads opening (before winter operation). Application of the methodological approach of Chukotka Autonomous District for calculating the amount of appropriations from the budget of the Republic of Sakha (Yakutia) for the maintenance, thorough overhaul and repair of highways of the corresponding category (by type of coverage) and winter roads with extended service life has a drawback in the absence of accounting the coefficient of territorial differentiation of cost of work (services) by municipal districts. Thus, the considered technique needs follow-up revision in order to be used in the Republic of Sakha (Yakutia).

In Krasnoyarsk Territory, the methodological approach is used to establish the standard of monetary costs for the maintenance, repair and thorough overhaul of highways according to the corresponding category (by type of coverage) and taking into account traffic intensity (number of cars per day). The distribution of standards for the 1st, 2nd and 3rd categories is applied to winter roads.

Krasnoyarsk Territory, like the Republic of Sakha (Yakutia), is characterized by a significant area of the region's territories and an impressive length of regional and intermunicipal roads. The regions determine the presence of intraregional differentiation and asymmetry in rise in goods and works (services) prices, depending on the municipal district transport accessibility. Thus, the inclusion of the territorial coefficient in the calculation, which takes into account the differentiation of the cost of road maintenance by municipal districts and urban districts, is reasonable and necessary.

In the Republic of Sakha (Yakutia), the normative standards for monetary costs for the maintenance, repair and thorough overhaul of roads of regional and intermunicipal importance of the 4th category, and not the 5th category as in other constituent entities of the Russian Federation, are established on the basis of Decree No. 352 of the Government of the Republic of Sakha (Yakutia) dated August 27, 2009.

In this methodology, winter roads are divided into two categories, but a reduction factor is applied for the northern regions that takes into account the seasonality: 0.96 for overland winter roads and 0.66 for ice winter roads. At the same time, 15 out of 36 municipal districts located in the Arctic and eastern zones of the Republic of Sakha (Yakutia) belong to the northern regions.

5 Conclusion

Based on the results obtained, it is recommended to apply following proposals in the rules for calculating monetary costs for the maintenance, repair and thorough overhaul of highways of regional and intermunicipal importance when determining the amount of appropriations from the budget of the Republic of Sakha (Yakutia):

1. The applied normative standard for the maintenance, repair and thorough overhaul of highways in the Republic of Sakha (Yakutia) approved in 2008, even taking into account the deflator index of investments in fixed assets, is two times lower than the average in Russia. It indicates the unreliability of this standard as for the real cost of work in the region. This leads to underestimated insufficient values of budget allocations. It is necessary to bring this normative standard to a level not lower than 1.68 than the average in Russia.
2. The presence of a significant share of winter roads of regional and intermunicipal importance justifies the need to include separate coefficients considering the differentiation of cost of work on winter roads in the rules for calculating the standards of financial costs.
3. In connection with the high intraregional differentiation of cost of construction and repair work, it is proposed to use an approach to the distribution of budgetary appropriations in 36 municipal districts, taking into account transport accessibility to nearby transport hubs, natural and climatic conditions and the type of road surface. Therefore, it is recommended to include the following additional increasing factors in the calculation rules: the coefficient of road natural and climatic zone and the coefficient of transport accessibility.
4. The deflator index of investments by years does not fully take into account the dynamics of cost growth rates for goods and services (work) in the region. Therefore it is proposed to use the deflator index of consumer prices for the region in the calculation rules for road maintenance, just as it is used in similar calculation rules for other constituent entities of the Russian Federation.

Acknowlegements

This work was prepared within the framework of the project on the state task of the Ministry of Science and Higher Education of the Russian Federation titled «The Patterns of Spatial Organization and Spatial Development of Socio-Economic Resource Systems of the Northern Region» (No. FSRG-2020-0010).

References

1. E.A. Latkina, T.F. Matyushenko, Roads and bridges **2(38)**, 1 (2017)
2. S.V. Fedotov, T.F. Matyushenko, Roads and bridges **2(18)**, 11-20 (2007)
3. V.V. Gasilov, M.A. Preobrazhensky, S.S. Zamchalova, M.A. Shibaeva, Science and technology in the road industry **2**, 10-12 (2002)
4. A.S. Eremeeva, News of higher educational institutions. Series: Economics, Finance and production management **2(12)**, 3-9 (2012)
5. V.K. Apelin, About inter-repair terms of capital repairs of highways in modern conditions of financing Roads and bridges **18(2)**, 20-29 (2007)
6. I.A. Kurbatov, Bulletin of the Moscow automobile and road state technical University (MADI) **3(22)**, 60-63 (2010)
7. T.P. Egorova, A.M. Delakhova, Proceedings of the International Conference "Aviamechanical engineering and transport" (AVENT 2018) Series:Advances in Engineering Research **158**, 123-129 (2018)
8. T.P. Egorova, A.M. Delakhova, Smart Innovation, Systems and Technologies **138**, 272-282 (2020)
9. T.P. Egorova, A.M. Delakhova, IOP Conference Series: Materials Science and Engineering **632(1)**, 012029 (2019)
10. V.N. Leksin, B.N. Porfiryev, Economy of Region, **3**, 641-657 (2017)
11. V.N. Borisov, O.V. Pochukaeva, Studies on Russian Economic Development **27(2)**, 159-165 (2016)
12. P.V. Gulyaev, Gornyi Zhurnal **12**, 23-26 (2018)
13. E.E. Grigoryeva, IOP Conference Series: Earth and Environmental Science **302(1)**, 012138 (2019)
14. V.V. Nikiforova, E.R. Romanova, E.E. Grigorieva, Gornyi Zhurnal **3**, 41-46 (2018)
15. V.V. Nikiforova, IOP Conference Series: Materials Science and Engineering **753(7)**, 072008 (2020)

FORMATION OF ECONOMIC SELF-DETERMINATION OF SMALL BUSINESS ENTREPRENEURS

Tatyana A. Terekhova[1], *Elena L.* Trofimova[2], and *Olesya B.* Bolshakova[3,*]

[1] Irkutsk State University, 1, K. Marx str., Irkutsk, 664003, Russia
[2] Baikal State University, 11, Lenin str., Irkutsk, 664003, Russia
[3] Chita Institute of Baikal State, 56, University, Anokhina str., Irkutsk, 672010, Russia

Abstract. The authors consider the economic self-determination of a small business entrepreneur as self-determination in the economic environment. The goal of economic self-determination is to achieve an optimal level of competitiveness, rational monetary attitudes and, as result, formation of economic behavior in order to achieve the set goals that contribute to the success in the professional activity. A group of 58 small business entrepreneurs took part in the survey. Preliminary, entrance testing was carried out with the use of the standardized methods, which was followed by correlation analysis. It allowed one to find out the relationship between the indicators of the scales. A group with a low level of competitiveness was identified using the standardized methodology "Attitude towards competition". As a result, the authors developed a program of group psychological counseling. The structure of the consultation process included the following stages: problem investigation, two-dimensional problem identification, planning, activity, assessment and feedback. After the group psychological counseling, the final control was carried out using the same methods and the effectiveness of this type of counseling was assessed using the Wilcoxon test. The results of mathematical and statistical analysis confirm the effectiveness of the developed program. Group psychological counseling allowed one to increase self-esteem, purposefulness, self-confidence, intensified the scenarios of economic behavior, changed the attitudes towards money and economic benefits which also ensured an increase in competitiveness indicators.

Keywords: Competitiveness, economic self-determination, group psychological counseling, small business entrepreneurs.

1 Introduction

Self-determination in today's economic environment is a key challenge for entrepreneurs.

Russian psychologists consider self-determination from the point of determinism, the integrity of activity, consciousness, consistency and personal development. M.G Yaroshevsky and A.V. Petrovsky defined self-determination as "a conscious act of defining and affirming one's position in problem situations" [9]. S.L. Rubinshtein pointed out that "self-awareness is an internal activity that gives a personality certainty and refracts outside influences" [12]; N.S. Pryazhnikov noted that self-determination is "a conscious process of analysis, acceptance and verification of the strength of one's position, as well as one's own ideas about oneself in difficult situations [10]. E.A. Klimov considers professional self-determination as one of the most important manifestations of human mental development, as a process of his inclusion in the professional community and, more broadly, in the social community [6]. According to A.L. Zhuravlev and A.B. Kupreychenko "self-determination as a search for methods of functioning and development can be simultaneously attributed to three main classes of mental phenomena, i.e. to be understood as a process, as a state (a certain cut of a process in a certain period of time) and as a property of a subject (a self-determining person or group is their special characteristic)" [15]. N.L. Ivanova believes that the process of self-determination of a person consists in choosing a certain profession, the formation of ideas about traits that help or hinder the performance of an activity [4].

The process of understanding self-determination plays an important role in the course of the personality formation of an entrepreneur. Russian researchers are convinced that a professional, unlike an ordinary performer, independently sets goals, determines the way and means to achieve them, and is also independently responsible for the consequences that have arise in the implementation of these goals.

Business psychology focuses on self-determination of the individual in social and professional terms. Subject is a person who has faced difficulties in choosing a life path. N.L. Ivanova argues that human life activity mainly depends on the results of choosing a life path and the ability to manage it [3]. In the course of her

* Corresponding author: terehovata@mail.ru

research, she found that self-determination is directly related to self-development, self-realization, a sense of self-confidence, as well as the ability for high-quality adaptation and activity. N.L. Ivanova draws attention to the fact that self-determination is a long-term process; however, the inner core helps the individual to maintain his unity and not lose himself in a constantly changing world [3].

Economic self-determination of an entrepreneur (ESDE) of small business is considered by us (the authors of the article) as self-determination in the economic environment. The status of the ESDE system, as a system of a specific class, with a built-in metasystem level, includes the system, subsystem, component and element levels.

The metasystem level of the entreupreneur's economic self-determination system is the unity of the personal (including linguistic and cultural), socio-economic and professional-activity metasystems [14]. Metasystems influencing the ESDE system are reflected at all its levels and primarily at the subsystem in particular, they determine the structure and content of current and potential subsystems of the phenomenon under study.

The goal of economic self-determination is to achieve an optimal level of competitiveness, monetary attitudes, which in turn form the economic behavior in order to achieve the set goals and contribute to the success in the professional activity.

R.A. Fatkhutdinov is the author of the theory of competitive advantage. According to this theory, competition is the process of a subject's leadership of personal competitive advantages in order to achieve goals, as well as victory over competitors in order to satisfy both subjective and objective needs in natural conditions or within the framework of legislation [2]

The main tool of a specialist's competitiveness is the availability of exclusive values, thanks to which the specialist has superiority over competitors and that in general constitute competitive advantages.

Competitiveness is one of the important structural components of economic self-determination, as well as one of the most important and key factors responsible for successful entrepreneurial activity. Competitiveness is considered by the authors as an integral characteristic, which includes the following properties and personality traits: high level of performance; striving for a high-quality end result; stress resistance, the ability to overcome difficulties; creative attitude to business, work; striving for professional self-improvement; the ability to make responsible, sometimes risky decisions; sociability, ability to cooperate, collaborate, co-create; the ability to master a new business quickly; the ability for self-education, self-realization, self-development, etc. [1, 8, 11].

2 Problem Statement

The problem of optional professional education is especially urgent for small business entrepreneurs in the context of the socio-economic crisis, aggravated by the course of the Coronavirus pandemic. It turned out that the methods used by entrepreneurs are clearly insufficient for business development in modern conditions. Retraining and advanced training for entrepreneurs are becoming fundamental for the development and maintenance of competitiveness at the current stage of development of small business in Russia.

The formation of economic self-determination of entrepreneurs is connected with 1) the development of the skill to take responsibility for his personal involvement in the activity; 2) understanding that verbalization and insight are not enough for sustainable changes.

The formation of entrepreneurs self-determination is associated with 1) practicing the ability to take responsibility for their own involvement in the activity; 2) understanding that verbalization and insight is not enough for full and sustained changes.

3 Research Questions

The subject of the survey is the formation of nuclear components (attitude towards competitiveness, attitude towards money) and peripheral components of economic self-determination (scenarios of economic behavior) of entrepreneurs in the process of psychological counseling.

1. The problem of the competitiveness of small business entrepreneurs is a central focus in our work. In our research we classify competitiveness as well as such subsystems as subjective characteristics of an entrepreneur, property, economic rationality, profit to the category of "nuclear", i.e. carrying meanings that contribute to the achievement of the goals of entrepreneurial activity [14].

2. The study of the monetary attitudes of entrepreneurs is the second direction in the process of the formation of economic self-determination. J. Keynes believes that the result of accumulation is the struggle between the motives of consumption and saving. The author emphasizes that a person begins to think about savings from a certain moment when his income begins to determine the level of an individual's well-being. In his opinion, an individual who is able to accumulate funds has such qualities as prudence, enterprise, craving for an independent life, prudence, discretion, the desire to leave a legacy, craving for the best [5].

3. Consequently, the study of scenarios of economic behavior of small business entrepreneurs determines the third direction of our research.

4. The forth direction is to determine the possibilities of group psychological counseling in the formation of the economic self-determination of entrepreneurs.

4 Purpose of the Study

The purpose of the survey is to develop and test a program of the formation of economic self-determination in the process of group psychological counseling, to prove the effectiveness of the program and show its impact on increasing the competitiveness of entrepreneurs, reduction the severity of negative monetary attitudes, and consolidation of the rational scenarios of economic behaviour.

5 Research Methods

A sample of 58 small business entrepreneurs aged from 25 to 40 (56 % of women, 44 % of men) took part in the study of economic self-determination. At the stage of the preliminary survey the respondents were tested with the help of three standardized methods: attitude to competition [1], attitude to money [13], scenarios of economic behavior [7]. The processing was carried out using the methods of mathematical statistics of the IBM SPSS Statistics 17 program.

According to the results of the survey, a group of entrepreneurs with low indicators of competitiveness was identified. A psychological counseling program based on the works of Yu. E. Aleshina and R. Kochyunas, D. Corey was developed for this group of the participants of the survey.

The training program includes 9 group consultations. Each consultation lasts 90 minutes.

The consultation process included the following stages: problem research, two-dimensional problem definition, problem identification, planning, activity, assessment and feedback (Table 1).

Table 1. Program of group psychological consulting of entrepreneurs

Title of the block	Aim	Psychological methods, procedures and techniques	Number of meetings
Analyzing the problem	Acquaintance. At this stage the contact is established and mutual trust is achieved.	Analytical (conversation, observation); procedures and techniques: provision of information; reflective listening, paraphrasing and summarizing; approval and calming, verbal and non-verbal communication techniques.	1
Two-dimensional problem identification	To characterize accurately the client's problem, find out cognitive and emotional aspects.	Analytical (conversation, observation); techniques of verbal and non-verbal communication; reception paraphrasing, encouragement to deep research of the problems that have arisen, analysis of feelings, statements.	1
Alternatives identification	To find out and openly discuss possible alternatives of the ways to solve the problems related to the development of economic self-determination of the entrepreneurs.	Analytical (conversation, observation); procedures and techniques: reflective listening; reflection of content (paraphrasing and generalization), confrontation.	1
Planning	Critical assessment of the chosen alternatives. Making a plan of the realistic problem solving.	Analytical (conversation, observation); procedures and techniques: providing information; paraphrasing and summarizing.	1
Activity	Consistent implementation of the problem-solving plan. Psychological counseling. As well as independent use of the acquired knowledge, skills and abilities for the development of economic self-determination.	Analytical (conversation observation), developmental (training exercises), modeling (role-playing games).	5 2 weeks
Assessment and Feedback	Assessment of the current level of goal achievement and generalization of the achieved results.	Analytical (conversation and observation), testing.	1

It should be added that the entire process of psychological counseling was carefully structured. The stages of counseling were distinguished, the results were assessed and information on the counseling process was provided. We discussed the results and formulated conclusions with the clients after each stage and then proceeded to the next stage of consulting. Different techniques of verbal and non-verbal communication were used during the process of psychological counseling.

The main tasks that the authors faced in the process of group psychological counseling:

1. Generating an atmosphere of respect and trust. One of the keys of successful counseling is helping a client to feel acceptance, understanding and sympathy from the counselor psychologist.

2. Providing participants with all the information about the group process and on how to get the most out of the group work. Explaining the goals of the group, guiding the group members into the structure of the session and informing them about the necessity of active actions. Discussing their expectations with clients and encouraging them to make assumptions about who and what the group can benefit from.

3. Creating a system for assessing the problems of group members. In this work, there was a multimodal assessment that included such areas as behavior, feelings, thinking, interpersonal relationships, etc.

4. Consolidating the skills of new behavior acquired by the members of the group and keeping control of them so that they gradually become aware of every, even a very small achievement.

5. Training the group members to take responsibility for their own involvement in different types of activity, both within the group and outside it. Helping clients understand in the process of psychological counseling that verbalization and insight are not enough for meaningful and sustainable change. Expanding their repertoire of adaptive behaviors, group members experiment with their own behaviors during the session and while doing their homework.

6. Helping team members prepare for the completion of the work. Supporting clients if it is necessary when there is time for discussing their reactions, integrating the gained experience and practicing the acquired skills necessary for effective functioning outside the group.

6 Findings

We determined the level of respondents' competitiveness using the "attitude to competition" methodology. According to the results of the study, 48% of respondents showed a low level of competitiveness. Therefore, a preliminary empirical study and further development of a program for the formation of economic self-determination of small business entrepreneurs was carried out in a group of respondents with a low level of competitiveness.

The results of the study of scenarios of economic behavior are presented in Figure 1.

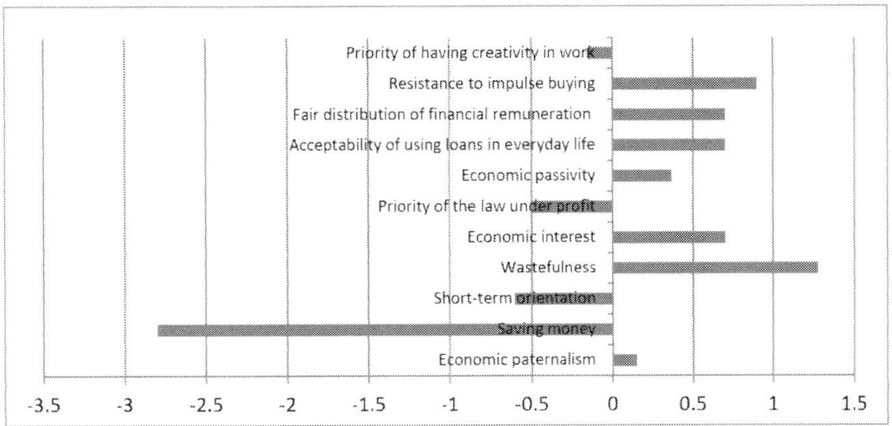

Figure 1. Results of the study of scenarios of economic behavior

Figure 01 shows the prevailing scenarios of economic behavior: "saving money", "wastefulness", "short-term orientation in economic behavior", "economic passivity", "economic paternalism". This type of behavior can be characterized as a dismissive and irrational attitude towards money, underestimation of the importance of time perspectives in doing business, an orientation towards making quick profits and achieving short-term prospects. At the same time, there are such scenarios as "economic interest", "acceptability of using loans in everyday life", "fair distribution of financial rewards" and "resistance to impulse buying" that indicates the involvement and orientation of the respondents to entrepreneurial activity. The "priority of law over profit" scenario indicates a preference for compliance with the legality of activities. An interesting indicator, although not high, was "the priority of having creativity in work". This result indicates the importance of focusing on the process, and not only on the result (profit) in one's activity. We can say that entrepreneurs who are not interested in competition are passive in their professional activities and do not focus on obtaining and increasing their profits. An uncompetitive entrepreneur is more inclined to focus on solving short-term and everyday tasks with a guaranteed, although not high, profit.

Figure 2 presents attitudes towards money of respondents with a low level of competitiveness.

Taking into account the low level of attitude towards competition, we can assume that dissatisfaction with their financial situation provokes an increased focus on money and related topics. Money is an emotionally meaningful object for them and it takes up their thoughts and attention. On the one hand, they carefully control their finances, but on the other hand, such entrepreneurs willingly take risks due to the high importance of money. They value money and often overestimate its value. In such cases, money can be a symbol of independence and power as well as an instrument for controlling others.

Further, a correlation analysis was carried out to study the relationship between competitiveness, attitudes towards money and scenarios of economic behavior of entrepreneurs (tables 2, 3).

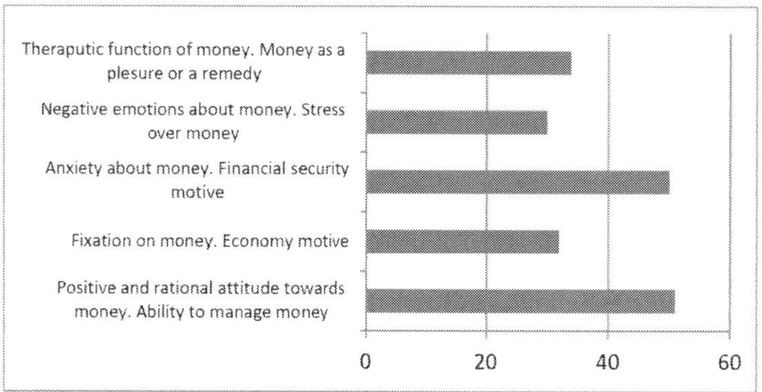

Figure 2. Results of the study of the attitude to money of entrepreneurs with low indicators of attitude to competition

Table 2. Program of group psychological consulting of entrepreneurs

Scales	Correlations		
	Fixation on money. Saving motive	**Anxiety about money. Financial security motive**	**Negative emotions about money. Stress over money**
Competitiveness	**0.248***	**0.490****	**-0.613***
Competition readiness	0.015	**-0.441***	-0.350
Defending your own interests	**0.468***	0.400	0.018
Taking into account the interests of a partner	-0.089	0.152	-0.136
Stiff competition	**-0.643***	-0.019	-0.151
Competitiveness	**0.248***	**0.490****	**-0.613***
** Correlation is significant at the 0.01 level (2-tailed)			
* Correlation is significant at the 0.05 level (2-tailed)			

Correlation analysis showed significant interrelation between almost all indicators of competitiveness, except for "taking into account the interests of the partner" and monetary attitudes.

Thus, the "competitiveness" scale correlates with the following scales: "fixation on money, the motive for saving" ($r = 0.248$, $p = 0.05$), "anxiety about money" ($r = 0.490$, $p = 0.001$), as well as with scale "negative emotions towards money" ($r = -0.613$, $p = 0.02$). That is, entrepreneurs who demonstrated a low level of attitude to competition show negative emotions towards money, and experience tension and anxiety about money. The higher the level of competitiveness, the sooner they will be guided by the motive of financial security and saving money in their activities. In this case, there is a desire to save money in case of unforeseen circumstances in order to reduce financial risks. The scale "defending one's interests" also correlates with the motive for saving ($r = 0.468$, $p = 0.06$).

The indicator of "readiness to compete" has an inverse relationship with "anxiety about money" ($r = -0.441$, $p = 0.03$), i.e., the lower the readiness to compete, the higher the anxiety about money. Here we can say that there is a frequent anxiety about money. When you need to invest a large amount or make an expensive purchase, they take a long time to make a decision, often get nervous, worried.

The "Stiff competition" scale has an inverse correlation with the "fixation on money, economy motive" scale ($r = -0.643$, $p = 0.013$). In conditions of stiff competition, the subjects are not focused on saving money but the lower the readiness for tough actions, the more they value money, often overestimating its significance.

There are no interrelations between competitiveness indicators and monetary attitudes such as a rational attitude to money and getting pleasure from money in this sample.

The analysis of correlations between indicators of attitude to competition and scenarios of economic behavior also demonstrated some significant connections (Table 3).

The lower the level of competitiveness of entrepreneurs, the more they demonstrate economic passivity ($r = 0.622$, $p = 0.034$) and wastefulness ($r = 0.588$, $p = 0.027$). A positive correlation was found between the "Willingness to compete" scale and the choice of long-term or short-term orientation of economic behavior ($r = 0.604$, $p = 0.022$) and the priority of profit or law. So, an entrepreneur with a low readiness for competition is focused on quick results and profit, as well as on the priority of the law over profit ($r = 0.724$, $p = 0.033$).

Correlations were also revealed between the scales "Taking into account the interests of a partner" and "Economic paternalism" ($r = 0.544$, $p = 0.044$). Entrepreneurs showing a low level of partner orientation tend to demonstrate a narrow perspective, limited interests, seeking patronage and a system of additional benefits, subsidies and payments for themselves.

The scales "Stiff competition" and "economic activity/passivity" ($r = 0.632$, $p = 0.015$) have a positive correlation, it means that if competition decreases, economic passivity also grows.

Table 3. Program of group psychological consulting of entrepreneurs

Correlations					
Scales	**Economic independence/ economic paternalism**	**Long-term/ short-term orientation on economic behavior**	**Economy/ wastefulness**	**Priority of profit over the law/ Priority of the law over profit**	**Economic activity/ Economic passivity**
	-0.062	0.139	**0.588***	-0.480	**0.622***
Competitiveness	0.127	**0.604***	0.360	**0.724***	-0.368
Competition readiness	0.085	0.100	0.445	0.082	0.209
Defending your own interests	**0.544***	-0.112	-0.040	-0.085	-0.273
Taking into account the interests of a partner	-0.037	-0.167	0.050	-0.385	**0.632***
**. Correlation is significant at the 0.01 level (2-tailed)					
*. Correlation is significant at the 0.05 level (2-tailed)					

The final stage of the psychological counseling program was aimed at processing and evaluating the effectiveness of the program and was based on the results of the data obtained before and after the group psychological counseling. At this stage, exit psychological testing was carried out. It also included the discussion of the results of the participants' independent work on the formation of economic self-determination.

For this purpose, the average values of the indicators of the attitude to competition were calculated before and after the training (Figure 03).

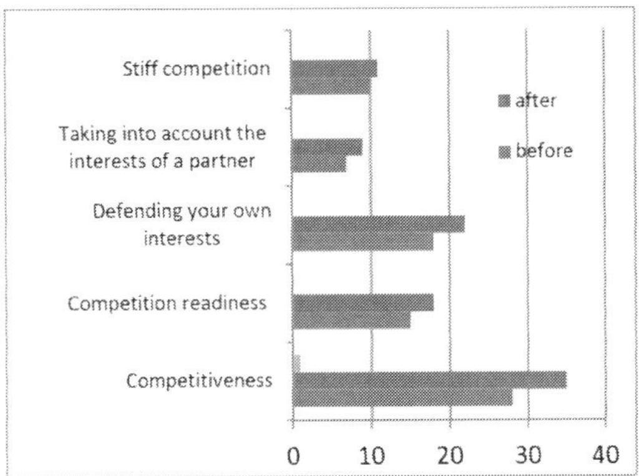

Figure 3. Comparison of results before and after group psychological counseling in terms of attitudes towards competition

Figure 3 shows that the indicators of "attitude to competition" have increased. These changes are due to the fact that during the program the emphasis was placed on the development of positive self-esteem and such qualities as self-confidence, purposefulness, communication, independence, responsibility, determination.

The entrepreneurs managed to form an attitude that competition is a stimulating factor for development and promotion both in personal and professional activities, and as the test results show, entrepreneurs began to assess their competitiveness more positively (the value is above average).

The results of mathematical and statistical analysis according to Wilcoxon's criterion confirm that there are significant differences in the scales of "competitiveness" ($p = 0.024$), "defending one's interests" ($p = 0.023$), "taking into account the interests of a partner" ($p = 0.014$) and "stiff competition" ($p = 0.020$). No significant differences were found on the "competition readiness" scale.

As for the monetary indicators, we can also see changes of all the indicators (Figure 04).

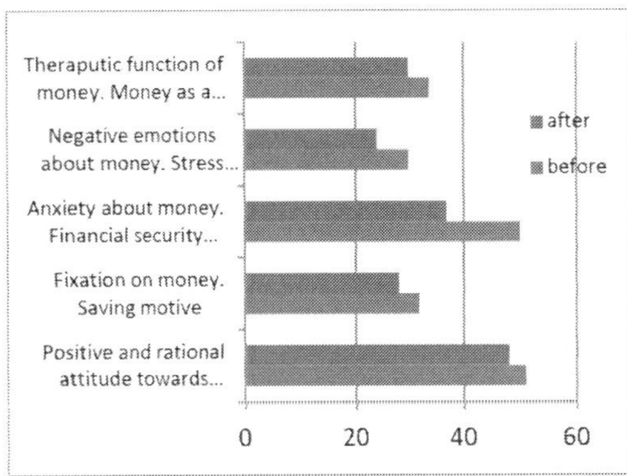

Figure 4. Comparison of the results by monetary attitudes before and after group psychological counseling

We can see a decent on all scales, this can be characterized as a decrease in anxiety about money, there is no fixation on money and a change in the motive for saving to the motive for financial security, thus the neutral or positive attitude in relation to money was formed.

The Wilcoxon signed-rank test showed the differences on the scales "fixation on money" ($p = 0.010$), "anxiety about money" ($p = 0.042$), and "negative emotions towards money" ($p = 0.049$), questionnaire "Relationship to money." No significant differences were found on the scales "positive and rational attitude to money" and "therapeutic function of money, money as pleasure".

The results of scenarios of economic behavior also demonstrate changes (Table 4).

Table 4. Results of the test "Scenarios of economic behavior" before and after group psychological counseling

	Mean												
scenario	**-3**	**-2,5**	**-2**	**-1**	**-0,5**	**0**	**0,5**	**1**	**1,5**	**2**	**2,5**	**3**	**scenario**
Economic independence				-1,8		*0,07*							Economic paternalism
Saving money	*-2,9*							1					Saving time
Short-term orientation (perspective) in economic behavior					-0,6					2,1			Long-term orientation (perspective) in economic behavior
Economy							0,7		1,2				Wastefulness
Economic indifference							*0,7*		1,4				Economic interest
Priority of the law under profit					*-0,5*	-0,3							Priority of profit under the law
Economic activity				-1,4		*0,36*							Economic passivity
Refusal to use loans in everyday life							*0,7*		1,2				Acceptability of using loans in everyday life
Equal distribution of financial reward							*1*		1,3				Fair distribution of financial reward
Addiction to impulse buying							*1*			2			Resistance to impulse buying
Priority of having creativity in work					-0,4	*-0,07*							Priority of the amount of reward

Significant changes can be seen in the scenarios "saving money". It changed to the polar one "saving time", "economic paternalism" was replaced by "economic independence", "orientation to the short-term perspective in economic behavior" changed to "long-term orientation (perspective) in economic behavior", and "economic passivity" was replaced by "economic activity". Entrepreneurs see the prospect of developing their business and are ready to invest into the future in order to increase their income. They are ready to act and be active.

In comparison with the indicators before testing, entrepreneurs now prefer saving their time rather than money, this approach indicates the value of their own time and effort, rational use of time and financial resources.

As for the "wastefulness" scenario, its indicators decreased towards "economy". The indicators of the scenarios "economic interest," "acceptability of using credit in everyday life," "fair distribution of financial rewards," "resistance to impulse purchases," and "priority of creativity in work," increased. The indicator of the scenario "priority of law over profit" changed its value to the neutral. And it is quite interesting.

The Wilcoxon signed-rank test showed significant differences in the indicators of the "Scenarios of Economic Behavior" practically for all scales, except three of them "Priority of profit under the law – priority of the law under profit", "Economic activity – passivity", "Distribution of financial rewards by equality – by fairness". Significant changes took place according to the following scenarios: "Economic paternalism" was replaced by "economic independence", "saving money" was replaced by "saving time", "wastefulness" decreased towards "economy". The scenarios "acceptability of using loans in everyday life", "short-term versus long-term orientation in economic behavior", "non-impulse buying", "priority of creativity in work" and "economic interest" have significantly increased.

Thus, the results of mathematical and statistical analysis show positive changes in the scales of competitiveness, scenarios of economic behavior, and attitude to money and it proves the effectiveness of the program of group psychological counseling.

7 Conclusion

The economic self-determination of a small business entrepreneur is viewed from the perspective of a metasystem approach.

The system of economic self-determination of an entrepreneur is influenced by more than one metasystems. The main ones are personal (with built-in linguo-cultural), socio-economic and professional activity.

Competitiveness is viewed as one of the "nuclear" subsystems in the structure of economic self-determination of an entrepreneur and the key factor that contributes to business success.

The results of the empirical research showed that 48 % of the surveyed small business entrepreneurs are characterized by the low level of competitiveness.

Entrepreneurs with a low level of competitiveness do not find links between the ability to manage money rationally and competitiveness. The lower the level of competitiveness, the sooner money evokes negative emotions, but at the same time there is no desire to save money and ensure financial security, and the more inherent a strategy of waste. These respondents are probable experiencing financial difficulties and cannot afford to use their money for pleasure they mostly spend it on basic urgent needs.

Entrepreneurs with a low level of competitiveness are more likely to show economic passivity, tend to be law-abiding and normative, and choose more stable types of business, with government support or other guarantees of regular income. Such entrepreneurs are characterized by an orientation towards short-term scenarios of economic behavior, quick results, and long-term prospects are not considered attractive and profitable.

Formation of the components of economic self-determination of small business entrepreneurs is possible in the process of psychological counseling, using analytical, developmental and modeling technologies.

Group psychological counseling made it possible to increase self-esteem, purposefulness, self-confidence, indicators of competitiveness, develop effective scenarios for economic self-determination and change attitudes towards money.

The results of mathematical and statistical analysis show positive changes in the structural components of economic self-determination and prove the effectiveness of the developed program.

Acknowledgments

The work was supported by grant 19-013-00430 A of the Russian Foundation for Basic Research.

References

1. O.S. Deineko. *Economic psychology*. (Publ. house of St. Peter. Univ., St. Petersburg, 2020)
2. R.A. Fatkhutdinov. *Management of the competitiveness of the organization* (Eksmo Publ. House, Moscow, 2015)
3. N.L. Ivanova. New perspectives of the business psychology. Organiz. Psychol., **10**, 208 (2020)
4. N.L. Ivanova, V.A. Shtroo, N.V. Antonova. *Business Psychology* (Yurait, Moscow, 2016)
5. J. Keynes. *The general Theory of Employment, Interest & Money* (Yurait, Moscow, 2018)
6. E. Klimov. *Psychology of labour, engineeraing psychology and ergonomics* (Publ. Center "Acad.", Moscow, 2015)
7. N. Lebedeva, A. Tatarko. Relationship of values and assessments of situations of social and economic behavior. *Mat. of 11 Int. sci. and pract. Conf. Economic psychology: topical theoretical and applied problems* (Irkutsk, 2010), pp. 96–102
8. K. Mikulchik. Assessment of the competitiveness of the personnel of the enterprise. Young sci., **5**, 370–372 (2016)
9. A. Petrovsky, M. Yaroshevsky. *Fundamentals of theoretical psychology*. (INFRA, Moscow, 1998)
10. N. Pryazhnikov. *Social and professional self-determination of personality in humanitarian and economic education: perspectives and development* (Publ. Cent. "Rus. Econ. univ.", Moscow, 2015)
11. S. Reznik. *Foundations of personal competitiveness* (INFRA, Moscow, 2019)
12. S. Rubinshtein. *Fundamentals of General Psychology* (Publ. House "AST", Moscow, 2020)
13. M. Semenov. *Questionnaire for studying the attitude of schoolchildren to money* (CAS Print. Cent., Omsk, 2009)
14. T. Terekhova, E. Trofimova (2019). Economic self-determination of an entrepreneur in the context of a metasystem approach. News of Irk. State Univ., Ser. Psychol., **30**, 68–82. DOI: http://doi.org/10.26516/2304-1226.2019.30.68
15. A. Zhuravlev, A. Kupreychenko. *Economic Self-Determination: Theory and Empirical Research.* (Publ. house Inst. of Psychol. RAS, Moscow, 2007)

Digitalization of transport sphere: relevance and main directions

Polina V. Kharitonova [1,*], *Natalia V.* Grokhotova[1], and *Evgeniya V.* Bogdanova[2]

[1] Bratsk State University, 40, Makarenko str., Bratsk, Russia
[2] Baikal State University (Bratsk branch), 14, Karl-Marx str., Bratsk, Russia

Abstract. The object of this study is the program of digitalization of the transport sector for the purpose of stable and effective development of the Russian economy. The subject of the research is the development of directions for the implementation of the digital economy in road transport. The working hypothesis of the study involves the construction of a systematic understanding of the role of digitalization in the transport sector in the context of improving the country's economy through modern IT technologies. In the existing theory and practice, the idea of the role of digitalization in the transport sector for improving socio-economic conditions needs to be clarified and supplemented. The theoretical basis of the research was the work of both domestic and foreign scientists devoted to the role of digitalization, including in the transport sector. The information base was the data of the Federal State Statistics Service, legal acts of Russia, information obtained at all-Russian and international conferences devoted to the problems of introducing the digital economy in various spheres of life. The article presents its own vision of the concept of "digital economy", substantiates the relevance and significance of the process of digitalization of the transport sector. The main directions of introducing the digital economy into the provision of services for the transportation of goods and passengers by road are proposed, the implementation of which will ensure stable and effective development of not only the transport industry, but also the country's economy as a whole.

Keywords: economy, road transport, digital economy, digitalization, industry economics

1 Introduction

Back in 2016 Russian President V.V. Putin set the vector for the development of the national economy, predetermining cardinal changes for the near future, setting the issues of its digitalization as one of the main guidelines. In modern conditions, information is becoming the highest form of knowledge, and the issue of digitalization is acquiring both theoretical and practical significance

In his 2018 Government Address, the President of the Russian Federation proposed a strategy for the development of regions and settlements, which will help improve the quality of life of people. The President noted that modern communications should cover the proposed process through and through.

It is obvious that it is not possible to implement the chosen strategy without modern digital communications. In this light, in our opinion, one of the main tasks requiring a priority solution is the improvement of activities, increasing the efficiency and coherence of the work of transport enterprises through the active introduction of the digital economy.

About the digital economy. Let's consider the terminological component. The concept of "digital economy" appeared relatively recently – in 1995. American economist N. Negroponte was one of the first to raise the question of the need to move the economy to a new level through information and communication technologies [7]. The term "digitalization" itself is widely used in modern conditions, but not everyone fully understands its essential component. After all, digitalization does not mean a global transition to advanced technologies and the ability to be successful and competitive in the market. This is a more complex concept based on a basic change in approaches to solving social and economic problems.

Today the so-called "digital economy" is becoming popular. At the universities of Russia, this direction of training is gaining relevance [9]. In simple terms, the essence of the digital economy is to optimize the turnaround time. Interest in the digital economy is growing every day, and this is evidenced by the number of reports presented at international conferences on world development [1, 2, 6]. Most countries consider digitalization as a priority area for the development of the national economy. It is digitalization that many researchers began to call "digital dividends", which will allow one in the long term to ensure the stable functioning of all sectors of the economy, improve the well-being of citizens, their level of education and development, the quality and transparency of public services provided, and develop other aspects of social and social life [5].

* Corresponding author: hpv83@mail.ru

2 Problem Statement

Due to the large length of the territory of Russia, the significant remoteness of its subjects, the transport industry plays one of the defining roles in the national economy, which confirms the importance and relevance of the task of digitalizing the management of the transport sector.

Currently, Russia is far from being the leader in the development of the digital economy. This is explained by many factors, such as low GDP growth rates and its volumes, lag in the development of IT technologies, etc. Nevertheless, in recent years, some large companies have made significant progress in the field of digital technologies. An example is the Russian multinational company in the information technology industry Yandex.

3 Research Questions

The study raised the following questions:

1. What are the approaches to the definition of the concept of "digital economy"?
2. How is the digitalization process going in the transport sector?
3. What areas of implementation of the digital economy in road transport can be considered the most promising?

3 Purpose of the Study

The purpose of the study is the development and addition of theoretical and methodological approaches to the digitalization of the transport sector based on IT technologies in relation to economic processes.

4 Research Methods

The present study is based on the systematic approach, for the implementation of which a logical approach, analysis, abstraction, classification and comparison techniques were applied.

5 Findings

The emergence of the digital economy in the 90s led to a large number of scientific publications on this topic. Discussions were unfolding in relation to the interpretation of the concept itself. At a seminar on December 20, 2016, the World Bank gave the following definition of the digital economy: "... the paradigm of accelerating economic development ...". A number of researchers understand the digital economy as a virtual environment that complements reality [5].

We cannot fully agree with this definition, since all actions performed in the economic environment are associated with production, distribution, exchange and consumption, and the appearance of these actions is not associated with the creation of computers. It seems to us that the digital economy can be viewed from two points of view. Firstly, as a mechanism of digital technologies (remote control in transport, distance learning, distance medicine, teleshopping, etc.). Secondly, like all economic production using IT technologies.

Currently, there are quite a few ways to develop the digital economy, and all of them are aimed at deep integration of IT technologies in relation to economic processes. Digitalization is focused on the consumer, segment, quality. The market can be changed to the maximum due to: information, which becomes the main resource; the Internet with its unlimited sites for the development of economic relations; competition; development of small and medium-sized businesses.

A number of researchers believe that the digital economy is an economy that is reflected in numbers [3]. In our opinion, this way of reflecting information is a requirement of the time, when in order to solve economic problems it is necessary to quickly find the relevant information and analyze it. In digital form, these actions can be performed as conveniently and quickly as possible. In addition, our modern life today is inextricably linked with the Internet. If you think about it, then in our hands we often take a phone not to make a call, but to access various sites and applications on the Internet. Citizens are increasingly using Internet sites, as it is very convenient, and sometimes much more profitable.

However, it should be noted that the digitalization process is not proceeding at the same pace in all industries. For example, in industries such as mechanical engineering or agriculture, we are not seeing such a rapid transition to digitalization as in the trade sector. This is due, first of all, to the peculiarities of the functioning of industries, their scale, the difficulties of introducing advanced technologies, the availability of sufficient financial resources, the feasibility and return on investment in the development of IT technologies.

Digital economy in the transport sector. Moving significant volumes of various goods, servicing the circulation process, carrying out passenger transportation, transport meets the needs of all other industries, plays the role of a "circulatory system" and therefore is one of the most important sectors of the economy of any state. For Russia, with its long length, given the current state of transport routes, the issues of the advanced development of the transport industry are significantly increasing. Our country is characterized by the presence of vast air, land and sea territories where it is possible to successfully develop unmanned vehicles, logistics using robotics, and use IT technologies. This is exactly what the President of Russia spoke about in his 2018 Address to the Government of the Russian Federation, outlining the vector of development towards the digitalization of the transport sector.

Transport as a sector of the national economy is a structural link of the entire economy as a whole. Therefore, the digitalization of the transport sector is directly related to the development of the digital economy at the national level [8]. The most obvious example of the transition to digital format in the transport sector is the synthesis of digital data between suppliers and buyers of transport services. These are currently the most popular electronic tickets and references.

The Program proposed by the President of the Russian Federation covers a wide range of issues related to measures for the transition to digitalization, with the platforms used for these purposes, with performance indicators and other significant aspects.

The transition to digitalization in the transport sector should unite all interested participants in this process: involve the spheres of science, education, entrepreneurship and the state. Only close interaction of these participants will allow us to create and build an effective and modern platform for digital transport management. The scientific community is already proposing effective mechanisms for the implementation of digitalization programs in the transport sector. The education sector must prepare competent specialists. The business sector is able to timely identify, indicate emerging problems and difficulties, which will ensure their prompt resolution or elimination. At the state level, it is necessary to develop appropriate regulatory and legal acts to ensure information security. The first steps in this direction have already been taken. So since 28.01.2021 the Federal Law of 31.07.2020, No. 258-FZ "On Experimental Legal Regimes in the Sphere of Digital Innovations in the Russian Federation", comes into force. The introduction of this regime involves the application of special regulation to its participants for a certain period of time, which cannot exceed three years. Decree of the Government of the Russian Federation No. 1750 dated 28.102020 approved the list of technologies used in these regimes.

Thus, the implementation of a systematic approach will make it possible to create on the basis of our country a powerful and unified system of digitalization of the transport environment, which ultimately will help to improve the quality of life of the population and create conditions not only for the efficiency of economic activities of transport companies, but also for the effective functioning of the national economy in the whole. In our opinion, one of the main results of the introduction of the digital economy into the transport sector can be the achievement of equilibrium in the transport services market, when the supply in volume and quality level fully corresponds to demand.

Transport, as a set of all types of communication routes, vehicles, technical and other devices that ensure the process of moving people and goods for various purposes, is classified into different types by the medium of movement, by purpose, by the energy used and other criteria. The most common is land transport, including road transport. The 21st century is characterized by the active development and improvement of motor vehicles, high travel speeds, and the development of new types of transport. These trends are confirmed by the analysis of the dynamics of registered vehicles in the Russian Federation for the period from 2007 to 2019, which are objects of transport tax in accordance with the Tax Code of the Russian Federation. During the analyzed period, there is an increase in all vehicles by more than one and a half times, the number of ground vehicles increased by 56.88 % and as of 01.01.2020 amounted to 57,650 thousand units. The share of land vehicles in the total number is quite high and ranges from 97.22 to 98.83 %. The share of road transport accounts for about 80 % of the total number of registered land vehicles, which confirms the relevance of digitalization issues in this area.

Table 1. Dynamics of vehicles registered to entities and individuals according to established order (according to the Federal Tax Service)

Indicators	2007	2010	2013	2016	2019
Total vehicles, thousand units	37 796	43 395	52 390	54 744	58 331
Growth rate in relation to 2007	-	1.1481	1.3861	1.4484	1.5433
Land vehicles, thousand units	36 747	42 868	51 714	54 051	57 650
Growth rate in relation to 2007	-	1.1666	1.4073	1.4709	1.5688
Share of land vehicles, percent	97.22	98.79	98.71	98.73	98.83

Compiled by the authors according to the Reports in form No. 5-TN (consolidated for the Russian Federation) for 2007, 2010, 2013, 2016, 2019. Retrieved from: http://www.nalog.ru

Digitalization in road transport. Today, assessing the development of digitalization in road transport, we can already talk about positive trends and results. This is, first of all, the widespread use of electronic travel documents, ordering transport services via the Internet, the possibility of their non-cash payment and other advantages. For example, a functioning GLONASS system provides users with ample opportunities: for business, it allows you to track the movement of industrial vehicles, monitor the implementation of tasks set by the employer; for citizens it makes it possible to plan a route of movement without waiting for transport at stops for a long time. Modern platforms allow tracking the movement of a vehicle through a smartphone, and all these advantages create more comfortable conditions for consumers of transport services.

Proceeding from the fact that any vehicle is an object of danger, in modern conditions of high speeds, car accidents are one of the factors of increased mortality among children and people of working age. According to the statistics of the Russian Federation, the largest proportion of all car accidents is accidents involving light vehicles. In 2019, this share was 83 %. Road accidents involving freight transport account for 8 % of the total number of accidents committed. Note that in recent years, there has been a positive trend in reducing the number of road traffic accidents (Fig. 1). But nevertheless, the issue of safety on Russian roads is still relevant.

An important role in ensuring road safety in modern conditions is assigned to the installation of photo and video recording objects on the roads. Work in this direction is being actively promoted not only in the European part of Russia, but throughout its entire territory and in regions remote from the federal centre.

Created in order to collect payments for compensation for damage caused by heavy vehicles to public roads, the Platon system, in addition to purely fiscal ones, can provide a solution to a number of other problems. For example, by registering and tracking freight traffic, it is possible to assess the congestion of roads, and the introduction of registration of the severity of goods transported and their hazard would help inform road users online in order to prevent and ensure their safety.

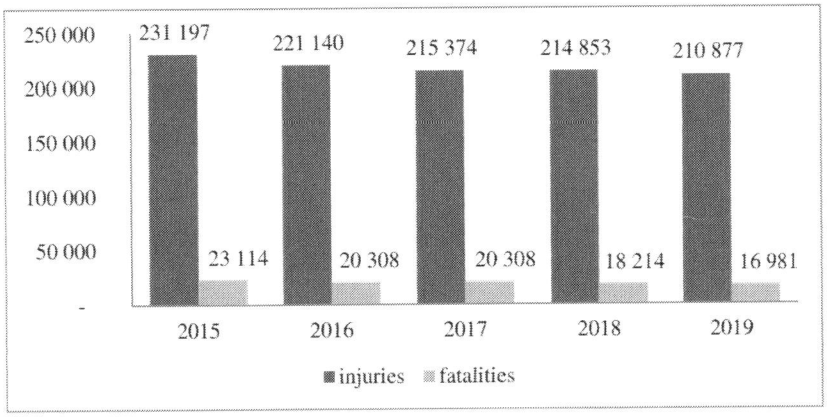

Figure 1. Dynamics of injuries and fatalities in road accidents, people

Nevertheless, at present in Russia the issue of quality and safety of the roadway is still relevant. This problem can also be solved with the use of digitalization. This is stated in the Order of the Government of the Russian Federation dated April 28, 2018 No. 831-r "On approval of the development strategy of the automotive industry of the Russian Federation for the period up to 2025". This document invites, using digital platforms, all road users to send wishes, photos and videos of roads that do not meet the quality and safety requirements. There are similar Internet resources in the regions of the country and are already actively used by caring Russian citizens.

It is gratifying that the partners GLONASS and Platon are currently working on the development of a single digital platform specifically for road transport. Using public-private partnerships, this platform will create additional digital services that help reduce operating costs for both motorists and carriers. It should be noted that this platform will be one of the solutions of the Safe and High-Quality Roads Program, contributing to the development of Russia's transport strategy until 2030.

An intelligent transport system is planned to become a real innovation in the near future. The American company Tesla has already introduced self-driving cars to the market. According to forecasts, ground-based drones will soon become the most anticipated innovative find in the field of digitalization in road transport.

6 Conclusion

Digital economy is a completely new look at economic relations around the world; it is a deep integration of IT technologies in relation to economic processes.

It is quite difficult to keep up with the new ideas of the leading countries in the development of the digital economy. Many opportunities are open for Russia to digitize processes in all spheres of activity, especially in transport. It is necessary at the state level to resolve issues of legal regulation in the field of digital innovations, to stimulate investments in the development of IT technologies and their active subsequent implementation in all sectors of the national economy, and especially in the transport sector.

In our opinion the main directions for the implementation of the digital economy in road transport are:

- firstly, ensuring comfortable conditions and efficiency in the provision of transport services to consumers;
- secondly, creating opportunities for regular monitoring and control of the compliance of the roadway with the established requirements for safe operation;
- thirdly, the solution of the problem of ensuring safety by means of remote control of observance of traffic rules by all its participants, with mandatory immediate punishment of the identified violators.

Investments in infrastructure complexes, competition between technological products and education of citizens are the main levers that will accelerate the process of digitalization of the country's economy.

The ideas for the development of digitalization in the transport sector presented in the article, in our opinion, will contribute to the creation of civilized relationships between the spheres of science, education, entrepreneurship and the state, which will ensure the implementation of the transport strategy of the Russian Federation for the period up to 2030.

References

1. L.V. Bolshedvorova, V.V. Kosyakova. Information culture as a factor in the development of the information society. In: *Current issues of the region's economy: analysis, diagnosis and forecasting* (pp. 270–273) (Stimul-ST, Nizhny Novgorod, 2016)

2. E.G. Grudistova, D.A. Pastuhova, A.M. Slinkov, O.K. Slinkova, T.N. Melnikov. Study of self-actualization needs of Russian students as a factor of competitiveness in the labor market. Espacios, **40(26)**, 1–11 (2019)
3. L. Kaverzina. Restructuring of the regional investment and construction complex. MATEC Web of Conf. Investment, Construction, Real Estate: New Technologies and Special-Purpose Development Priorities, **212** (2018)
4. L. Kaverzina, G. Kovalenko, I. Dudina, O. Belskii. Cost efficiency assessment of automated quality control of precast structures. MATEC Web of Conf. Youth, Science, Solutions: Ideas and Prospects, **143** (2018)
5. L. Kaverzina, P. Koshevoi, V. Dorofeeva. The improvement of company planning at the building enterprise. IOP Conf. Ser. Mat. Sci. and Eng., **667** (2019)
6. I.I. Lomov, M.Y. Vakhrusheva. Automation of the process of selling and accounting of insurance products in commercial banks. Probl. of socio-econ. develop. of Siberia, **2(32)**, 64–70 (2018)
7. N. Negroponte. *Being digital.* (Hodder&Stoughtong, London, 1995)
8. A.M. Patrusova. Modern information technology assesses the effectiveness of investment IT projects. Modern technol. System anal. Model., **2(42)**, 92–96 (2014)
9. A.M. Patrusova, M.Y. Vahrusheva. Process management: Some implementation aspects. IOP Conf. Ser. Mat. Sci. and Eng., **753(8)** (2020)

Assessment improvement of road accidents on highways of the Far North as the basis for sustainable regional development

Dmitry A. Skorobogatchenko[1,*], *Vitaly S.* Borovik[2], *Vitaly V.* Borovik[3], *Anton Y.* Zhabunin[4], *Roman R.* Chugumbaev[5]

[1] Volgograd State Technical University,
[2] Volgograd Scientific and Technological Center,
[3] State Owned Public Enterprise of Volgograd region «Directorate of highways»,
[4] Volzhsky Polytechnic Institute,
[5] Russian Transport University,

Abstract. The paper substantiates the need to assess the road safety in the specific conditions of the Far North by means of the analysis of the factors in the "driver-car-road-environment" system. The authors suggest a methodology for assessment of road traffic accidents, which makes it possible to take into account a wide range of factors affecting road accidents. In particular, the simulation takes into account the characteristics of the driver, technical condition of the vehicle, road conditions, weather and climate. Adaptive neural networks based on fuzzy inference systems are used as a tool for road safety assessment. The authors mention the results of statistical studies on a number of variables of the "driver-car-road-environment" system, which make it possible to form membership functions in the fuzzy inference system. The final part of the paper presents the practical results of road safety assessment for various categories of drivers in different road conditions in one of the largest cities in the Far North.

1 Introduction

The Far North has as natural reserves up to 90% of coal deposits, significant oil and gas reserves, almost the entire volume of explored rare earth metals and diamonds, half of the iron ore deposits, 80% of forest resources and more than 60% of fresh water reserves of the Russian Federation [1]. At the same time, at present, the transport system of the region is characterized by extremely uneven development and a poor use. The lag in transport infrastructure development results in an increase in the cost of transportation of goods and passengers. Thus, the share of transportation costs in the value of the final product of the Far North and the Arctic zone is up to 60%. The dependence of the regions mentioned above on supplies is more than 90%, and the cost of road transport in the Far North is 5-10 times higher than in the central part of the Russian Federation and 1.5 times higher than that in similar areas of other countries of the world [2].

Thus, the key condition for the implementation of Russia's strategic plans to involve huge natural raw materials and other economic resources in the economic turnover is the effective functioning of the road transport system of the Far North and the Arctic zone. A significant problem affecting the efficiency of the road transport complex in the regions of the Far North is the road accident rate, which has serious demographic, economic and social consequences. Despite the fact that the number of road accidents on highways of the Far North is insignificant due to the low population density and, as a consequence, insignificant motorization, in cities the percentage of road accidents remains quite high (Figure 1). More than a quarter of people killed on the roads are the most active people of the working age, and the situation makes a serious demographic problem.

In addition to direct losses associated with death and injury of people, when assessing the socio-economic damage from road accidents, the damage to the owners of motor vehicles, loss of road maintenance services, loss of cargo owners, police costs and medical institutions, costs of enterprises and institutions whose employees became participants in road accidents, costs of social security agencies, as well as a number of additional indirect costs are calculated independently.

One of the mesures reducing the number of accidents and the severity of their consequences is the development of organizational steps in the field of traffic safety, as well as automated systems for assessment and accident prevention, which have significant features in the Far North.

* Corresponding author: dmitryskor2004@gmail.com

https://doi.org/10.1051/shsconf/202111200007

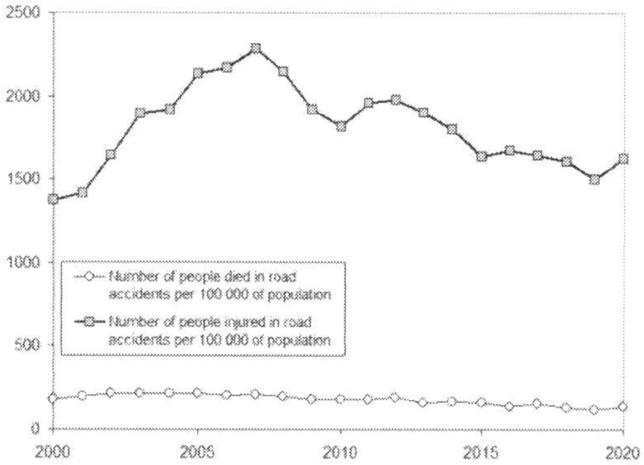

Figure 1. Analysis of the accident rate in the Far North in dynamics, 2000-2020.

The accident rate in the North is determined significantly by the actual condition of pavement, characteristics of speed regime, time and seasonality. Consequently, the definition of traffic safety in the Far North must be carried out taking into account road conditions [3]. It should be noted that the accident rate in the North significantly depends on the behavior of the driver and the technical condition of the car, as well as the transport and operational state of the road, peculiarities of natural and climatic impact on both road conditions and the operation of the car and, in particular, on the behavior of drivers and pedestrians [4]. At the same time, if the number of accidents caused by technical problems of vehicles and unsatisfactory transport and operational state of road decreases every year, then the number of accidents connected with the violation of traffic rules by drivers is steadily increasing.

Consequently, at present, the following issues connected with the assessment of road accidents in a complex socio-economic system are of particular relevance and they are presented as objective factors which are becoming urgent: design parameters and road condition, traffic intensity of vehicles and pedestrians, road construction with facilities and means of regulation, time of year, hours of the day. The subjective factors, characterizing the features and condition of drivers and pedestrians, as well as their violation of established rules are also taken in consideration.

The aim of the study is to develop an automated system for a comprehensive assessment of road safety in the framework of the "driver-car-road-environment" system.

2 Analysis of methodological approaches to problem solution

Since the assessment of road safety in the "driver-car-road-environment" system aims to take into account the maximum possible number of parameters for each of the factors that form the system as a whole, the most common integrated approaches are presented in the following paper [5]. As an example of the implementation of automated safety assessment systems based on a comprehensive assessment of "driver-car-road-environment", we can mention the following study [6], which presents a family of computer simulation models, many of which formed the basis of official documents introduced into practice. A specificity of this direction is the ability to obtain the quantitative assessment of the totality of factors that have the greatest impact on road traffic situation within a particular road segment.

The weakest link in the "driver-car-road-environment" system is the person. In this regard, a significant part of the approaches to automated safety assessment is based on considering "driver-car-road-environment" system as a biotechnical system with the identification of psychophysiological characteristics of drivers [7].

There are a number of papers that, although they analyse traffic safety in the framework of "driver-car-road-environment" system, they focus on the reliability of individual parameters of the system. So in the following paper [8], a complex of simulation models for accident rate assessment based on road parameters, road condition and the quality of drivers is described. There are a number of studies assessing the reliability of vehicle mechanisms and the reaction of drivers when assessing road accidents [9].

Both the classical approaches of the probability theory mathematical statistics and modern Big Data technologies are used as tools for the use of automated traffic safety assessment within the "driver-car-road-environment" [10]. The method of hierarchy analysis and expert assessments are widely used in the design of "driver-car-road-environment" assessment systems [11].

3 Safety assessment model in the Far North

The general structure of the "driver-car-road-environment" system in terms of road safety assessment is described in sufficient detail in a variety of scientific studies. It is substantiated that the assessment of the accident rate in the "driver-car-road-environment" system is well formalized using the method of expert assessments. However, the use of this technique is difficult, due to a large volume of input data, and it is difficult to isolate within them patterns that simplify the construction of the rule base. Therefore, as an implementation tool, a set of instruments of fuzzy neural networks of the ANFIS type was chosen, which, through training, automatically generate a rule base. The details of the functioning and principles of the network are presented in the paper [12].

Let us analyze the features of the variables that form the basis for the formation of term sets of variables for the conditions of the Far North.

All researchers agree that changes in a driver's performance are influenced by age, which affects the mental and emotional state of the driver [13]. At the same time, among young and middle-aged drivers, the risk of road accidents for men is significantly higher than for women, while among older drivers the opposite tendency prevails, i.e. the risk of road accidents for older women is higher than for men of the same age. The study of individual characteristics of driver behavior is shown in Figure 2.

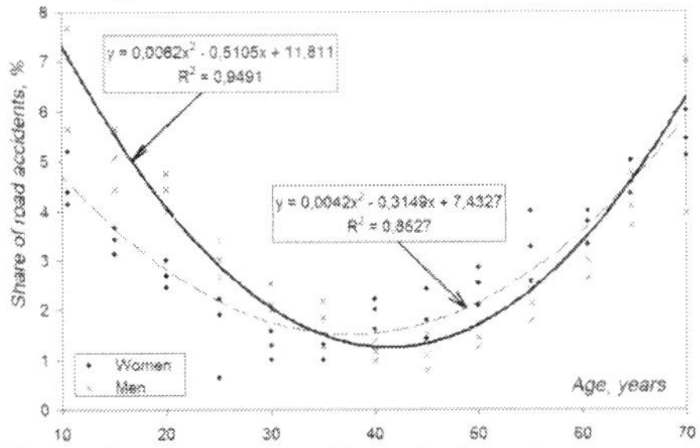

Figure 2. Analysis of accidents in dependence on age and gender of the driver.

The safety of the factor "car" in the Far North is largely influenced by weather and road conditions. In general, when describing the reliability of this factor, the efficiency of the vehicle safety systems was taken into account, with respect to time between failures in time. The results of observations of vehicle failures are shown in Figure 3.

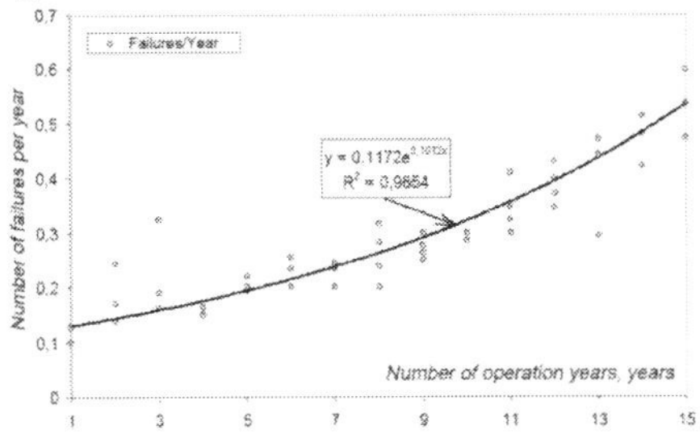

Figure 3. Reliability analysis for «car» factor.

When assessing the "road" factor, it is advisable to use a system of variables defined by the concept of "road conditions". In this case, the most significant variables that determine traffic safety in the "driver-car-road-environment" system are the intensity of traffic flow and the vehicle speed (Figure 4).

The environment or meteorological conditions are the main external factors affecting all other elements of the "driver-car-road-environment" system. Weather and climate in the Far North have a significant impact on the general psychophysical state of both the driver and the pedestrian.

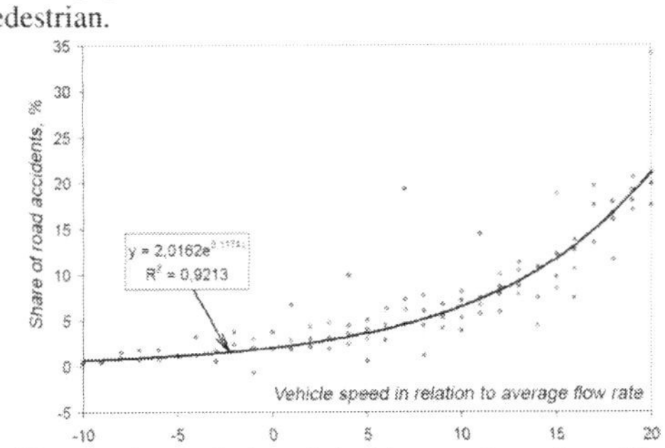

Figure 4. Analysis of incidence of road accidents depending on the change in vehicle speed in relation to average flow rate

There are a lot of scientific studies devoted to the analysis of changes in the consumer qualities of roads and road structures under the influence of weather and climate. And if the influence of pavement condition on road safety is well studied, then the effect of temperature and pressure changes on the driver's behavior had to be developed by forming the "driver-car-road-environment" system of variables (Figure 5).

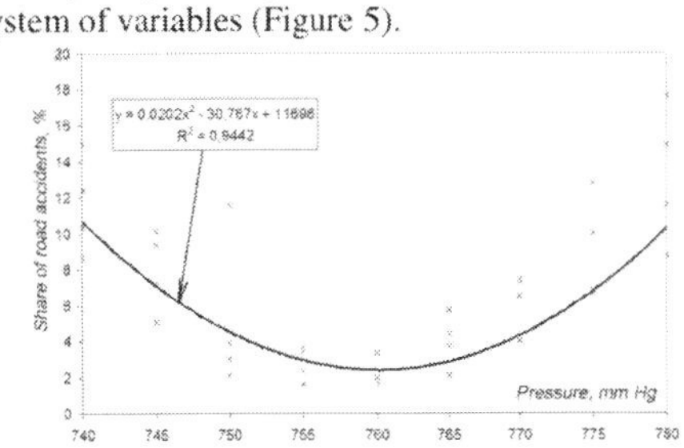

Figure 5. Influence of atmospheric pressure on accidents caused by the driver.

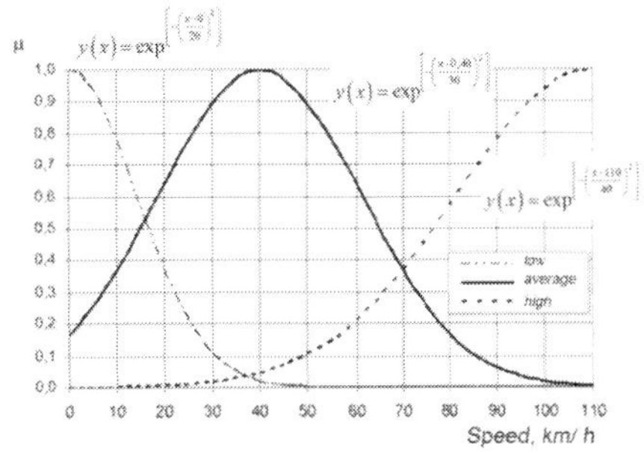

Figure 6. Membership functions used as the example of the variable "speed"

Taking into account the dependencies of the type presented in Figures 2-5 and similar ones, not presented in the paper by the authors for reasons of article volume growth, for all variable factors "driver", "car", "road", term sets for linguistic variables are identified. As an example of the formation of term sets before normalization, Figure 6 presents data on the variable "vehicle speed". The general structure of the input variables is shown in Figure 7.

On the basis of the formed term sets, an artificial NNN was developed, which is a parallel computing system consisting of a large number of elementary information processing units i.e. neurons that accumulate experimental knowledge and provide them for further processing (see Figure 7).

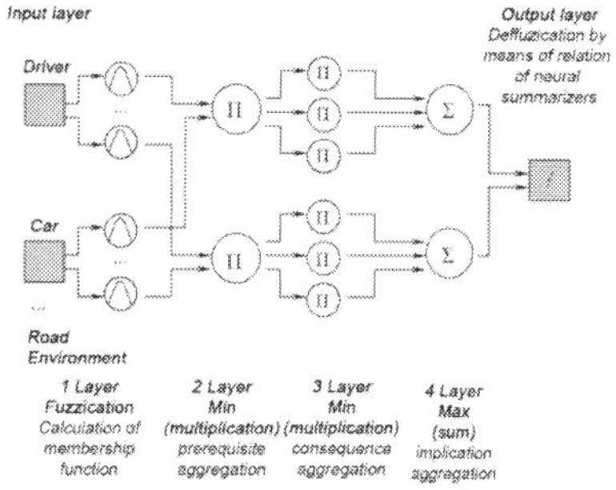

Figure 7. General diagram of the NNN structure, according to the assessment of traffic accidents on highways

The general structure of input layer variables is shown in Figure 8.

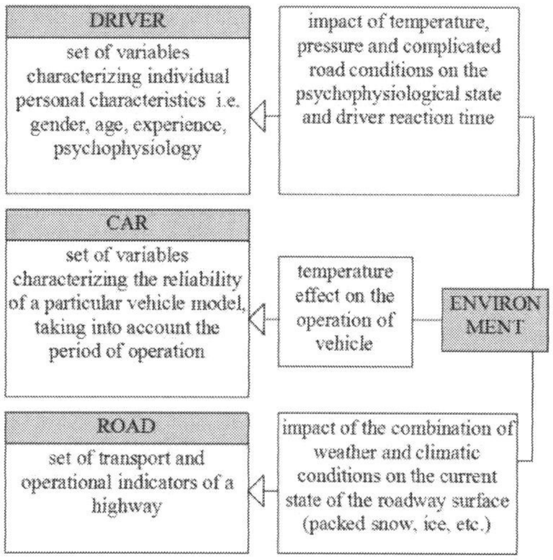

Figure 8. The set of input variables in the NNS, according to the assessment of traffic accidents on highways

Fuzzification is carried out on the first layer of the NNN, according to the assessment of road traffic accidents.

$$\mu_A^{(k)}(x_i) = \frac{1}{1 + \left(\frac{x_i - c_j^{(k)}}{\sigma_j^{(k)}}\right)^{2b_j^{(k)}}} \qquad (1)$$

where k - number of membership functions (k=1...M); x_i - current value of the i-*th* input variable after normalization; j - number of variables (j=1...N); $c_j^{(k)}$, $\sigma_j^{(k)}$, $b_j^{(k)}$ - parameters that determine, respectively, the center, width and shape of the k-*th* membership function of the j-*th* variable.

In the second layer, the values of variables x_i are aggregated in accordance with the formula:

$$w_k = \prod_{j=1}^{N} \frac{1}{1 + \left(\frac{x_i - c_j^{(k)}}{\sigma_j^{(k)}}\right)^{2b_j^{(k)}}} \qquad (2)$$

The parameters w_k (k = 1 ... M) calculated in this way are simultaneously sent further into the 3rd layer (for the multiplication by weights) and into the fourth layer for the calculation of their sum in the neuron f_2.

The third layer calculates the centers for k-rules using the formula:

$$y_k = p_{k0} \qquad (3)$$

where p_{k0} can be viewed as the center of the membership function c_k.

After that the aggregation of consequence is made by means of the algebraic multiple carried out: $w_k \times y_k(\mathbf{x})$.

The fourth layer is presented by two neurons: f_1 and f_2 performing the aggregation of results:

$$f_1 = \sum_{k=1}^{M} w_k \times y_k(\mathbf{x}) = \sum_{k=1}^{M}\left[\left(\prod_{j=1}^{N} \mu_A^{(k)}(x_j)\right) \times c_k\right] \qquad (4)$$

$$f_2 = \sum_{k=1}^{M} w_k = \sum_{k=1}^{M}\left[\prod_{j=1}^{N} \mu_A^{(k)}(x_j)\right] \qquad (5)$$

The fifth layer is presented by a single neuron performing defuzzification:

$$y(x) = \frac{f_1}{f_2} = \frac{\sum_{k=1}^{M} w_k \times y_k(\mathbf{x})}{\sum_{k=1}^{M} w_k} = \frac{\sum_{k=1}^{M}\left[\left(\prod_{j=1}^{N} \mu_A^{(k)}(x_j)\right) \times c_k\right]}{\sum_{k=1}^{M}\left[\prod_{j=1}^{N} \mu_A^{(k)}(x_j)\right]} \qquad (6)$$

The NNN learning algorithm can be conditionally divided into two stages. Initially, the parameters of the center of the output membership functions in the third layer are subject to training. Further, using the gradient descent method, the parameters of the membership functions of the first layer are adjusted.

4 Conclusions and discussion

Analyzing the presented dependencies, it is obvious that the accident rate changes during the week having the peaks in the morning and on Saturday. Also, the diagrams clearly show the difference in accident rates for men and women of different ages. Particularly noteworthy one is the tendency of an increase in the accident rate among drivers with significant experience, caused by traffic conditions (nighttime and negative temperatures).

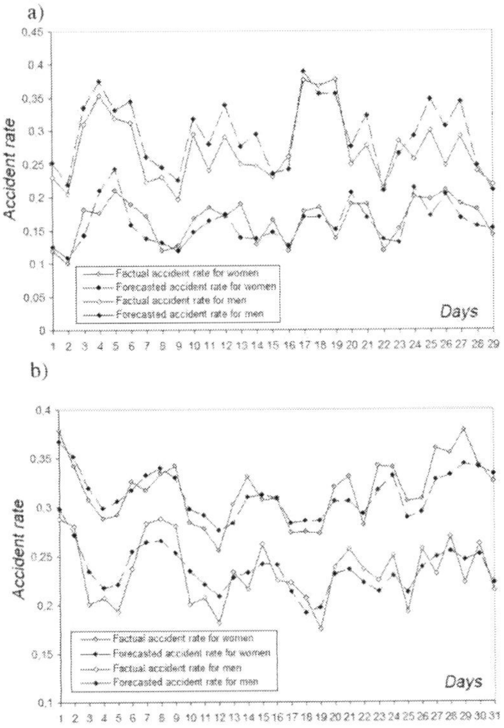

Figure 9. Forecast of the accident rate on the roads of Murmansk: a) for drivers under 25 in the daytime, June; b) for drivers 35-50 years old, in the nighttime, February.

In general, it can be concluded that the authors have developed a model for accident rate assessment taking into account the individual characteristics of the driver, traffic conditions and the specifics of the impact of weather and climate. As a result of training the NNS underlying the model, sufficiently accurate results of the accident rate assessment were obtained (error is less than 10%). The model makes it possible to take into account the peculiarities of the regions of the Far North and, in particular, the influence of road conditions, weather and climate on the behavior of drivers and the operation of vehicles when assessing the accident rate. The practical use of the dependencies suggested by the authors will make it possible to develop measures to reduce the accident rate on the roads of the Far North, and, therefore, to increase the efficiency of the road transport complex, which will directly contribute to the social and economic development of the region.

References

1. I.V. Samsonova, A.B. Yarlykapov Economic policy, **4**, 13-18 (2016)
2. A.V.Shpak, V.A. Serova National interests: priorities and security, **17**, 31-36 (2011)
3. J.J. Rolison, S. Regev, S. Moutari, A. Feeney Accident Analysis & Prevention, **115**, 11-24 (2018)
4. Haque M. Ohidul, Haque Tariq H. Transportation Research Part A: Policy and Practice, **118**, 594-607 (2018)
5. V.A. Korchagin, S.A. Lyapin, V.E. Klyavin, V.V.Sitnikov Fundamental research, **6-2**, 251–256 (2015)
6. V.M. Eremin CAD and GIS systems of automobile roads, **1(2)**, 90-93 (2014)
7. A. Bennajeh, S. Bechikh, L.B. Said, S. Aknine *Transactions on Computational Collective Intelligence XXX. Lecture Notes in Computer Science*, **11120** (2018)
8. A.V. Skrypnikov, E.V. Kondrashova, S.V. Dorokhin, A.G. Chistyakov Bulletin of Voronezh State University of Engineering Systems, **3(61)**, 77-83 (2014)
9. S.N. Lazarev, N.A. Oreshyn, D.A. Mamleev Management of road safety activities: state, issues, ways of improvement, **1(2)**, 278-284 (2019)
10. H. Wang, XY. He, LY. Chen, et al. J. Comput. Sci. Technol, **35**, 875-888 (2020)
11. E.S. Voevodin, E.V. Fomin, K.V. Pulyanova, A.M. Askhabov, A.S. Kashura, N.V. Golub Bulletin of Irkutsk State Technical University, **5(136)**, 240-250 (2018)
12. D.A. Skorobogatchenko Bulletin of Volgograd State University of Architecture and Civil Engineering. Series: Construction and Architecture, **45(64)**, 188-198, (2016)
13. S. Ghaemi, S. Khanmohammadi, M.A. Tinati, International Journal of Control, Automation and Systems, **10**, 517-528 (2012)
14. D.A. Skorobogatchenko, A.V. Erohin Bulletin of Volgograd State University of Architecture and Civil Engineering. Series: Construction and Architecture, **36(55)**, 174-181 (2014)

Development of management abilities of motor transportation companies' leaders

Nadezhda I. Chernetskaya[1,*], *Elena A.* Kedyarova[2], *Vera V.* Monzhievskaya[3], and *Margarita Yu.* Uvarova[4]

[1] Irkutsk State University, 30-14, Mayakovsky Str., Angarak, Russia
[2] Irkutsk State University, 39-8, Komsomolskaya Str., Irkutsk, Russia
[3] Irkutsk State University, 42-2, Rosa Luxemburg Str., Irkutsk, Russia
[4] Irkutsk State University, 27-312, 25th of October Str., Irkutsk, Russia

Abstract. The paper presents main ways of development of management abilities of motor transportation companies' leaders. Summary of modern investigations devoted to development of management abilities in made. Peculiarities of management abilities is motor-transportation branch are shown. The study devoted to personal traits of motor transportation leaders that can be a factor of their better quality and higher effectiveness at work is done and its results are presented. It was proved that several personal traits give the basis of a better quality at work of motor transportation leaders: activity, communicativeness, persistence, stress tolerance, purposefulness, and absence of the role conflict at work. These traits and qualities can be taken as a basis when we decide to develop management abilities of motor transportation leaders. At the particular motor transportation company in Irkutsk (Siberia region) where the investigation was done the majority of leaders have these traits and qualities at much lower level than it is required. As a result, the developing program for them was proposed. This program included the description of conditions on which the developing work can be done effectively and the content of the developing work itself. The proposed and tested program consists of four blocks including developing, modelling and correcting psychological technologies that taken in common had shown a significant shift of management abilities of motor transportation leaders.

Keywords: management abilities, transportation companies' leaders, personal traits, qualities, role conflict

1 Introduction

The XXI century is the time when not only high technologies, but also highly qualified leaders in all branches are required. Some decades ago, high qualifications meant mainly the education and the experience in the branch, but today it is not the way – the personal traits and qualities and good management abilities among them are the main point of high qualification. Management abilities and personal traits and qualities related to them give a significant reason for promotions and for professional success. In the transportation branch that grows and changes every day effective leadership is one of the most important issues to study and to develop.

The problem of development of management abilities was taken into consideration and was studied many times [1–3], but not in relation to personal traits and qualities and not in the field of motor transportation branch as much as it is required and as we have shown in our previous publications [10, 11].

2 Problem Statement

The applied point if this research is related to the particular case in Irkutsk: a general manager of the motor transportation company made a request upon the investigation of his employees that do management work and hold managerial duties. He characterized them as highly qualified specialists in their field but not effective enough leaders. These managers mostly understand that they are not quite good at their leadership at work, they have a fear to lose their posts and, as a result, they are under stress almost all the work-time. Recruiting new managers is not the best solution in this case because this branch has very specific demands upon employees and it is not easy to find many qualified motor transportation specialists who are ready to come into the company. The same problem is stated constantly in publications such as: Druker P.F. (2004), Erina S.I. (2009), Gerasimov B.N. (2017), Gornakova N.S., Uvarova M.YU. (2018), Ogorodnikov A.Y., Przhlenskaya I.B. (2019), Pfeffer J. (2015), Savelyev I.I., Mekhdiev S.Z. (2017), Zhuravleva I.I. (2017). Researchers confirm that nowadays leaders are highly qualified, but have poor leadership and management abilities.

So the request turned into a study of personal traits and qualities of motor transportation company leaders that determined their leadership's effectiveness, and as a result the program of development of management abilities of these employees was prepared and tested.

* Corresponding author: cherna@yandex.ru

3 Research Questions

1. The first point of the study was to make a theoretical analysis of the researches done upon the subject of the study and to see what developmental methods are used for leadership development.

2. The second point of the study was to find personal traits and qualities that can be taken as indexes of management abilities of motor transportation company leaders.

3. The third and the main point of the study was to prepare a program for motor transportation company leaders to develop their management abilities in a short time period.

4 Purpose of the Study

The aim of the study was to make a research of management abilities of motor transportation leaders, to prepare a developing program for them and to prove its effectiveness.

5 Research Methods

The following empirical methods were used: perceptive detection of stress-tolerance, diagnostics of communicative and management abilities with the 'KOS-2' test, diagnostics of leadership orientations by T. Santalayen, diagnostics of the role conflict by S. Erina.

The following mathematical and statistical methods were used: Wilcoxon Rank-Sum Test, Mann–Whitney U-test.

6 Findings

6.1 Indexes of management abilities and their correlation

The first index of management abilities was stress tolerance, and it appeared to be low in the sample of the motor transportation company leaders. 17 % of them have a type A stress tolerance, they are very motivated to success and use all possible means to achieve it but also they are constantly unsatisfied by themselves and by their life circumstances, so sometimes they can be aggressive and not patient. They are highly active; they have a high verbal indexes and very strong subverbal signals that can cause problems both with their colleagues and with administration. Only 42 % of leaders have a high level of these qualities. 4.17 % have an extremely low level of these qualities. These test subjects felt uncomfortable during testing, were suspicious about the research especially about the point whether this testing was related to administrative check, they wanted to finish the testing as soon as possible. They were not interested about the results of the testing; they did not have questions about it and were disagreeing about some point of the research.

Leadership orientations' testing has shown that the majority of leaders (more than 75 %) were oriented to the task. That orientation was good from the point of view of corporative culture, corporative values of this motor transportation company. This company is very orientated to the value of time, success and motivation. But also 17 % of leaders of the company did not show any orientation in leadership. It means that these leaders cannot set tasks; they do not have strong goals and cannot organize collective work of other employees.

The role conflict testing let us define the group of leaders (8.3 %) that demonstrated a high level of role conflict. They are very much under stress about being leaders and about their responsibility at work. They want to be leaders and they mostly like their job but they hate the responsibility and they suffer from work stress. It makes their everyday work stressful and hard for them mainly because of constant necessity to control them.

After counting averages of management abilities' indexes, we defined two groups of motor transportation company leaders: group 1 – leaders with high management abilities and group 2 – leaders with low management abilities (later this group was also divided into two samples – experimental and control groups). Results of the comparison are presented in figure 1.

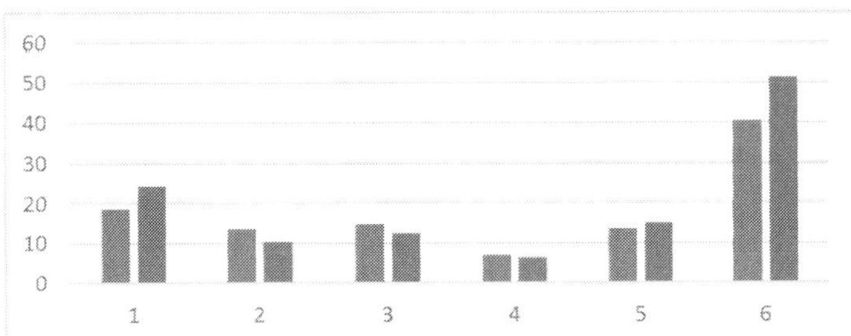

Legend: blue – group 1, red – group 2
1 – stress tolerance indexes of A-type
2 – communicative abilities
3 – organizational skills
4 – orientation for people
5 – orientation for work tasks
6 – role conflict indexes

Figure 1. Ratio of indicator of management abilities of motor transportation company leaders of groups 1 and 2 (averages)

6.2 Developing program to increase management abilities of motor transportation company leaders

The program was developed on the basis of the results of the diagnostics described above. The program consisted of four parts. Each of them was focused on a particular index of management abilities of motor transportation leaders. It is presented in table 1.

6.3 Efficiency marks of the developing program

The program has shown the high level of efficiency upon all indexes of management abilities – all of them at a certain point are higher in the experimental group than that in the control group.

By means of Wilcoxon Rank-Sum Test we found out that changes that took place in the experimental group after the developing program were statistically significant for the perceptive level of stress tolerance, communicative and organizational skills, role conflict level. The indexes are presented in table 2.

Table 1. The program of development of management abilities of motor transportation company leaders

No.	Time	Topics	Comments
		Part 1 Leadership and goal setting	
1	10:00 – 10:30	Introduction	Greetings, presentation of topics
2	10:30 – 11:00	Mini-lection	Management abilities
3	11:00 – 11:15	Coffee break	
4	11:15 – 12:05	Leadership and management	Leaders and manager terminology, developing leadership exercises
5	12:05 – 12:30	Mini-lection	'Why it is important to set priorities to achieve goals'
6	12:30 – 13:00	Goal setting	Goal setting exercises
7	13:00 – 14:00	Lunch break	
8	14:00 – 14:30	Mini-lection	SMART-tasks
9	14:30 – 15:00	Training	Technologies of working with SMART-tasks
10	15:00 – 16:00	Conclusion	Feedback and concluding exercises
		Part 2 Communication	
1	10:00 – 10:30	Introduction	Greetings, presentation of topics
2	10:30 – 11:00	Mini-lection	Communicative and organizational skills
3	11:00 – 11:15	Coffee break	
4	11:15 – 13:00	Communication development	Communication training
5	13:00 – 14:00	Lunch break	
6	14:00 – 15:15	Teaching movie	Watching the movie 'Convince and win'
7	15:15 – 16:00	Conclusion	Discussing the movie and the feedback
		Part 3 Forming the role of the leader	
1	10:00 – 10:30	Introduction	Greetings, presentation of topics, introductory exercise
2	10:30 – 11:00	Mini-lection	Self-esteem
3	11:00 – 11:15	Coffee break	
4	11:15 – 13:00	Confident leadership	Training of self-confidence
5	13:00 – 14:00	Lunch break	
6	14:00 – 15:00	Conclusion	Concluding exercises and the feedback
		Part 4 Stress tolerance	
1	10:00 – 10:30	Introduction	Greetings, presentation of topics, introductory exercises
2	10:30 – 11:00	Mini-lection	'How to avoid distress'
3	11:00 – 11:15	Coffee break	
4	11:15 – 13:00	Stress tolerance	Exercises to increase stress tolerance, learning and training relaxation techniques
5	13:00 – 14:00	Lunch break	
6	14:00 – 15:10	Stress tolerance	Training techniques with self-control and relaxation
7	15:10 - 1600	Conclusion	The feedback and conclusion

Table 2. Significance indexes of differences between management abilities of motor transportation company leaders before and after the developing program on the basis of Wilcoxon Rank-Sum Test

Significance	Stress tolerance	Communicative skills	Organizational skills	Orientation for people	Orientation for work tasks	Role conflict
Z	-2.214b	-1.826c	-2.070c	-1.841b	-1.826c	-2.207b
Asymptote 2-ways significance	0.027	0.068	0.038	0.066	0.068	0.027

Legend: b – on the basis of positive ranking; c – on the basis of negative ranking

As it can be seen from the data, the given perceptive level of stress tolerance of the A-type was lower in the experimental group while in the control group with time it even was higher. Maybe due to the tasks leaders of this group were working on while others were in the experiment made their stress tolerance even higher than at the very beginning.

Communicative and organizational skills in the experimental group got higher while in the control group they stayed at the same level as at the very beginning, organizational skills on some reason was even lower than they were before. It gives us a reason to assert that our developing program was effective for the experimental group leaders.

Comparing the results in experimental and control groups, we can also assert that the average level of orientation for people in the experimental group got lower while the average level of orientation for work tasks got higher. The leaders of the experimental group became more confident in their decisions; they are not that much under the fear of being wrong as they were before the program. In the control group, the orientation for people did not change and the orientation for work tasks was even lower that in was before which means they were still very much more worried about relationships and about approval than about their work. The role conflict level in the experimental group became lower.

All the indexes are presented in table 3.

Table 3. Significance indexes of differences between management abilities of motor transportation company leaders before and after the developing program in experimental and control groups on the basis of Mann–Whitney U-test and Wilcoxon Rank-Sum Test

Stat. test	ST	ST	Com	Com	Org	Org	OP	OP	OWT	OWT	RC	RC
Mann–Whitney U-test	17.000	12.500	4.500	3.000	16.000	9.000	17.000	15.500	14.500	11.000	15.000	8.000
Wilcoxon Rank-Sum Test	38.000	33.500	25.500	24.000	37.000	30.000	38.000	36.500	35.500	32.000	36.000	29.000
Z	-0.160	-0.889	-2.189	-2.423	-0.324	-1.478	-0.166	-0.419	-0.565	-1.139	-0.481	-1.604
Asymptote 2-ways significance	0.873	0.374	0.029	0.015	0.746	0.139	0.868	0.675	0.572	0.255	0.630	0.109
Precise significance	0.937b	0.394b	0.026b	0.015b	0.818b	0.180b	0.937b	0.699b	0.589b	0.310b	0.699b	0.132b

Legend: ST – perceptive level of stress tolerance; OP – orientation for people; Com. – communicative skills; OWT – orientation for work tasks; Org. – organizational skills; RC – role conflict. b – not corrected for correlation

7 Conclusion

After analyzing the theoretical material around the point of management abilities the main definition of it was specified: management abilities are complex abilities to set goals, develop strategies and plans but also abilities to communicate and organize people of different social, gender, educational and psychological categories to make them work in the same vector.

Studying peculiarities of work of motor transportation company's leaders, we concluded that their work is full of stress and high tension; it is multifunctional and can be done only when they are focused of work tasks.

Taking management abilities' development as its improvement we created and realized a program for motor transportation company leaders. This program was built on the basis of psychological conditions of management abilities of leaders; personal traits and work orientations were taken as main conditions above the others. During the study we divided our leaders' sample into two parts – with high management abilities and with low management abilities and the second one were later used for an experimental and control groups.

After mathematical analysis we concluded that among other indexes used to estimate the effectiveness of the program the following indexed got significantly better: stress tolerance, communicative and organizational skills, and the role conflict at work. It gave us a reason to assert that our developing program was effective and can be used in similar transportation companies to make management abilities of their leaders better.

References

1. J. Arevalo, R. Laud, M. Johnson. Managerial skills, mindsets, and roles: advancing taxonomy to relevancy and practicality. J. Acad. of Manag. Proc. (2014)
2. K. Blanshar. *Leadership: going to the highest points of success* (Piter Press, St. Petersburg, 2011)
3. S.A. Bobinkin. Correlation between management abilities and individual traits of personality in work or line leaders. St. Peter. Univ. Bull., **12(3)**, **Part II**, 17–21 (2009)
4. S.I. Erina, B.S. Lisovenko. Question of creativity in the situation of role conflict. Costr. Univ. Bull., **15**, 114–117 (2009)
5. B.N. Gerasimov. Game modelling of the process of development and working out strategic decisions in organizations. Econ. and busin.: theory and pract., **4(1)**, 57–64 (2017)
6. N.S. Gornakova, M.Yu. Uvarova. Development of stress tolerance of station duty officers in

psychological counselling. *Proceedings of XVII scientific conference* (Irkutsk, April 2018, Irkutsk State University) (2018), pp. 196–201

7. A.Y. Ogorodnikov, I.B. Przhlenskaya. *Basics of management* (Publ. House Rusines, Moscow, 2018)
8. J. Pfeffer. *Leadership: Fixing Workplaces and Careers One Truth at a Time* (HarperCollins, New York, 2015)
9. I.I. Savelyev, S.Z. Mekhdiev. *Personnel management: basics of management in the aspect of organizational behavior.* (Rusline, Moscow, 2017)
10. E.A. Sukneva, E.A. Kedyarova, N.I Chetnetskaya. Development of leadership qualities of future civil employees. Pedag. Image. **13(2(43))**, 240–250 (2019)
11. M.Yu. Uvarova, V.B. Rekoslavskaya. Study of management abilities of civil servants at motor transportation companies. *Material of the XVIII scientific conference* (Irkutsk, 25-26 April 2019, Irkutsk State University) (2019), pp. 253–257
12. I.I. Zhuravleva. Forming and developing management. Busin. Ed. in econ. of knowl., **1(6)**, 38–44 (2017)

Training of specialists for transport construction in the context of digitalization

A Zhygadlo, *K* Poleshchenko, and *Yu* Kodentseva*

Siberian State Automobile and Highway University (SibADI), 5, Mira Avenue, Omsk, 644080, Russia

Abstract. The main goal of the study is to determine the direction of transformation of "SibADI" FSBEI HE as a center of attraction for all concentrated knowledge, skills and abilities in the development of the road industry, within the framework of the development conditions established by the scientific and technological revolution. This takes into account synchronization of all efforts on the part of scientific communities and industrial partners. Today, the Siberian State Automobile and Highway University faces the task of transforming from an industry-specific university with an accumulated scientific base. It includes transport construction, into an entrepreneurial University with commercialization of scientific, technical and educational and results in the following areas: "Creation and testing of intelligent composite materials"; "Creation of automated complexes for modeling multicomponent new road materials, including those of technogenic soils"; "Informational modeling of road objects with data transmission system"; "New methods of road diagnostics and defect recognition"; "Digital control, monitoring, forecasting systems"; "Intelligent logistics and design of network road configurations, as well as the development and transfer of entrepreneurial competencies into educational content". The specificity of training of engineers requires reaching a qualitatively new level, which makes it possible to enter the number of international recognized specialized universities, to solve ambitious tasks.

1 Introduction

A distinctive feature of the world economy is rapid development of science and technology. Emergence of end-to-end technologies indicates onset of a new stage of development in the economy. The challenges we face today determine the directions of transformation of scientific, educational institutions and technology companies to synchronize their efforts in achieving the goals set by the fourth scientific and technological revolution.

In addition, new entrepreneurial models become an element of such associations, which allow creating the effect of trust between the participants in the system, which leads to the emergence of cross-industry cooperation. Therefore, creation of a new mechanism for interaction and further development between science and industry through new models in the form of our proposed Competence Transfer Center on the basis of "SibADI" FSBEI HE will allow solving a role of a guide to new markets emerging at the intersection of industries.

Currently, the trend is to create a single research space in scientific communities based on the solution of the internal market, in which researchers, scientific knowledge and technologies will move freely, and with the help of which the technological base will be strengthened, competitiveness will increase and the ability to meet challenges collectively.

The tasks that "collaboration" solves are to increase the efficiency of research systems of the scientific community, find the optimal balance between cooperation and rivalry in interaction, remove barriers in the labor market for researchers, create conditions for the free exchange of knowledge and technologies, as well as the development of international cooperation. The development of such forms of interaction is also considered in the papers [1, 2, 3].

2 Materials and methods

Transfer of competencies, taking into account emergence of a stack of end-to-end technologies, brings us to network forms of cooperation that form ecosystems around large scientific schools, allowing to use the potential of external scientific and industrial schools, research institutes, as well as science-based small and medium-sized businesses in order to solve the problems of the ecosystem formed around the scientific community. The creation of such an ecosystem in the form of the competence transfer center on the basis of "SibADI" FSBEI HE is considered in this article for preparation of highly qualified specialists who develop transport infrastructure in the context of digitalization. At the same time, the whole range of advanced

*Corresponding author: si_kodentseva@sibadi.org

technologies and approaches ensure high competitiveness of transport infrastructure facilities and structures that meet triad of requirements of the modern global market associated with reducing decision-making time, for the implementation and execution of projects, as well as for bringing products, facilities and structures to the market.

Therefore, "SibADI" FSBEI HE will have to transform from an industry-specific university with the accumulated scientific base for transport construction into an entrepreneurial University with commercialization of scientific, technical and educational results in the following areas: "Creation and testing of intelligent composite materials"; "Creation of automated complexes for modeling multicomponent new road materials, including those of technogenic soils"; "Informational modeling of road objects with data transmission system"; "New methods of road diagnostics and defect recognition"; "Digital control, monitoring, forecasting systems"; "Intelligent logistics and design of network road configurations, as well as the development and transfer of entrepreneurial competencies into educational content".

3 Results and discussion

In the point of view of public policy, technology transfer is of particular strategic importance, affecting the socio-economic development of business entities. Taking into account the search for a regional agenda for the digital transformation of regions, this task as the transfer of competencies, mainly based on new technologies, also becomes an important task for the subjects of the Federation.

Historically, in 1930 SibADI was founded not by chance in the Omsk region, because the predominant physical and geographical location of the Omsk region allows developing and transferring technologies for the road construction industry in Siberia and the Far East with coverage of the Far North and transmission to the CIS countries. Development of social and economic condition of the region in terms of providing infrastructure, pace of development of buildings and structures in the region, ensuring availability and quality of transport and logistics services in the field of cargo transportation at the level of needs of the development of the regional economy is closely related to the participation of SibADI in these issues. Common tasks today are to solve the problems of reducing the outflow of the population and especially talented youth, creating comfortable studying conditions for them and further employment in the region.

To solve these goals, it was proposed to create on the basis of "SibADI" FSBEI HE the competence transfer center, the purpose of which is to create effective interface for interaction between scientific communities and interested industrial partners in developing the scientific potential of the industry and improving the training of highly qualified personnel for transport infrastructure.

The tasks of the center are development of strategic partnership in the system "University-industrial partner"; development of competencies of employees of enterprises, business structures and university employees in the field of technological management, technological entrepreneurship and technology brokerage; assistance in the formation of a personnel reserve of road enterprises, as well as assistance in the employment of graduates in the road infrastructure.

The idea of the Center is that during the formation of the competencies the future graduates, who are at the same time participants in scientific and educational entrepreneurial projects, in addition to mastering engineering skills and abilities, shall receive additional competencies of technology brokers, technology managers and technology entrepreneurs in order to provide intellectual and personnel support at all stages of development innovative technologies.

The focus of this Center is related to the current agenda of innovation and technological development of road enterprises according to the following existing problematic aspects:

- the lag between the pace of development of competencies among university graduates and engineering personnel and the pace of development of innovative technologies;
- lack of mobile infrastructure for educational training in the field of technology entrepreneurship;
- lack of universal models and effective mechanisms for forming project teams (communities of practice) in breakthrough areas of technological regional development.

The need for such transformation is due to the global trend of modern development - the formation of the modern innovative knowledge economy (the hyper-competitive, global, knowledge-driven economy of the 21st century) in the context of the rapid development of technology, total computerization and automation, globalization and hypercompetition, constantly accelerating changes. The task of creating and developing the innovative knowledge economy, high technologies and science-intensive industries is becoming generally recognized.

The task is to create an "economy that generates and applies knowledge-based innovation", not generate "innovations" for their painful introduction into the economy. In this situation, universities face the choice: either to disappear as an educational institution, or to become active players in the "innovation field", both within the global or regional level. A necessary condition for embedding universities in global or regional innovation systems is acquisition of additional competence – entrepreneurial one, which inevitably leads to restructuring (transformation) of any university into the entrepreneurial university.

In the global innovation policy, the "triple helix" model (Henry Etzkowitz, Loet Leydesdorf) is widely spread, according to which universities are reorienting to the entrepreneurial model, in which the commercialization of scientific results and technical and technological developments plays a significant role.

The main feature of the entrepreneurial university is its focus on innovation. In addition to the traditional types of university activities: educational process and scientific research, project activities, as well as the entrepreneurial function in the form of technological entrepreneurship, are applied. Small and medium-sized innovative businesses build their development namely on the instruments of technological entrepreneurship - drivers of economic growth.

The difference of technological entrepreneurship is concluded in the fact that the creation of new products is directly related to the use of the latest scientific and technical knowledge, as well as technologies. The specificity of technology entrepreneurship is as follows:

- concentration on the discovery of new opportunities and the training of unique entrepreneurs in various areas of professional activity who can realize them;
- focus on project activities and project investment;
- ability of specialists to identify, select and develop new features of products and technologies (scientific and technological groundwork) in order to create and increase the value of an innovative product or technology.

The role and importance of technological entrepreneurship in the process of creating added value is steadily increasing. Consequently, there is the growing need for personnel capable of carrying out the relevant activities.

In addition, the problems of personnel training in the field of technology management and technology brokerage are becoming more acute.

The main benefits of the Center and such cooperation are the following:

1). Strengthening the role of industrial partners and universities in the development of regional and national economies.

2). Saving available funds in training of specialists;

3). Access to high-tech equipment;

4). Possibility of conducting internships for teachers at enterprises and improving the qualifications of employees of enterprises in universities;

5). Increasing the competitiveness of universities and industry enterprises.

Forms and models of strategic partnership are determined based on the goals, directions and resources available to the parties. Among them, two main areas of interaction can be noted.

First, the enterprise and the university are two parties of the educational process. The university is a manufacturer, and the enterprise is a consumer of specialists. Therefore, the degree to which the quality of training of specialists meets the wishes of the employer depends on the effectiveness of the feedback between them, and, consequently, the demand for a university graduate in production.

Secondly, the enterprise and the university, in accordance with global and domestic trends, become partners in the field of technology commercialization. At the same time, universities are increasingly trying to engage in technology transfer. Classification of existing forms and models of partnership between universities and enterprises is given below.

Existing models and interactions between Universities and industrial partners:

I. Educational programs.

• "Basic Department" and training of specialists under the order of the corporation;

• "Technology penetration";

II. Orders from industrial partners.

• "Business contracts";

• "Runaway Research Institute";

"R&D orders from the state bodies" (218 Resolution, technological platforms, programs for innovative development of state corporations);

III. Development of new businesses (licenses, patents, IIP).

New forms of interaction between universities and industrial partners:

- independent platforms for cross-industry communication and industry foresight (independent associations of open innovation, clusters);
- creation of joint laboratories/research institutes and a "belt" of companies around the university;
- interaction of universities and companies;
- independent technology transfer centers.

Thus, the creation of the Competence Transfer Center will allow combining new forms of interaction between the University and industrial partners into one structure at the University and increase the efficiency of its work due to the transparency of management and synchronization of its activities under the direct requests of the industry.

The specificity of the Center's activities is determined by its main target function - the function of a "system moderator" that contributes to the formation and transfer of new competencies, mainly to support the joint R&D activities of the university and industrial partners required at TRL3-TRL6 stages, as well as the competencies required for the commercialization of scientific and technical developments.

Therefore, the first "track" of the Center's activities should be focused on advanced formation of competencies necessary for the development of scientific and educational entrepreneurial projects by the type of REC projects, only being implemented both by individual faculties and by inter-faculty teams.

The second "track" should be focused on the formation of intermediaries-entrepreneurs with design, innovation and intellectual competencies - technology brokers.

The tech broker is a scientist entrepreneur who is located at the interface between the academic world (developer, scientist) and the market (project packers, marketers). The main feature of the tech broker is that he is "deeply involved in the technology", in contrast to the packer-marketer and technology entrepreneur, who, rather, will only have superficial understanding of technological issues.

The functions of the tech broker turn out to be similar to the functions of the packer-marketer due to the fact that competencies are added with technical and technological knowledge that is combined with knowledge of product sales in the research area.

The competencies of the tech broker include the following:

• Organization of marketing and analytical work for the "packaging" of scientific technological groundwork.

• Working with roadmaps: ability to read company roadmaps and ability to draw up own research roadmaps.

• Competence in the field of "product management", development of project and commercial proposals.

• Creation of the system of innovative communications with scientists and industry representatives.

• Ability to accumulate competencies and use them for the further commercialization of R&D results.

The first step towards organizing educational forms of training of the tech brokers that are strategically important for both enterprises and universities could be a training program, including intensive trainings and tests, which would identify personnel who are inclined and able to work at the interface between university and industry.

The objectives of the Competence Transfer Center are the following:

1. To contribute to the achievement of partners' goals by developing their professional competencies in those areas of the economy, additional knowledge and skills in which they need to solve specific production and business problems.

2. To assist in the development of a system for improving professional competencies, attracting leading practicing domestic specialists to cooperation.

The tasks of the Center are the following:

1. To form a single communication platform based on the Transfer Center competences based on the exchange of experience and best practices.

2. To provide industrial partners with access to the best techniques and practices in the commercialization of their scientific and technical developments.

3. To provide methodological support to industrial partners at the level of preparation of analytical and information materials and their scientific and technological groundwork to prepare them for commercialization.

4. To develop the competencies of partners in the convenient mode for them, providing them with the opportunity to receive up-to-date information in the format convenient for them - through webinars.

In the future, the functions of the Center, in addition to the function of forming the competencies of participants in scientific and educational entrepreneurial projects, corresponding to the competencies of tech brokers, technology entrepreneurs and technology managers, can be expanded in accordance with both the needs of partner enterprises and the need to solve the strategic tasks of transforming SibADI into the entrepreneurial university.

4 Conclusion

This Competence Transfer Center complies with the current agenda, which determines the need for partnership between the university and industrial partners to reach a qualitatively new level, dictated by the need to intensify regional economic and technological development.

Today, neither industrial partners nor the university can exist without each other. The period of mutual disappointment has passed. New general and large-scale tasks have emerged that require mutually beneficial partnerships. New tasks require systematic, well-defined work in the most significant problem-oriented areas.

Analysis of foreign and domestic experience has shown that the minimum necessary (starting) conditions for the creation of entrepreneurial universities are:

• availability of partnerships with enterprises;

• availability of mechanisms for systematic support and stimulation of project-oriented and entrepreneurial activities contained in the innovation ecosystem of the university and allowing developers to overcome administrative barriers to the commercialization of innovations;

• availability of innovative and competent leaders at the university;

• availability of innovative and competent teams;

• availability of innovative scientific and educational projects (entrepreneurial projects).

In general, the expected results of the creation of the Competence Transfer Center will be determined by its positioning in the innovation ecosystem of the university, namely:

- as an element of the innovation ecosystem of the University, providing transfer of competencies based on the functioning of partner networks;

- as a coordinator of the activities of project teams (communities of practice) consisted of scientists, engineers, specialists and managers;

- as an analytical center that provides joint activities of the University and industrial partners to create a portfolio of scientific and technological groundwork for the commercialization of scientific and technological developments.

In the short and medium term, the activities of the Competence Transfer Center can be focused on the implementation of the SibADI transformation program to the entrepreneurial university. In addition, the activities of the Center can be strategically effective in coordinating joint work with industrial partners.

References

1. D. Yu. Znamenskiy *Upravlenie* **7** **(3)** 12-20 (2019) doi: 10.26425/2309-3633-2019-3-12-20
2. A. O. Grudzinskiy, L. N. Zakhrova, Z. M. Saralieva *The management staff of the University*, 24-32 (2014)
3. K. Belyakov, A. Gavrilov, A. Ishenko, T. Poselova *Technology transfer in the digital economy* (Moscow: NATT, 2020)
4. M. V. Akulcheva, O. V. Sedova *Economic and socio-humanitarian studies* **2** **(2)** (2014)
5. I. V. Rozhdestvensky, O. I. Rozhdestvensky, A. Yu. Tarshin *Innovations* **11** **(205)** (2015)
6. A.N. Link, D.S. Siegel, M. Wright *The Chicago Handbook of University Technology Transfer and Academic Entrepreneurship* (University of Chicago Press, 2015)

Socio-ecological discourse in education for the sustainable development of the North regions

Galina Egorova[1*], *Amina* Nijazova[1], *Aram* Vartanjan[2], *Iuliia.* Gibadullina[3], and *Olga* Demidova[4, 5]

[1] Surgut State Pedagogical University, 10/2, Street 50 years to AULYCL, Surgut, 62842, Russia
[2] Yugra State University, 16, Chekhov street, Khanty-Mansiysk, 628012, Russia,
[3] FSAEI HE University of Tyumen, 6, Volodarskogo St., Tyumen, 625003, Russia
[4] FSAEI HE National Research Tomsk Polytechnic University, 30, Lenin Avenue, Tomsk, 634050, Russia,
[5] FSAEI HE National Research Tomsk State University, Lenin Avenue 36, Tomsk, 634050, Russia

Abstract. The article deals with a new position of socio-ecological discourse in educating a younger indigenous generation for the sustainable development of the North regions. The concept "socio-ecological competence" has been disclosed as a key personal quality and a specific behavioral model of a younger generation. This concept provides integrity, interaction, interpenetration of cognitive, pragmatic, emotional, problem-solving stimuli, providing successful gap overcommig between knowledge, skills, and abilities of ecological and social focus, their practical application in the society, which a younger indigenous generation follows. The material for studies is a model of developing socio-ecological competence in the younger indigenous generation in the North regions. The theoretico-methodological analysis of the model has been grounded on the leading ideas of the sustainable development of the regional communities. Essential principles of socio-ecological discourse in educating a younger indigenous generation for the sustainable development of the North regions, KhMAD – have been disclosed. A model has been developed and tested, as well as content modules, development levels of the socio-ecological competence have been determined, basing on the principle of the socio-ecological discource parameters fulfillment. It has been proved that social and ecological discource is a key direction for the sustainable development of every district.

1 Introduction

The North of the Tyumen region is considered a strategic real estate; a stand-by and reference edge in Russia. Northern territories in Russia take 66% of the whole area, where more than ten million citizens live, who contribute to the achievements of the region's economy. Northern territories are a powerful production resource, yielding more than 42 % of GDP[1, 2].

With the account of the conceptual focus on a long-term social-economic development (2025 year), Northern territories aim at innovations in society and economy development which creates favorable living conditions for the population of the North [4].

However, with the account of a faster development of the Northern territories, several contradictions appear among:

1. high intensity of the industrial development and rapid reduction of the lands for indigenous people for the traditional household;
2. fast resource development and resource depletion as a traditional support of indigenous people;
3. intensive technologies' development of the Northern territories and unemployment among indigenous people.

Along with the socio-economic contradictions there are problematic socio-ecological issues, which orient the regional community to conceptuality and rationality of their solution.

Let's note negative trends in the development of the indigenous people in the Tyumen North:

- high level of innovations in the society and weak social security of the indigenous people;
- high level of technology and antropocentric nature of the region demands high sociolo-ecological competence of every indiginous citizen.

It should be noted that the key sector of the economy in the region is oil and gas extraction and processing. It is not a secret that the fieldwork is conducted by integrated enterprises (more than 85 names), which requires new methods of recording technogenity, protective measures for social and ecological welfare.

These contradictions and modern conditions necessiate increasing the level of socio-ecological competence and thinking among younger generaion of indigenous people in Northern Territories. This fact is undeniable as the contemporary education of the younger indigenous population suggests separation of a person from traditional life that influences socialization, ethnicity and environmental friendliness of a person.

* Corresponding author: egorovagi@list.ru

2 Problem statement

These contradictions conditioned a set of the following problem objectives. What is the subject matter of the social and ecological discourse for the sustainable development of the region? What are the possibilities for socio-ecological education of a younger indigenous population in the North, preparation to the life activity in the regional community of the North? The solution of these issues has a significant contribution into the consistancy of the region's development.

It is important to understand the concept "sustainable development" from several positions. From the point of view of the social discourse, sustainable development in the region is understood as possibility of students' further living in the region, as they are subjects of the education. From the point of view of the ecological discourse it is disclosing and positioning ecological knowledge with the account of the value components in the regional area, which is important for upbringing a younger indigenous generation [10].

For complex study of the objective of the article let's specify the contents of the social and ecological discourses, built into the model of socio-ecological education for indigenous people for the sustainable development and living of the North Regions.

3 Discription of the subject of the article

During literature review more than 200 definitions of the concept "sustainable development" have been analyzed. It allowed determining the core of the concept, based on the integrative interaction of sociality, cost effectiveness, environmental friendliness – as characteristic trends of human life and activities [7]. Special mentioning should go to the definition which presents: progressive development and absence of irreversible changes in biosphere, preservation of non-renewable resources, equal possibility for younger indigenous people [8]. The key invariant core focuses on such parameters as: anthropocentricity (survival factor, humanity development) and bio-sphere centricity (preserving bio-geosphere as an existancial basis). According to such researchers as K.YA. Kondratiev and K.S. Losev) [9], to achieve sustainable development it is necessary to solve such issues as: relation of the society and nature (socio-ecological problem); natural depletion as the activity base (eco-economical issue); enhancing the level of living of indigenous people (socio-ecological problem).

Current understanding of the concept "sustainable development" aims at a special self-organization of a younger generation, a society for understanding the importance of the socio-ecological discourse, its efficiency, safety for living [5,11].

Socio-ecological discourse of educating a younger indigenous people has been and is widely discussed [10]. Categorial-conceptual field of the concept "social-ecological discourse" has been determined and proved. In our research we talk about the dialectal dependance of the internal structure of the concept "socio-ecological discourse" we study it within the parameters of the dialectal dependance "system – environment– process – result". The universal nature of the dialectal dependence has been proved by the fact that it is manifested in all spheres of activity (animate and inanimate nature, cognition, education, including parameters of methodology, content, principles of cognition and education).

Socially-oriented discourse is aimed at preserving and stability of some systems (society and nature), elimination of distructions and conflic situations in the regional comunity. We pay special attention to the parameters of fair welfare division, preserving cultural herritage by the younger generation, developing interconnection between the generations of indigenous people.

The key result of the social discourse is to orient every representative of a young indigenous generation to understanding society and culture as spheres of regional life, their monitoring and increase.

Ecological discourse is aimed at provision and preserving biosphere, nature (its internal and external parameters). Biosphere as a sphere of life and stability includes regional spheres, as dynamic systems, which constantly recover independently and improve (industrial, urban, educational and so on). Herewith, every active environment is ecologized with the account of ecological laws and principles [4].

From this perspective let's note conceptual ideas, created on the principles of ecologization and socialization, which provide methodological grounds for developing main blocks of the model for socio-ecological education of a younger indigenous generation for the sustainable development of the North region.

Basing on the conceptual ideas (U.V. Bromley, A.P. Gudyma) we stress the integrity of the model for socio-ecological education including the parameters of the environment as a combination of interactions, which provide the fulfillment of the idea of the sustainable development in the region, combination of the conditions, content, preparing indigenous population to life activity. Socio-ecological educational environment integrates scientific achievements of different spheres, fulfills them in practical activity and gives an opportunity to include content areas of the sustainable development of the region, both into subject areas and practical activity of teachers and students. It is the first conceptual foundation.

The second conceptual foundation is formed by integrating ideas disclosing functionally important aims for socio-ecological content [6]. Conceptual importance of socio-ecological content is based on the invariant and variative part of the whole education within the sustainable development of the region: a) invariative features of socio-ecological content are incorporated into enlarged modules of the subject areas; b) the variative part operates by independent enlarged modules of ecological and social content including ideas of the sustainable development of the region into the system of elective and optional courses.

The model of socio-ecological education for a younger indigenous generation for the sustainable development of the North regions is transformative and managative at the same time. Its resulting parameter is socio-ecological competence.

Let's note the fact that socio-ecological competence as a special model of behaviour is one of the concepts of a new scientific trend – sociology and pedagogical ecology. Combining and integrating cognitive, pragmatic and value-based parameters, it provides overcoming the gap between ecological and social knowledge, skills and experience and their practical application in the system of rules, and norms of behaviour which a younger indigenous generation should follow.

The features of socio-ecological competence characterising it as a separate concept are principles of regulating interactions and behaviour of a young person in the systems "person - society – nature" [3]. Socio-ecological competence reflects scientifically grounded scope of knowledge, skills and experience, those general requirements for younger generation with the account of theoretical and practical experience.

In general, socio-ecological knowledge is determined by the regularitites of ecology and socialization, personality development. Socio-ecological skills (competences) are understood as the combination of successive actions, based on knowledge of theory and practice of ecological and social content. Mostly these are meta-skills – skills to think ecologically and act socially, analyze facts and events from the ecological and social point of view.

The content of the modules is determined by creating the social and ecological constructs in the minds of younger indigenous generation when studying subject areas. Let us represent fragments of developed and approved modules.

Module 1. Biosphere and its role in the life of the indigenous population.

Biosphere regularities and biodiversity of the Northern nature. Domiant parameters of the Northern environment. Ecology, society as systems and factors of life activity. Socio-ecological competence and tolerance. Aboriginal fishery. Climatic and chemico-technological changes for commercial activities. Ellimination causes and methods. Siberian wildlife parks and their role in the life of indignous people and biodiversity of the life system.

Module 2. Sociocultural sphere and its role in the life of indigenous population.

Social factors and regularities. Number of indigenous Northern people and their role in the life of society. Forecast and perspectives of demographic situations in Northern regions. Health, education and culture as factors of sustainable living of an indignous citizen. Global challenges of culture and health care. Indigenous culture and its preserving. Indigenous activities. Deer breeding, fishing, hide manufacture. Socialization of indigenous people. Polyculture of a younger generation.

Module 3. Ecosphere and its role in life of indigenous people.

Resource components of ecosphere. Renewability and non-renewability of resources. Natural resources as reserves in life activity of indigenous people. Global warming impact on the climate of the North. Resources depletion in the conditions of the region's sustainable development. Younger indigenous generation and ecological crises. Water as a natural resource. Water consumption challenges. Northern land and possibilities of soil resources. Preservation and renovation of the soil. Retrofit, erosive destructions. Oil spills and their impact on indigenous lands.

Module 4. Energy sphere in life activity of the indigenous population.

Natural energy and ways of its utilization. Ways to increase resources production and utilization. Thermal power. Variety of sources based on coal, oil and gas. Pollution and its main causes. Criteria and assessment parameters.

Module 5. Healthy lifestyle of a younger generation as a factor of the sustainable development of indigenous generation.

Health (physical, psychological, emotional, general), risk factors for the younger generation. Living environment and its parameters. Harmful and toxic substances and their influence on the alive systems. LOC of diverse effects (air, water, soil). Health and environmentsl pollution. Attitude to environmental pollution for health maintanance. Nutrition and health of a younger generaion. Emulsifies, coloring agents, alkaloids, as harmful sources for the health of a younger generation.

Harmful addictions (illnesses) of a younger generation. Drug-taking. Alcohol addiction. Nicotine addiction. Sources and ellimination methods.

Module 6. Management of sustainability of a younger indigenous generation.

Management and types of environmental management. Environmental management in the North. Legislation and regulating within the area of indigenous population. Topical importance of the sustainable development of indigenous population in the North. Historical and modern discource of developing parameters for the sustainable development. Legal documents, reflecting sustainable development of the indigenous population. Rationality of the scenario about the development of indigenous population as a northern community.

The process of implementing the model of socio-ecological education of a younger indigenous generation for the sustainable development of the region (KhMAD – Yugra) is carried out in the system of life-long education and on the bases of laboratories of Ethnic education (Surgut State Pedagogical University). Integrity of socio-ecological education has been provided by the inclusion of content modules into the educational process vertically and horisontally.

Practice-oriented classes in the tune of activitites of indigenous population were of great importance. The classes were on traditions, culture, special technologies based on natural raw materials. Let us note some themes represented in the elective disciplines.

"Birch bark and operation techniques with it"- a practical class about the work of Khants with natural material.

"Fur clothing and footwear by Wei" - a practical class about creating necessary things by the indigenous population.

"Technological peculiarities of wood works (log canoe)" - a practical class about the work of Mansi with natural material.

"Technology of song creating"- a practical class about the peculiarities of the the song by Nenets Vipinemya – about Amanitaceae mother for keeping a family, tribal lands, deer.

Dealing with integrated concepts (region, surrounding, culture, society, tribe, natural conformity and other), students estimated modern condition and the role of their younger generation in the sustainable development of an indigenous younger generation. Being not indifferent to the issues under study , students joined the scientific work on the themes demanded during the model implementation. Talking about the fields of scientific research, let us note some variants of the themes: "Strategy for the sustainable development of my ethnos", "Studying natural objects for increasing their resistance to technogenic environment", "Health of a younger generation and harmful habits", "Socio-ecological factors and rational life activity", "Estimating the technogenic and ecological nature of oil and gas processing enterprise for the sustainable development of the region", "Studying the ways to reduce man-caused risks for the northern nature", "Estimating and value of natural areas as specific territories", "Studying local communities for developing civic consciousness with the account of autonomy".

4 Methods used during the implementation

Table 1. Indicators and parameters of socio-ecological competence for educating a younger indigenous generation

Indicators	Subject-matter and appearance	Parameters of socio-ecological competence
Subject bases (humanitarian, scientific, trends)	Developing integrated ideas about different spheres of activity (social, natural, ecological) on the synthesis, analysis of the concepts, generality, personal enrichment	knowledge-representations, knowledge-images (socio-ecological view)
Conceptual	Conceptual importance of understanding social, ecological problems, cause-effect analysis	World-view judgements, conceptual importance of the socio-ecological thinking
Integrational	Integrting facts, theories while creating unified scientific world view	Integrated knowledge, skills, worldview
Active	Cognitive, research, and practical activity	Metasubject competences, experience of practical activity

The problems stated in the article have been solved with the following methods of research: theoretical (literature review on the topic of research, studying scientific materials and publications); empirical (pedagogical observation, testing, questionaire survey, comparative analysis), mathematical statistics methods.

Model implementation brought the parameters of socio-ecological competence to high levels, which was the most important goal of the research. During the experiment representatives of a younger indigenous generation (Khanty, the Mansis, the Tartars) developed mental skills and abilities "think — act — think — think practicaly".

12.0% of students in the experimental group and 38% of students in the control group had a low level of competence. At the high level there were students of the experimental group with the results 28.0%. The effectiveness of the control group was rather low, only 16.00%. The results at the intermediate level were 3% higher than in the experimentl group.

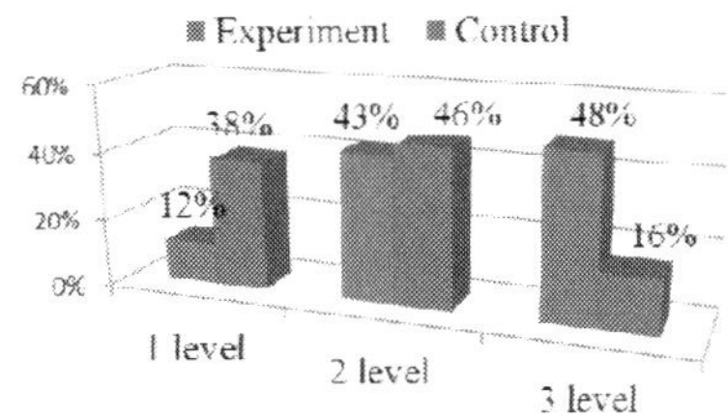

Figure 1. Level expressiveness of socio-ecological competence for students of a younger indigenous generation

Let us note that during the experiment there was a positive dynamics of socio-ecological competence parameters. Along that, there were positive trends in personal development of every student – a representative of the indigenous population. Students expressed their opinions on the following issues: socio-ecological issues reveal contemporary character of problematic areas, study natural and man-made environment; reveal the role and importance of the ethnos in region's activity as a specific oil-gas system; disclose polycultural facts of interethnic cooperation between the Kanty and the Mansis.

Positive responses prove the importance of the model for developing socio-ecological competence of a younger indigenous generation in the region. Besides, the efficiency of the model has been provided and connected with certain types of activity and conditions, they are:

1) choosing an individual development pathway of socio-ecological competence, implementation methodology and technology, educators' professionalism;

2) understanding that sustainable development of the region is a factor and condition of development of the indigineous population;

3) motivation and interest of the younger indigenous population to the possibility to develop socio-ecological competence not only in theory but in practice;

4) development of socio-ecological competence is built on the principles of face-to-face communication, visualization with traditions and culture of the indigenous population.

5 Conclusion

A younger generation of indigenous people of Northern territories lives in conditions which require a high level of socio-ecological competence as an integral personal characteristics, providing a high level of socialization.

The article solves the determined problems. It discloses the pecularities of the social and ecological discourse for the sustainable development of the region. The possibilities of the content modules have been shown for the personal development of younger indigenous people of the Northern territories and their training for life activity in the regional community.

The scientific character of the work is conducted with the methodological model validation, choice of content modules, methods and ways of their implementation.

High practical character of the model is connected with involving younger generation into the activity where natural, cultural, artistic materials are used. It is highly demanded in the life of indiginous population.

References

1. A.G. Aganbegyan, Bull. of the IE RAS. **1**, 7 (2018)
2. O. Bogomolov, The world of transform. **1**, 34 (2015)
3. U.V. Bromley *Ethnical social processes: theory, history, modernity*, 332 (1987)
4. V.I. Bulatov, *Analytical review*, SPSTL SB RAS, 116 (2012)
5. A.P. Gudyma *Social-Philosophical and Ecological Aspects of the Sustainable Development of the Scanty Northern Peoples*, 343 (2000)
6. *Euro-Asian Integration: problem points and ways of development. Project "Social-political dimension of the processes of Euro-Asian integration"*, **3** (2017)
7. G.I. Egorova, N.I. Loseva, O.A. Ivanova, B.M. Chabarova, O.M. Demidova, *Advan. in Intell. Sys. and Comp.*, 45 (2019)
8. G.I. Egorova, *Devloping noospheric ideas in the pedagogical science and educational practice*, (2019)
9. K.Ya. Kondratiev, K.S. Losev, Bulletin of RAS, **72**, 505 (2002)
10. *Conception of the Sustainable Development of the Scanty Northern Peoples of the Khanty-Mansiisk autonomous district – Yugra till 2025year*, 122 (2020)
11. I.F. Vereshagin, K.S. Zaikov, A.M. Tamitskii, T.I. Troshina, F.KH. Sokolova, N.K. Kharlampieva and others.; edited by N.K. Kharlampieva *Ethnical and national processes in the Arctic: tendencies, challenges and perspectives*, 325 (2017)

SHS Web of Conferences 112, 00011 (2021) https://doi.org/10.1051/shsconf/202111200011
Northern Sustainable Development Forum 2020

Innovations in Russian tourism: ethno-tourism and sharing model

Anna N. Polukhina[1,*], and *Oksana S.* Agalakova[2]

[1] Volga State University of Technology, 3, Lenin sq., Yoshkar-Ola, Russia
[2] Vyatka State University, 111, Lenin str., Kirov, Russia

Abstract. This paper aims to shed light on the prospects of ethno-tourism cluster development in Russian regions. The provided methodology is applicable for the study of the cluster approach in the field of tourism. The focus is made on good practices of ethno-clusters in tourism in the Russian regions. There are some scientific methods in this research: methods of the system and comparative analysis, a technique of assessment of resources and capacity of regions, methods of administrative diagnostics, expert estimates, a deep interview, methods of social research will be applied. Tools of a research include (depending on a solvable task): dialectic, formal and logical, concrete historical, structurally functional and statistical methods. Moreover, authors research a new economy model – sharing – and its influence of tourism area. The main results are the challenges and prospects for ethno-tourism development are defined and supported by Mari El case study. The factors exerting influence on ethno-tourism development include tourism potential of the region; attitude of ethnos and local authorities to tourism; availability of the necessary infrastructure and qualified personnel and new information technology. The conditions facilitating success of the tourism cluster development are analyzed. These include, first of all, the development of the target programme for tourism development with the emphasis on ethno-tourism on the level of local authorities; secondly, investment in tourism infrastructure development; thirdly, the sharing model will contribute to establish the ethno-tourism cluster. The research is of practical importance for modernization of domestic tourism programmes in order to boost regional economy.

Keywords: ethno-tourism, cluster, regional tourism, sharing model

1 Introduction

Modern dynamics of crisis phenomena, their scale, lost opportunities for modernization and restructuring of Russian economy aggravated by the gap in the quality of living across territories and regions of the Russian Federation represents a serious problem for the country's development. The majority of Russian Federation areas face a range of socio-demographic, economic, infrastructural, administrative problems which act as the limiting factors for sustainable development of the country in general.

Tourism is one of the most dynamically developing spheres of economy. The development of tourism assumes the involvement of organizations from different fields of economy and long-term cooperation with them [4]. It also stimulates networking and the accelerated implementation of innovative technologies, methods and models of management. The sharing economy is an economic model based on acquiring, providing, or sharing access to goods and services. This model is currently considered as one of perspective innovative models. The model highlights the preference for individuals under certain conditions to rent or borrow goods rather than buy and own them, reducing, thereby, owning costs. Sharing economy services are rapidly developing and wining the market of the USA and other developed countries, thus affecting also the markets of developing countries. Sharing economy in Russia is a new and rapidly developing phenomenon.

The Russian Federation has entered the Federal Target Program on Tourism Development until 2025. The main aim of the program is commercialization of the Russian tourist product on the national and international markets via the development of the tourist and recreational complex in Russia and improvement of the quality of tourist services. As wrote M. Porter, the cluster approach is regarded as the basis for this change [16], i.e. concentrating on a single lot the enterprises and organizations involved in the development, production and sale of a tourist product, as well as related business. In the framework of the program it is planned to establish recreational tourism clusters in most prospective regions. The clusters presume the construction of modern and comfortable hotels, camping sites, cafes and restaurants, sports, recreational and trading centres, parking lots and other infrastructure units. The major financing is expected to come from

* Corresponding author: Poluhinaan@volgatech.net

private investors based on competition of investment projects. The Federal Budget will partly cover investors' expenses in credits and loans. It is planned that the regions shall be partly subsidized (co-financing scheme) for the construction of infrastructure.

Nowadays there is a hierarchy of the tourist regions in terms of scale: tourism and recreation zone, tourist macro-region, tourist meso-region, tourist neighbourhood, tourist object. In different parts of the globe, in different countries the potential tourist regions vary in terms of tourism attraction. Thus, attraction of the region may be obtained due to a number of factors: nature, culture and history, society and economy, ethnic-confession or religious factors.

The record shows that there is a close link between the tourist regions of various levels and tourism and recreation complexes [20]. By the latter, we mean a complex establishment which includes healthcare and recreational institutions, servicing institutions and related industries (retail enterprises and catering facilities, social, cultural and sports institutions, sightseeing objects, traffic network, souvenir production, etc.). Thus, A.Yu. Aleksandrova argues that the most typical features of tourism: vastness of interfiled ties, fragmented structure, dominance of small and middle-sized businesses, non-material character of the tourist product and some others make the application of a cluster approach desirable for tourism development [1]. At the moment the urge for development and the principles for tourism clusters establishment in Russia are approved by the Government.

Ethnographic tourism (ethno-tourism) is defined as a type of cognitive tourism aimed at visiting ethnographic objects in order to learn the culture, architecture, traditions of people, ethnos that are living or used to live on the visited territory. Ethnic differences, uniqueness of the place and the culture are more and more often becoming the local tourism band in the world practice. Such old crafts as winemaking, woven work, fishing and pottery have become additional attractions for tourists. This is one of the ways how new tourist destinations are created, for example, in France, Spain and some Scandinavian countries. Shift in tourist preferences lead to the development of a new model of tourism market, which involves ethno-tourism landscape, which preserves traditional lifestyle with all its characteristic features: cuisine, architecture, etc.

Another good example of ethno-tourism development is from Canada, the country, which focused on tourism as the priority direction of economic development. In this country tourism helps the government solve the problems of aboriginal organizations and northern territories, while aborigine people get the chance to restore and preserve their cultural traditions. The aboriginal tourism industry in Canada is presently estimated at $596 million employing 14.000–16.000 aboriginal people (Report ATAC, 2015). The business includes aborigine specific tourist trails, hotels and camping sites, restaurants with the national cuisine in Indian reserve lands, combined clusters, cultural centres, interactive museums-villages, totemic souvenirs [6]. The development of sharing services is in rush development, especially for ethno-tourism cluster. For example, the Italian Touristic Recourse – VoomaGo [24] web-site provide the information not about tours on reasonable prices, but about ones, that carefully thought-out multi-day tours with immersion in local culture and traditions. The issues of preserving traditional culture, and at the same time, the local population employment are relevant for many regions of Russia.

2 Problem Statement

In many Russian regions there are large-scale ethno-tourism projects. In order to capitalize ethno-cultural potential it is very important to create recognizable and attractive regional brands. The major problem in Russia nowadays is that although regions we tend to create locally recognizable brands, they cannot often link them into a cluster and promote in a proper way [6]. Mari El Republic may be considered as a potential regional tourist cluster. The unique character of the Mari territory in terms of history, culture and ecology can surely make the place a desirable tourist destination. In other words, the regional tourism cluster should be based on the unified concept. We believe, that this unified concept is the concept of ethno-tourism (preserving traditional religious belief of Mari people – paganism), that should act as a basis for the programme of tourism development in the region.

With the development of information technologies, society has returned to the practice of sharing things and services, which is not limited only the personal contacts communication but has grown to the scale of the whole world. This phenomenon was called the sharing economy – collaborative consumption – and was included in the list of ideas to change the world in the near future by Time magazine. The concept of collaboration consumption was proposed by Rachel Botsman and Rogers Rogers in their book «What's Mine Is Yours: The Rise of Collaborative Consumption» [5]. Speaking at the TED conference, Botsman called the sharing economy a new socio-economic model that revolutionizes our consumption of goods and services. Botsman's forecast comes true since millions of consumers use Airbnb, the BlaBlaCar app, Uber, eBay, and other sharing economy services. Online platforms that allow people and companies to share their resources have already created a global market with a volume of $ 15 billion and the prospect of growth to $ 335 billion by 2025. In tourism, the sharing economy model is now widely used, especially in the field of hotel and transport services. But authors think that the sharing model will o establish the ethno-tourism cluster in Russian regions.

3 Research Questions

3.1 Short characteristics of successful ethno-tourism clusters in Russian regions

Ethno-tourism development depends on a number of factors: tourism potential of the region, potential of the area, attitude of local community and local authorities to this kind of tourism, availability of the necessary infrastructure and personnel qualification. In places where the efforts join, there are obvious positive results. The best-developed sites in terms of ethno-tourism are located in Yamalo-Nenets and Khanty-Mansiysk Autonomous Districts of Tyumen Region, Krasnodar and Stavropol Territories in Russia.

Ethnic direction for the Russian tourism is not a new issue. For example, ethnographic open-air museum Torum Maa in Khanty-Mansiysk was founded back in 1987. It is a popular visiting place for both local residents and tourists [22]. In Krasnodar Territory due to ethno-tourism development, all interested people can get to know their national culture, as well as immerse themselves into the traditions of other Kuban peoples. In Anapa region there is another unique park called Dobrodea. The park includes ethnographic exposition "Kazache podvorie" (Cossack habitation), which is represented as a Cossack museum telling the visitors about the history of Cossacks in the Kuban [7]. And finally, not very far from Moscow, close to Borovsk there is one of the largest ethnographic parks in Russia called Ethnomir.

In 2018, the Ministry of Culture of Russia released a collection of "Best Practices of Ethnographic Tourism," which included 129 ethno-tourist objects. A list of 10 most popular ones was highlighted, which included Cossack village "Ataman" (Krasnodar Territory), Park Museum "Ethnomir" (Kaluga Region), Finno-Ugric Ethnopark (Komi Republic), Rural Park "Okolitsa" (Tomsk Region), Ethnopark "Nomad" (Moscow) [21].

3.2 Establishing ethno-tourism clusters in Volga-Kama territory

The Perm Territory is a significantly large area inhabited by numerous nations. In Byrma village the visitors were welcomed with bread and salt and Chak Chak (a sweet dessert, made of honey and pastry) [7]. The two dishes are the traditional welcome symbols of Russian and Tatar people. Another ethnographic route is a cognitive tour "Visiting Sylvenskie Mari people". This tour combines the elements of rural, ethnographic and environmental tourism. The programme also includes learning about household activities: vegetable lot, garden, shed, fenced area for livestock, bathhouse.

However, one of the most famous ethno-tourism events in the Perm Territory is the festival of ethno-futuristic Finno-Ugric culture KamWA which traditionally finishes on the first Sunday of August in the historic and ethnographic nature reserve "Khokhlovka" [9, 18].

In Komi Republic close to Syktyvkar there is the first stage of the large Finno-Ugric ethno-cultural park. The first stage of the park, as envisioned by the architects, is bound to bear the cultural and cognitive meaning.

The Republic of Tatarstan was among the first Russian regions (after Moscow and Saint Petersburg) which started investing into its image as an ethno-tourism region from 1990s. Thus, the Kazan Kremlin is considered the monument of the Khan's period, even though only ground it from that time; the mosque itself has been erected on the ancient ground place.

In our research we referred to Mordovia experience of Ethno-tourism development. The major rivers of the Republic are Sura, Moksha and Insara, which are not the desired tourist destinations. There are over a thousand of historic and cultural monuments on the territory of the Republic, 52 of which are acknowledged on the Federal level. There are currently 7 centres of national culture working on the territory of the Republic. Such centres are especially important for domestic intraregional tourism development. They can attract tourists through organizing holidays and festivals in the centre of national culture, such as the festival of national cultures Shumbrat or Finno-Ugria.

3.3 Challenges of tourism cluster development (Mari El case study)

Mari El Republic may be considered as a potential regional tourist cluster. The unique character of the Mari territory in terms of history, culture and ecology can surely attract tourists. Our opinion is that ethno-tourism (preserving the traditional religion of the Mari people, paganism) should become the central idea for the Mari El Republic which should define the program of tourism development in the region.

There are ethno-cultural monuments almost in every region of Mari El Republic. Gornomariysky region (Mountainous Mari Region) and Kozmodemyansk are famous for their historical potential and tourist resources, particularly for cognitive tourism. Emelian Pugachev (Russian peasant leader, head of the peasant rebellion of 1773–1774) passed this place together with his troops. There is well-known evidence to this fact, such as Pugachev's oak. The town wooden architecture erected in the 19th century still appeal to tourists' eyes. There are a lot of ancient archeological objects in the region. We believe there is an opportunity to significantly develop the potential of ethnographic tourism resources in Mari El by introducing myths of Mari and Russian peoples, contemporary animation experience.

We shall refer to the use of Mari mythology for innovative product development. For example, one of the most colourful solar holidays for Mari people is called Surem (Сÿрем). According to the literature, on this day people exiled evil spirits from the village. The main idea of the holiday is to drive out demons. We believe that the concept of tourism development in Mari El Republic should be developed based on the local authentic culture, thus, it would be unwise to concentrate on one culture tradition only, be is Mari, Tatar or Russian, as in this area we face the multiculturalism model which has been developed for centuries in the Volga region. It is necessary to acknowledge the mutual interrelation and impact of cultures and myths of these peoples onto each other.

The model of ethno-tourism development includes a number of constituent elements, such as certification of the regions in the Republic based on the presence of ethnographic monuments, real business-ideas and projects, search of investors with the focus on tourism infrastructure including accommodation, catering and main transport routes.

3.4 Sharing model: methodology and practice

Let us turn to the study of the economy of joint consumption. Collaborative consumption, which is a set of transactions that are usually conducted using the Internet, primarily online platforms that effectively match the needs of some people with the capabilities of others [5, 12]; is becoming widespread in a number of industries, facilitating the exchange of goods or services between economic agents for certain benefits, both financial and any other [3, 10, 14].

A co-consumption economy is defined as an economic model in which users share assets that they underutilize for material and intangible benefits [5]. The economy of joint consumption in practice represents activities for the principle of P2P (Peer-to-Peer) for receiving joint access to goods and services which are coordinated by online services at the level of communities [8], and connects sellers and buyers by means of digital platforms. In the scientific environment, several conceptually different approaches to determining the economy of joint consumption can be distinguished, each of which reflects a certain aspect of this phenomenon [19]:

- collaborative consumption (Botsman, Rogers, 2010);
- sharing economy or sharing economy (Belk, 2014; Hartl at al., 2016; Hawlitschek at al., 2016);
- access-based economy (Bardhi, Eckhardt, 2012; Böckmann, 2013; Strahilevitz, Myers, 1998);
- peer-to-peer economy (Venkatesh at al., 2012);
- on-demand economics (De Stefano, 2015);
- commercial sharing systems (Kramer, 2011);
- co-production (Hibbert, Horne, 1996);
- co-creation (Lamberton, Rose, 2012; Postigo, 2003);
- product-service systems (Moeller, Wittkowski, 2010);
- online volunteering (Piscicelli at al., 2015);
- anti-consumption (anti-consumption) (Ozanne, Ozanne, 2011).

Sharing economy development is extremely important for social and cultural integration in tourism (Guttentag, 2015; Oskam, Boswijk, 2016; Katsoni, 2017; Zhu et al., 2017; Katsoni, Segarra-Oña, 2019). And this is more important for most guests than a cooked breakfast or a stocked fridge. The sharing services in tourism are experiencing explosive growth. For example, A VoomaGo [24] relies on carefully thought-out multi-day tours with immersion into the local culture and traditions. This is especially important for promoting tourism in regions with depressed economic and social development. It can also serve as a basis for the development of an innovative model for the development of sustainable tourism in the region.

4 Purpose of the Study

The paper is aimed at the development of the ethno-tourism cluster in Russian regions on the basis of the sharing model.

5 Research Methods

When choosing approaches and methods for studying a phenomenon or object, the assumption regarding the nature of the studied object is decisive. With regard to the study of the proposed topic, it is important to emphasize the network nature of the subject of the study.

First, tourism is a fragmented industry, geographically distributed, with many stakeholders (local communities, specialized businesses, non-profit organizations, local administrations, etc.) contributing to the creation of the tourist product and the general impression of it. At the same time, as a key unit of the analysis for understanding of all tourist system in general it is considered to be a tourist destination [23]. This, according to definition of the World Organization of Tourism [25], can be considered as network of the specialized companies, the organizations and communities concentrated in the limited geographical territory.

Second, sharing economy is a set of transactions that are typically conducted using online platforms that match the needs of some people with the capabilities of others [5, 10–12]. It is an innovative economic model based on the advantages of information and communication technologies and a network coordination mechanism, which allows creating situations of mutual gain within the network and increasing its overall efficiency by combining insufficient resources and competencies of each individual network member.

In turn, it is generally accepted that the study of networks in the economy is clearly interdisciplinary [2]. In addition, the study of sustainable development also involves a combination of different scientific disciplines. Today, a tendency has emerged for interdisciplinary study of the theoretical and methodological aspects of the transition to sustainable development of territories.

Accordingly, a cross-disciplinary approach is key for this study. The study is based on theories and concepts developed at the junctions of economics and other scientific disciplines, including the new institutional economic theory (NIET), the theory of entrepreneurship, the theory of innovation, the theory of spatial development, the network approach in management, marketing relationships.

6 Findings

According to the author's mind, the model of ethno-tourism development includes a number of constituent elements, such as certification of the regions in the Republic based on the presence of ethnographic monuments, real business-ideas and projects, search of investors with the focus on tourism infrastructure including accommodation, catering and main transport routes. The next stage of creation of ethno-touristic cluster will be development and implementation the digital platform for regional tourism development to promote the sharing economy services examples of the regions. The virtual platform, created on the basis of the Moodle system (for example), will allow participants to create a tourist territorial socio-cultural network (TTSCN) – a network of interaction between representatives of stakeholders, socio-cultural organizations, representatives of municipal and regional authorities, tourism entrepreneurs, representatives of museums and exhibitions, etc. for the development of local self-government, improving the welfare of the local population in the context of the concept of sustainable development.

The free-access resource aims at creating an integration network of interaction between producers and consumers of cultural, historical and nature tourism services.

Moreover, academic staff (University researchers of regions) will be assist in the development of a draft program to support local small entrepreneurs and promote local competitive products and services, based on the history and culture of the region (including food and drinks, crafts, performing arts, agricultural products, etc.), conducting the survey among the policymakers, stakeholders, organized civil society, public and cultural organizations representatives, regional and municipal authority representatives, tourism entrepreneurs, museums and exhibition representatives on the results experience of the economy of "collaborative consumption".

Evaluation of the results of creating a socio-cultural model of integration of tourism stakeholders on the principles of the sharing economy in terms of creating new jobs in the region, promoting the potential of cultural and historical tourism and creating new models of socio-cultural leisure practices (recreation ideas) bases on the healthy lifestyle ideas, environmental friendliness, cultural identity preservation. Collaborative consumption economy is a phenomenon that describes cultural, technological, and economic transformations brought by innovations in information and communication technology. The sharing economy is defined as an economic model when the users share the underutilized assets for material and non-material benefits. Collaborative consumption economy presents a P2P (Peer-to-Peer) activity for sharing products and services that is coordinated by community-based online services, and connects buyers and sellers through digital platforms. Thus, the collaborative consumption economy is the complex of virtual platforms to facilitate the exchange processes between users. The impact is that the creation of virtual platform will allow one to generalize and promote the cultural, historical and nature tourism resources of the region, directly connecting producers and consumers of tourism services without intermediaries.

7 Conclusion

The use of ethno-tourism concept for the regional tourism development in Volga-Kama region will definitely make the region more attractive for investors on the one hand and stir interest in authentic culture of people residing in the area, on the other one. However, in order to achieve success in using the historian and cultural objects for tourism it is deemed necessary to develop a target programme of tourism development on the level of regional authorities with the focus on ethno-tourism; more carefully develop the necessary infrastructure. We believe that in order to create ethno-tourism cluster in any Russian region including the Mari El Republic it is first of all necessary to define the key concept.

As a result, the implementation of the sharing model in the ethno-tourism cluster will lead to the creation of new jobs in tourism sector, increase the level of the region economic development and strengthen social stability, offering the possibility of creating own business for the representatives of the local population, including the exposed population. It will also have a beneficial effect on the cultural heritage of the region. Thus, there should be cooperation between the local entrepreneurs, investors and local authorities. One or several large-scale tourist operators should also be involved in order to promote the product on the national tourism market. Establishing a tourism cluster (with sharing model) in the region is bound to significantly facilitate regional tourism development.

References

1. A.Yu. Aleksandrova. Tourism clusters: contents, boarders, mechanism of functioning. Contemp. Probl. of serv. and tour., **1**, 51–61 (2007)
2. R. Baggio, M.Yu. Sheresheva. Network approach in economics and management: interdisciplinary character. Bull. of Mosc. Univ., **6(2)**, 3–21 (2014)
3. R. Belk. You are what you can access – sharing and collaborative consumption online. J. of Busin. Res., **67(8)**, 1595–1600 (2014). DOI: 10.1016/j.jbusres.2013.10.001
4. M. Beni. *Globalização do Turismo: Megatendências do Setor e a Realidade Brasileira* (Aleph, São Paulo S.P., Brazilian, 2003)
5. R. Botsman, R. Rogers. *What's Mine is Yours: The Rise of Collaborative Consumption* (Harper Business, New York, 2010)
6. O. Chizhova. *The potential of Russian territories in ethno-tourism is huge* (2011). Retrieved from: http://www.ratanews.ru/news/news_30092011_3.stm
7. Ethnographic tourism (ethno-tourism) (2016). Retrieved from: http://www.rusotdih.ru/publications/show15/
8. J. Hamari, M. Sjöklint, A. Ukkonen. The sharing economy: Why people participate in collaborative consumption. J. of the Associat. for Inform. Sci. and Technol., **67(9)**, 2047–2059 (2015)
9. *Kamwa: livestream* (2020). Retrieved from: http://www.kamwa.ru/2020

10. V. Katsoni. An investigation into the sharing economy phenomenon in the Greek tourism industry in the accommodation sector. J. of Tour. and Reg. Develop., **7**, 25–35 (2017)
11. V. Katsoni, M.Yu. Sheresheva. The sharing economy in hospitality and tourism industry. Vest. of MGU, Ser. 6 Economy, **1**, 71–89 (2019)
12. P. Marshall. *The sharing economy*. Inform. J. SAGE Busin. Res. (SAGE Publ., Inc., 2015)
13. M.Yu. Sheresheva, M.A. Bek, N.N. Bek, E.V. Buzulukova et al. *Methodology for researching network forms of business organization* (Publ. house of the Higher School of Econ., Moscow, 2014)
14. S. Miller. First principles for regulating the sharing economy. Harvard J. on Legislat., **53**, 149–150 (2015)
15. *Report ATAC* (2015). Retrieved from: http://aboriginalcanada.ca/corporate/wp-content/uploads/2015/04/REPORT-ATAC-Ntl-Ab-Tsm-Research-2015-April-FINAL.pdf
16. M. Porter. *Competition*. (Williams, Moscow et al., 2001)
17. M. Porter. *On Competition*. (Harvard Busin. School Press, Boston, 2001)
18. *Salted ears, zyryane and erzyane – Finno-Ugric ethno-tourism on the march* (2015). Retrieved from: http://finugor.ru/node/22336/
19. *A.A. Shalaeva, V.A. Rebyazina, M.M. Smirnova. The economy of joint consumption as a new business model. In: Innovations of marketing models of Russian companies in the era of digital transformation (SPbMEU, St. Petersburg, 2018)*
20. N.V. Shabalina. *Introduction into tourist cauterization* (2012). Retrieved from: http://bzm.su/articles/article-05
21. *The Ministry of Culture has compiled a list of places for lovers of traditions and everyday life of different peoples.* Retrieved from: https://iz.ru/817972/aleksandra-rykova/put-v-tygydym-nazvany-10-luchshikh-obektov-etnoturizma-v-rossii/2020
22. Yu. Timofeev. *Russian regions are just at the beginning of their way in ethno-tourism development* (2011). Retrieved from: http://www.ratanews.ru/news/news_6052011_1.stm, in Russian
23. N. Vanhove. *The Economics of Tourism Destinations* (Elsevier Butterworth-Heinemann, London, 2005)
24. *Voomago*. Retrieved from: // https://voomago.com
25. *Indicators of Sustainable Development for Tourism Destinations*. A Guidebook (World Tourism Organization, Madrid, 2004)

Strategies of developers to housing supply within framework of project financing

Olga V. Grushina[1,*], and *Natalya N.* Shelomentseva[2]

[1] Baikal State University, 11, Lenina ave., Irkutsk, Russia
[2] Irkutsk State University, 20, Gagarin boulevard, Irkutsk, Russia

Abstract. The problem of providing comfortable and affordable housing for Russian citizens as a national goal has gained an impetus to solution after the introduction at the legislative level of project financing for housing construction using escrow accounts. The application of this new method of financing to the decision-making algorithm of a developer requires mathematical modeling of the economic ratios of the main indicators of the developer's activities, taking into account the interests of other subjects in the housing construction sector: banks and population. The approach proposed by the authors is aimed primarily at expanding the supply in the housing market. We have formalized the possible sale strategies of the developer using a matrix approach in varying prices and the number of meters sold at different stages of construction, as well as interest rates for bank crediting, depending on the degree of escrow accounts accumulation. Testing the model has shown that developers should face serious problems related to increasing the cost of replenishing working capital for the construction process, which will inevitably be included in the price per square meter of housing and paid by the consumer (population). At the same time, banks become the main controllers and beneficiaries of the project financing. The model developed and tested by the authors will allow construction organizations to overcome the first difficulties of project financing application and competently and consistently make a choice of a strategy for offering housing to buyers, taking into account the variability of the economic parameters of all parties.

Keywords: project financing, supply in the housing market, costs, working capital requirements, escrow accounts.

1 Introduction

The milestone transition from various schemes of share participation to project financing in housing construction has been widely discussed in the scientific community, especially in the last 3 years. Nevertheless, the perspectives of the project financing have been estimated earlier for some large projects in the Russian Federation, while abroad this method of financing has a well-established long-standing practice [3, 9, 11].

The project financing itself can be impersonally characterized as a special way of financing "a large and viable project, which is financially and legally separated from the activities of a company, where a money lender considers the cash flows and economic income from this project as a source of credit repayment" [20]. In other words, the project financing features the separation of the construction object from the activities of the developer as an enterprise, the targeted use of credit under the control of the bank and its repayment from income obtained from the sale of this object.

The main declared goal of the project financing is to solve once and for all the problem of defrauded homebuyers with the help of not state, but banks. The latter are embedded in the hitherto short link "developer-housing equity holder" as a third actor, last but not least (last in number, but not in importance). The main consequence is that the role of banks is dramatically increasing many times over, driving out the functions of the state. According to the prophecy of R. Hilferding, financial capital becomes the ruler of people's lives [7]. Apparently, this goal is not declared. Banks that take on the risks of equity holders according to the previous scheme get the main benefits of the project. Hitherto free funds of equity holders are transformed into the borrowed moneys of the bank by a clever movement of escrow accounts.

The noble social purpose of providing citizens with affordable and comfortable housing casts a shadow on the super-profit of the financial sector and the further aggravates the crisis in the real sector of the housing construction industry. It should be emphasized that money on the escrow accounts is "frozen" for developers, but not for banks. "The minimum risk-free rate is equal to the key rate of the Central Bank and amounts to 7.75 %, more profitable instruments being also available. If to add 9–11 % on mortgages and credit interest of the developer, it becomes clear that the banks

* Corresponding author: olga7771972@mail.ru

can gain about 23 % from the same house under construction. According to predictions of the experts, when the entire construction sector will be completely dependent on the banks, the latter can earn about 1.5 trillion rubles annually [19]. Meanwhile, for the developer, the project financing will cost almost twice as much as the funds of equity holders [1], since "all risks are associated not with the project financing mechanism itself, but with the size of the interest rate at which banks will lend to developers" [12].

In the present paper, we do not set the goal to study in detail the monetary policy of the Central Bank of the Russian Federation. However, we agree with Ershov [2], Katasonov [10], Glazyev [4], et al. that the business loan of the state is the main factor of maximizing investment in fixed assets as well as an advantage in global competition. The developed countries, issuers of reserve currencies, successfully employed this strategy, while the Central Bank of the Russian Federation stubbornly refuses to use it. In our opinion, the under-monetization of the economy (40 % of GDP compared to 100 % or more in developed countries [4] and the high price of money hinders any good activity in the development of the domestic economy. The issues of the relationship between the established key rate, the lack of money in the state with ways to solve the problem of housing affordability were discussed in detail by us earlier [5, 6].

2 Problem Statement

The arrangement for using the escrow accounts established by Federal Law 214, January 1, 2019, formally prohibited construction companies to directly accept funds from equity holders. However, "from more than 4 thousand domestic construction companies 275 ones have escrow accounts" [15]. As, according to O.A. Mishchenko and V.I. Privalov, such a rate of opening accounts can lead an imbalance in the market and decrease the number of objects put into operation [15]. At the same time, the number of banks allowed to open escrow accounts and support projects amounted to 95 by June 1, 2019.

Solution of equity holder problems with the help of the project financing resembles a guillotine in the case of a headache: the interest rates cut off most small and medium-sized enterprises from the construction industry like a knife. We are talking about a radical change in the "landscape of the construction industry", when small companies will completely leave the market that results in "freezing" of the started housing construction projects [15]. "In the country as a whole, according to experts, it takes about 1 trillion rubles to finance new construction projects that do not fall under the old scheme of financing" [8]. This increases the leverage of the construction industry almost by 3 times and enhances the financial risks of the remaining companies.

The optimists of the project financing usually make the following arguments:

- only finished apartments will be presented on the market;
- risk of an increase in construction time will be minimized (its increase is absolutely unprofitable for the developer in new conditions);
- difference in the cost of housing at the excavation stage and finished housing will be reduced to almost zero;
- all risks are distributed between the developer and the bank, while the state and buyers do not risk at all;
- speculative factor of real estate purchase is neutralized;
- and, finally, there will be no significant increase or decrease in property prices [13].

Indeed, registration of ownership and actual implementation will be carried out only after the object is put into operation. However, nothing prevents the developer to initiate the accumulation of money in escrow accounts from the start of construction, manipulating the price per square meter of future housing. Therefore, the difference in the cost of housing at different stages of construction will not be zero at all. As a result, the speculative factor in the acquisition of real estate remains challenging.

The construction time can increase up to 6 months and even longer under circumstances beyond the control of the developer. Noteworthy, potential buyers do not have any more leverage in this case. All leverages have been given to the banks: their sanctions, one way or another, will force construction organizations to take it out on the consumers. Therefore, the last argument of the optimists sounds especially naïve. The reduction in the number of developers, monopolization of the housing market, a decrease in housing supply, high interest rates (the promise to reduce them to 8 % by 2024 [16] is simply a mockery) inevitably increase prices per square meter of new housing. The dispute can only be about how much will raise the prices.

As for the risks, if a bank goes bankrupt, they will inevitably come upon the housing buyers and, ultimately, on the state, when the latter insures escrow accounts for up to 10 million rubles (Article 12.2 of the Federal Law of 23.12.2003 N 177-FZ, revised on 28.11.2018, "On insurance of deposits in banks of the Russian Federation" (as amended and supplemented, entered into force on 01.01.2019). "Let us note that the risks of developers are covered not by the deposit insurance system, but by escrow accounts." [18]

We are deeply convinced that banks, as intermediaries, should not make money on providing citizens with a vital benefit, a roof over their heads. The obligation to provide citizens with housing is spelled out in the Constitution of the Russian Federation.

3 Research Questions

3.1 Methods for managing the working capital of a developer company in terms of project financing

These issues and problems of the project financing are actively discussed in scientific and public circles. We would like to pose another task that escapes everyone's attention: how will building organizations provide the urgent need for working capital during the construction period? In the strict framework of the new legislation, is there a choice to optimize the ratio of funding sources for working capital and a classic compromise between a sufficient level of profitability and the risk of a shortage of the working capital s, which entails a loss of solvency?

At first glance, there is no room for maneuver. The entire volume of working capital becomes credit for the entire construction period (minus the initial amount of the developer's own funds). However, the regulatory framework for the project financing of housing construction in Russia links the cost of funding sources for working capital with "conditional sales". We have proposed the term "conditional sales" to denote the accumulation of funds in escrow accounts.

In the present paper, we have analyzed the following possible approaches to crediting the need for working capital during the construction process: (i) the credit line is fully used at the beginning of the construction period; (ii) the credit line is used evenly during the construction period; (iii) the credit line is used as needed in the construction process. The first approach is strongly recommended by banks, since at the moment their software is not ready to use another credit line. Construction organizations are more accustomed to employ the working capital as production needs. The even application of the credit line during the construction period, from our point of view, is a compromise between two opposite positions (bank and construction organization). The weighted average interest rate on the credit line depends on the amount of money on escrow accounts, for example, as presented by JSC DOM.RF, founded by the Government of the Russian Federation. In the absence of money on escrow accounts, the rate is 12 %. When the escrow account reaches half of the credit line, the rate is 9 %. In case of the amount of money on escrow accounts coincides with the size of the credit line, the interest rate will be 6 %. and can be reduced to 3 % if the amount of money in escrow accounts exceeds the amount of the credit line 1.5 times.

Consequently, the availability of working capital and, first of all, the costs of servicing the source of their financing (credits from banks holding escrow accounts) will depend on the rate of money accumulation by future home buyers, that is, on the choice of a sales strategy. Both this choice and its marketing support completely depend on the developer.

As a result, 4 possible strategies for "conditional sales" of future apartments can be proposed.

Strategy I comprises active sales of apartments at the beginning of construction. Due to the accumulation of money on escrow accounts, the interest rate may decrease.

Strategy II consists of active sales of apartments in the middle of construction. In the initial period, money on escrow accounts are close to zero and the interest rate may decrease only at the end of the construction period.

Strategy III involves active sales of apartments at the end of construction. In the initial and middle periods, money on escrow accounts are close to zero and the interest rate does not change during the entire construction period.

Strategy IV is based on active sales during the entire construction period. It is possible to reduce the credit interest rate by several times due to the accumulation of money on escrow accounts.

3.2 Model for assessing the effect of the housing "conditional sales" strategy on the key subjects of the construction project financing

The interests of the economic subjects of housing construction are different. When modeling, we considered the interest of the major actors: construction organization (developer), bank, population (consumers).

Table 1 summarizes the key indicators that are relevant to the application of project financing methods and depend on the choice of a sales strategy and accumulation on escrow accounts.

Legend:

S is the area of the construction object;

n is the number of periods (quarters);

Q_t is the quantity of conditionally sold square meters in n-period (quarter) $t = \overline{1, n}$;

P_t is the price in t period (quarter);

C_t is the sum of costs for construction of the object in t -period;

NWC is net working capital;

K is the credit line (total amount of credits);

$З_t$ is the credit for interest charge in n-period;

r_t is the bank's interest rate in the n-period;

$Э_t$ is the amount of funds on escrow accounts in the t -period.

ROE is Return on Equity, %;

Ra is return on assets earnings before interest and taxes (EBIT)), %;

T is the adjusted profit tax rate, units;

(1-T) is the tax corrector, net income / income before taxes;

Kd is the estimated tax rate for borrowed capital, «debt price», %;

L is the borrowed capital, rubles;

E is the average of owned capital, rubles;

L/E is the leverage;

B_j is the income from conditionally sold square meters when implementing the j sales

C is the construction cost

$ПВ_{ij}$ is the interest payment to the bank for the i type of crediting and j sales strategy.

The main interests of the population in the framework of housing construction are:

1) a reasonable ratio between the cost and quality of housing;

2) the ability to purchase housing in accordance with income;

3) minimal risks related to purchasing a new housing.

Most of the above interests have not been satisfied at the moment, since other economic actors involved in the housing construction process do not fully take into account the interests of the population.

Table 1. Formalization of key indicators of the subjects of the project financing

Subjects and indicators	Formula
Developer:	
Sales profit	$\text{ПН}_{ij} = B_j - C - \text{ПВ}_{ij}$, $i = \overline{1,m}$, $j = \overline{1,l}$, $B_j = \sum_{t=1}^{n} P_t \cdot Q_t$; $C = \sum_{t=1}^{n} C_t$;
Effect of financial leverage	$EFL = (1 - T) \cdot (R_a - K_d) \cdot \frac{L}{E}$,
Return on equity	$ROE = (1 - T)R_a + EFL$.
Bank:	
Interest income from crediting the developer	$\text{ПВ}_{ij} = \sum_{t=1}^{n} З_t \cdot r_t$. The bank sets the interest rate depending on the amount of money in escrow accounts: $r_t = \begin{cases} r_1, & at\ Э_t < \propto_1 К, \\ r_2, & at\ \propto_1 К \leq Э_t < \propto_2 К, \\ r_3, & at\ \propto_2 К \leq Э_t < \propto_3 К, \\ r_4, & at\ Э_t \geq \propto_4 К. \end{cases}$
Consumers:	
Increase of cost of 1 m^2	$\text{УС}_{ij} = \frac{\text{ПВ}_{ij}}{S}$,

Compiled by the authors

4 Purpose of the Study

The purpose of this work is to offer tools that allow construction companies to expand their supply on the housing market through the rational use of their own and borrowed resources via the project financing with the escrow accounts.

5 Research Methods

The study is based on the method of economic and mathematical modeling, which includes several stages. The first stage comprises formulation of the economic problem and its qualitative analysis through understanding the available information on the topic under consideration.

The second stage involves the construction of a mathematical model via formalization of the economic relations of the problem in the form of specific mathematical dependencies and relationships.

The next stages relate to the mathematical analysis of the model and the preparation of the initial information to clarify generality of the model and to process information obtained from management accounting data of the developer.

At the stage of numerical solution, we have proposed an algorithm for the numerical solution of the problem, developed a software product, and calculated significant amounts of information.

The final stage includes the analysis of the numerical results and their application to reveal specific quantitative factors that influence economic processes. Also, the consequences of changes in the conditions of the development of economic subjects have been evaluated.

6 Findings

The proposed model has been tested by the example of activity of "Instroytech" company. The project envisages the construction of four residential buildings, including three 13-storey houses and one 10-storey house with 368 apartments of a total area of more than 24 thousand m2. The options of the strategy for selling apartments have been calculated assuming that a non-revolving credit line is used. We have assessed the amount of funds on escrow accounts on a quarterly basis for four sales strategies:

I – at the beginning of construction,

II – in the middle of construction,

III – at the end of construction,

IV – uniform sales throughout the construction period.

The average cost of construction of 1 m2 in the Irkutsk region is 51.900 rubles. In the initial period of construction, the selling price of 1 m^2 is 54.000 rubles. This price increases quarterly by 2 %. The construction is carried out at the expense of own and borrowed funds. The need for working capital in each quarter is calculated by the authors as a percentage of the expenditures for the implementation of the project according to the data "UniverStroy" company (Table 2). As seen from Table 2, that the greatest need for financing falls on the middle of construction.

Table 2. The quarterly need for funds of the construction company, expressed as a percentage of the project cost.

Quarter	1	2	3	4	5	6	7	8	9
Percentage of the project cost	9.69	11.57	14.95	11.07	13.38	13.01	13.33	3.32	3.94

Source: calculated by the authors using the data of «UniverStroy» company

The amount of the developer's own funds should be at least 10 % of the project cost, as stated in Federal Law 214 "On shared construction ...". When calculating, we used the size of the own funds at the level of 10 % of the project cost (90 % of borrowed funds), since in the case of shared construction, construction organizations have borrowed this sum from the population [15]. In our calculations, we employed the values ∝_1=1/2, ∝_2=3/4, ∝_3=1, 〖∝〗_4=1 1/2 and r_1=12, r_2=9, r_3=6, r_4=3.

6.1 From the perspective of a construction organization

Profit of a construction company before taxes for various sale strategies is presented in Table 3 (the value that is best for a construction company is highlighted in bold).

For a developer, the highest profit is provided with equal crediting during construction and active sales of apartments at the end of construction. With the same type of crediting and various strategies for sales of apartments, the difference between the highest and the lowest profit before taxes reaches 127.451.400 rubles.

If to compare the best strategy of the project financing with a similar strategy in shared construction, then the construction organization loses up to 41 % of profit.

Table 3. Profit of a construction company before taxes for various sale strategies using borrowed funds in the amount of 90 % of the project cost

Sale strategy / **Type of crediting**	**I**	**II**	**III**	**IV**
At the beginning of construction	-41 183 227	-39 120 666	-21 541367	-63 307 916
Evenly during the construction period	29 689 799	88 509 739	157 141 199	79 638 137
As production needs	29 184 520	91 842 385	125 102 807	72 290 865

Source: calculated by the authors

6.2 From the perspective of the bank

The interest that the bank will receive from crediting in the amount of 90 % of the project cost for the entire construction period is presented in Table 4 (the value that is the best for the bank is given in bold).

Table 4 shows that for a bank it is more profitable to grant a credit at the beginning of construction and fill escrow accounts at the end of construction. The data in Table 3, on the one hand, demonstrate the bank's income, and on the other hand, display how much money a construction company loses with the transition from shared construction to the project financing. The interest in the amount of 297.804.276 rubles is 23.63 % of the construction cost!

Table 4. Profit of a construction company before taxes for various sale strategies using borrowed funds in the amount of 90 % of the project cost

Filling escrow accounts / **Type of crediting**	**I**	**II**	**III**	**IV**
At the beginning of construction	119 121 710	212 717 340	**297 804 276**	238 243 421
Evenly during the construction period	47 648 684	85 086 936	119 121 710	95 297 368
As production needs	50 282 269	86 010 902	155 416 715	106 901 253

Source: calculated by the authors

6.3 From the perspective of the consumer

The construction company will include the interest for the use of borrowed funds in the cost of each square meter that will effect on the cost of housing for the consumers. The increase in the cost of $1m^2$ for various types of crediting and the time of money transfer to the escrow account ("conditional purchase") are presented in Table 5.

Table 5. Increase in the cost of $1m^2$ with various types of crediting and the time of transfer to escrow accounts when crediting a construction organization in the amount of 90 % of the project cost

Filling escrow accounts / **Type of crediting**	**I**	**II**	**III**	**IV**
At the beginning of construction	4 905	8 758	12 261	9 809
Evenly during the construction period	1 962	3 503	4 905	3 924
As production needs	2 070	3 541	6 399	4 401

Source: calculated by the authors

The data of Table 5 show, on the one hand, the increase in cost of 1m2 for a construction organization. On the other hand, Table 5 demonstrates how this fact will affect the price for the consumer, in whose interests such an increase is undesirable. In Table 6, the percentage of cost increase that the developer includes in the future housing price, is presented.

Table 6. I Increase in the cost of $1m^2$ with various types of lending and the time to transfer money to escrow accounts, %

Filling escrow accounts / **Type of crediting**	**I**	**II**	**III**	**IV**
At the beginning of construction	9.5	16.9	23.6	18.9
Evenly during the construction period	3.8	6.7	9.5	7.6
As production needs	4.0	6.8	12.3	8.5

Source: calculated by the authors

Our calculations prove that in case of project financing of housing construction, the credit load for a construction organization significantly affects the increase in the cost of $1m^2$ for the consumer, on average. In all options, this increase is about 10 %, but banks will certainly seek the most profitable options for themselves that leads to the rise in housing prices by 16–23 %.

7 Conclusion

With the help of the project financing in housing construction, our state intends to solve the problem of providing the population with housing [17]. The interaction of subjects of the project financing is full of contradictions. Risks of the developer risks are covered by filling escrow accounts. It is beneficial for the bank to cover the need for the developer's working capital in full at the initial stage, thus deforming the developer's balance sheets, threading its financial stability, and increasing the financial expenses. As in the case of shared construction, the consumer should immobilize money on escrow accounts at the excavation stage, losing the profit from the possible use of the accumulated financial resources in order to reduce the price per square meter. It would be preferable for the developer to sell all apartments after the completion of the project at the maximum price, but in order to reduce interest rates on the credit the developer should replenish escrow accounts with the only available way, i.e. to raise the price of housing. If not to employ the administrative resource to oblige the bank to credit the developer depending on the production need, lowering the interest with the replenishment of escrow accounts, then no compromise will be reached. Consequently, the project financing will trigger a pathological process: massive bankruptcy of developers, freezing of numerous housing projects, decrease in supply volumes, and the growth of in prices for the ready objects in 2–3 years due to developer credits. Ultimately, this growth will be attributed to the cost of construction and will significantly increase the selling prices of the "new housing" [14], according to our calculations, at least by 15–22 %.

The estimations of the effects of project financing of housing construction on the interests of its subjects (developer, bank and consumer) differ significantly. Despite the fact that economic actors interact with the aim of providing the population with affordable and high-quality housing, their positions are not equal. Consumers are the most deprived of rights. They cannot influence the price per square meter of housing under construction and bank credit rates. Meanwhile, the bank can agree with the developer to the prejudice of the consumer. Apparently, the problem can be solved via drastic changes in the monetary policy of the Central Bank of the Russian Federation and through state regulation of interest rates on mortgage and the developer credits within the framework of project financing to overcome lifelong financial slavery of most of the population in order to fulfill the basic living need, a need for a roof over the head.

References

1. E.A. Alekseeva. The advantages and disadvantages of the transition to project financing of housing construction. *Topical issues of law, economics and management: collection of articles of the XIX International Scientific and Practical Conference* (MCNS, Science and Education, Penza, 2019), pp. 231–233
2. M. Ershov. The 2008 Crisis: "A Moment of Truth" for the Global Economy and New Opportunities for Russia. Econ. issues, **12**, 4–26 (2008). Retrieved from: https://doi.org/10.32609/0042-8736-2008-12-4-26
3. J. Gathergood, J. Weber (2017). Financial literacy, present bias and alternative mortgage products. J. of Bank. & Fin., **78**, 58–83
4. S.Yu. Glazyev. *Leap into the future. Russia in new technological and world economic structures.* Collection of the Izborsk club (Book World, Moscow, 2018)
5. O.V. Grushina. Theoretical and methodological comprehension of the main aspects of capital deficit in the Russian Federation as one of the obstacles to achieving affordable housing prices. *J. Labour and social relations* (2011). Retrieved from: http://www.e-rej.ru/Articles/2011/Grushina.pdf
6. O.V. Grushina. *Strategy for ensuring the affordability of housing in the Russian Federation.* (BSU Publ. House, Irkutsk, 2017)
7. R. Hilferding. *Financial capital. The newest phase in the development of capitalism* (State product., Moscow, 1924)
8. S. Ishkov. *Temporary stress: how the introduction of escrow accounts will change the market* (2019). Retrieved from: https://pro.rbc.ru/demo/5c73af629a7947ef6ae2a42c
9. H. Jialin, R. Zhao. Housing boom, real estate diversification, and capital structure: Evidence from China. Emerg. Mark. Rev., **32**, 74–95 (2017)
10. V.Yu. Katasonov. *About the interest on the loan, the court, the reckless.* A textbook of modern problems of "monetary civilization" (Publ. House Res. Inst. of School Technol., Moscow, 2011)
11. H.L. Kim, N. Graeme. Real estate global beta and spillovers: An international study. Econ. Modell., **59**, 297–313 (2016)
12. A.Yu. Kovalenko. Risks of transition from equity participation to project financing. *Prospects for the development of science and education. Collection of scientific papers based on the materials of the international scientific and practical conference* (2017), pp 65–67
13. I.A. Kharitonova, T.O. Pinchuk. The reason for the cancellation of equity financing of residential construction and the transition to project financing. *ISTU Youth Bull.*, **8**, 135–140 (Publ. of Irkutsk Nat. Res. Techn. Univ., Irkutsk, 2018)
14. A.N. Larionova. On the issue of premature abandonment of shared construction and transfer of housing construction to project financing. Constr., econ. and manag., **3(31)**, 4–11 (2018)
15. O.A. Mischenko, V.I. Privalov. Bank ecosystem as a mechanism to reduce the credit risk for project financing. Econ. of stable develop., **1(37)**, 52–56 (2019)
16. A. Petrov. *Forecast of the Ministry of Finance for 8 % mortgages depends on the Central Bank rate* (2018). Retrieved from: https://rueconomics.ru/354140-prognoz-minfina-po-8-ipoteke-zavisit-ot-stavki-cb

https://doi.org/10.1051/shsconf/202111200012

17. N.N. Shelomenceva. To the problem of project financing in housing construction. Econ. of Constr., **6**, 22–31 (2018)
18. A.N. Shulekin. Analysis of changes in legislation in the field of housing construction: the transition from shared construction to project financing. Siber. Finan. school, **4(129)**, 34–37 (2018)
19. L. Taejun. Growth, financial development, and housing booms. J. Econ. Modell. (2017)
20. A.A. Tsygankova, D.I. Koptelova, K.A. Bareshenkova. Residential construction project management during transition to project financing and digital technologies. *Contemporary problems of project management in investment and construction sphere and environmental management. Proceedings of IX International Scientific and Practical Conference, devoted to the 112th anniversary of Plekhanov Russian University of Economics* (Rus. Univ. of Econ. named after G.V. Plekhanov, Moscow, 2019), pp. 226–232
21. N.V. Voronina, V.V. Serova. Project financing as a new mechanism of financing household construction objects. Electr. Sci. publicat. "Scientific Notes of PNU", **9(2)**, 55–60 (2018)

Products from mammoth Tusk in Yakutia in the 18th century (based on the materials of funerary monuments)

Aleksandra Prokopieva*, and *Aytalina* Sleptsova

Institute for Humanities Research and Indigenous Studies of the North of the RAS, 1, Petrovskogo St., Yakutsk, 677027, Russia

Abstract. The article is devoted to the study of products made from mammoth Tusk in Yakut burials of the 18th century. The work is based on literary and archival as well as on the collections of the museums. Analysis of products made from mammoth Tusk showed that they were not widely distributed among the Yakuts. This could be due to the fact that in the religious beliefs of the peoples of Yakutia the mammoth had a negative colour. Two movements of carving from a mammoth Tusk in the 18th century were revealed. The first movement was based on the traditions and techniques of local carving, and the second one was based on imitation of the Northern Russian school.

1 Introduction

Mammoth mining, import and export of raw materials from Russia and the manufacture of tusk products are becoming topical issues in the Arctic zone of Russia and Yakutia in particular. Both practical and theoretical aspects of this problem are being discussed. One of the interesting topics is the question: how traditional and primordial is the extraction of mammoth tusk and the production of household and artistic products from it? When discussing this topic in the context of the Yakut culture, they often turn to ethnographic materials and decorative and applied art, often without including archaeological research data in the source database. Meanwhile, at the moment, more than 300 funerary monuments dating from the 17th-18th centuries have been studied. Analysis of the accompanying inventory of the burials showed that among a large number of wooden, metal and bone products there are objects made of mammoth tusk. We assume that the number and nature of the products from the burials, the technique and quality of execution may indicate the level of mastering of the material by local craftsmen.

2. Materials and methods

According to R. I. Bravina, 69% of the total number of ground burials of the 15th - early 19th centuries. In Yakutia, there are items of accompanying inventory. They are more common in male and female burials (78% and 68%, respectively), less common in children's burials (57%) [1 p. 99]. The set of items of accompanying inventory usually consisted of a suit, personal and household items, hunting tools and weapons, horse equipment, religious items and funeral food. According to the composition of the inventory, the material and the quality of its execution, it is possible to make assumptions about the status of the buried person, his occupation and professional affiliation. To identify items made of mammoth tusk in the burials, we studied museum collections (Yakut State United Museum of History and Culture of the Peoples of the North named after Em. Yaroslavsky, Museum of Archeology, Ethnography and Higher School of NEFU), archival materials (field reports from the Scientific and Branch Archive of the Institute of Archaeology of the Russian Academy of Sciences, diaries, protocols of opening graves from the funds of the National Archive of the RS (Ya)) and published literature. As a result, the discovery of 14 items from the mammoth tusk in 13 burials from the total number of funerary monuments dating from the 17th-18th centuries was recorded and explored by archaeologists. The selection is based on the availability of an accurate definition of the material of manufacture of the item. All items are personal items and include the following items: combs, pipe heads, cane heads, spoons, belt buckle, and two sticks of unknown purpose. In the course of studying the museum collections, we found that not all the items were preserved in the museum collections, some of them are known only from the reports of researchers.

3. Results

3.1. Hairbrush and combs

Hair combs in many cultures were endowed with symbolic symbols associated with the magic of hair [2 p. 147]. At the moment, it is difficult to say whether they carried remarkable symbols in the context of the funeral rite of the Yakuts of the 17th-18th centuries, but wooden and bone combs were found in male and female burials. Among them, the combs from the mammoth task in the amount of 5 specimens are highlighted. The combs were

* Corresponding author: aleksandra.pkkyyha@gmail.com

found in the burials of Yelets I of the Megino-Kangalas district, Shaman Tree III of the Churapchinsky district [3 p. 102-103], Sangar of the Kobyai district [4 p. 73], Batyyalakh I and Koeboeng of Olenek district [5]. All combs are frame combs (the outside teeth are wider) of the bilateral type. The ridges have a trapezoidal shape, where the upper narrower part has rare large teeth, and the lower wider part has frequent small ones. The size of the combs varies from 6 to 7.5 cm in height, with the widest part from 6.6 cm to 7.5 cm; in the narrowest part from 6 cm to 7.2 cm. The number of large teeth is 18-20 pieces; and that small teeth are up to 60 pieces. Marking lines are drawn on the surface of the products, sometimes serving as decoration (Shaman Tree I). At the comb from the burial of Yelets I, the sides have figured faces, decorated with recesses and notches. Judging by the materials of the Alazey and Stadukhinsky prisons, these combs are of Russian origin [6, p. 46-48]. This is also evidenced by the Siberian materials [7]. Apparently, under the influence of Russian traditions, similar double-sided combs were common in the Yakut culture of the 18th-19th centuries. It should be noted that the stories related to the semantic status of combs in the Yakut culture are borrowings from Russian culture [8, p. 217].

The topography of the combs inside the coffin is only known in three cases. In the partially destroyed burial of Sangar in the Kobyai district, a comb was found inside a birch vessel located at the feet of the buried man. Inside the vessel we placed scissors and a copper ring in addition to the comb. In the rich burial of Elecha I in the Megino-Kangalas district, a traveling bag with a large amount of contents was found at the feet of the buried man. Among other things. It contained a wooden box for the storage of handicrafts, inside of which there were a birch mound, pieces of fur, white and black beads on a thread, a metal thimble and a mammoth tusk comb. In this context, it is likely that the comb was not used as a hygiene item, but as a tool for brushing tendon threads or horsehair. In the burial of Shamansky III in the Churapchinsky region, a comb was found on the left side of the thorax of the buried. The comb was in a wooden case and, judging by the tissue print on the surface of the case, was placed inside the caftan. According to the anthropological materials and accompanying equipment, all the burials with the combs in question were female.

3.2. The heads of smoking pipes

Smoking pipes, common on the territory of Yakutia of the 18th-19th centuries, are composite: a wooden chub made of two halves, a separate metal or bone head, a leather belt for winding the chubuk. In the ethnographic literature, these pipes are known as "belt pipes" and according to the typology of A.V. Shapovalov, they belong to the 4th or "American" type of smoking pipes in Siberia [9]. In the Yakut burials of the 18th century, pipes with a metal head are most often found. Mammoth bone heads were found in three burials: Shaman Tree I and III in Churapchinsky district and Ordegen I in Nyurbinsky district. All products are of the same type and are similar to metal ones, consisting of a bowl and a wide base - a leg. The height is 2.3-3.2 cm, the diameter of the bowl is 3.5-4.5 cm. The bowls are sculpturally carved and decorated to varying degrees with a carved pattern consisting of simple geometric shapes: lines (Ordegen I), dots and arched motifs (Shaman Tree I). The most elegant and delicate carving on the bowl from the group burial is the Shaman Tree I [3, p. 95]. The upper part of the body of the bowl is decorated with a line of a number of points, under it there are two girdles of two straight parallel lines. Immediately below the second girdle, closer to the bottom, the bowl is decorated with a repeating arched ornament with descending sprouts in the technique of relief carving. In general, decorative motifs and techniques are characteristic of Yakut carving of the 18th century, but the combination of smooth and relief carving in one product is not typical for the studied period. As with the metal-headed tubes, the tubes in question were found in the boot tops of the buried men.

3.3. A clasp

A mammoth tusk clasp was found in the male burial of Ordegen II of the Nurbin district. On the burial there was a caftan with asymmetrical shaped sides, smelling on the right side. At the same time, the lock was only at the top, on the waist a wide port side abruptly interrupted and formed a broken line, continuing straight down the center. The common edge of the caftan is similar to the stylish kaftan of Aba Wos, the head of the Meghini Yakuts, Jorgo Idelgyeva [10 pp. 115]. R.S. Gavrilyeva referred to the East Asian type of the Yakut and found similar elements with Mongolian and Manchurian shoulder clothes [11 pp. 16-17]. The upper part of the caftan was fastened on copper buttons, and at the belt on a lock of mammoth tusk type «togle» or «crutch». The clasp consists of two elements: a ring and a T-pin. The ring is capelin in section, the diameter of the outer side is 6.4 cm, and the diameter of the inner side is 3.8 cm. The pin has a stylized shape with a short base foot and slightly arched shoulders, the end of which is shaped like the heads of a horse. The height is 1.2 cm; the length is 5.6 cm. This type of clasp has not been seen before and requires further study.

3.4. The staff heads

Staffs are rarely found in Yakut burials and, according to anthropological data, their presence in a burial is related to the problems of the musculoskeletal system of the buried. The staff were made of wood, sometimes composite: a wooden base with a bone plate. Two burials known from field reports were found with mammoth tusk heads. In the first case, the article was found in the burial ground of Prince Boltoru Yraev in the burial ground of Onogoschut Unuoga, Upper Vilnius district, N.D. Arhipov. A wooden staff with a head from mammoth tusk was laid on the right side of the grave. There is no further detail on the plaque in the report, apart from the fact that it is decorated with longitudinal straight cuts [12]. A second copy was found in one of the women's aerial graves on Batyyalah (an influx of the r. Olenek), but apart from the mention in the newspaper article there is no more detailed information about the object [13].

3.5. Spoons

The accompanying material often includes dishes and utensils, funeral repast. Individual bone spoons and wooden spoons are often placed in low wide vessels or in travel bags. Sometimes spoons are made with special covers, which usually refers to imported metal spoons. Spoons were most often made of wood, less often bone. Spoons from mammoth tusk are only recorded in two cases and both in the Olenek Valley. The first spoon was discovered by H.V. Xenophontov in a ground male burial [1 pp. 277, para. 11]. The spoon was placed inside a wooden bowl laid to the right at the feet of the buried man. Another spoon was discovered in the aerial burial of the Yakut republican branch All-Russian Society for the Protection of Monuments of History and Culture (these letters in Romanized form are YaRO VOOPIK) [12]. Among the many iron tools, fragments of sheet iron and religious objects was a spoon of mammoth tusk with 4 grooves on the handle. The end of the handle is decorated with two parallel recesses, where two miniature bumps» [12] are formed. Based on the specifics of the accompanying inventory, A.N. Kopirin believed that the buried one could be a blacksmith. Unfortunately, both spoons did not survive in the museum collections [14].

3.6. Sticks

Two mammoth bone sticks, found in the rich grave of the legendary virgin deity Bolugur Ayyith by an expedition of the Yakut Museum under the leadership of S.I. Bolo in the Churapchin region, remain as items of unclear use. This is how the sticks are described in the grave-opening protocol: They are reminiscent of Korean or Chinese food sticks, but considerably thicker than them. The bottom of the stick is thicker than a normal student pencil. The opposite end of the stick is gradually tapered, semi-concave. Both round, length 22 cm» [15]. Sticks were found at the right elbow, next to a knife with pearl inlay. No photographs or drawings have survived, nor have they been found among museum objects. Were they imported or made by local craftsmen - question remains open.

4. Discussion

4.1 The emergence of mammoth mining and carvings

Mammoth bone began to appear in Western European markets as an equivalent of ivory from 17 to 17. [16]. In the 18th and 19th centuries. Russia was one of the main suppliers of tusks, which was connected with industrial development of new bone areas, especially in Yakutia. There are two views on the expansion of mammoth mining in the territory of Yakutia. According to the first position, the extraction of mammoth tusk and tusk carvings existed before the arrival of Russian explorers [17 p. 61; 3 p. 9]. And the other one says that these phenomena were caused by acquaintance with the Russian bone marrow case, in particular the holmogor carving [18 pp. 156; 5 pp. 8]. Let us look at these positions in more detail.

According to researchers of different times, the fact that mammoth mining was known to the Yakuts before the arrival of Russian explorers is supported by documents 17th century. Thus, N. Vereshaghin cited data of Smolensk voivode Musin-Pushkin of 1685 on the islands of the mouth of the Lena, where the locals hunted the walrus to hunt and sell its fangs [19 pp. 36]. H. Vereshhagin here combined the hunt for walrus tusks and mammoth tusks and anticipated the existence of this hunt even before the arrival of the Arctic explorers. Also quoted is a document according to which on June 1, 1652, the Yakut Voivode Mikhail Ladyshevsky was sent a letter «About gathering mammoth bones on the banks of the North Ocean» [20 pp. 19]. It should be noted that the documents refer to the time after the arrival of Russian explorers in the territory of Yakutia and cannot serve as solid evidence in support of this assumption. N.A. Alexeyev believed that bone products had similar motifs to wood carving, which had long been a well-known art of the Yakuts. In his opinion, the Yakuts knew the bone-cutting craft before the arrival of the Russians and northern peoples of the basin of the middle current of the river Lena widely used mammoth bone in household in 17th century, because they did not know iron [17 p. 61].

However, a more common view is that the mammoth bone has become one of the trades of the peoples of the North with the appearance of Russian immigrants in 17th century. S. Shelekhov in his work mentions that the mining of mammoth bones began in the second half of XVII century [21 pp. 24]. According to D.N. Anouchin, the mammoth bone had been mined in Siberia by 17th and by the second half of 18th centuries; bone carving was widespread [22 pp. 8]. S.V. Ivanov, M.A. Sergeyev assumed that among the Russian immigrants were bone-cutting craftsmen who brought this craft to Yakutia [18 pp. 156]. According to V.H. Ivanov and L.I. Yakunina, the extraction of mammoth bone in Yakutia became a target of fishing only in the 18th century, and until the mid-1670s. Mammoth bone was not mined and traded in Russia on the whole [23 p. 14; 3 p. 7]. Having considered the data of written sources, archaeological excavations and information about the mining of mammoth bone, V.H. Ivanov came to the conclusion that the time of emergence of bone-cutting art in Yakutia is the 18th century [23 pp. 13-15].

The same opinion is shared by art critic Z.I. Ivanova-Unarova, highlighting stylistic features of Yakut craftsmen in decorative design of products and in material processing techniques [24]. The study of Yakutia's bone-cutting case of the 18th century is based on Yakut's work of mammoth bone from the collection of the State Historical Museum: the bilateral combs, the 1764 class and the 1799 Casquet-Teremok. The 1743 comb in the center depicts lions with human faces and a unicorn surrounded by plant ornaments. Such a story is more common among Holmogor masters of the early 18th century. This is due to the fact that the main customers of mammoth bone products were wealthy segments of the urban population, especially visitors from Central Russia. On the watch stand (podchasnik) of 1764 you can see the plot of shaman's camlania, on the top there is a heart pierced with arrow, which was also characteristic of North Russian cutters. About the same item V.H. Ivanov wrote that the watch stand can already be considered as an independent work of the Yakut bone carver, while the Yakut master imitated the Khomogorsk Kostorores in the manufacture of the 1743 comb. In most of the works of this time, the substance prevails over decorative processing, which also distinguishes the Yakut masters from the Holmogor masters. The products are schematic, have simplified forms, are characterized by their brevity and strict decorative design [23 pp. 16-18]. In addition, immigrants from Russia - representatives of the authorities, industrialists, merchants, exiles - brought with them products made of bone of Russian cutters, which contributed to the rapid development of bone-cutting art in Yakutia. The movement of artistic carvings from tusks was most likely developed in the urban environment - the place of the intersection of cultures and development of trade.

4.2. Cost of raw materials

It is known that mammoth bone mining in the territory of Yakutia was mainly carried out in the Verkhoyansk and Kolyma districts, from where it was exported in unprocessed form to Yakutsk, and then to other cities and countries. Unfortunately, we do not have data on the value of mammoth bone products in the 18th century, but information on the value of raw materials has been preserved. So in the yasak book we find the following: the residue of the year 726 tithe payment collection - ... > 2 poods of mammoth bone in sutunkas - 2 roubles; sent with Osipa Ushnitsky and comrades the part not yet paid in Irkutsk - to give - 6 pounds of walrus bone and 7 pounds of mammoth bone - 1 rub. 25 cop. to Aleksey Zinoviev and comrades not yet paid in Irkutsk - 27 pounds of mammoth bone - 50 cop.» [25 s. 57]. In 1799, Tunguska merchant Osip Shumakov, while searching for the bones of a mammoth, discovered a corpse. He sold the tusks of this mammoth for 50 roubles, each weight was about 6 poods, in 1803 to the merchant Roman Boltunov, who sold them in turn to merchants going to Moscow for 21 roubles per pood [22 pp. 11].

The attitude to the mammoth of the local population. According to the Yakuts (like many other northern peoples), the mammoth was associated with the underground world [26] and considered to be the spirit-master of the water. The Yakuts envisioned him as a bull with huge horns and gradually transformed his image into a cold ox with the breath of which comes winter. In the Yakut society, the image of the winter bull and its taboo activities are still alive. In some places it is forbidden to watch the ice drift, as the river is freed from the ice, the souls of the dead are released in winter and, together with the ice and parts of the body, the cold ox carries them to the Arctic Ocean [27]. Mammoths were thought to live permanently underground, and coming out of the ground, smelling the air, immediately died [28 pp. 110]. Similar representations were made in the course of Hunts, Samoyeds, Yakuts, Evenks, in areas where most of the remains of mammoths are found. For example, in 1799, during mammoth bone mining, one of the Evenks discovered a mammoth carcase. When he told the house, they reacted with fear and said that the beast had already shown up in the same place and that the entire family of the Evenk who found it died shortly after [22 pp. 9-11]. According to the same text, the Evenki, on the other hand, tried to hide the corpses of mammoths for fear that the officials would come to them and force them to work, transport these corpses for hundreds of miles on the reindeer, and the local could lose a large number of reindeer [22 c. 11]. In the Chukchi extracting of mammoth bone was banned, despite the high cost. It was believed that the person who found the bone of the mammoth «should pay the ransom to the «spirit», having made some sacrifice [29 pp. 38]. Thus, at the end of the 17th and 18th centuries, the peoples of Siberia, including the Yakuts, had representations of a mammoth of a religious nature, explaining the origin of tusks.

5. Conclusion

The following conclusions can be drawn from the study. Due to industrial exploitation of the northern territories of Yakutia, the extraction of mammoth tusk became one of the traditional crafts of the Yakutia peoples by the 18th century. The above data indicate that mammoth tusk products were very rarely encountered in the context of Yakut burial monuments. It is not possible to speak of the variety of objects made of tusks. This may indicate that up to the end of the 18th century, mammoth tusk products were not widely used within Yakut society. This could be due to the negative connotation of the mammoth in the religious beliefs of the Yakutia peoples.

The objects found in the Yakut burials are substantially different from the works of the Yakut masters from the collection of the State Historical Museum of the same period. In our opinion, this shows the existence of two movements of mammoth tusk carving in the 18th century. The first is based on traditions and techniques of local carvings on bone and wood and has an original character. Local craftsmen produced household items for the local population. The second direction should be designated as the nascent bone-cutting art based on imitation of the northern Russian school. Masters of this direction created artistic works in European style and gradually developed their own style.

6. References

1. R.I. Bravina, V.V. Popov, *Funeral and memorial rites of the Yakuts: monuments and traditions (15-19th centuries)*. (Nauka, Novosibirsk, 2008). (in Russian).

2. N.I. Gagen-Torn, *Women's clothing of the peoples of the Volga region. Materials for ethnogenesis.* (Chuvashskoe gosudarstvennoe izdatel'stvo, Cheboksary, 1960) (in Russian).

3. E. Kryubezi, A. Alekseev, Eds., The world of ancient Yakuts: the experience of interdisciplinary research (based on materials of the Sakha-French archaeological expedition (Izdatel'skij dom SVFU, Yakutsk, 2012) (in Russian).

4. A.P. Okladnikov, *Lena antiquities* **1.** (Izd-vo AN SSSR, Yakutsk, 1945) (in Russian).

5. R.I. Bravina, N.D. Archipov, Funeral monuments of population of the pool of the Olenek river Yakutia (17-19th centuries) *Echoes of an Arctic Odyssey: the fate of ethnic cultures in the research of Northern scientists. Collection of materials of the All-Russian scientific and Practical conference with international participation*, 85-91 (2019) (in Russian).

6. A.N. Alekseev, *The first Russian settlements of the 17-19th centuries in the north-east of Yakutia* (Nauka, Novosibirsk, 1996) (in Russian).

7. I.V. Balyunov, *Vestnik NGU. Seriya: Istoriya, filologiya*, Novosibirsk, **15**, 149-157 (2016) (in Russian).

8. *Yakut fairy tales. Saha ostuorujalara*, (Yakutskoe knizhnoe izdatel'stvo,Yakutsk, 1967) (in Russian).

9. A.V. Shapovalov, *Tobacco and smoking pipes in Siberia in the 17-18th centuries (according to archaeological sources)* (Omsk, 1999) (in Russian).

10. M.M. Nosov, *Clothes and jewelry of the Yakuts of the 17th-18th centuries* (YAkutsk, **1**, 1955) (in Russian).

11. R.S. Gavril'eva, *Clothing of the Sakha people of the late 17th-18th centuries* (Nauka. Sib. predpriyatie RAN, Novosibirsk, 1998) (in Russian).

12. N.D. Arhipov, Report on the work of the archaeological and ethnographic squad of YSU in the summer of 1967 NOA IA RAN. F.1. R.1. Op. 1. D. 4255. (in Russian).

13. N. Arhipov, *In the Land of Yuryan Hosuun* **2** (1985) (in Russian).

14. A.N. Kopyrin, Report of the archaeological and ethnographic expedition of the Yakut republican branch of All-Russian Society for the Protection of Historical and Cultural Monuments for 1980 NOA IA RAN. F.1. R.1. Op. 1. D. 8608. (in Russian).

15. Protokol No.1. Description of items from the burial of a yakut women "Bolugur Ajyyta", NA RS(YA). F. 1407. Op.1. D. 38. (in Russian).

16. I.M. Potravnyj, A.V. Protopopov, V.V. Gassij, Arktika: ekologiya i ekonomika, **1(37)**, 109-121 (2020) (in Russian).

17. N.A. Alekseev, *Overview of the development of Yakut bone carving* (YAkutsk, 1966) (in Russian).

18. S.V. Ivanov, M.A. Sergeev, Sibirskie ogni, **4**, 155-168 (1956) (in Russian).

19. N.K. Vereshchagin, Why did mammoths become extinct, (Nauka, Leningrad, 1979) (in Russian).

20. V.L. Priklonskij, Sibirskij sborni, **1**, 24 (1893) (in Russian).

21. S. Shelekhov, *About the beginning of the development of the Siberian trade* (Avgusta Semena, Moskva, 1846) (in Russian).

22. D.N. Anuchin, About the restoration of a mammoth for an Anthropological exhibition (Lavrova, Moskva, 1879) (in Russian).

23. V.H. Ivanov, Yakut bone carving (Nauka, Moskva, 1979) (in Russian).

24. Z.I. Ivanova-Unarova, Izobrazitel'noe iskusstvo Urala, Sibiri i Dal'nego Vostoka, **1**, 122-133 (2019) (in Russian).

25. Acts of the archives of the Yakut region*:* (s 1650 g. do 1800 g.) Izvestiya YAkutskoj Oblastnoj uchenoj Arhivnoj Komissii ; 1(1) (1916) (in Russian)

26. A.R. Fedorova, A.A. Sleptsova, CHelovek i kul'tura, **6**, 164 - 170 (2019) (in Russian).

27. A. Prokop'eva, Field materials of the author. (2014)

28. I. Ides, A Brand, *Notes on the Russian Embassy in China (1692-1695,* (Glavnaya redakciya vostochnoj literatury izdatel'stva, Moskva, 1967) (in Russian).

29. V.G. Bogoraz, Chukchi (Leningrad, 1939) (in Russian)

Ethnocultutal identity of the indigenous people of the Arctic (a case study of Anabar district in Yakutia)

Viktoriya Filippova[1,*], *Liliya* Vinokurova[1], *Yana* Sannikova[1], *Natalya* Zakharova[1], and *Akulina* Mestnikova[2]

[1] Institute for Humanities Research and Indigenous Studies of the North, Siberian branch, Russian Academy of Sciences, #1 Petrovskogo str., Yakutsk, 677027, Russia
[2] Federal State Budget Educational Institution of Higher Education "Churapcha state institute of physical education and sports", #2a Sportivnaya str., Churapcha, 678670, Russia

Abstract. The article presents the preliminary results of a comprehensive study on the social anthropology of the cultural landscape of the Khatango-Anabar region as an integral historical and cultural space, formed as a result of the interaction of man and nature, local ethnic communities and social institutions. The purpose of the article is to study the regional ethno-cultural identification of the population living in one of the Arctic regions of Yakutia - Anabar. The field research, including interviews, a sociological survey, and a psycholinguistic experiment revealed indicators for determining the ethnocultural identity. Three main criteria of ethnocultural identity of residents of Anabar district were identified. These are territorial, cultural and linguistic ones. It was found that the unifying indicator of the population living in this area is regional identity as Anabarians.

1 Introduction

The ethnocultural diversity of the Russian population is studied within various social humanitarian disciplines. Despite a significant number of Russian and foreign publications, most of the works are aimed at studying Russian identity; issues of regional and local identity remain outside the scope of research [1].

The relevance of this article is due to the insufficient study of regional ethnocultural self-identification. Assuming that ethnocultural identity is based on language, culture, historical past, the research supported by the RFBR (project 20 09 00257A) was conducted to study indigenous peoples of the Arctic living in Anabar national (Dolgan-Evenk) ulus (district) of the Republic of Sakha (Yakutia).

The choice of the region was determined by the Arctic natural landscapes and ethnic composition of the population: the Dolgans, Evenks and Yakuts are the indigenous ethnic groups. Adaptation to the harsh local conditions and interaction of different cultures and languages have formed an ethnocultural identity of the inhabitants of this area.

Anabar ulus (district) of Yakutia is a territory with a unique experience of interethnic interaction, economic and cultural adaptive practices. A number of gaps in the historical-geographical and cultural-anthropological studies of the region should be eliminated.

The sources were unpublished materials found in the state and regional archives, current archives of institutions and organizations; the published sources are data of statistical records of the population, reports and information posted on the official websites; field materials collected in Anabar during the expedition in September 2020, sociological survey and psycholinguistic experiment results.

2 About Anabar district

2.1 Landscape

Anabar ulus lies beyond the Arctic Circle, in the permafrost zone and tundra landscapes. The harsh climate is softened by the Laptev Sea, which washes the region in the north. The ulus has got its name from the Anabar river that flows into the Anabar Bay. The relief is flat: the region lies on the North Siberian lowland. However, there are ridges of low hills in the northeast and in the south. In the wide valleys of meandering rivers, there are many channels and lakes. There are many lakes in the northern part of the ulus. It borders on Krasnoyarsk Territory, Bulun and Olenek Districts of the Republic of Sakha (Yakutia).

The nature of Anabar is represented by two landscape zones - tundra and forest-tundra. In these conditions, as a result of long historical ties between the Tungus, Yakuts and Russians, the Dolgan nationality formed [2].

2.2 Population of Anabar District

Anabar is a place of residence of two ethnic groups: Dolgans and Evenks. Unlike the Tungus-speaking Evenks, the Dolgan ethnos, whose origin is associated with the ancient Evenk clans Dulgan, speaks the language of the Yakut subgroup of the Turkic language family of the Altai language community. The ethnonym

* Corresponding author: filippovav@mail.ru

with the root dol/dul is widespread in the Tungus-speaking environment [3], and the Dolgan language, in addition to the Evenk component, stores "elements of the Yakut language of the period of the first waves of their migration to present-day Yakutia" [4].

The main ethnic settlements are two rural local communities: Saskylakhskoye and Yuryung-Khainskoye. These communities are characterized by intra-district identification as Evenks (Saskylakh village) and Dolgans (Yuryung-Khaya village).

In 2020, the number of residents in Saskylakh was 2,420 people, in Yuryung-Khaya - 1,216 people. The share of the Saskylakh people is dominated by the indigenous small-numbered peoples of the North - 69.3%. Their distribution is as follows: Evenks - 30.9%, Dolgans - 27.7%, Evens - 10.3%, Yukaghirs - 0.4%. The share of Yakuts is 25%. In the ethnic structure of Yuryung-Khaya Dolgans are dominant (74.2%) ; the share of Yakuts is 19.7%, the share of Tungus-speaking representatives (Evenks and Evens) is only 2.5%.

The population is polyethnic, which gives particular interest to the study of ethnocultural identification.

2.3 Traditional Economic Activities

Deep connections of the traditional nature management of the indigenous peoples of the North with the geographical landscape are one of the most important areas of research in Russian science. Undoubtedly, along with the gradual disappearance of traditional nature management, the accumulated historical experience of ecological culture goes away, as a result, the social structures that form the traditional society disintegrate.

Tundra deer breeding, fishing and hunting have been natural economic activities for the natural conditions of this Arctic territory. In the district, deer are used as a means of transportation for migrations, hunting fur-bearing animals, fishing. In summer, they collected wild plants (dove and cloudberry), bird eggs, killed wild geese and ducks, and in winter they killed partridges [5].

The main direction of the traditional economy of the indigenous population of Anabar is domestic reindeer herding, which plays a leading role in the socio-economic development of the region. As of 2019, the total number of deer was 18,050. Today, the reindeer herding companies are Arktika and MUE n.a. Hero Spiridonov, nomadic tribal communities Bolshoy Begichev, Sedeme and Balyksyt. In the ulus, three farms have fishing plots: the Uele tribal community, the N. Ye. Androsov MMUP and A. Koryakin who is an individual entrepreneur; all 14 farms of the ulus are engaged in hunting: 3 municipal enterprises, 5 nomadic tribal communities and 4 individual entrepreneurs in agriculture [6].

In general, the indigenous population managed to preserve and continue to develop the traditional sectors of the economy and land management, with the exception of the Arctic fox fishery. Today, the local population has managed to preserve reindeer pastures (about 72.7% are used) and hunting grounds distributed among the farms. Additional protection for traditional occupations is the registration of the territory of the Anabar ulus as a territory of traditional nature management and the presence of specially protected natural areas [7].

3 Indicators of Ethnocultural Identity of Anabar District

Ethno-cultural identity is the awareness of one's belonging to a particular ethnic community based on the common territory of residence, stable features of culture and language. Thus, there are three main criteria for ethnocultural identity: territorial, cultural and linguistic. Modern culture is characterized by differences between ethnic groups which causes an identity crisis, and the process of preservation is observed ; the growth of ethnic self-awareness caused by social transformations is the basis for ethnocultural identity development [8].

There are many areas that address the issues of ethnic identification based on various methodologies.

Based on the interviews of Anabar residents, the sociological survey and the psycholinguistic experiment, we identified the following criteria of ethnocultural identity.

3.1 Territorial Identity

N.A. Mamontova said that earlier the name of a genus group was used to reveal the approximate circle of social connections of the interlocutor. In order to indicate this ability of the Evenks, the author uses the concept of geographical model [9].

During the interviews, many residents pointed to their family and ancestral affiliation. Based on the interviews and published sources, it was established that the Saskylakhs and Yuryung-Khains are descendants of the following clans: Khangalas, Khatygyn, Omoldon, Beti, peasants, Dulgan, Baraakh, Sedeme, Aryyaat, etc. Many ethnonyms are used in the place names of Anabar ulus. The residents distinguish by surnames belonging to a particular genus. For example, the Vinokurovs are the Khangalas, the Androsovs are the Khatygyns, the Tuprins belong to the Sedeme clan, etc. The territorial-clan self-identification was studied by local researchers, which indicates the importance of this type of self-identification.

Settlement is the most important aspect of sociocultural self-identification. The results of the study indicate the division of the inhabitants by the settlements of the ulus: Saskylakhs and Yuryung-Khains. The old-timers mentioned that earlier the clothes helped to determine where a person is from: "We used to wear shoes with a straight seam (pattern) - lapaika, the Saskylakh people called us "lapayki". Similar markers like clothing or a dialects denote the ethnocultural identity of local groups [10]. In the ulus we did not reveal the division into residents of the periphery and the center, which may be due to their "equivalent" functions: Saskylakh is the regional center; in Yuryung-Khaya, there is a river port that supports the entire ulus.

However, the population of the ulus, like the majority of the population of the northern regions of Yakutia, emphasize their extreme northern geographical position in relation to the inhabitants of the central part - "I will reach diono". The isolation of the region is perceived as a matter of course, reflecting the area of settlement of the Dolgans in this area and nowhere else. This was also reflected in the respondents' answers; only 10.4% indicated the importance of history and territory in determining nationality.

The general regional self-identification of "Anabars" is characteristic both of the indigenous inhabitants and of those who do not consider themselves either Saskylakh or Yuryung-khain. One of the regional types of self-identification is the Arctic identity due to a close connection with nature, perception of the Arctic as a homeland.

3.2 Reindeer herding and hunting culture

Ethnocultural identification is associated with the traditional economy of the indigenous peoples; the leading occupation has been domestic reindeer herding. In this regard, we observe the persisting figurative and real "reindeer world" of the Anabarians.

Economic and everyday culture is similar to the one of Tungus nomads. In the vocabulary, there are traces of the presence of hunting, fishing and reindeer breeding culture and the absence of cattle breeding culture typical of the Turkic-speaking peoples. When analyzing the reindeer-breeding vocabulary of the Anabar Dolgans, it was revealed that 35.7% of reindeer-breeding terms are Evenk in origin. The reindeer herding vocabulary is developed, which confirms the fact that reindeer husbandry remains the main economic activity and occupies a significant place in the Dolgan culture. The analysis of vocabulary revealed the predominance of words associated with the names of deer, sex and age characteristics of these animals (27.9%), their common names (14.5%) and habits (13.1%). The share of other words related to reindeer husbandry is insignificant and ranges from 4% to 9%: the names of deer diseases (4.9%), games and competitions related to reindeer husbandry (4.9%), reindeer in pack-riding and narten transport (6.5%), reindeer pastures and forage (8.2%), names of deer by the shape of the antlers (9.8%) and names of the deer by their external characteristics (9.8%).

The results of the free associative experiment conducted in Anabar ulus in 2020 confirmed these assumptions. 150 people of various nationalities took part in the survey. The questionnaires were developed in the Yakut language. Consider the images of cow and horse, which reflect the pastoral culture and the deer image related to the reindeer herding culture.

The image of cow is associated with an animal producing milk similar to a horse which lives in the central mainland. The image of horse is associated with a strong animal, cattle similar to the cow. Such similar reactions indicate that the Anabarians are familiar with these concepts, but they are not the realities of their everyday life.

The image of deer is diverse and can be seen especially clearly in the answers given by the respondents from Yuryung Haya, where the reindeer culture is dominant. For example, the sacred white deer, which is distinguished by its beauty and purity, is a traditional food. Some words from the reindeer herding terminology - deer road or a deer footprint, new sledges, nets, lasso. A single response - reindeer-riding - was given to the "star" stimulus, which suggests some kind of individual experience related to night travel.

Thus, reactions to the image of the deer are present in different areas of life, which indicates the presence of this concept in the life of the Anabarians.

3.3 Language as a Cultural Indicator

Ethnocultural identity is based on language, culture, and historical past – these are the unifying features that create a unique image of an ethnic group. A sociological study conducted in Anabar ulus in October-November 2020, in which 201 respondents from Saskylakh and Yuryung-Khaya took part, revealed that the primary marker of ethnic identity is a traditional culture. To the question, "What determines your nationality?" the respondents' answers were distributed as follows: culture and traditions – 60.2%, native language – 26.9%, history and territory – 10.4%.

Based on the sociological research data, we can say that in Anabar, the role of culture and traditions in the process of ethnic identity is the most important ethnic identification criterion. The survey results proved this: 96.7% of Evenks, 71.8% of Dolgans, and 92.3% of Evenks indicated Yakut as their native language, which is 83.1% of all respondents. The respondents indicated two and sometimes even three languages as their native languages. The majority of respondents also named Yakut as the language of family communication (94.5%). The Yakut language is most often used in the forest (taiga) (50.3%), in their own locality (69.7%), in the district center (54.2%), and in Yakutsk (49.8%). However, this fact does not affect the ethnic identity of the peoples inhabiting this area: 51.2% of Dolgans, 23.4% of Sakha, 14.9% of Evenks. 6.5% of Evenks indicated their own nationality.

According to the results of the sociological survey conducted in Anabar, more than half (53.2%) of the respondents said that they observe the traditional rites and customs in everyday life; 26.9% chose the option "sometimes," and 10.9% do not observe them. Traditional festive ceremonies are held mainly in Yakut (79.6%), Russian (35.3%), Dolgan (34.3%), and Evenki (17.9%) languages, and the option of holding events in a mixed language is also common (22.9%). The most preserved elements of traditional culture include national food (70.1%), everyday clothing (57.2%), tools, hunting gear (45.8%), and transport (16.4%).

Thus, based on the sociological data and analysis of the current research, we can say that a decline in the importance of linguistic identity and traditional culture (consisting of ritual practices, national holidays, traditional food culture, and national clothes) remains one of the core components of the ethnic and cultural identity of Arctic peoples as a special kind of metaethnoregional identity character. The uniqueness of the traditional cultures of the Arctic peoples of the North reveals the diversity of local cultures (Dolgan, Evenki, Even, Yakut, Russian) that have emerged on the geo and ecocultural landscape of the Khatango-Anabar region. This represents a single environment of ethnic groups in the context of interethnic and intercultural relations, which will be the subject of further research of ethno-social and cultural processes in this region.

4 Conclusion

Thus, the ethnocultural identity of residents of Anabar is determined by three criteria:
- territorial, containing family and clan historical memory of the region of residence and including the settlement affiliation of the inhabitants of two municipalities - Saskylakh and Yuryung-Khaya and positioning as a regional community - "Anabarians" and "residents of the Arctic";
- linguistic, not dominant for ethnic identification, when the language of communication for the majority of the indigenous population (Dolgans, Evenks, Yakuts, Evens, etc.) is Yakut ; native languages are taught at school (Dolgan and Evenki), and the official language is Russian;
- cultural, fueled by Arctic traditional activities and their sustainability, elements of the traditional way.

Acknowledgment

The article was carried out within the framework of the grant "Khatango-Anabar region in the XX-early XXI centuries: anthropology of the cultural landscape" of the Russian Foundation for Basic Research, grant No. 20-09-00257A

References

1. V. A. Tishkov, *Herald of the Russian Academy of Sciences.* **89(4)** 408-412 (2019). DOI: https://doi.org/10.1134/S1019331619020229
2. V. I. D'yachenko, Hunters of high latitudes. Dolgans and northern Yakuts (2005) [In Russian].
3. G. M. Vasilevich, *Sovetskaya ethnografiya* **4** 34-49 (1946) [In Russian].
4. N. M. Artemyev, *Dolgan language: structural and semantic analysis.* PhD abstract, (2001) [In Russian].
5. *Anabar District: A history. Culture. Folklore* (2005) [In Russian].
6. Municipal program "Development of agriculture and food of Anabar District for 2020-2024» (2019) [In Russian].
7. V. V. Filippova, L. I. Vinokurova, Ya. M. Sannikova, N. E. Zakharova, A. E. Mestnikova, *Nauchnyi dialog,* **11,** 495-508(2020). DOI: 10.24224/2227-1295-2020-11-495-508 [In Russian].
8. N. R. Krasovskaya. *The Authority*, **4** (2020) DOI: https://doi.org/10.31171/vlast.v28i3.731 [In Russian].
9. N. A. Mamontova. Identity and language of the IlympicEvenks in the XX-XXI centuries. PhD abstract, Moscow (2013) [In Russian].
10. N. I. Shietova, *Tomsk journal of linguistics and anthropology*, **2**, 160-170 (2020).

Development of a system of transport-interchange hubs in urban transport network

Oleg V. Moskvichev[1], and *Svetlana A.* Leonova[1,*]

[1] Samara State Transport University, 18, Zavodskoy highway, Samara, Russia

Abstract. The paper considers the issues of the formation of a system of transport interchange hubs (TIH), associated with the determination of the location of TIH in order to improve the quality of the provided transport services. The model of the optimal choice of TIH system is presented in the form of a linear Boolean programming problem. The paper gives the factors that reduce the average travel time around the city due to the formation of a system of transport hubs. In order to solve the problem of the choice of TIH locations, the approach based on the definition of "effective interchange hubs" is proposed. The determination of effective interchange hubs is carried out according to the following algorithm. First we found a list of non-stop (direct) rides. Then, we consider the rides for which there is no direct route or for which there is such a route, but this revealed connecting route will take less time. These hubs are efficient. Next, a list of effective interchange hubs is determined and the amount of passenger traffic at each hub is specified. Using mathematical methods to solve integer programming problems, it is possible to select such efficient interchange hubs, in which it is advisable to create transport interchange hubs. These scientific results can be used in programs and projects for the development of urban transport systems in the development of basic requirements for the formation of transport hubs.

Keywords: transport interchange hub, urban transport network, efficient interchange hub, optimal choice, linear Boolean programming problem, passenger traffic

1 Introduction

Recently, the problem of the development of urban transport infrastructure in cities and megalopolises, including transport interchange hubs (TIH), has become increasingly important. A special place in the formation of a modern urban transport system is assigned to transport and interchange hubs. Urban transport hubs are formed at the intersection of various types of public passenger transport. The number of transport hubs and their locations determine the state of the entire transport system of urban public transport. Optimal placement of TIH increases the attractiveness of public urban transport ensuring fast, safe and comfortable travel through interchange hubs.

In the scientific literature, the choice of the locations of TIH in the network of urban public transport is based on the use of the expert assessment method [11, 14]. It is also proposed to create transport and interchange hubs on the basis of railway stations, passenger depots and metro stations. According to this principle, transport hubs have been formed in super-large cities of Russia [3, 11, 12, 14, 15] and foreign countries [5, 10].

However, in many cities, high-speed off-street transport (urban rail transport, subway) is insufficiently developed, which must also be taken into account when designing additional transport infrastructure, including TIH. It is necessary to conduct a comprehensive analysis of the state of the road network and the demand for urban public transport based on the construction of quantitative models. The important characteristic of a transport hub is the amount of passenger traffic.

In order to solve the problem of the choice of the location of TIH, it is necessary to select the type of model. The analysis of the literature on the development and use of mathematical models for the description of urban passenger transport processes [1, 3, 8, 9, 12, 13, 15] shows that most of them use a simulation tool. For example, N.A. Kalyuzhny [2, 3] considers the VISUM software package and Citraf software package, which the author chose because of the simplicity and availability of modification. The use of this tool allowed identifying some properties of the network of urban passenger transport, which indirectly indicate those network hubs that can become TIHs. In this case, only the hubs formed around the stations of high-speed off-street transport are considered.

The criterion of the influence of the waiting time delay revealed by the author can be considered only as one of the factors determining the priority of the construction of the transport-interchange hub, but there are a number of other factors that can influence the choice of the location of the transport-interchange hubs. It is necessary to note that the conclusions drawn on the

* Corresponding author: moskvichev063@yandex.ru

basis of simulation modeling of the urban transport system are not universal in nature and it is incorrect to apply them for the network of any city. Moreover, the network parameters that have to be pre-entered into the model require large statistical measurements of passenger traffic in a particular network. As a result, further scientific research and substantiation of the issues of the choice of the location of urban transport hubs in order to improve the urban transport system require further study.

2 Problem Statement

For the accurate assessment of the state of the urban transport system and its development, including the development of urban transport and interchange hubs, it is required to perform a system analysis of the processes occurring in the urban passenger network, based on criteria expressing the main purposes of the system and build models that express the target function as the degree of achievement of the main purpose.

Therefore, in order to characterize the quality of the urban transport system, the value of the average statistical time of one ride is taken. The optimal location of urban transport hubs in order to improve the quality of the provided transport services is formally expressed as an objective function

$$\bar{T} = f(X) \rightarrow \min,$$

where X denotes a certain variable subset of TIH, which is a variant of constructing a TIH system for the entire urban passenger network.

It is obvious that in the general case, during the choice of the optimal system (set) of TIH, there are always restrictions on the choice of X. Therefore, the options of X must belong to some admissible set G, that is, $X \in G$. Thus, in general form, the problem of optimal choice of the TIH system can be written as

$$\bar{T} = f(X) \rightarrow \min \quad (1)$$

$$X \in G \quad (2)$$

Consequently, the simulation of the processes occurring in urban passenger transport should describe the objective function (1) and explicitly express the constraints (2). The urban public transport network is a graph that includes subgraphs of all types of urban public transport and a subgraph of transfers. The network graph includes stops and routes of urban public transport (Figure 1).

The connectivity of the graph is taken into account during the solution of the problem of the choice of the location of TIH. It is known that not all dependencies between the parameters of the graph model can be described in the form of analytical dependencies. In addition, we should expect a model with an extremely large number of variables and constraints, which makes the search for a solution labor-intensive or almost impossible. As a result, it is important to propose an approach that will allow solving the problem of the optimal choice of locations for urban transport hubs.

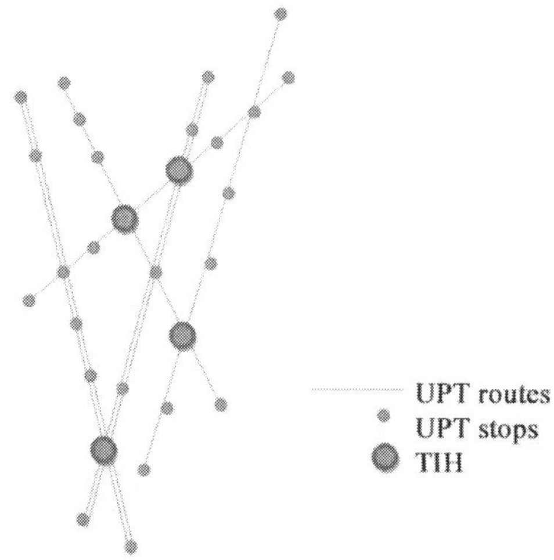

Figure 01. Network graph of urban public transport (UPT)

3 Model of the optimal choice of the TIH system in the form of a mathematical programming problem

In order to solve the problem of optimal choice of the location of transport hubs for urban transport network, a possible mathematical model should be considered. The network of urban public transport is a graph with stops and routes of urban public transport (Figure 1). In the first assumption it is considered that any passenger route within a straight line is the best and no choice is required. If the starting and ending stops are on different lines, then a transfer is required at the stop, which will be called a transport hub. i – the number of the initial stop of the passenger's route, j – the number of the final stop of the passenger's route, k – the number of a possible interchange hub. There can be several numbers. We can assume that for the studied network, from any stop i, one can get to stop j through at least one possible interchange hub k. Since the network topology is given, it is possible to specify all possible routes in advance from i to j through possible interchange hubs k.

It is necessary to note the factors that reduce the average travel time of a passenger around the city due to the formation of a system of transport hubs.

1. Direct routes from i to j are replaced by interchange routes if travel time decreases.
2. Transition to TIH from one route to another is carried out faster by the value Δt_{TIH}.
3. New interchange routes appear from point i to point j with a change at the lth hub due to the construction of a transport hub with $t_{ilj} < t_{ij}$.

Further, the shortest paths in time from i to j are found using well-known algorithms [4]. Then the whole set of possible triples of numbers (i, k, j) becomes known and the value K is the maximum possible number of interchange hubs.

It is obvious that rational routes with more than one change are possible. Then such routes can be specified by a sequence of numbers (i, k, l,... r, j). In addition to the network graph, the main characteristics of the passenger traffic are specified:

1. The average number of passengers traveling from i to j (for the base period), c_{ij} – the matrix of correspondences [6].
2. Average travel time from i to j along a straight line t_{ij}.
3. Average travel time from i to j through the transport hub k t_{ikj}.
4. Average travel time from i to j through interchange hubs k and l t_{iklj}.
5. The total number of passenger rides during the base period – N.
6. The total number of stops of the transport network – n.
7. The maximum allowable volume of passengers at the kth interchange hub for the base period is D_k.
8. The minimum allowable volume of passengers at the kth interchange hub for the base period is d_k.

All studied characteristics are obtained using statistical surveys, using prediction of future values.

All stops of the city network are numbered so that any stop of any line has its own unique number.

It is necessary to formulate the problem of optimal selection of the locations of TIH based on the condition of a given passenger flow and in order to reduce the time spent on the transfer of all passengers in the system of urban public transport. Further, we consider a possible mathematical formulation of the optimization problem according to the criterion T. For this, it is necessary to formulate the objective function and the limitations of this problem. A change in the value of T under the above mentioned assumptions can be carried out only by the choice of a TIH. The values x_{ikj} and x_{iklj} are considered as controlled variables, which are equal to 1 if we go from i to j through the hub k or kl, and are equal to 0 otherwise.

If there is a direct path from i to j, then it is considered to be identical, $x_{ij}=1$ for any i and j. It is important to note that the considered model assumes that there is a rational path from i to j, which lies on the same line, but the ride can be performed through interchange hubs k or kl, since $t_{ikj}<t_{ij}$ (for example, if it is possible to use a faster transport).

The average travel time of passengers around the city can be expressed as follows:

$$\overline{T} = \frac{1}{Nn^2}\Big(\sum_{i=1}^{n}\sum_{j=1}^{n} c_{ij}t_{ij}x_{ij} + \frac{1}{K}\sum_{k}^{K}\sum_{i=1}^{n}\sum_{j=1}^{n} c_{ikj}t_{ikj}x_{ikj} + \\ + \frac{1}{KL}\sum_{k}^{K}\sum_{l}^{L}\sum_{i=1}^{n}\sum_{j=1}^{n} c_{iklj}t_{iklj}x_{iklj} + ...\Big) \rightarrow \min \quad (3)$$

Here, it is taken into account that each time t_{ij} is calculated by the number of passengers who spend this time. From the above mentioned formula, it can be seen that more than two transfers can be taken into account. Next, the restrictions on the choice of variables x_{ij}, x_{ikj}, x_{iklj} should be formulated.

The transportation from i to j with one transfer is carried out only through one selected interchange hub

$$\sum_{k}^{K} x_{ikj} = 1, \forall i, j; (i \neq j) \quad (4)$$

The transportation from *i* to *j* with two transfers is carried out only through one selected pair of transport hubs

$$\sum_{k,l}^{K} x_{iklj} = 1, \forall i, j; (i \neq j) \quad (5)$$

The volume of the kth interchange hub (out of all possible K hubs) is limited by the value D_k

$$\sum_{i=1}^{n} c_{ik}x_{ik} + \sum_{j=1}^{n} c_{kj}x_{kj} + \sum_{i=1}^{n}\sum_{j=1}^{n} c_{ikj}x_{ikj} + \\ + \sum_{i=1}^{n}\sum_{j=1}^{n} c_{iklj}x_{iklj} \le D_k, \forall k = (1, K) \quad (6)$$

In the problem (6), passengers leaving k, arriving at k and transferring to k are taken into account. On the other hand, there may be a limitation on the fact that the transport hub must be large enough, for example, to cover the cost of additional construction of the hub infrastructure. Therefore:

$$\sum_{i=1}^{n} c_{ik}x_{ik} + \sum_{j=1}^{n} c_{kj}x_{kj} + \sum_{i=1}^{n}\sum_{j=1}^{n} c_{ikj}x_{ikj} + \\ + \sum_{i=1}^{n}\sum_{j=1}^{n} c_{iklj}x_{iklj} \ge d_k, \forall k = (1, K) \quad (7)$$

We get the problem (3–7) of linear Boolean programming, which, in principle, can be solved using standard computer programs [4, 7]. In the algorithm for the solution of this problem (3–7), various options of transport hubs among the allowable ones will be sorted out, but since the objective function is minimized, the optimal transportation option will use the allowable number of interchange hubs from the condition of minimizing the objective function (3). Therefore, in this statement, the question of the optimal number of transport hubs K^* is solved automatically.

The number of routes passing through the interchange k (value b_k) is determined.

Let the solution of problem (3–7) be found. These are arrays of zeros and ones $(x_{ikj}, x_{iklj})^*$ (asterisk marks all values obtained as a result of the solution of problem (3–7)). Then

$$b_k = \sum_{i=1}^{n}\sum_{j=1}^{n}(x_{ikj}) + \sum_{i=1}^{n}\sum_{j=1}^{n}(x_{iklj}), (k = 1, K) \quad (8)$$

The vector b_k gives all the necessary information about the solution of the initially posed problem. First, the number of transport hubs in the solution is determined by the number of nonzero components of this vector

$$K^* = \sum_{k}^{K} \delta(b_k), \quad (9)$$

where $\delta(b_k)=1$, if $b_k>0$ and $\delta(b_k)=0$, if $b_k=0$.

We obtain the capacity of interchange hubs B_k (the total number of passengers who visited the kth interchange hub) using the following expressions:

$$B_k = b_k = \sum_{i=1}^{n}\sum_{j=1}^{n} c_{ikj}(x_{ikj}) + \sum_{i=1}^{n}\sum_{j=1}^{n} c_{iklj}(x_{iklj}), (k = 1, K) \quad . (10)$$

B_k shows the number of passengers transferring at the kth hub.

Further, the complexity of the solution of the problem (3–7) and the ways of its simplification are estimated. Considering this problem, it is believed that only single transfers are possible (with the increase in the number of transfers, the complexity of the solution increases dramatically). In this problem, only the variables x_{ikj} are varied. Their number is determined by the number of encountered connecting routes.

The considered problem of the optimal choice of the TIH system is a multidimensional problem of linear Boolean programming and, therefore, can be solved only approximately by one of the enumeration methods. It is necessary to propose an approach that allows simplifying the solution of this problem and obtain practical options for its solution.

4 Purpose of the Study

The purpose of the work is to form a system of TIHs based on the optimal choice of TIH locations in urban transport network.

5 Research Methods

In order to solve the problem of the choice of TIH locations, an approach based on the definition of "effective interchange hubs" is proposed. Such hubs are candidates for the role of transport hubs.

In efficient interchange hubs, passengers are transferred in case of their travel by the fastest routes, which are determined using graphical analytical methods.

Further, using mathematical methods for the solution of integer programming problems (the solution of the knapsack problem, the branch-and-bound method), the locations of the transport hubs are determined.

Thus, the problem of the optimal choice of locations for urban transport hubs is solved in two stages.

1. At the first stage, for each transfer route, there are transport hubs r that provide the shortest path from i to j through the transport hub r

$$r = \arg\min_{k}(t_{ikj}) \quad (11)$$

It is necessary to note that in the future, the expression for one transfer will be denoted by the number k, although k can be considered a vector of transfers, and then (11) is also valid for the case of many transfers.

The hubs of the urban transport network that satisfy the problem (11) are efficient interchange hubs. It is obvious that now only effective interchange hubs should be considered as allowable for the solution of the problem of the choice of TIH hubs. Processing this stage radically narrows the range of feasible solutions and simplifies the algorithms for the solution of the problem. A set of effective interchange hubs is found R = (1,2,3,..., r, ..., N_r), where r is the number of the effective interchange hub, N_r is the total number of interchange routes. In the case of two or more transfers, the transfer vector (k, l,... q, p), which has its own number, can also act as a transfer.

2. At the second stage, a Boolean variable is considered: $x_r = 1$, if the rth effective hub is TIH; $x_r = 0$, otherwise. Then the set of TIHs is determined by a vector of zeros and ones of dimension N_r.

$X=(x_1,x_2,\ldots,x_N)$, $X \in R$.

At the same time, during the choice of the optimal X^*, the optimal number of TIHs in the project n

$$\sum_{r=1}^{N_r} x_r^* = n \quad (12)$$

Thus, it is possible to simplify the description of the mathematical model through the introduction of the concept of effective interchange hubs and a vector that determines the solution of a much smaller dimension.

6 Findings

The following aspects are considered as the initial data for the studied problem of finding effective interchange hubs: a matrix of interstop correspondence [6, 15], a graph of the urban transport network indicating routes, stops and possible interchange hubs, travel times between all public transport stops on all city routes [7].

The search of efficient transport hubs is carried out according to the following algorithm. A table of stops and the names of these stops are created. For each route, the numbers of stops (a chain of stops i and j) through which this route passes are recorded. If a bus / trolleybus / tram follows the same route in the forward and backward directions, it is recorded as a sequence of stop numbers in the same direction. If the route is circular, then it is recorded as a sequence of stop numbers recorded twice in order to accurately determine the length of the route, the time and the number of stops that should be traveled when traveling from i to j. After that, there is a list of non-stop (direct) rides from i to j.

Then we consider rides from i to j for which there is no direct route or it exists, but the revealed connecting route will take less time. These interchange hubs are called "efficient" interchange hubs. Next, we consider rides from i to j with two interchange hubs. These are routes for which there is no one-transfer route, or routes that are faster than routes with one-transfer route.

For the practical implementation of the above mentioned calculation algorithm, a computer program "Efficient Transfers" is developed, which allows obtaining many effective interchange hubs for the network of urban passenger transport [7].

The correspondence matrix $C^0=\{c^0(i,j)\}$ determines the average predicted value of the passenger traffic moving from stop i to stop j over a certain period of time. This time includes the standard waiting time for public transport. For routes with transfers, c_{ikj} is the number of passengers traveling from i to j and changing at hub k. The number of passengers who follow from i to j with two or more transfers $c_{ikl\ldots pj}$.. It is considered that k is the set of numbers of transport hubs. It is necessary to determine the largest efficient transport hubs in terms of the number of passengers. Some of these hubs will become TIHs in the future.

The creation of TIH is aimed at the improvement of the convenience of passenger service during transfer and requires significant costs for redevelopment of the transport hub area, construction of convenient crossings and additional infrastructure (commercial areas, waiting rooms, etc.). The development of a TIH system should entail a reduction in the time spent by passengers on total service. The choice of transport and interchange hubs is primarily related to the number of passengers who transfer, since the main criterion for the optimal urban transport network requires minimizing the average time of one ride.

The average time of one ride with a transfer can be expressed as:

$$T = \frac{1}{\sum_{i,k,j} c_{ikj}} \sum_{i,j} c_{ikj} t_{ikj} \rightarrow \min \quad , \qquad (13)$$

where c_{ikj} –the number of passengers traveling from i to j, transferring to k;

t_{ikj} –the time spent on passengers transfer from stop i to stop j through k.

In the problem (13), it is noted that for all i and j the transfer at the transport hub k is fixed. Based on it (13), we can conclude that the average time of one ride is determined by significant passenger traffic. That is, the more passengers will be processed through the formed system of transport and transport hubs, the greater the obtained reduction in time.

Having determined the set of effective transport hubs (11), it is possible to calculate the amount of passenger traffic at each hub c_r.

$$c_r = \sum_{i,j} c_{i,r,j} + \sum_i c_{ir} + \sum_j c_{rj} \quad , r = 1,2,\ldots R \quad , \qquad (14)$$

where $\sum_{i,j} c_{i,r,j}$ – the number of passengers changing at the rth interchange hub (TIH);

$\sum_i c_{ir}$ – the number of passengers arriving at TIH;

$\sum_j c_{rj}$ – the number of passengers leaving TIH.

Since formula (13) expresses the criterion of the average time of one transfer ride and its values c_k are the frequency of occurrence of the values of the time of one ride t_k, it turns out that the values of c_r determine the main criterion for the creation of transport hubs. The amount of passenger traffic at each efficient hub provides information on where to set up TIHs.

7 Conclusion

Thus, the model of the optimal choice of the TIH system considered in this paper in the form of a linear Boolean programming problem and the approach based on the choice of efficient interchange hubs allow solving the problem of the optimal placement of TIHs in the urban transport network. Efficient interchange hubs are hubs in the network of urban public transport where passengers are transferred when traveling around the city, while choosing the optimal route based on the criterion of the minimum travel time. These very hubs lie at the intersection of several routes and determine the location of urban transport and interchange hubs.

The results of this research can be used in programs and projects for the development of urban transport systems in the determination of basic requirements for the formation of transport hubs.

References

1. B. Fan, Y. Yang, L. Li. Integrated optimization of urban agglomeration passenger transport hub location and network design. URASIP J. on Wirel. Communicat. and Network., **168** (2018)
2. N. Kalyuzhny, L. Losin. A method of mathematical modeling for transport hub establishment in Saint Petersburg. Transport. Res. Proc., **36**, 245–251 (2018)
3. N.A. Kalyuzhny. Justification of the priority of the location of transport interchange hubs in the structure of the agglomeration by the method of mathematical modeling. Bull. of Civil Eng., **5(64)**, 142–148 (2017)
4. V.G. Karmanov. *Mathematical Programming* (Science, Moscow, 1986), 288 p
5. P. Kristersson. The role of public transport interchanges in regional planning. Reg. Magaz., **285(1)**, 16–17 (2012)
6. S.A. Leonova. On the formation of a system of transport interchange hubs in the Samara city district. Railway transp., **8**, 58–61 (2019)
7. S.A. Leonova. Choice of locations for transport hubs of the urban passenger transport network. Transp. of the Urals, **4(63)**, 101–105 (2019)
8. Z.-C. Li, W. Lam, S.C. Wong. Modeling intermodal equilibrium for bimodal transportation system design problems in a linear monocentric city. Transport. Res., **46, Part B** (2012)

9. I. Ludan, E. Maiorov, J.D.M. Santana, O. Saprykin. Integrated approach to building a microscopic city model. *In Proceedings of the 5th International Young Scientists Conference on Information Technologies, Telecommunications and Control Systems* (2018), p. 8
10. A. Monzón, S. Hernández, F. Di Ciommo. Efficient urban interchanges: the City-HUB model. Transport. Res. Proc., **14**, 1124–1133 (2016)
11. A.A. Shagimuratova System analysis in determining the priority directions for the development of transport hubs of railway transport. Urban plan., **2(42)**, 63–71 (2016)
12. E.V. Sherbina, N.V. Danilina, D.N. Vlasov. City planning issues for sustainable development. Int. J. of Appl. Eng. Res., **10(22)**, 43131–43138 (2015)
13. C. Sun, X. Chen, H M. Zhang, Z. Huang. An Evaluation Method of Urban Public Transport Facilities Resource Supply Based on Accessibility. Hindawi J. of Advan. Transport., **11,** (2018)
14. D.N. Vlasov. *Transport hubs.* Monograph, 2nd ed. (MSSU, Moscow, 2017), 192 p
15. M.R. Yakimov, A.A. Arepyeva. *Transport planning. Features of modeling traffic flows in large Russian cities.* Monograph (Logos, Moscow, 2016), 280 p

Healthcare financing in Russia: mechanism and assessment of regional budget opportunities

Tatyana V. Faiberg[1,*], and *Irina N.* Shcherbakova[1]

[1]Baikal State University, 11, Lenin str., Irkutsk, Russia

Abstract. In the difficult economic conditions in Russia, caused, as in the all countries, by the consequences of the coronavirus pandemic, the search for additional government revenues is becoming especially relevant. All possibilities of cutting budget expenditures are being considered, including the optimization of the number of civil servants. The deputies of the State Duma of the Russia over the past month have been actively discussing the effectiveness of the functioning of state extra-budgetary funds and expressing ideas about the abolition of the Pension Fund of Russia and others, with the simultaneous return of functions back to the federal budget. It is important to reasonably assess the effectiveness of the financial mechanism of each extra-budgetary fund in Russia, including the Federal and Territorial Compulsory Medical Insurance Funds. To this end, it is necessary to comprehensively consider the mechanism of financing health care in the country within the framework of the current budget-insurance model. This article reveals the features of the financial mechanism in the healthcare sector in Russia, primarily in terms of methods for the formation of financial resources. In general, the Russian health care system is financed by the country's budgets, Compulsory Medical Insurance Funds, voluntary medical insurance and paid medical services. In accordance with this, the budget, insurance and market methods for the formation of financial resources in health care are identified. The article describes each method; the effectiveness of the current financial mechanism was assessed and recommendations for its improvement were proposed.

Keywords: Health, financing, budget expenditures on healthcare, financial mechanism

1 Introduction

In the context of the coronavirus pandemic, all countries began to pay special attention to such economic sector as healthcare. The current epidemiological situation exacerbates the issue of the effectiveness of the health care system, the quality of medical care. Each resident involuntarily compares the availability of medical services in different countries, which directly depends on the chosen funding model and on the well-functioning financial mechanism in the healthcare sector. All this determines the relevance of the organization of financing of medical institutions and development prospects in this area.

2 Problem Statement

In Russia, the healthcare financing process has many shortcomings in the financing mechanism for medical institutions. In particular, these include the existence of redundant links in the chain of the process of transferring funds to medical institutions, which affects the final amount of funding for medical services; the complexity of bringing funds to institutions; the availability of different ways of obtaining and using financial resources by medical organizations. In this regard, it is necessary to analyze the organization of the financial mechanism in the Russian healthcare system and determine the ways to improve it.

3 Research Questions

The research analyzed the practice, studied and systematized the scientific literature on the financing mechanism of economic sectors and, in particular, the healthcare system in Russia.

3.1 Content of the «financing mechanism»

In Russian practice, the content of the «financing mechanism» is considered both from a broad and narrow point of view.

In a broad point of view, the financing mechanism is identified with the concept of «financial mechanism» developed in Russian financial science. This is «a set of types and forms of organization of financial relations, specific methods for the formation and use of financial resources and methods of their quantitative determination» [5, 7]. This concept was first formulated in the 70s of the twentieth century in the papers of M.A. Pessel in relation to corporate finance [6]. In the papers of V.P. Ivanitsky [3] V.I. Samarukha [8], the

* Corresponding author: faibergtv@mail.ru

financial mechanism was considered in relation to sectors of the economy. In our opinion, this concept applies to other areas of finance, in particular, healthcare.

Further, Yu. M. Berezkin has offered financial-engineering interpretation of the concept of «financial mechanism» [2].

For the purposes of this study, the financing mechanism will be considered from a broad perspective as synonymous with the financial mechanism: a financial mechanism refers to the way in which a business, organization, or program receives the funding necessary for it to remain operational [11].

From a narrow point of view, the financing mechanism is the transfer of funds to enterprises, institutions, and the population.

3.2 Features of the Russian healthcare financing mechanism

The papers of such Russian economists as O.V. Emelyanova, N.V. Krivenko, E.S. Malysheva, I. Sheiman, S. Terentyeva, A.I. Tsvetkov, N.I. Yashina, K.S. Yashin and others are devoted to the problems of healthcare financing, assessing the effectiveness and comparing models of healthcare financing.

Most studies focus on particular issues of organizing healthcare financing in Russia, the most relevant of which are assessing the effectiveness of healthcare financing models [4, 10, 12], features of health care financing at the federal level [1, 13] and in the regions [9].

From our point of view, the issue of simplifying the financing mechanism for medical services (possibly by reducing the «intermediate links») in order to increase the availability of medical care, to rapidly increase the volume of medical services in a pandemic, that is, to increase maneuverability, while improving the quality of medical services is of principle.

4 Purpose of the Study

The purpose of the study is to highlight the peculiarities of the Russian healthcare financing mechanism, primarily in the methods of generating financial resources, to assess the effectiveness of the current mechanism, and to develop recommendations for its improvement.

5 Research Methods

The depth of research, the validity of scientific results, the reliability of conclusions and recommendations were ensured by the use of general scientific methods: observation, comparison, synthesis of theoretical and practical material, using methods of economic and statistical analysis (grouping, comparison, generalization).

6 Findings

The financing mechanism in healthcare in Russia is quite complex. It has developed as a result of reforms not only in the healthcare sector, but also as a result of budget reforms (introduction of performance-based budgeting; Treasury budget execution; procurement systems; changes in the organizational and legal forms of state (municipal) institutions, etc.), focused primarily on borrowing foreign experience, the so-called «best practice».

Of all the elements of the financial mechanism, the methods of forming financial resources, methods of their transfer and quantification have significant features in healthcare.

The basis of the financial mechanism in health care is the financial relationship between recipients of medical services and medical organizations in the process of redistribution of financial resources that mediate the provision of medical care (Figure 1).

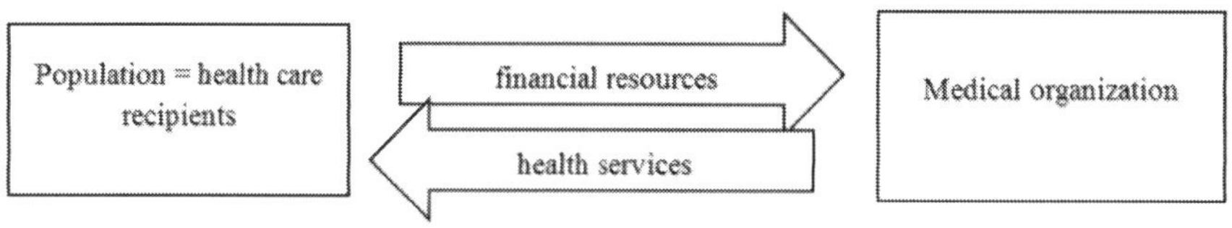

Figure 1. Healthcare Financial Relationship Diagram

In world practice, there are different methods of transfer of financial resources depending on the chosen model of financing of health care: budget model, insurance model or budget-insurance model.

In Russia, a budget-insurance model of financing health care has developed, despite attempts to implement an insurance model. In my opinion, in Russia there is a «quasi-budget-insurance» model, or rather a budget model financing of health care, because it historically developed in the early twentieth century and proved its effectiveness. Before 1917, there was no public health care system in Russia practically and private medical practice dominated. Therefore, attempts to complicate the existing mechanism of financing healthcare institutions and to introduce insurance reduce the efficiency of financial resources and the availability of medical services.

The analysis of the financing mechanism of the state healthcare system in Russia makes it possible to identify the following methods of generating financial resources, depending on funding sources:

- market method (method of exchange relations);
- actually financial (budgetary) method;
- insurance.

In Russia, the market method arose in the 90s of the twentieth century and means direct transfer of money to medical institutions as payment for medical services. At the same time, medical services can be provided not only by public, but also by private organizations.

The budgetary method of generating financial resources in healthcare involves the redistribution of funds received from taxes into the country's budget system in favor of medical institutions under the Program of State Guarantees of Free Medical Care (Figure 2).

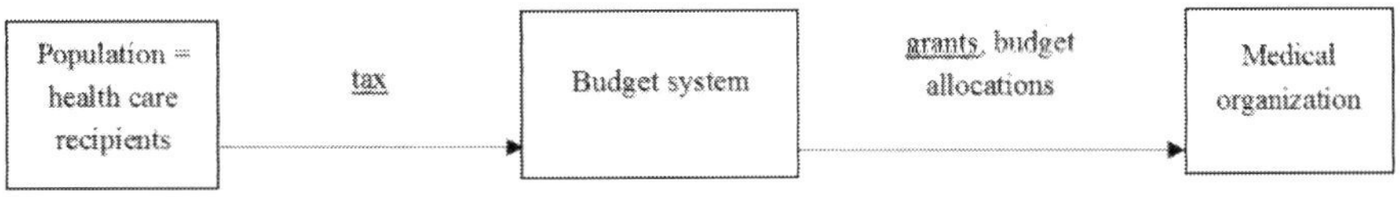

Figure 2. Budgetary method of financial resources formation in Russian healthcare

Budgets of all levels of budget system participate in the process of budget financing of institutions: Federal, regional and local budgets. The main role is assigned to regional budgets, since almost all health institutions in the country are regional owned.

The form of transferring funds to medical institutions may vary depending on the legal form of the institution. State budgetary and autonomous institutions receive grants (for the fulfillment of a state assignment; for budget investments; for other purposes), and government - budget allocations. At the same time, budgetary and autonomous institutions have freedom in the disposal of income from paid services, while government institutions irrevocably transfer them to the budget.

The insurance method of generating financial resources in the healthcare of Russia involves the compulsory and voluntary medical insurance.

Extra-budgetary Compulsory Health Insurance Funds were created in 1993 in Russia. This is the only system of extra-budgetary funds in Russia that has a two-level structure: the Federal Compulsory Health Insurance Fund and the Territorial Compulsory Health Insurance Funds in the regions of the Russian Federation (Figure 3).

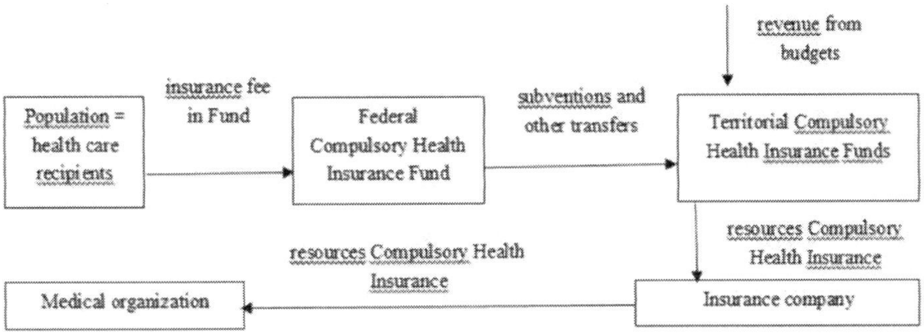

Figure 3. Insurance method of financial resources formation with compulsory medical insurance in Russia

The borrowed foreign experience of insurance medicine, aimed at improving the quality of medical services, and explaining the emergence of insurance companies in the mechanism of financing health care, did not radically change the medical care system, but increased the costs of maintaining funds, insurance companies, created new problems, negatively affected the availability of medical care.

The amount of funding for medical institutions depends on the number of services provided and the per capita financing rate, which, according to the law, should take into account all the costs of medical care. However, in fact, in the face of a constant budget deficit, it is not possible to finance all expenses, as in any other branch of the social sphere. Therefore, in practice, the financing rate is calculated taking into account the real capabilities of the budget, that is, not «from needs» but «from opportunities». In these conditions, the costs of maintaining funds and insurance companies only exacerbate the shortage of funds.

The costs of maintaining the Compulsory Health Insurance Funds in 2019 amounted to 15.19 billion rubles.

Insurance companies keep a percentage of the total cash flow for doing business, the amount of which is fixed annually by the laws of the regions «Budget of the Territorial Compulsory Health Insurance Fund». In each region, this percentage is different. For example, in the Irkutsk region in 2019, it amounted to 1 %. At the same time, the total cash flow from the Territorial Funds in 2019 amounted to 2 265.98 billion rubles. Thus, according to the most modest estimates, insurance companies formed income only due to budget cash flow in the amount of 22 billion rubles, not counting fines and sanctions imposed on medical institutions for poor-quality, untimely and/or incomplete medical care received by insurance companies (44.5 billion rubles were transferred fines in 2017).

It seems doubtful to integrate private insurance companies in the financial mechanism of the Russian state health system, which are aimed at making a profit, as a controller of the quality of medical services. Our conclusions are indirectly confirmed by the opacity of the statements of insurance companies on compulsory health insurance. In our opinion, not the desire to develop the voluntary health insurance sector by these companies is associated with receiving cash flow for compulsory health insurance.

The insurance method for the formation of financial resources with voluntary medical insurance in Russia is schematically shown in Figure 4.

This method of generating financial resources depends on the efforts of insurance companies to conclude insurance contracts for voluntary medical insurance, the established amount of insurance premiums and made insurance payments to medical organizations.

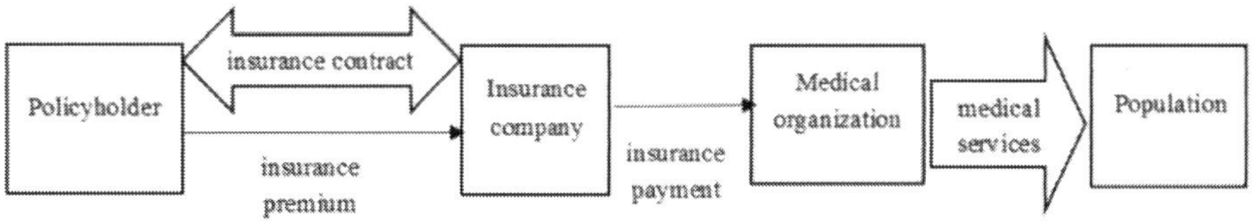

Figure 4. Insurance method of financial resources formation with voluntary medical insurance in Russia

According to the Bank of Russia, the volume of the voluntary medical insurance market in terms of the amount of contracts concluded in 2019 amounted to 136.1 trillion rubles. However, the volume of insurance payments is only 126.5 billion rubles with paid contributions of 180.65 billion rubles. Thus, the net income of insurance companies on voluntary medical insurance amounted to 54 billion rubles.

Thus, 1/3 of the cash flow by the insurance method of generating financial resources from insurance companies in Russia is provided by compulsory medical insurance, while the costs of attracting customers are minimal.

The analysis of the insurance method in the healthcare financing mechanism of Russia showed a focus not on the development of the country's health care system, not on improving the quality of medical services, but on diversifying the Russian economy at the expense of insurance companies, whose stability in crisis is guaranteed by constant cash flow from extra-budgetary funds. From our position, insurance companies should actively develop voluntary medical insurance, and not claim a share in the cash flow of compulsory medical insurance.

Methods of using financial resources in health care depend more on the legal form of medical institutions; the participation of the Federal Treasury in financing the institutions; the applied procurement system than on the sources of their funding.

The ratio of healthcare financing sources in Russia is clearly shown in Figure 5.

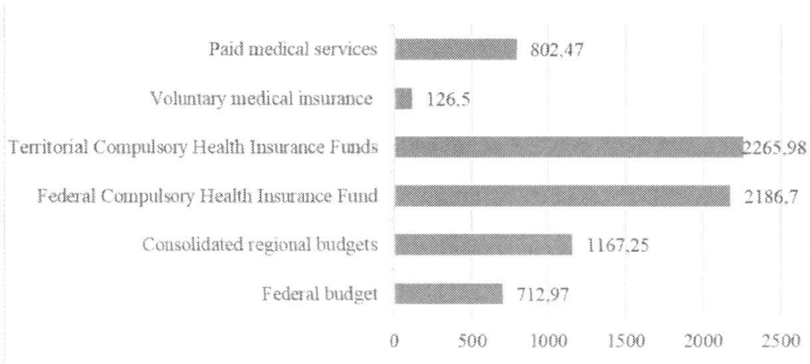

Figure 5. Sources of healthcare financing in Russia in 2019, billion rubles

Figure 05 clearly shows that the dominant cash flow passes through the compulsory health insurance funds (61 %). At the same time, a fairly high share of the Russian budgetary system in health care financing remains (26 %). Analyzing the structure of budget financing for healthcare, we can conclude that the regional budgets dominate, through which 1.163.36 billion rubles were transferred in 2019 (62 % from all budgets).

It should be noted that despite the introduction of single-channel funding in the compulsory health insurance mechanism, still not all expenses of medical institutions for the provision of medical care under the «Program of State Guarantees of Free Medical Care» are covered by the funds of compulsory medical insurance funds. For example, costs of diagnosing and treating coronavirus infection. The problems associated with financing the procurement of equipment and medical supplies for coronavirus infection were resolved not through health insurance, but through budget funding (by the providing grants for other purposes).

Thus, it is impossible in the system of compulsory health insurance to quickly increase the volume of medical care and, accordingly, financing without losing interest to insurance companies.

In this regard, we consider it expedient and economically justified to return to budgetary financing of healthcare in Russia with the transfer of most of the spending powers to the regions of the Russian Federation. They have the majority of medical organizations under their control and can quickly reallocate scarce resources within budgets according to financial policy priorities.

7 Conclusion

In the article, we identified the features of the financial mechanism in the healthcare sector in Russia, primarily in terms of methods for the formation of financial resources. The characteristics of the market, insurance and budgetary methods are given; the effectiveness of the current mechanism is evaluated and recommendations for its improvement are offered.

The main recommendations may be the transition to budgetary financing of healthcare in Russia; simplification of the insurance mechanism of compulsory health insurance by optimizing cash flows between Federal and Territorial Compulsory Health Insurance Funds, as well as between Territorial funds and insurance companies. Insurance companies should develop voluntary health insurance, and not apply for a percentage of the compulsory insurance health cash flow. The current situation - the coronavirus pandemic - presents broad opportunities for the development of voluntary health insurance, subject to the development of quality insurance products. From our position, insurance companies, along with medical institutions, should be responsible for both timeliness and accessibility of medical care.

References

1. E.V. Ageeva. Financing of national systems health care: search of new approaches. Econ. and entrepren., **3(92)**, 1276–1281 (2018)
2. Yu.M. Berezkin. Two approaches to understanding the concept of financial mechanism: accounting-financial and financial-engineering. Izv. of the Irkutsk State Econ. Acad., **2(82)**, 27–32 (2012)
3. V.P. Ivanitsky. *Formation and development of a financial mechanism based on the distribution of monetary accumulations of industry (theory and methodology)* (Irkutsk State University, Irkutsk, 1984)
4. N.V. Krivenko, A.I. Tsvetkov. Efficiency of Funding Healthcare for Ensuring the Economic Security of a Region. Econ. of the reg., **3**, 970–986 (2018)
5. E.V. Markina. *Finance.* Textbook (KNORUS, Moscow, 2014)
6. M.A. Pessel. *Financial and credit mechanism for intensifying public production* (Finance, Moscow, 1977)
7. V.M. Rodionova. *Finance.* Textbook (Finance and statistics, Moscow, 1995)
8. V.I. Samarukha, N.S. Permyakova. *Development of the financial mechanism of the corporation in conditions of risk and uncertainty* (Baikal State Univ. of Econ. and Law, Irkutsk, 2001)
9. T.V. Shchyukina, S.N. Plyuta. Improvement of health care financing in the region. Univ. Bull., **5**, 48–88 (2019)
10. I. Sheiman, S. Terentyeva. International Comparison of the Effectiveness of Budget and Insurance Models of Healthcare Financing. Econ. policy, **6**, 171–193 (2015)
11. G. Wiesen. *What is a Financial Mechanism?* (2020). Retrieved from: https://www.wisegeek. com/what-is-a-financial-mechanism.htm.
12. N.I. Yashina, O.V. Emelyanova, E.S. Malysheva, N.N. Pronchatova-Rubtsova. A methodology methodology to evaluate the efficiency of public health financing in Russia. Finan. and credit, **3(771)**, 693–708 (2018)
13. N.I. Yashina, K.S. Yashin, E.S. Malysheva, O.I. Kashina, N.N. Pronchatova-Rubtsova. *Features of state financing of health care in the context of economic reform* (Publ. House of RADONEGE, Nizhny Novgorod, 2019)

Components of the competitiveness of Arctic university graduates as an object of social and labor relations

Olga Ivacheva[1,*], *Varvara* Zarovnyaeva[1]

[1]NEFU Road Construction Faculty, 13, Krasilnikov street, Yakutsk, 677007, Russia

Abstract. The state of the transport infrastructure is one of the most problematic areas in the Arctic. Taking into account the strategic position of the Arctic territories, the growing interest of foreign states in Russian natural resources and communications, transport support for existing and planned industries, the training of qualified specialists should promptly have developed within the framework of the national program for the development of the Arctic territories. The implementation of the program requires a special approach to the training of competitive graduates of the transport complex. Large-scale production projects will require hundreds of engineering and technical specialists: engineers, transport workers, logisticians, builders, etc. The components of competitiveness of a university graduate have been determined, which constitute general requirements for future transport specialists working in the conditions of the Arctic.

1 Introduction

The relevance and significance of this study are determined by the fact that the development of the Arctic transport system, including air, water and land communications, is one of the strategic priorities of the Russian state policy in the Arctic. The tasks of developing Arctic natural resources are directly related to the need for radical improvement and the creation of new transport hubs, new roads, new vehicles, and improved infrastructure, which will help improve the living conditions of the local population, solve social problems, and increase the level of transport accessibility of Arctic settlements.

The length of public roads in the Arctic zone of Russia is 108.9 thousand km, of which only 38.4% (41.8 thousand km) is improved. In completely Arctic regions this figure is 0.6% of the total length of the road network. Thus, the successful solution of the problems of the transport complex in the circumpolar territories, including cargo transportation, the operation of road transport enterprises, the construction of new and reconstruction of old roads, will largely depend on the level of training of specialists - engineers in the field of road transport who are competent and competitive and in demand in the Russian and regional markets.

Within the framework of the State Program of the Russian Federation "Development of Education for 2018-2025" (Decree of the Government of the Russian Federation No. 1642 of 26.12.2017), the main goal of higher professional education is to train the qualified employees of the appropriate level and profile, competitiveness in the labor market, who are responsible and fluent in their profession and capable of effective work in the specialty at the level of world standards [9].

Graduate competitiveness is a new trend in Russian higher education. According to the definition of the famous manager of the twentieth century, Paul Drucker, "it consists of the competitive advantages and distinctive features of the employee". R I Platonova points out, "The competitiveness of a specialist of a regional university," is determined by the content of the educational order of the state and the regional community and the requirements of the educational standard and regional labor markets. Competitiveness is formed in certain socio-cultural conditions generated by social and cultural traditions, and the mentality of the peoples of the region" [5]. In the education system, competitiveness is defined as the sum of the competitiveness of an educational institution, the personal and professional advantages of its graduates.

The purpose of the study: to determine the components of the competitiveness of a graduate in the transport industry.

Research objective: to identify the specific features of the formation of a competitive specialist in the field of motor transport for work in the Arctic zone. To achieve this goal, empirical research methods were used: observation, and survey.

The reform of engineering education in the field of motor transport will be accompanied by an active search for means to intensify the process of forming the competitiveness of future specialists, which is confirmed by the results of scientific research (V I Andreev, L M Mitina, D V Chernilevsky). The training of engineering specialists in the field of motor transport in various aspects has become the object of scientific research of Russian scientists of different generations (A N Kara, A A Nikolaeva, P A Fatkhutdinov, N N Kolobkov, Yu A Dmitrieva, etc.).

* Corresponding author: ivacheva.o.a@mail.ru

2 Scientific and theoretical approaches to the formation of a competitive specialist in the university

A wide range of positions of scientists on the competitiveness of specialists gave grounds to identify the following approaches to clarify this essence:

1. General scientific approach determines the content of professional abilities and the development of a competitive personality (A A Angelovsky, T V Bendas, E F Zeer, V Ya Zinchenko, E I Mashbits, L M Mitina, C A Novoselov).

2. Management approach developed within the framework of the theory of management by foreign (I Ansoff, A Weissman, F Woodcock,) and domestic (V A Dyatlov, V V Travin, P A Fatkhutdinov) researchers.

3. Pedagogical approach reveals the competitiveness through priority, core qualities of the individual (V I Andreev, S A Borisenko, N V Borisova, E F Zeer, O I Polkina, S M Shirobokov, N V Fomin, D V Chernilevsky, etc.).

4. Psychological approach: Competitiveness is interpreted as the ability to anticipate, update and use all the opportunities for development (L M Mitina); development of criteria and levels of competitiveness formation (H H Sidorova, N V Fomin, N L Hristolyubova); conditions for the formation of students' competitiveness (Y V Andreeva, R Y Akhmetshin, I V Darmanskaya, H H Sidorova, F R Tuktarov).

5. Acmeological approach studies the competitiveness of a specialist in terms of achieving a high level of quality of professional activity and creativity (K A Abulkhanova-Slavskaya, A A Derkach, A A Bodalev, N V Kuzmina, E A Klimov); aspects of the manifestation of the competitiveness of specialists in various fields of activity (A R Kilba, K A Krasilnikova).

6. The sociological approach understands the competitiveness as the basis for the stability of society, based on the conscious subordination of the individual to the norms and requirements of culture by virtue of their own benefit (N M Mukhamedzhanova), a package of abilities and personal qualities that ensure successful socialization in society and active adaptation in the labor market (A Y Flier); education that is formed during a person's life, due to social impact (O V Dushkina); the ability to see a long-term perspective (J Naisbit).

7. Economic approach (P A Fatkhutdinov, T B Marushchak, L D Medvedeva, S I Sotnikova, I V Virina, A F Stepus, E A Galuzo, T I Bezdenezhnykh, A G Shatokhin, E V Pirogova, P A Krysin, etc.) develops the competitive principle between legal entities, goods, services and the ability of the subject (participant) of the competition to be more attractive to consumers in comparison with analogues due to the presence of a number of competitive advantages. According to Dmitrieva Y A, the following components identified:

- motivational component (system-forming in relation to all other components of competitiveness), which determines the list of abilities and qualities to be developed, in accordance with the system of goals and values of the individual;
- the qualification component that determines the educational and professional activities of the graduate;
- a personal component that reflects the characteristics of the competitive behavior of a university graduate when interacting with potential employers (university teachers, practice manager, mentor, etc.);
- the business component that characterizes the competitive advantages of the graduate and the strategy of behavior in relations with the employer.

3 Components of the competitiveness of the Arctic University graduate

As part of our research, we will clarify the components of the competitiveness of a graduate student and a young specialist.

The second level of the competitiveness pyramid can be specified as follows: the competitiveness of an individual —> the competitiveness of a graduate student —> the competitiveness of a particular specialist in the field of motor transport. Graphically, the pyramid of a specialist's competitiveness can be represented in the form of Fig. 1.

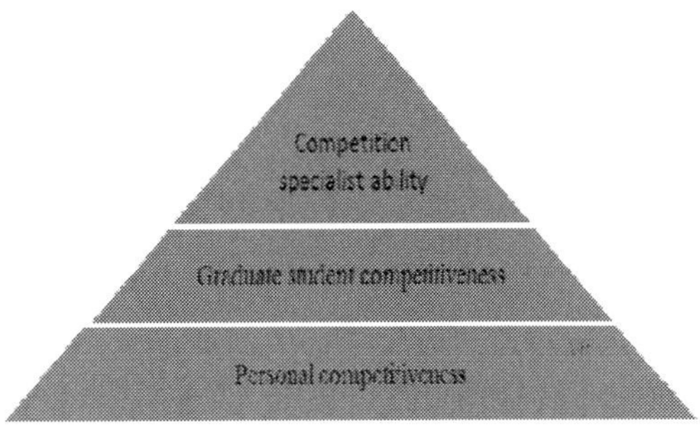

Fig.1. The pyramid of a specialist's competitiveness

We have determined that the competitiveness of a specialist is the unity of his motivational, qualification, personal and business characteristics that ensure a high level of theoretical training, possession of practical skills of professional activity, free orientation in related fields, the ability to adapt to the professional environment, compliance with the requirements of the employer and teamwork.

A N Kara clarifies that the specificity of a specialist as an object of social and labor relations in the system of competitiveness, having a specific product - the ability to competent and productive work-consists in the long-term nature of its possible competitiveness, so the process of achieving it and ensuring it is also quite long. At the same time, when considering the concept of "competitiveness of a university graduate", it is necessary to take into account the objective features due to the nature of the intermediate position of the graduate, since on the one hand, during the training process, they were exposed to the university (compliance of the graduate with the requirements of the Federal State Educational Standard of Higher Education), and on the other hand, to the requirements of the labor market and a specific employer. Additional factors influencing the formation of the future specialist's competitiveness are the conditions prevailing in the labor market, as well as the degree of adaptation to the graduate himself and his ability to effectively interact with the employer and teachers.

As a graduate of the university masters, professional knowledge, skills, skills of work in the specialty, and ways of communication, their adaptation to the workplace, ideally after a certain period of time (on average up to 3 years), the competitiveness of the graduate is transformed into the competitiveness of a specialist.

In her work, Nikolaeva A A proved that the content structure of the student's competitiveness included a combination of 3 components:

- cognitive, consisting primarily of knowledge and skills in a specific professional field, including knowledge of modern information technologies, knowledge of foreign languages, native and foreign culture, the ability to navigate a huge amount of changing information and quickly process it;
- emotional, reflecting the value orientations of the individual, his attitude to himself, his life, his chosen cause, interaction with people in joint activities, his future and the future of his country;
- activity-based, related to the professional self-determination of the individual in specific life circumstances on the basis of individual and professional capabilities.

These components in the structure of the competitiveness of the student's personality are interrelated and interdependent, each of them, regardless of the other two, has no independent significance. This allocation is appropriate to identify the dynamics of the development of the competitiveness of the individual in the context of higher education.

Based on the description of each of the selected scientists' components of competitiveness, we can assume that the emotive component of competitiveness according to A A Nikolaeva corresponds to the motivational, cognitive - qualification and activity business component allocated by Dmitrieva Y A.

It should be noted that we are talking about a strong interdependence of the components of competitiveness of graduates in relation to each other that, in our opinion, is truer.

The analysis of the scientific base devoted to the problem of assessing and improving the competitiveness of a specialist has shown that individual researchers have formulated their own definitions for understanding the concepts of "employee competitiveness and personnel competitiveness".

Therefore, it can be said that the degree of compliance of the characteristics of the labor activity of employees with the requirements imposed on their work by employers, competitiveness in the labor market is determined not only by the qualitative characteristics of the labor force, but also by the conditions on the labor market.

4 Pedagogical conditions affecting the increasing competitiveness of future specialists in the transport industry in modern

Table 1. Rating of factors affecting the training of a competitive specialist in the transport industry.

№	Factor name	Ball
1	Competence and professionalism of teachers and employers	5.0
2	Direct participation of employers in the educational process	5.0
3	An interesting and meaningful program training	5.0
4	Comfortable, supportive and friendly environment	5.0
5	Practice organization quality	5.0
6	The level and quality of the material base of the university and the enterprise, the equipment of the educational process	5.0
7	High-speed and stable internet	5.0
8	Provision with modern simulators, computer and office equipment, library funds	5.0
9	The possibility of self-realization and all-round development of the student personality	5.0
10	Application of new teaching methods and technologies focused on modern educational and digital technologies	4.7
11	Human relationships in the link "teacher-student", "employer-trainee"	4.7
12	Professional development, internships employees	4.3
13	Planning and control	4.3
14	Conducting seminars, conferences to exchange knowledge and experience	4.3
15	Having a good scientific foundation of the training unit	4.3
16	Availability dining room	4.0
17	Dormitory availability	3.7

In recent years, the modernization of Russian vocational education is associated with its integration into the international educational community, for training specialists in the modern economy and improving the quality of education. On the basis of an internal assessment, the factors influencing the training of a competitive specialist in the transport industry and the quality of student education are identified and ranked. Quality education at all its levels is achieved and will be effective only with the active participation of all stakeholders of the educational organization, including the students themselves, their families, and employers. The introduction of competence-based educational standards, leveling the system of training students, measures to strengthen the links of universities with production and the scientific sector, the allocation of leading universities, designed to become methodological and scientific centers in their regions - all this contributes to solving the problems of training personnel taking into account the needs of the Russian economy and global competitiveness of domestic technical education [7].

The formation of fundamentally new technological environment enterprises in the transport industry requires an influx of qualified specialists who have modern engineering and technical knowledge and advanced production technologies, who understand the trends in the development of the transport sector, and have regional practice-oriented training.

An expert group consisting of the educational part, teachers Faculty of Road Transport and employers identified, evaluated and ranked factors affecting the quality of student learning and the competitiveness of graduates. So, 13 main factors were identified, presented in table 1, of which 10 are the most significant. According to ten factors that have the highest degree of influence on the quality of the educational process, a self-assessment of the work of the faculty was made with the ultimate goal of training a competitive specialist in the transport industry, increasing the rating of the main factors affecting the quality of student education and the training of a competitive specialist in the transport industry. This highlights the most relevant aspects of educational, informational, social and psychological nature.

Currently, special requirements are imposed on the content of educational programs, on the implementation and level of mastering by students of practice-oriented competencies to increase the competitiveness of graduates of the transport and road industry [1]. The faculty has concluded about 30 long-term agreements with various organizations in all areas of training and levels of education - practice bases, where key professional competencies are formed, incl. skills in design, research activities, experience of practical training in regional conditions, opportunities for further employment [2]. The largest basic partners of the faculty are AO Mechel, GAZ Group (Nizhny Novgorod), AO Polyus Gold, FKU Uprdor Vilyui, PJSC Doriss, LLC Investdorstroy, AO RIK Avtodor, LLC Rechnoy port of Yakutsk ", JSC" Kranservice ", MUP" YAPAK ", YIPK and other potential employers who, in modern conditions of production, as well as an educational organization, have become interested and responsible partners in the implementation of high-quality practice-oriented training.

5 Conclusion

Thus, the components of the competitiveness of a university graduate constitute and determine the general requirements for future specialists in the field of motor transport, laying the basis for the concept of a competitive specialist-motor transport worker working in the Arctic are:
- readiness to work in innovative and extreme modes;
- practice-oriented training in regional conditions aimed at high-quality mastering of technology and professional competencies;
- high adaptability to changes, positive perception of innovations and technologies of transformation in professional activity;
- subject-personal attitude to active intellectual activity and interaction with students, teachers and colleagues;
- professional and personal readiness for the increasing intensity of work;
- need for training, self-education, professional retraining, professional and personal improvement;
- willingness and ability to participate in decision-making at different levels (in study groups, projects, work collectives, professional associations, etc.);
- orientation to maintaining social and personal health, including a culture of thinking, work behavior, nutrition, consciousness, etc.

Modern global changes in the world show that it is the field of education that determines the priority direction in the training of competitive specialists.

References

1. Valishin P.R. Scientific opinion **3**, 59-63 (2016)
2. Ivacheva O.A., Zarovnyaeva V.I. Vestnik NEFU series "Pedagogy and Psychology" **4**, 28-35 (2019)
3. Kara A N, Kuznetsova E Y *Questions of economic theory. Macroeconomics* 1 (37), 46-49 (2011)
4. Dmitrieva Y A *Improving the level of competitiveness of a university graduate in the specialty "Personnel management" in the labor market*, dissertation Candidate of Economic Sciences (Moscow, 2010)
5. Platonova R I *Formation of a competitive specialist in the system of regional higher pedagogical education (by the example of pedagogical universities in Yakutia)*, Dissertation for the degree of Doctor of Pedagogical Sciences. (Moscow, 2011)
6. Decree of the Government of the Russian Federation No. 1642 *The State Program of the Russian Federation "Development of Education"* for 2018-2025 (2017)
7. Strategy for the development of the Arctic zone of the Russian Federation and ensuring national security for the period up to 2020. Retrieved from: http://n-west.ru/wp-content/uploads/2012/01/Strategiya-razvitiya-Arktic heskoy-zonyi-RF.doc

The paradoxes of Russian railways' loan policy

Darya A. Dinets [1,*], and *Anatoly S.* Merkulov [1]

[1] Irkutsk State Transport University, 236/B, Baikalskaya str., Irkutsk, Russia

Abstract. Russian Railways is gradually increasing its loan debt, which reflects the general trend of the economy. But, taking into account the objective economic state of railway transport and its low profitability, the increase in the share of borrowed funds reduces the company efficiency. This contradicts the purpose of using the loan, the purpose of which is to increase the return on equity. While analysing the results of the transport industry, we regarded it as an infrastructure industry, whose goal is primarily to provide macroeconomic indicators, and the problems of making profit recede into the background. An important factor in the development of the Russian economy in consideration of its geographical location is the development of transport networks, in which the predominant (at present and in the future) one is railway transport. However, its technical state requires significant modernization, investments far exceeding its own resources. The state, as a sole shareholder of Russian Railways OAO, does not have the necessary financial resources to handle the modernization, so Russian Railways is turning to the capital market, even reducing profitability. Private capital is willing to invest its capital only in separate modernization projects that can provide an acceptable rate of return for them. This explains the borrowing of capital for modernization at a price far exceeding the return on equity. At present, economic losses are expected to be set off by future revenues and mainly by the multiplicative effect of the economic growth of the national economy and the resolution of geopolitical problems. The purpose of the article is to try to explain the loan policy of Russian Railways.

Keywords: borrowed funds, long-term loans, short-term loans, bond loans

1 Introduction

Since 2003, with the beginning of the reform of railway transport, its status has changed. The railway became a shareholder company, but the state owns 99.99 % of the shares. The reform assumed that the main carriers would be private companies, but the experience of 2004–2006 showed that it is not always expedient from a macroeconomic point of view. One of the fundamental provisions of the reform is the idea that Russian Railways should gradually become an organization granting the right to use the railway structure rather than engage in transportation as the main activity.

One of the provisions of this reform is the technical modernization of infrastructure. The railway transport modernization process has increased dramatically since 2016. If we allocate the railway infrastructure as a separate economic management facility in order to reduce the cost of transportation process by the partial financing of the infrastructure from budget funds, it does not guarantee the actual reduction of tariffs. Tariffs will be supplemented by various surcharges that take into account the interests of the private carrier. This is what hinders the development of reform as a goal of strengthening the commercialization of the transportation process by railway transport.

At present, the role of loans is gaining ground due to the following reasons: firstly, the amount of monetary capital, which is not involved directly in the real capital movement, is increasing. This leads to the reduction of capital price. Secondly, there is growing uncertainty of economic processes, which often drives the need for additional investments, which cannot be covered by their own funds. Thirdly, the increase in financial uncertainty requires borrowed funds to maintain repayment ability. Fourthly, borrowing volumes in the consumer sector are increasing sharply. These phenomena are descriptive of the properties of the new stage in the development of capitalism: the growth of the importance of financial capital, when the credit sphere from the conditions of the real sector development turns into the dominant sphere of development and operation of capital. This raises the question of how the traditional terms of borrowing and credit granting have changed by the example of an individual organization. The answer to this question provides an opportunity to explain the seeming paradox of the loan policy of Russian Railways OAO.

2 Problem Statement

The volumes and stages of the proposed investments in the modernization of the main technical and

* Corresponding author: dinets_da@irgups.ru

technological base were determined depending on the economic situation in Russia. 2.44 trillion rubles were invested from 2010 to 2019, and it is still expected to invest up to 4.4 trillion rubles by 2022. In 2019, Russian Railways' in-house resources in the form of profit and amortization amounted to only 19.4 % of the funds invested in modernization. Most likely, the State, as the sole shareholder, on the basis of the established priorities, which are advocated by large industries of the first and second redistribution, cannot invest the missing amount from the National Welfare Fund. Private investors are willing to invest only in individual highly profitable projects or those that are needed as elements of their logistics chain. The insufficiency is supposed to be covered by borrowed funds.

The return on equity for the period of 2012–2019 was about zero. It makes one wonder whether it is possible to form the loan policy of this organization by analogy with a commercial one. Naturally, this poses the question about the effectiveness of borrowing.

3 Research Questions

3.1 Analysis of the long-term borrowed funds utilization efficiency

Let us set ourselves the goal of determining the effectiveness of the utilization of long-term borrowed funds. For this purpose, we will use the data given in Table 1, where the dynamics of the growth of the share of long-term debt in revenue is clearly visible. Their share increased 7.5 times.

Table 1. Analysis of the dynamics of long-term loans and revenues, billions of rubles

Indicators/period of time	Code	2019	2018	2017	2016	2015	2014	2013	2012	2011	2019 to 2011
Borrowed funds	1410	1143.0	997.5	934.8	829.1	900.4	715.7	579.2	359.5	263.2	4.34
Revenue	2110	1848.1	1798.4	1697.6	1577.5	1510.8	1401.7	1376.6	1366.0	1288.3	1.43
Total due interest	2330	72.1	55.9	55.2	64.2	64.5	34.7	25.2	17.2	17.6	4.09
Borrowed funds for 1 ruble of revenue		15.84	17.85	16.93	12.91	13.96	20.63	22.99	20.95	14.92	1.06
Net profit (loss)	2400	53.46	18.36	17.50	6.50	0.32	-44.08	0.74	14.11	16.82	3.18

Source: [11, 12]

The increase of 4.4 times in long-term debt resulted in the increase of 1.2 times in revenue and 1.3 times in cargo turnover. Investments increased only 2.2 times during this period. There is a replacement of in-house and public sources of investments for borrowed funds. A large part of the borrowed funds does not act as capital, which leads not to growth, but, on the contrary, to a decrease in the return on equity. A vicious circle is created: the lack of in-house funds, the necessity to use borrowed funds for investments, and their growth leads to a reduction in their in-house sources.

Debt on borrowed funds includes not only interest on the loan, but also repayment of credit resources.

The main conclusion that can be drawn from Table 02 is that dependence on borrowed funds has increased dramatically. For each ruble of the revenue obtained, the company paid 28 kopecks in 2012 to repay the financial debt. In 2019, it paid 66 kopecks. Although there was a slight decrease in the cost of credit resources during this period, which reflects the general economic trend of the Russian financial market. However, despite the increase in financial risks, the International rating of Russian Railways is stable. This can be explained by the fact that the railway holding is state-owned and the risks will be repaid at the expense of public funds [3].

Another conclusion from Table 2 is an acknowledgement of the negative ratio of the dynamics of net profit and the total interest paid. They grow at a ratio of 3.18 to 4.34.

What would be the profit without the use of credit resources?

During the calculation, we take into account that the interest to pay relates to costs, which reduces the taxable profit and thus the net profit increases by 20 % of the amount of interest accrued.

The accrued net earning for the period from 2011 to 2019 amounted to 83.7 billion rubles, and the total interest paid amounted to 909.7 billion rubles, which is 10.7 times more. State funding for the same period amounted to 2,216 billion rubles. Thus, the financial sector (both national and foreign) increased their income by 909.7 million rubles. It is obvious that taxpayer funds have become a source of profit for private investors. Much of the shortage in the financial resources of the state holding is compensated from state sources. Isn't this financial market a vacuum cleaner that sucks out sources of development from the real sector?

From the standpoint of managing financial flows, it is advisable to analyse the ratio of short-, medium- and long-term borrowed funds, on the one hand.

3.2 Short-term loan policy analysis

What is the content-related difference between short-term loans, on the one hand, and medium-term and long-term loans on the other hand? Analysing this problem, it is necessary to determine the criteria for categorizing loans to a particular group. On the surface, it is quite simple. Banking statistics, carried out by the Bank of Russia, counts loans up to 1 year as short-term, loans up to 3 years as medium-term, and all others are long-term. However, these criteria are formal. Generally, "time" is a very under-researched concept in the economics.

Perhaps the definition of the loan urgency criteria should be based on the concept of loan. If the loan is used as a means of business development, i.e. as capital, the sign of which is self-reproduction, it is a medium- and long-term loan. In this case the price of the loan is tied to the return rate for the industry where the funds received on credit are invested.

The use of loans to maintain repayment ability should be seen as financing not capital, but money in their payment function. Here the invested value does not increase, but creates conditions for growth. This is a short-term loan. In this case the price of this loan has no objective basis. Only supply and demand determine the price of a short-term loan, as the use of these funds is not aimed at developing or making a profit, but preventing bankruptcy of the business. In accounting, this form of loan is presented as accounts payable.

The amount of short-term loans (calculations on short-term loans and borrowings) increased 2.8 times. There is no clear correlation between the dynamics of short-term loans and revenue.

Table 02. Analysis of the dynamics of short-term borrowing funds and revenue

Indicators/period of time	2019	2018	2017	2016	2015	2014	2013	2012	2011	2010
Short-term borrowing funds, billions of rubles.	334.5	263.67	140.74	91.91	88.49	140.51	36.96	60.17	120.69	42.8
Borrowed funds, total	1477.5	1261.2	1075.5	838.2	988.9	856.2	616.1	323.4	597.8	306.1
Share of short-term debt, %	22.60	20.90	13.10	1.10	8.95 %	16.40	6.00	18.60	20.20	14.0
Revenue, billion rubles	1848.1	1798. 4	16975	1577. 5	1510. 6	1401.7	1376.6	1366.0	1288.3	1848. 5
Share of short-term debt in revenue, %	5.53	6.82	120.62	17.16	17.07	9.98	0.04	22.7	10.67	5.53

Calculated: [1, 5, 6].

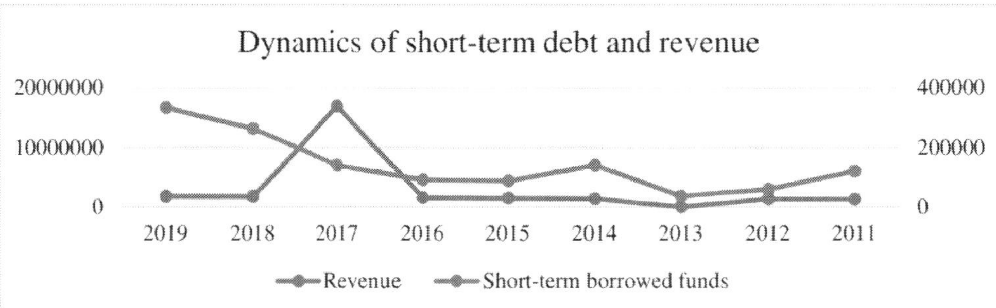

Figure 01. Short-term debt and revenue dynamics

Table 03. Analysis of dynamics of short-term borrowing funds and debts receivable, billion rubles

Indicators/period of time	2019	2018	2017	2016	2015	2014	2013	2012	2011	2011 to 2019
Accounts payable	334	264	141	92	88	141	37	60	121	3
Accounts receivable	139	138	124	122	135	123	57	60	86	2
Payables-receivables ratio	2.41	1.92	1.13	0.75	0.65	1.14	0.64	1.01	1.40	2

Calculated: [12].

Therefore, short-term borrowings are not associated with the changes of working capital, which is usually linked to revenue. Interest payments on this type of financial debt have little influence on prime cost. It is not possible to prove a correlation with long-term investments. It should be borne in mind that a significant part of accounts receivable is the result of late transfer of funds by the state. Thus, in 2016 accounts receivable related to decisions of federal and regional authorities amounted to 148.1 billion rubles [13]. Does short-term debt compensate for the working capital shortage due to the growth of accounts receivable?

Creditors' borrowed funds (except for 2016, 2015, and 2013) exceed accounts receivable. Moreover, this excess increased from 1.4 times in 2011 to 2.41 times in 2019, which reduced the saleability of Russian Railways OAO.

3.3 Analysis of Russian Railways' borrowings from the perspective of urgency

Of interest is an analysis of the debt policy of Russian Railways OAO in relation to the loan structure by maturity. The data is presented in Figure 2.

The general trend is to increase loans received for more than 3 years. However, each year the structure is formed according to the financial situation. This indicates a long-term policy on the structure of the loan portfolio according to due dates and ability to adapt to changing conditions. This structure increases the paying capacity of the organization, more in accord with investment policy, although it increases costs. It is important for Russian Railways to tie the loan periods to the life cycle of implementable investment projects.

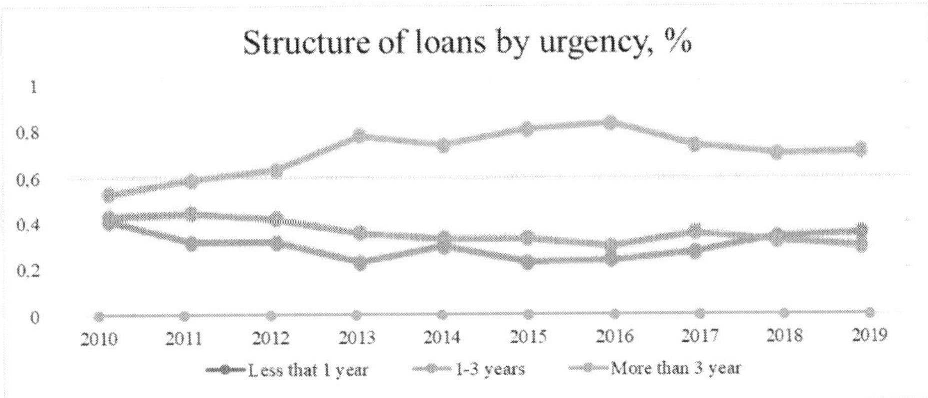

Figure 02. Dynamics of borrowed funds of Russian Railways OAO from the position of urgency . Calculated: [11, 12].

3.4 Analysis of the debt portfolio structure by types of instruments

Data for the analysis of the correlation of bank financing, domestic and foreign bond loans in rubles is presented in Table 5. The transfer of euros, Swiss francs and pounds into rubles was carried out at the time of the issue.

The dynamics of bond loans tends to grow. The share of bank borrowings in 2019 decreased from 36.0 to 18.0 % when compared to 2010. This is explained by the increasing role of the stock market in long-term financing.

Table 04. The dynamics of the debt portfolio by type of instruments, in billions of rubles

Indicator name/period of time	2010	2011	2012	2013	2014	2015	2016	2017	2018	2018 to 2010
Bank financing	107.28	166.32	169.3	29	76	0	18	103	226	2.1
Eurobonds	149	82.026	156.9	237	398	488	637	430	406	2.7
Local bonds	41.72	129.65	86.73	340	369	481	490	527	612	14.7

Calculated: [2, 4, 9].

The main advantage of bond loans is that it is necessary to repay loans at the end of the period. The bond loans are of a long-term nature. This most fully corresponds to capital-intensive and long-term projects, as the source of bond repayment appears only with the beginning of the business activity. Banks, on the other hand, usually require regular repayment of debt before the project is developed. Bond loans more correspond to the nature of a loan as a source of recapitalization. Moreover, the number of individuals buying bonds increases. In 2020 bond loans have become more diverse. "Green" and perpetual bonds have become available. Their peculiarity consists in the fact that, in compliance with the law, the issuer has the right to unilaterally decide not to make payments by the bond coupons. In the case of Russian Railways' decision not to pay coupon bond yield, the Russian Ministry of Finance provides compensation for the forgone income to investors. The mechanism of the compensation of forgone income is in the process of being agreed upon.

In mid-May acting Prime Minister Andrei Belousov (who is also the head of the Russian Railways Board of Directors) signed an instruction allowing Russian Railways OAO to place "perpetual" bonds worth 370 billion rubles. In terms of the approved bond program, the company will be able to make several issues of perpetual bonds, as well as in foreign currency (within the approved ruble limit) [10].

Let us analyze the correlation of the dynamics of ruble and foreign exchange loans to Russian Railways OAO. Borrowings in currency are appropriate when using these funds abroad, being able to accurately predict the exchange rate of the national currency.

We will try to analyze the borrowed funds received by Russian Railways in foreign currency. According to the loan policy, during the period 2010–2019 lending in foreign currency changed little and was 35 % on average, although in 2019 it was only 31.0 %. However, it is not indicated anywhere that foreign exchange borrowing should be related to foreign exchange earning amounts. Foreign exchange loans not tied to foreign exchange earnings are very risky because of volatility. In 2018, the total revenue of Russian Railways from foreign activities amounted to 471.8 million rubles, having increased by 33.0 %. This suggests that the risks associated with currency changes are decreasing [12].

Foreign currency loans are made in the form of bonded Eurobond loans. In 2010 bonds worth \$1.5 billion were issued at 5.74 % interest and with the period of repayment in 2017. This amount was equivalent to 45.5 billion rubles (at a rate of 30.4 rubles). The total payout was \$2.1 billion. To pay off this debt, it was necessary to purchase dollars for the sum of 223 billion rubles. The value of the bond loan issued in the national currency for the sum of 45.5 billion rubles would have cost Russian Railways only 75.0 billion rubles. There is no financial explanation of the advantages of bonds issued in foreign currency in the official sources of Russian Railways.

4 Purpose of the Study

Loan policy is determined by the economic state of the railway industry.

Table 01 data shows the steady pace of economic development. The volume of freight transportation increased by 30.0 %. The GNI grew by 23 % in the same period. The volume of rail transport turnover depends basically on the advance of the economy, but due to the growth of the share of services and mainly due to the increase in transportation distances, the railway industry can achieve better turnover.

Table 05. Some economic characteristics of Russian Railways OAO

Indicator/ periods of time	2010	2011	2012	2013	2014	2015	2016	2017	2018	2019	2019 to 2010
The amount of long-term debt, trillion rub.	0.26	0.36	2.60	0.58	0.72	0.90	0.83	0.93	1.00	1.14	4.3
The amount of freight transportation, trillion t. km	2.01	2.13	2.22	2.2	2.3	2.316	2.34	2.49	25.9	2.6	1.3
Revenue, trillion rub.	1.3	1.3	1.4	1.4	1.4	1.5	1.7	1.8	1.6	1.64	1.2
The amount of investment, trillion rub.	0.31	0.40	0.48	0.47	0.39	0.42	0.38	0.48	0.53	0.68	2.2
Return on equity, %	2.52	0.78	1	0	0	0	0	-1	0	0	-2.5
Share of long-term debt in revenue, %	0.5	0.6	1.19	1.65	2.36	3.92	3.92	3.09	3.11	7.0	

Calculated: [5, 11]

Revenue increased by 20 %. This points to the fact that there is a relative reduction in transportation costs, which is a positive result. When assessing revenue, it should be taken into account that Russian Railways is a natural monopoly and transportation tariffs are regulated by the state.

The growth of investments 2.2 times shows that all the stakeholders are serious about the implementation of the modernization program. And since the recoupment of infrastructure facilities is objectively very low, the return on investment is low, too. This is shown in the indicators of return on equity. The low return on equity accounts for a fact that, on the one hand, there is high capital intensity and, on the other hand, there is a much higher increase in long-term debt, since the return obtained on additional investments is paid as interest. Loan policy of Russian Railways for 2020–2022 is characterized by the increased activity of Russian Railways in the loan market. It is expected to borrow 297.3 billion rubles in 2020, 382.35 billion rubles in 2021 and 399 billion rubles in 2022. The increase in borrowed funds in 2020 is a consequence of the pandemic. State resources have been reduced and replaced by borrowed funds.

This will provide an opportunity to fulfill plans for investments in modernization. It is expected to spend up to 819 billion rubles in 2020, 842.1 in 2021 and 863.7 billion rubles in the coming year. The average loan rate is 7.7 %. The volume of loans in foreign currency during this period is 100–150 billion rubles, which will keep the share of foreign currency borrowings at the level of no more than 35.0 [10].

5 Research methods

Railway transport belongs to the infrastructure industries, the peculiarity of which consists in the fact that at some historical stage this industry, being socially necessary, however, cannot be profitable. This changes the way its effectiveness is evaluated. Natural indicators that characterize its contribution to the national economy development have top priority. In this case, the financing of infrastructure development can be provided out at the expense of society, and not only through investment at the expense of in-house funds. The option may be to finance investments at the expense of the state, and to cover current costs from the in-house sources. The multiplier effect, obtained by the process of developing businesses using infrastructure, will exceed the cost of the state's infrastructure financing [7].

The article uses the accounting records of Russian Railways as the most reliable source of information. Information is often distorted in the media because of the amateurism of journalists.

The amount of borrowed funds is taken from the balance sheet, line 1410. Therefore, this does not include the accounts payable submitted by line 1520.

An accounting report was used to obtain information on the amount of long-term debt, line 5110.

Revenue is in accordance with data taken from the financial results report (earnings and losses), line 2010. Net profit is also obtained from this accountability form, line 2400.

Data on planned investment amounts are obtained from the "Investment program disclosure form (about the investment program projects) and reports on their implementation of Russian Railways" for the period of 2014–2016."

The bulk of the information on the quantitative parameters of bond issuance is taken from the website of RUSBONDS of the Interfax Group and from the document "Borrowing policy and debt securities" for different years, posted on the website of Russian Railways. This document presents the structure of loans by various parameters. Payments on loans are most accurately presented in the cash flow report, line 4123.

The complexity of the use of statutory records is related to changes in indicators and accountability forms.

6 Findings

The management of Russian Railways pursues an active loan policy, increasing the share of financial liabilities.

However, this reduces financial results due to the increase of the price of loan resources on the efficiency of their use. But given that the problem of increasing the contribution to the macroeconomic indicators of the long-term development of the national economy is put in the forefront before the Russian railway industry, this kind of a debt policy can be considered necessary. It can be improved by replacing foreign exchange loan resources with local bonds by increasing the share of government financing.

7 Conclusion

The transportation per se should be carried out largely by private companies. Prospects for the development of the economy in Russia which is spatially very extensive and oriented to large flows of raw materials for export face the limited capabilities of transport networks. The structure of domestic and international transportation is focused on railway transport that requires modernization. Modernization requires significant investments from Russian Railways. Due to insufficient in-house funds and severe budgetary restrictions, all these years the management of Russian Railways pursues an active loan policy. The paradox of the loan policy of Russian Railways is explained by the economic role of railway transport in Russia, which is to ensure the effective functioning of the overall public capital of the Russian economy.

References

1. *Balance sheet* (2011). Retrieved from: https://www.tsum.ru/upload/iblock/133/133dcd68a82b90ca1746ba51835dc7c9.pdf
2. *Borrowing Policy* (2019). Retrieved from: https://old-ir.rzd.ru/static/public/ru?STRUCTURE_ID=65
3. *Company ratings* (2020). Retrieved from: https://company.rzd.ru/ru/9474
4. *Debt portfolio structure* (2014). Retrieved from: http://ar2013.rzd.ru/ru/investor-relations/loan-portfolio 11)
5. *Federal Statistics Service* (2020). Retrieved from: https://rosstat.gov.ru/storage/mediabank/3TZ2EcWc/gruz-ob.xls
6. *Financial results analysis, Russian Railways annual report* (2020). Retrieved from: https://ar2019.rzd.ru/ru/performance-overview/financial-results/income-expenses-analysis
7. A.S. Merkulov, O.G. Terekhova. The problem of financing transport infrastructure and how to solve it. *Transport Infrastructure of the Siberian Region. Materials of the 7th Int. Sci. and Pract. Conf., dedic. to the 355th anniver. of the found. of the city of Irkutsk*, vol. 4 (IrGUPS, Irkutsk, 2016), pp. 104–110
8. *Russian Railways OAO: accounting and financial analysis* (2020). Retrieved from: //https://www.audit-it.ru/buh_otchet/7708503727_oao-rossiyskie-zheleznye-dorogi
9. *Russian Railways placed the issue of perpetual bonds for 30 billion rubles* (2020). Retrieved from: https://www.rusbonds.ru/nwsinf.asp?id=526570
10. *Russian Railways raised the loan plan in 2020-22 in contrast to increased investments* (2019). Retrieved from: https://company.rzd.ru/ru/9401/page/78314?id=18198515)
11. *The 2014 Annual Report o3f the Russian Railways OAO* (2015). Retrieved from: https://ar2018.rzd.ru/ru/performance-overview/financial-results/debt-policy
12. *The 2019 Annual Report of the Russian Railways OAO* (2020). Retrieved from: https://ar2019.rzd.ru/ru/performance-overview/financial-results/income-expenses-analysis 6
13. *Working capital management* (2016). Retrieved from: https://ar2016.rzd.ru/ru/financial-results/working-capital

Regarding the infrastructure modernization of the Arctic settlements using the example of Chokurdakh village

Albert Kardashevskii[1,*], *Anna* Shishigina[1], *George* Volkov[1], and *Evgeny* Gribchatov[1]

[1]North and Arctic Interdisciplinary Research Division of Arctic Scientific Research Centre, Academy of Sciences of Sakha Republic, 22 Kurashova st., Yakutsk, 677000, Russia

Abstract. The article discusses the infrastructure development of the Arctic settlements through the example of Chokurdakh village in the Allaikhovsky district of the Sakha Republic (Yakutia) which is on the basis of a project approach linked to ongoing and planned projects for the development of territories and sectors of traditional natural resource use. The results of a sociological survey and an analysis of the actual state of infrastructure and social facilities on the basis of which projects on improvement of the living conditions of the population are proposed. It includes creating a favourable economic environment and creating new lines of economic and commercial activity in the Arctic settlements, using local natural and human resources are presented in the article.

1. Introduction

The existing large Arctic settlements on the territory of the Sakha Republic were formed during the formation and development of the Soviet state for the implementation of priority state tasks - the study and development of the natural wealth of the Arctic, mining, border protection, the formation of transport hubs, etc. At the same time, the costs of operating the infrastructure were of secondary importance.

After the collapse of the USSR, the economic turmoil of the 90s and the change in priorities for the development of the state, many city-forming enterprises went bankrupt, and there was a significant outflow of population from the North and the Arctic, which led to the closure of many industrial villages. So, according to researchers, 136.3 thousand people or 13.3% lived in the Arctic economic zone of the YASSR in 1986 and in 2017 - 68.5 thousand people or 7.1% of the total number of residents of the Sakha Republic [1]. According to preliminary estimates, the population of the 13 Arctic regions of the republic as of January 1, 2019 was 67674 people, which is 7.0% of the total population of the republic [2].

The village of Chokurdakh, Allaikhovsky district, like many other Arctic villages of Yakutia, for the reasons mentioned above, fell into decay. The outflow of the population and the appearance of empty apartment buildings led to a change not only in the architectural appearance of the village, but also in the patterns of energy and heat supply. According to the data for 01.01.2020, the population of Chokurdakh was 2077 people [3].

The decrease in the population is primarily due to migration outflows associated with a sharp decrease in the quality of life in the macro-region, including the level of wages that do not compensate for the poor quality of life, the departure of the able-bodied population to Yakutsk and other areas, the reduction of jobs in the village.

2. Results and Discussion

2.1 Social research

2.1.1 Content and directions of the survey

In order to determine the problematic issues of the life of the population in the village of Chokurdakh, a sociological survey was conducted from 03.08.2020 to 10.09.2020. The full study involved 115 participants, including 47 men and 68 women representing different categories of respondents (age, income level, profession, etc.). The sample amounted to 5.5% of the target audience, while 90% of the surveyed residents were natives of the Allaikhovsky district. The main focus of the study was on determining the satisfaction of the population with living conditions and the current state of infrastructure and social facilities, as well as plans to change their place of residence, move to other settlements of the Sakha Republic or regions of the Russian Federation.

2.1.2 Interpretation of results

According to the results of the survey, 15.7% of respondents expressed their readiness to move, and the reasons for this decision were the tension in the labor market, the lack of development prospects, leisure and unsatisfactory infrastructure. These data point to the possibility of securing the population while creating

* Corresponding author: agkard@mail.ru

acceptable living conditions and new jobs. This fact is indirectly confirmed by statistical observations, according to which the population of the village has remained stable over the past 3 years.

An important role in a sustainable population is played by 70% of respondents having their own housing in apartment buildings. 20% live in rented housing and 10% in official housing. An important indicator of the sustainability of the population of the village is the fact that 55% of respondents are married and 70% of respondents have children, which in a compartment with the presence of their own housing indicates settling. In addition, 60% of respondents classify themselves as middle-income citizens, which indicates the presence of permanent and sufficient sources of income for living in difficult Arctic conditions.

At the same time, 50% of respondents rate the living conditions in the village as satisfactory, 25.0% - unsatisfactory, 20.0% - good and only 4.0% of respondents rate excellent.

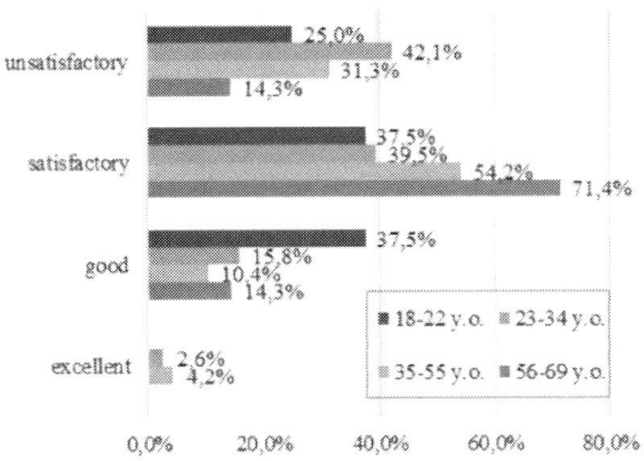

Figure 1. Satisfaction of the population of Chokurdakh with living conditions.

The population connects its future fate with the village and is interested in its development. The most pressing requests of the population according to the results of the survey are high-quality housing, the construction of roads, sidewalks, playgrounds, public spaces and accessible high-speed Internet (Figure 2).

These needs of the population are largely characteristic of all the Arctic and northern settlements of the Sakha Republic, where there is a great deterioration of real estate and engineering infrastructure, as well as practically no modern public spaces for recreation.

The improvement of public spaces for the residents of the village of Chokurdakh is primarily associated with the construction of playgrounds, street lighting, sidewalks, drainage systems and landscaping.

According to the majority of respondents, 60%, the population of the village is not active in solving existing problems and the reason for this behavior is the lack of initiative and focus only on the authorities. This position indicates the lack of understanding of the population of its role in solving the pressing problems of urban economy. At the same time, as the practice of implementing federal and regional urban improvement programs shows, the population directly living in the territory and striving for positive changes initiates most successful projects. In addition, similar data were obtained by researchers from other Arctic regions with a relatively developed infrastructure, where an assessment of the current state of the urban environment is recorded as moderately unsatisfied. Estimates of residents of small cities and single-industry towns do not fundamentally differ from estimates of residents of a large city - the regional centre [4].

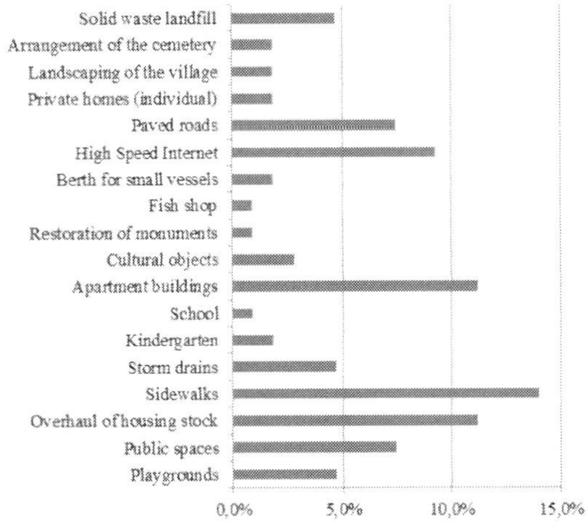

Figure 2. Priorities of the population in the development of the village of Chokurdakh.

2.2 Projects and mechanisms for their implementation

Analysis of modern large projects of development of resources of the Russian Arctic conducted by researchers in terms of spatial effect, which they generate, shows the presence of three main spatial effects of projects: localization effect, district effect, inter-corporate effect [5].

In the context of an acute budget deficit and an unstable economic situation in the energy market, it is a very difficult task to predict the development of remote Arctic and northern territories. The main framework for the implementation of the state policy of the Russian Federation for the development of the Arctic zone is the state programs.

The organization of management processes on the basis of strategic planning and the tasks, goals and priorities formed in the planning process involves the implementation of projects and programs for the development of territories. This takes into account financing from various sources: both from budgets of different levels, and using the funds of departments and private investors, as well as the option of public-private partnership [6].

A key advantage of the village of Chokurdakh, as well as other Arctic settlements of the Republic of Sakha

(Yakutia), is the opportunity to participate in the implementation of special tools to ensure the socio-economic development of the Arctic. It includes the state program of the Russian Federation "Socio-economic development of the Arctic zone of Russia until 2025", the Strategy for the socio-economic development of the Arctic zone of the Republic of Sakha (Yakutia) for the period until 2035 (Decree of the Head of the Republic of Sakha (Yakutia) dated 14.08.2020, No. 1377).

In these conditions, based on the current state of the infrastructure and the needs of the population of the village, development projects were formulated in relation to the state and regional programs being implemented.

2.2.1 Renovation of the housing stock in Chokurdakh

The project is aimed at designing and building new residential and social buildings, improvement facilities and recreation zones, bringing to a satisfactory state the level of improvement of the courtyard and public areas of the village.

The main mechanisms for the implementation of this project are consistent and mutually complementary state and municipal programs and investment projects of economic entities implemented on the territory of the Allaikhovsky district of Yakutia.

Currently, the following programs are being implemented in the district:

1. State program of the Republic of Sakha (Yakutia) "Formation of a modern urban environment in the territory of the Republic of Sakha (Yakutia) for 2020-2024 years". At the initial stage of the program implementation, the master plan is developed and the urban planning documents of the village of Chokurdakh are updated. The size of financing of the first stage is 1.8 million rubles.

2. Regional program for the overhaul of common property in apartment buildings in the village of Chokurdakh. The planned volume of investments within the framework of the regional program of overhaul of common property in apartment buildings in 2020-2035 amounts to 173.755 million rubles.

3. The republican targeted program "Resettlement of citizens from the emergency housing stock 2019-2025", which includes 27 apartment buildings in the village of Chokurdakh or 311 residential premises with a total area of 15101.31 sq. m. Under this program, the improvement of housing conditions will affect 800 residents, which is 38.7% of the population of the village. The amount of financial resources for 2019-2025 provided for the implementation of the program in the village of Chokurdakh is 1747.765 million rubles (in 2019 prices).

The marginal cost of construction (acquisition) of 1 square meters of the total area of housing under the resettlement program for the conditions of the village of Chokurdakh is: stone - 118 870 rubles, quick-to-produce - 91 131 rubles, wooden - 79 258 rubles. According to the Ministry of Construction and Housing and Communal Services of the Russian Federation (Order No. 351/pr "On the standard cost of one square meter of the total residential area for the second half of 2020 and indicators of the average market value of one square meter of the total residential area for the constituent entities of the Russian Federation for the third quarter of 2020" dated 29.06.2020) the average market value of 1 sq. m of the total housing area in the Sakha Republic for the 3rd quarter of 2020 is 83,191 rubles. Accordingly, the increase in construction costs for the conditions of the village of Chokurdakh, taking into account the complex transport scheme and the limited construction season, according to approved standards, is 42.8% compared to the average market value of 1 sq. m of the total housing area in Sakha Republic.

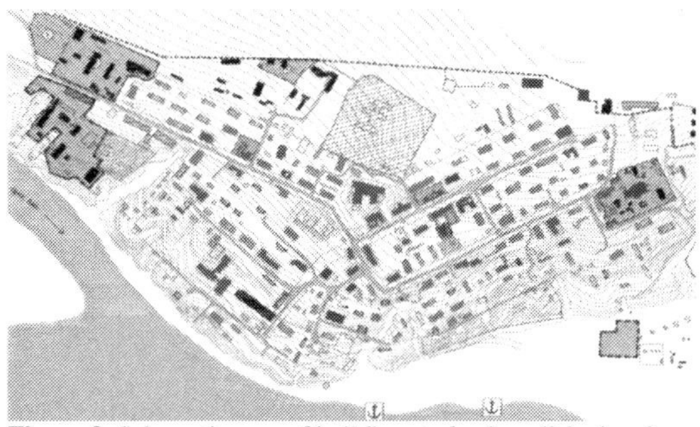

Figure 3. Schematic map of buildings to be demolished under the program "Resettlement of citizens from the emergency housing stock 2019-2025".

The real cost of building residential buildings in a stone version in the village of Chokurdakh, subject to the delivery of materials from Yakutsk, will be within 125-132 thousand rubles per sq. m. That does not meet the requirements of the program, and residents of inaccessible Arctic areas find themselves in limited conditions for choosing the structural part of future housing. For them the most durable and reliable type of buildings becomes practically inaccessible - stone buildings on a reinforced concrete foundation.

Uneven deployment of production facilities for the production of construction materials in the territory of the Sakha Republic significantly increases the cost of construction due to high tariffs for transport services [7].

A significant increase in the cost of building stone buildings is due to the high transport costs for the delivery of the initial components of the concrete mixture - cement, crushed stone, sand. Moreover, the largest size is occupied by mineral fillers crushed stone and sand. The transportation scheme for the delivery of materials and the absence of locally developed quarries of building materials (ASG, crushed stone, construction sand, etc.) significantly increases the cost of construction, especially stone buildings on reinforced concrete foundations of various types tape, slab, pile. In addition, for the construction of intra-quarter roads and sidewalks, engineering communications and improvement facilities (children's playgrounds and sports grounds), OPI such as river sand, sand and gravel mixture, limestone or gravel are used in large volumes.

The lack of accessible developed deposits of building stone, sand, clay, etc. in the Arctic regions forces many developers to look for alternative solutions to stone construction. In recent years, in the Arctic regions, technologies of frame house building have been actively introduced using light steel thin-walled structures and mineral wool insulators. But the text of annex 2 of the republican targeted program "Relocation of citizens from the emergency housing stock for 2019-2025" explicitly states that "It is not recommended to build houses and purchase housing in houses made of light steel thin-walled structures (LSTK), SIP panels, metal sandwich panels..."

The problem of the lack of local building materials in the conditions of inaccessible territories of the North and the Arctic is limiting not only in housing construction, but also in the improvement of the yard areas of settlements, the construction of paved roads and bridge crossings.

The optimal solution to this problem for the conditions of the village of Chokurdakh is the organization of small production of building materials based on local mineral raw materials with orientation and interaction with existing and planned state and municipal programs for the development of territories.

The implementation of this project will be facilitated by the existing system of training specialists in the field of construction materials science with extensive practical experience at the North-Eastern Federal University [8].

2.2.2 Small production of building materials and products

Developed deposits of construction raw materials are mainly grouped in the central, southern and western part of Yakutia [9].

The prerequisites for organizing the production of building materials are the Hara-Tas igneous rock (andesite-basalt) deposits located 17 km east of Chokurdakh, suitable for the production of building materials. The forecast reserves of raw materials according to the estimates of the State Committee of the Republic of Sakha (Yakutia) for geology and subsoil use are 2065 thousand m^3, while the average annual production levels can reach 60 thousand m^3. In addition, in the area of Oyotung (26 km from the village of Chokurdakh) has reserves of building stone with certain book reserves of 13 thousand m^3 suitable for crushed stone production.

Sources of raw materials for the production of building materials can be deposits in the nearest areas along the Indigirka riverbed. According to the Register of the existing licenses for subsoil use on subsoil plots of local value on the navigation channel of the Indigirka River in Abyysky district the extraction of construction sand and sand-gravel mix by large enterprises of the region-state enterprise "Road of the Arctic" and Public Joint-Stock "Lena River United Shipping Company" is conducted.

Difficult conditions in the development of existing deposits of common minerals require a large amount of investment and such projects cannot be carried out by small enterprises, especially if they are engaged in the implementation of short single projects, such as the construction of low-rise apartment buildings or the improvement of a limited area (sidewalks, playgrounds, intra-quarter roads, etc.).

An integrated approach to solving issues of landscaping and housing construction involves attracting and coordinating the work of many multi-directional enterprises and institutions, which significantly complicates the work process and increases risks. The most suitable option for the organization of work on the improvement of the village from the point of view of management and cost optimization is the creation of a joint venture within the framework of a municipal-private partnership, whose functions will be transferred to work on projects for the improvement of the village and housing construction.

Currently, the Municipal Unitary Enterprise Industrial Transport and Construction Complex "Polyarnik" (Municipal Unitary Enterprise PTSK "Polyarnik") operates on the territory of Chokurdakh, on the basis of which a joint venture can be organized with the involvement of interested economic entities. Directly producing building materials in the Allaikhovsky district belongs to the production cooperative of the nomadic ancestral community "Allaikha" (PC KRO "Allaikha"). This organization has established the production of insulation from polystyrene foam in the form of boards of various sizes. Today, KRO Allaikha PC has the necessary production areas and equipment to organize the production of other types of construction materials.

In accordance with the estimated calculation in the next 5 years for the implementation of plans for the demolition of emergency and the construction of new housing with a resettlement program from the emergency housing stock for 2019-2025, the need only for crushed stone for the arrangement of vertical planning, reinforced concrete foundations and public roads of local importance will be more than 42 thousand m^3.

The delivery of material in such volumes increases the cost of construction products by 62-66%, which ultimately increases the estimated cost of finished housing in the stone version.

The purpose of the proposed project is to develop new areas of economic activity, as well as optimize costs for the construction of residential and public buildings in the village of Chokurdakh. The main mechanism for implementing the project is public-private partnership with the creation of a consortium or cluster of enterprises operating in the construction sector.

The initial stages of the project implementation include the following works: justification, registration, preparation of permits (licenses) and development of a field of common minerals in the territory of the Allaikhovsky district.

The next stage of the project implementation is the organization of small production for the manufacture of construction rubble, concrete blocks, paving tiles and curb stones. The project provides for the development of a quarry of building stone and the organization of its processing in accordance with the requirements of GOST 8267-93 "Crushed stone and gravel from dense rocks for construction work. Specifications" with a total capacity of 10 thousand m^3 per year.

On the basis of the obtained crushed stone, it is proposed to organize the production of concrete wall and partition stones according to GOST 6133-99 "Concrete wall stones. Specification" with a total capacity of 2100 m3 per year and workshops for the production of mortars and concrete in accordance with the requirements of GOST 7473-2010 "Concrete mixtures. Specification" with total capacity of 20 m^3/h.

The project provides for the modernization of the existing PK KRO "Allaikha" workshop for the production of polystyrene foam to increase the size and assortment of products. After the modernization of the production line, the following types of products will be additionally produced: polystyrene foam shells for utilities and tanks, bulk insulation in the form of pellets, extruded polystyrene foam for road construction, inserts for heat-efficient three-layer wall blocks.

The implementation of the project is planned in stages with synchronization with resettlement programs from the emergency housing stock, the formation of a modern urban environment and the overhaul of apartment buildings for 2020-2025. The total projected amount of financing is 38.7 million rubles.

3. Conclusion

The implementation of interconnected projects to renovate the housing stock and organize the production of building materials in the village of Chokurdakh will not only create additional jobs, increase the income part of the municipality's budget, but also provide a qualitative jump in the diversification of the region's economy and improve living conditions.

The total amount of investments envisaged for the implementation of all projects from the federal, regional and municipal budgets and funds of economic entities will amount to about 2 billion rubles. This, of course, is the largest in the post-Soviet history of the region and a unique chance for the Arctic settlement to change the existing situation and transition to sustainable development of the territory.

The implementation of these projects will:

- ensure the transition to a qualitatively new level of construction of housing, administrative buildings and public spaces, transform the architectural appearance of the village and create modern infrastructure to ensure quality housing, work and recreation;
- modernize the engineering equipment of buildings and improve their energy efficiency;
- create additional (seasonal) jobs for residents of the village during the improvement of territories, construction and repair of buildings (estimated the need for seasonal labor personnel is determined at 25-30 people (handyman, mason assistant, painter, carpenter, road worker, welder, etc.) annually during the construction season (April-October).

In general, the sustainable development of Arctic territories without the creation of high-quality social and housing infrastructure is impossible [10]. Therefore, new types of production that contribute to the development of the region should become priority measures on the path to sustainable development of the village of Chokurdakh along with the development of traditional industries.

References

1. I. Nikolaeva, S. Pavlova, Innovation and Investment **1**, 111-114 (2018)
2. E. Danilova, Innovation and Investment **12**, 260-265 (2020)
3. Territorial body of the Federal State Statistics Service for the Republic of Sakha (Yakutia), Retrieved from: https://sakha.gks.ru/folder/32348
4. O. Russova, T. Smak, I. Tarasov, Arctic and North **41**, 236-247 (2020)
5. A. Pilyasov, E. Putilova, Arctic and North **38**, 21-43 (2020)
6. I. Bachurinskaya, Bulletin of Altai academy of economics and law **12-3**, 5-12 (2019)
7. O. Burenina, A. Andreeva, M. Savvinova, INNOVATIONS IN SCIENCE **5 (66)**, 69-73 (2017)
8. A. Mestnikov, S. Semenov, D. Vasileva, Fundamental research **12-1**, 80-84 (2017)
9. Federal State Budgetary Enterprise "A.P. Karpinsky Russian Geological Research Institute" Retrieved from: https://vsegei.ru/ru/info/gisatlas/dvfo/yakutia/
10. T. Ediseeva, Journal of Economy and business **5-1 (63)**, 173-176 (2020)

The fishery in the northern regions of the Far East of the USSR in the context of late Soviet modernization

Nonna Platonova [1,*]
[1] Far Eastern State Transport University, 47 Serysheva street, Khabarovsk, 680000, Russia

Abstract. The article describes the development of the fishing industry in Kamchatka, Sakhalin, and Magadan regions in 1965-1985. In a general historical context, the article pays attention to the modernization processes in the fishery of the northern regions of the Far Eastern territory, emphasizing general trends and differences in their dynamics. The special importance and level of development of the material and technical base of the fishing industry is noted. The article reveals the working conditions of ship repair enterprises and the reasons for the financial deficit of ship repair. Statistical data on the accident rate of fishing vessels of the branch administrations of Kamchatka, Sakhalin, and Magadan are analyzed. The results of the introduction of 200-mile economic zones that complicated the fishing activities of enterprises and the fleet of the fishing industry in the northern regions of the Soviet Far East are shown. It was found that social and economic experiments in the late Soviet period and modernization, which was carried out from above by the political elite, did not eliminate disproportions in the development of the fishery. Territorial remoteness, difficulties of timely material and technical supply and ship repair and maintenance problems were difficult-to-overcome challenges which accelerated the systemic crisis in the fishery.

1 Introduction

The Far Eastern basin occupies one of the leading positions in the development of the fishing industry and the fishery industrial cluster of the Russian Federation. The Far Eastern Basin accounts for the bulk of fish catch and fish production, which is determined by the territorial access to biological resources of the Okhotsk, Bering and Japan Seas, and the northwestern Pacific Ocean; the historically traditional nature of fishing; a significant share of labor resources. For many regions of the Far Eastern Federal District (FEFD), it acts as a town and village-forming component. In addition, the proximity of markets of the neighboring countries of the Asia-Pacific region, which have a powerful economic potential, reveals multifaceted prospects for international cooperation, influences the government strategy for the development of the FEFD fishery industry. The historical retrospective of the results of late Soviet modernization of the fishing industry of Sakhalin, Kamchatka, and Magadan regions is of scientific interest.

2 Historiography, problem and purpose

In Soviet historiography, the history of the fishing industry of the Far Eastern territory was represented by articles [1], but the studies by A.T. Mandrika, which were published in the Soviet and modern periods, are of special interest. Analytical assessments and conclusions of this author reveal the regional specifics of the fishing industry, show its influence on the national economy of the Far Eastern territories [2].

At the latest stage, many problems of the history of the development of the Far Eastern region were rethought. The research on the socio-economic development of the fishing industry, its material, technical and personnel support has become actualized, which is reflected in the works by Far Eastern authors [3 – 5].

However, there are no special studies that describe the state of the fishing industry in the northern regions of the Far East in the late Soviet period at the latest stage.

The **subject** of the article is the development of the fishing industry in the northern regions of the Far East in the context of late Soviet modernization (1965 – 1985).

The **purpose** is to study the development of the fishing industry in the northern regions of the Far East through the analysis of archival and statistical data, factors and methods that determined the dynamics of the industry in the context of the late Soviet modernization.

Research methods: problem-chronological, statistical analysis, and comparative.

3 Results and discussion

In the economy of the Soviet Far East, the fishing industry, being the basic sector, played an important role in the food supply of the country. The Far Eastern fishery basin supplied more than ⅓ of the all-Union catch of fish and sea animals. Sectoral enterprises produced 6.1 and 15.9 million cans of crab and salmon ; 41% of fish products were exported. Only Japan

*Corresponding author: nonnaplaton@mail.ru

received 81% of red caviar and 56% of seafood. In particular, the fishing industry of the Sakhalin region supplied many types of finished products and raw fish to Singapore, Japan, as well as to France and Belgium, receiving various imported consumer goods, specialized machinery and equipment [6, F. 399. Op. 3. V. 3. D.1315. P. 2; F. 399. Op. 3. V. 3. D.1435. P. 107 – 108; F. 399. Op. 3. D.1483. P. 69 rev.].

The dynamics of the progressive development of the fishing industry was determined by the Pacific seaports in Magadan, Petropavlovsk-Kamchatsky, Kholmsk, Nevelsk, etc., which were centers for fish processing with an annual cargo turnover of about 60% An important role was played by their proximity to rich fishing areas, for example, to the ports of the Kamchatka region. Thus, the distance from the Bay of Lavrov, located in the northwestern part of the Olyutorsky Bay in the Bering Sea, to the fishing area of the Pribylov Islands was 870 miles, while from Vladivostok to the same islands 2500 miles [7].

The coastal strip of the Sea of Okhotsk (about 2.5 thousand km long) and the rich fish resources of the Amur river with its tributaries predetermined the specialization of a number of northern regions (Okhotsk, Tuguro-Chumikan, Nikolaev, Ulcha, Vanin and Sovetsko-Gavan) in the fishing industry [8, F.R-353. Op. 10. D. 350. P. 5].

In the second half of the 1960s, the fishery industry was well-coordinated under the jurisdiction of the industrial-territorial Main Directorate of Fisheries of the Far Eastern Basin "Dalryba" (Vladivostok). It included the territorial administrations "Primorrybprom", "Sakhalinrybprom", "Kamchatrybprom", "Magadansky", "Nizhneamursky", "Okhotsk" trusts, the Department of Whaling Flotillas and "Vostokrybholodflot". In May 1970 the Department "Dalmoreproduct" was created [5]. A significant contribution to the development of the fishing industry was made by the Pacific Department of Reconnaissance and Research Fleet (TURNIF) and the Pacific Research Institute of Marine Fisheries and Oceanography with offices in Kamchatka, Sakhalin, Magadan and Chukotka.

The economic reform of 1965, which initiated the modernization processes, revealed miscalculations of self-supporting experimentation, emphasizing the problems of the material, technical and personnel potential of sectoral enterprises. Heads of the territorial departments solved complex strategic tasks, which were aimed at expanding the fishing areas, developing new stocks of biological resources in the open parts of the Okhotsk and Bering Seas, and industrializing the production processes, increasing the output of fish food products. For the party elite of the USSR, it was clear that the fishing enterprises and the fishing fleet of the Sakhalin, Kamchatka, and Magadan regions were not ready for the forthcoming reorganization. In this regard, the USSR State Planning Committee developed new working conditions whoch should be implemented in January 1, 1968, but by that time only accompanying documents had been prepared and standards for working capital had been developed. In the following years, the process of curtailing economic reform began, the transformation of socio-economic stability into stagnation intensified, so since the mid-1970s, the growth rates in the fishing industry of the Far Eastern territory, although they remained high, nevertheless, the economic crisis affected them. The introduction of market elements was incomplete [6, F. 4372. Op. 66. D. 107. P. 182 – 184].

Another factor that hindered the economic reform in the fishing industry was the uneven development of regional administrations. The Dalryba Main Directorate made biased administrative decisions, thereby creating serious difficulties. In particular, the planned targets for daily fishing for the large freezer fishing trawler (BMRT) Primorrybprom were lower than for similar vessels of Kamchatrybprom by 8% in 1963, 32% in 1964, 34 % in 1965, and 13% in 1966.

The sectoral specificity of the remote northern regions actualized the problems of timely fuel supply for ships. However, in 1965 Primorrybprom received 35% more fuel for the export of finished products, while the Kamchatka Administration received 32.5% less fuel. There were also unreasonable disproportions in the provision of fishing vessels with a receiving and transport fleet: while waiting for the delivery of finished products, the downtime for one BMRT of the Primorsky Department was 8 days, for Kamchatka - 22 days. Repeated failures in the provision of refrigerated vessels were the result of financial losses of the coastal fish processing enterprises of Kamchatka region. In 1966, the fish processing plant in Anapka suffered losses in the amount of 470 thousand rubles.

These "manifestations of parochial tendencies" on the part of the central administration were desscribed in a letter addressed to the department of light and food industries of the Central Committee of the CPSU by Orlov, the first secretary of Kamchatka party committee, who suggested making changes in the fisheries management system by creating the regional Main Directorate subordinate to the USSR Ministry of Fisheries. His views were supported by the first secretary of the Sakhalin regional committee P.A. Leonov.

However, these plans were not implemented, since for the top party leadership, reorganization of the fish industry seemed economically inexpedient. However, the powers of territorial production departments were expanded, the deployment of the fleet was optimized, the volumes of centralized financing increased, etc [9, F. 5. Op. 58. D. 110. P. 24, 29, 32].

The governement strategy for the accelerated comprehensive development of the Far East, which had ambiguous consequences, contributed to significant economic transformations in the fishing industry. On the one hand, trawl fishing, carried out in the coastal zones of the Tatar Strait, replaced oceanic fishing, which significantly increased the volume of fishing on Sakhalin and Kamchatka (Figure 1).

An important result was the construction of an active oceanic fishing base located in the north of the Khabarovsk Territory in the city of Sovetskaya Gavan. Capital investments in its material and technical development reached 24 million rubles. New jobs were created, which increased the influx of people to the area.

https://doi.org/10.1051/shsconf/202111200020

The finished products, the annual volume of which exceeded 20 million conventional cans of canned food, were supplied to various regions of the USSR and for export. The basis of the base was the fishing fleet with special fishing vessels, where the processing of raw fish was carried out, and in the working village of Lososin there were specialized workshops for the production of finished products. Raw materials for processing were partially supplied by Vladivostok and Khabarovsk, separate fishing collective farms, tin containers came from Kholmsk and Nakhodka. In general, the opening of the active oceanic fishing base marked a new trend in the development of the fishing industry throughout the Far East of the USSR [8, F.P-137. Op. 14. D. 2254. P. 75].

However, on the other hand, this led to some degradation of coastal fisheries, which particularly affected the settlements in the fishing areas of the Lower Amur, the Amur estuary and the western coast of Kamchatka [10].

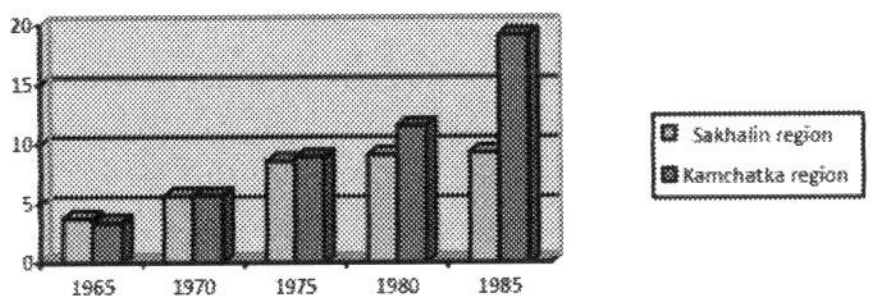

Figure 1. Indicators of fishing on Sakhalin and Kamchatka in 1965 – 1985 (In million tons) [12, F. 17. Op. 150. D. 465. P. 7].

In the late Soviet period, the main vector of the government economic policy was focused on strengthening the material and technical base of the fishing industry ; therefore, the fishing fleet was provided with new vessels. As a result, in the 1970s, Sakhalinrybprom's fleet consisted of 33 BMRTs, 47 Medium Fishing Freezer Trawlers, 80 Medium Fishing Trawlers, 40 Fishing Seiners, 70 Fishing Freezer Seiners, 8 floating bases and factories. Between 1976 and 1985, the volume of government support for the regional fishing industry amounted to 53 million rubles [11, F. 540. Op. 1. D. 728. P. 4].

"Kamchatrybprom" possessed 257 fishing and fish processing vessels which had modified equipment for the production of food products [12, F. 17. Op. 106. D. 404. P. 2]. By 1985, the fishing industry numbered 288 vessels which emphasized the oceanic nature of fishing in Kamchatka. In 1966-1970, the volume of capital centralized investment amounted to 302 million rubles, in 1971 - 1975 - 342 million rubles, in 1985 - 445.7 million rubles. [9, F. 5. Op. 41. D. 151. P. 10; 12, F. 17. Op. 145. D. 426. P. 4]. Fleet modernization not limited by the remoteness of the navigation areas and climatic conditions, had a positive effetc on the extraction and processing of fish raw materials in a mechanized way [Figure 2].

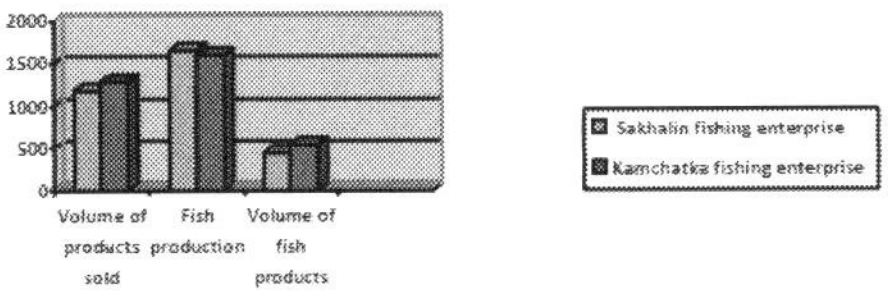

Figure 2. Performance indicators for the fishing enterprises of Sakhalin and Kamchatka regions in 1966 – 1970 (million rubles; thousand tons) [13, F.P-68. Op. 54. D. 203. P. 112].

The technical modernization was more complicated in Magadan region, where until 1969 the fishing industry did not have its own floating facilities. With the support of the local party leadership, Magadanrybtrest received 48 vessels, and until 1985 this number did not change. The problem of an acute shortage of vessels worsened because of oceanic fishing. The existing vessels were not able to cope with the volume of fish processing and supplied raw materials to the coastal enterprises of Magadanrybtrest, many of which became unprofitable. Their number was rapidly decreasing [12, F. 17. Op. 103. D. 601. P. 44; 9, F. 5. Op. 43. D. 118. P. 5].

Against this background, the capacity for repair and maintenance of vessels increased slowly, which led to an increase in imbalances in the dynamics of development of the fishing industry. During the period under review, the financial deficit of ship repair has constantly increased [Figure 3].

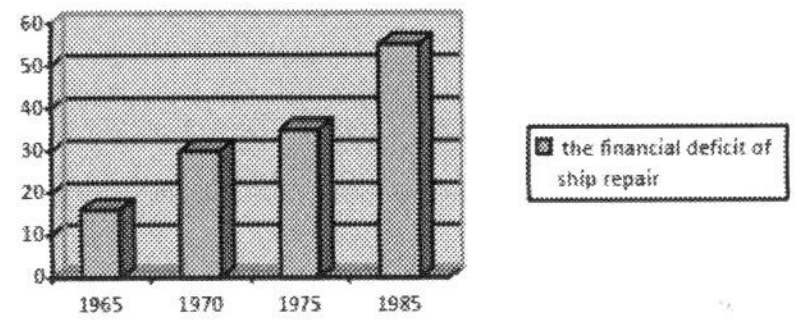

Figure 3. Dynamics of the financial deficit of ship repair in 1965 - 1985 (million rubles) [6, F. 399. Op. 3. V. 2. D. 726. P. 7].

Far Eastern ship repair enterprises did not use their capacities rationally, working in one shift. They had a lot of unidentified equipment ; auxiliary operations were mechanized by 40-50%, and there were no qualified personnel. Quite often, the intensification of industrial processes was replaced by slogans and all sorts of workers "initiatives" initiated from above, which resulted in emergency work. About 52% of the vessels were out of repair in the third decade of the month. Whole-day and intra-shift downtime, loss of working time were significant.

For example, in the 1970s at the shipyards of Kamchatka region the cost of equipment stored in the warehouses amounted to 3003 thousand rubles, more than 40% of machines and mechanisms were used for 20 or more years. The replacement of obsolete equipment began only in the early 1980s. The ship repair base of Sakhalinrybprom did not have the technical capabilities (components, mechanisms, equipment) to provide the mining, processing and transport fleet in a timely manner, so more than 60% of the preventive repair of the BMRT was carried out outside the Sakhalin region, thereby significantly increasing its cost [12, F. 17. Op. 150. D. 465. P. 127].

The poor material and technical base of enterprises hindered the possibilities of accelerated, mobile repair of ships, which led to an increase in downtime. In 1975, the Kamchatrybprom vessels were at sea for 33% of the fishing time, while the target was 43%; in 1985 - 9 thousand ship days and 13%, respectively; the vessels of Sakhalinrybprom were at sea for 32 thousand ship days (46%) [12, F. 17. Op. 154. D. 525. P. 19]

With the growth of crisis phenomena in the economy by the mid-1980s, the workers felt the lack of technical means for sorting fish and processing raw materials; therefore, fishermen were often forced to dump part of the catch overboard. Most fishing and processing vessels remained unprofitable. This caused accidents. For example, in 1971-1975, 114 accidents occurred in the Kamchatrybprom fleet with a damage amount of 13.5 million rubles; in 1976-1980, 157 accidents occurred with a damage amount of 4824.6 thousand rubles ; in 1981-1985, 134 accidents occurred with a damage amount of 1115.9 thousand rubles. A high rate of accidents was observed in the Sakhalinrybprom fleet, where in 1980-1985 20 shipwrecks and accidents were registered, technical losses amounted to 137.8 thousand rubles ; in Magadanrybprom, these figures were 7 and 48.2 thousand rubles, respectively [12, F. 17. Op. 145. D. 424. P. 14].

In general, by the mid-1970s, a period of stabilization began in the fishing industry. The total number of industrial enterprises was more than 200, including large seaports, shipyards, and the fleet. Fish was peocessed by 47 enterprises located along the Far East coast. The fishing industry included 4 fish processing plants and 1 fish factory in Magadan region; 12 fish factories in Kamchatka; 9 fish processing plants, 2 fish and 1 agar factories on Sakhalin; 2 fish processing plants and 3 fish factories in Khabarovsk and 11 fish processing plants and 1 agar plant in the Primorsky region [5].

Nevertheless, the economic situation in the second half of the 1980s was complicated by difficulties caused by the introduction of 200-mile economic zones the consequences of which were serious restrictions on fishing activities and significant changes in fishing areas. For this reason, fishermen of Kamchatka region did not add 62.5 million rubles to the gross output (94.8%), in Magadan region, this figures were 5.3 million rubles or 89.4%, respectively. In fact, in all the fishing regions of the Soviet Far East, many fish processing enterprises and floating canneries were idle due to the lack of raw materials. The fishing industry, its material and technical base, infrastructure needed time and money for restructuring and adapting consumers to new types of finished products.

As a result of the joint activities of the regional party and economic leaders of the Far Eastern territory, measures were taken to change the coastal fishing and processing base. For example, in connection with a significant reduction in the catch of coastal fish species, the coastal plants began to use raw materials from more remote areas (South and North Kurils). They processed new species of fish for canned food (saury, mackerel, etc.) and eliminated the factor of production seasonality [12, F. 17. Op. 150. D. 999. S. 8; 6, F. 399. Op. 3. V. 3. D.1435. P. 155].

4 Conclusion

The fishing industry of the northern regions of the Far Eastern territory developed in a national context, being an important component of the national economic system. State regulation of the industry was expressed in the formed vertical of management, the establishment of uniform purchase prices, centralized financing.Along with the reform, the centralized system of material and technical supply limited the economic independence of enterprises, negatively affecting their performance. At the same time, ministries continued to set economic indicators. In such conditions, the enteprises were not able to restructure the production process in accordance with the consumer market conditions and continued to supply raw materials in large volumes. The gaps between the growth rates of the capital-labor ratio of sectoral enterprises and labor productivity and between the dynamics of fixed assets and the total volume of output were gradually increasing.

The traditional fish-raw material base became depleted, new areas and fishing objects were being developed extremely slowly. Measures for fish conservation and fish farming were ineffective. The dynamics of modernization was influenced by the disproportions between the level of development of the coastal economy and needs of the fleet, the lagging of ship repair facilities, processing plants and industrial infrastructures. The material and technical base of the fishing industry of Kamchatka, Sakhalin, and Magadan regions needed special equipment that was not produced in the Far East, which aggravated the growing economic and economic crisis.

References

1. Yu. M. Kolesnik, *Local history notes* **4** 42-49 (1973).
2. A.T. Mandrik, *Pages of the history of the fishing industry of the Soviet Far East. Part 1 (1923 - 1958).* (Vladivostok, 1974).
3. V.G. Svidersky, *From the history of fishing in the Far East.* (Vladivostok: RIO Primoruprpoligrafizdat, 1990).
4. A.T. Mandrik, *The history of the fishing industry of the Russian Far East (1927 - 1940).* (Vladivostok: Dalnauka, 2000).
5. N.M. Platonova *Industrial and civil complex of the Far East of the RSFSR (1965 - 1985). The experience of historical development.* (Moscow: Ethnosocium, 2013).
6. Russian State Archives of Economics (RSAE).

7. P. Ya. Baklanov, Far Eastern region of Russia: problems and prerequisites for sustainable development. (Vladivostok: Dalnauka, 2001).
8. State Archives of Khabarovsk Region (SAKR).
9. Russian State Archives of Contemporary History (RSACH).
10. L.V. Volkov 2016 Institutional aspects of the development of the fishing industry of the Far East *Regionalistics*. **3 (6)** 56 - 68 (2016).
11. State Archives of Sakhalin Region (SASR).
12. Russian State Archive of Social and Political History (RSASPH).
13. State Archive of Primorsky Region (SAPR).

The role of the state in ensuring digital transformation of banking sector

Yulia B. Bubnova [1,*]

[1] Baikal State University, 11, Lenin Str., Irkutsk, Russia

Abstract. In the age of digital technologies, the main factor of competitiveness in the financial market is "digital capital". Using developments in the field of information technology, banks change the format and methods of providing their products and services, participate in the provision of services that are not typical for them. In such circumstances, banks have to completely transform their business and build it on the basis of platforms. Changes in the operating conditions of the financial market entail changes in the rules of conduct on it and approaches to its regulation. Investing in technologies, creating and developing ecosystems require significant financial investments from banks, which not all banks can afford. This research analyzes the possibilities of using digital capital by the most successful Russian banks in building and developing ecosystems. The main models of banking ecosystems used in Russian practice are systematized. The necessity of state participation in financial support of the digital transformation of the banking sector is justified. The steps taken by the Bank of Russia in this direction are evaluated. The article summarizes the Bank of Russia's efforts to develop approaches to financial market regulation to ensure information security and prevent financial isolation. Steps taken by the regulator to ensure an equal access to financial services for all citizens, equal opportunities for competition of various financial market players.

Keywords: banks, financial market, ecosystem, digital technology, regulation

1 Introduction

The Russian banking sector is the largest user of digital technologies and the main driver of digital transformation of the entire economy. The financial services market has recently become very attractive for many participants, as it is becoming more and more technological. Services that still recently were provided only by credit organizations are becoming popular among companies that use modern mobile and Internet technologies. Telecom operators are responsible for instance issues. Customer creditworthiness assessment (online scoring) is carried out by specialized companies, for example, "Direct Credit" and "Gridfore". In the sphere of crediting of the enterprises of small and medium business, banks compete with technologically advanced platforms: "City money" and "Moulding". Many financial services are provided by such companies as "UK FinEx Plus" and "Cindicator", "Optiacs", "Unicombay", "Simple Invest", etc.

The emergence of a new format of banks – non-banks – in the last decade also poses a threat to the functioning of banks in the traditional format. The long and successful work of "Tinkoff Bank" confirms this. Among the newcomers who have proven themselves in the banking market, we can mention "Roketbank" – one of the first mobile banking services in Russia, created in 2012. Since 2015, "Touch Bank" has been operating in an online format as a retail direction of "OTP Bank". "Modul Bank" has been operating in a new format since 2016. "Tochka Bank", as a branch of "FC Otkritie", is also a representative of non-banks. The interesting project in the field of a new format of banking services in messengers is "TalkBank" – a resident of the SKOLKOVO Innovation center.

In the works of many domestic [1, 3, 5, 7, 16] and foreign authors [14, 15], the influence of technology on the financial sector in general and the banking sector in particular is revealed. Digitalization has significantly affected the marketing activities of banks, the ways of organization and conducting business, changed the structure of banks' expenses, reducing management, organizational expenses, costs associated with storing and transporting cash, the costs of receiving, processing and storing information, and provoked an increase in the scale of operations. Many banks, using their established customer bases and having professionals in the field of information technology, try to create customized products and services for both existing and potential customers, create and develop ecosystems.

Building an ecosystem and developing the necessary technologies to implement such solutions is not an easy task for many banks, as it requires highly qualified personnel, strong technical support and huge investments in technology. So, Sberbank has spent $1 billion on buying assets for its ecosystem for the last three years

* Corresponding author: 23bal@mail.ru

 https://doi.org/10.1051/shsconf/202111200021

(about 3 % of its profit during this time). For many banks, with a reduced financial result, such investments are simply impossible. But there are some options here too. If the Bank cannot afford to create such technologies on its own, it can join the existing ecosystem. The model options may be different. Many small regional banks have followed the latter way. Joining groups of larger credit organizations, they occupy their own niche there and develop due to the help of the parent company.

On the one hand, this increases concentration in the banking sector, which can lead to less competition, and, consequently, is the reason for a decrease in the quality of services. However, according to research conducted by "Ernst & Young", an international leader in the field of audit, taxation, transaction support and consulting (hereinafter – E&Y), the Russian banks occupy leading positions in the implementation of new technological solutions. In the field of digitalization of their business processes, Russian banks are not inferior, and in some areas they are ahead of their American and European counterparts. Domestic credit organizations benefit from flexibility and ease of use, including online banking systems, contactless payment services, and so on.

Banks with state participation are the driving force behind such changes. The infusion of public investment in this sector has given impetus to the development of technological financial services. It worth noting that for a long time our banking system has been developing based on the experience of Western banks. But in the terms of sanctions, our authorities paid more attention to the experience of forming the Chinese banking system. In China, the share of state-owned banks (banks financed by the government) is very high. When forming its banking system 30 years ago, China did not abandon state-owned banks, but gave them the authority to develop and finance technological projects. The result is obvious today. China is the largest economy and one of the most high-tech powers. It was his experience that provoked the creation of the FINTECH Corporation by the Bank of Russia and the infusion of a huge amount of funds to support technological projects in the financial sector. Nowadays, the largest banks (mainly with state and foreign participation) show significant success in the field of technological financial services.

The research conducted by the E&Y company in 27 global markets confirms that Russia is one of the world leaders in the penetration of FINTECH services (figure 1).

The level of its penetration is 82 %, which is significantly higher than the global total, only China and India (having 87 %) passed ahead. Compared to 2017, the index has grown almost doubled.

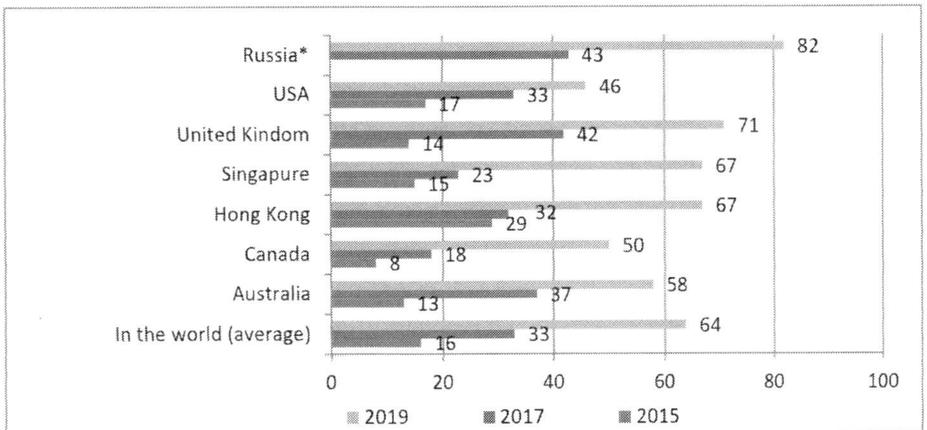

Figure 1. The comparison across countries of the level of financial technology penetration in 7 markets in the period of 2015–2019, % [17]

* In Russia in 2015, the level of penetration was not determined

2 Problem Statement

The introduction of digital technologies saves transaction costs, creating new potential, and at the same time new requests and requirements to the market, accelerates business. As a result, the life span of not only the product, but also the company is shortened. So, in the Standard & Poor 500 rating, the life span of large corporations has been reduced from 60 years to 18. And according to Forbes analysts, over the past fifty years, only in the United States of America, the life cycle of large companies on average has decreased five times – from 75 to 15 years x4ъ.

Yesterday's competitors of banks can become partners. This is what happened to companies that provide e-money transfer services, which eventually became subsidiaries of banks, and the banks themselves began to transfer funds without opening an account.

It should be noted that a new high-quality product, a more technological process provide only a temporary competitive advantage. In order to remain competitive and maintain their position in the market (or just stay in the market), banks have to be able to apply a new more effective model, change it and refine it in time. All this forces banks to constantly change and rebuild and it means to be innovative.

Such changes require large financial investments that only large banks can make. In this regard, the Bank of Russia should create equal opportunities for all financial market participants to switch to the digital format and ensure legal regulation of new banking business formats.

3 Research Questions

Changes in the operating conditions of the financial market entail changes in the conditions of competition in it and the rules of its regulation.

1. Today for the banking sector the displacement factors of competitiveness take place in the direction and initiative of its employees and prospective strategic planning. The Bank is more interested in how to serve its customers at a higher qualified level, offering a more technologically advanced service that allows you to quickly satisfy as many needs of its customers as possible.

2. To make a profit from the use of digital technologies, banks have to radically change their business models and be very mobile. A new business model based on the network effect can be a platform that involves its joint use by all partners. Today, the largest Russian banks are beginning to widely use open interfaces, create and develop ecosystems. This business model makes the market more mobile and transparent. But its functioning can become the basis for price discrimination.

3. In maintaining and developing a competitive environment in the financial market, the role of the regulator that sets the rules and monitors their implementation is important. Moreover, we need financial support from the state in implementing large-scale projects that will allow us to modernize the entire financial services sector.

4 Purpose of the Study

The purpose of this study is to justify the increasing role of the state in ensuring the digital transformation of the banking sector. The Bank of Russia's efforts should be aimed at maintaining a competitive environment in the financial market, providing financial support for the introduction of new technologies, and transforming the banking business. Special attention should be paid to the development and implementation of new rules of conduct for financial market participants, and also monitoring their compliance.

5 Research Methods

In the course of the research, such methods as analysis and synthesis, the grouping method and the graphical method were used. The analysis and synthesis were used in the consideration of the concept of digital capital, in generalizing the practice of forming banking ecosystems, activities carried out by the Bank of Russia to maintain a competitive environment in the financial market, the introduction of new technologies and the transformation of banking business. The grouping method was used to identify the main ecosystem models. The graphical method was used to visually compare countries by the level of financial technology penetration.

6 Findings

The penetration of digital technologies into all spheres of human activity requires banks to develop new products and services, new methods and models of interaction with customers. The source of income for the Bank and its competitive advantage is knowledge of the problems and needs of customers, the technological ability to quickly and efficiently create and provide services. Analysts of the international consulting company "McKinsey & Company identified the resources needed to develop new products and services in the digital economy as "digital capital" [2]. According to some authors, this form of capital includes both traditional assets (servers, routers, web portals and basic software that allows working on the Internet) as well as intangible assets (unique developments that contribute to the development of digital technologies, opportunities provided by technologies such as Big Data that contribute to the growth and expansion of business, etc.) [8] defines digital capital as "a set of internalized abilities and abilities" (digital competencies), as well as "external resources" (digital technologies) that can be historically accumulated and transferred from one field of activity to another. The contribution of digital capital to the creation of a new product is more significant than labor [11].

Another important factor in the competitiveness of banks is the creativity of employees. In the context of digital transformation and the digital economy, it is no longer enough to improve human capital to generate super-profits. An important factor becomes the formation of creative capital [12], the possession of which brings a stream of superprofits in the implementation of creative ideas.

To make a profit from the use of digital technologies, banks have to radically change their business models and be very mobile. A new business model based on the network effect is a platform. Multidimensional integration based on a digital platform is a complex process of forming network connections between market entities and allows rational using the potential of interacting entities [13]. Banks are beginning to make extensive use of open interfaces, create and develop ecosystems. A study conducted by the consulting company "Accenture" on 120 major banks confirms this trend: 88 % of the studied banks are aimed at creating partner ecosystems.

Models for building ecosystems can be different. For example, associations of banks within open banking platforms. Banks include their products and services in combined catalogs together with other banks via API (application programming interfaces). An example of such a model can be the transfer system by phone number of Sberbank (number 900), which in 2018 was joined by 4 more banks – "Tinkoff Bank", "Sovcombank", "Proinvestbank" and "SDM-Bank", and also 4 companies – "Svyaznoy-Euroset", "MTS", "WebMoney" and "Moby.Money". This competitive advantage of Sberbank has become the basis for price discrimination, which consists in limiting the ability of customers to transfer funds within the Bank. The amounts transferred in excess of the established limits will be charged a 1.5 % commission.

In order to minimize transaction barriers, the Bank of Russia has launched a fast payment system service that allows individuals to transfer money instantly by mobile phone number to themselves or to others in another bank. The absence of the need to enter the recipient's card details and the speed of transfer ensure high of transfers' availability. The Bank of Russia's setting of tariffs for money transfers using this system ensures their low cost. This made it possible for banks that have enabled this service to compete in the field of transfers of funds from individuals with Sberbank.

Another version of the Bank's ecosystem that has become widespread in Russian practice is the creation of a referral platform. This is the principle used by credit organizations within a financial group. All major banks such as "Sberbank", "VTB Bank", "Raiffeisenbank", "Alfa-Bank", etc. are the main credit institutions in the consolidated group. When the client contacts the Bank, he is redirected to other providers of services of his interest that are members of this group. For instance, an application for issuing a car loan received by "Sberbank" will be satisfied with the participation of a subsidiary bank "Setelem Bank" that specializes in car loans. Leasing services for clients within the consolidated groups are provided for-by: Sberbank – "Sberbank-leasing", "VTB Bank" – "VTB-leasing", "Alfa-Bank" – "Alfa-Leasing", etc.

In addition to financial services (banking, insurance, investment), banks are beginning to introduce non-financial services: food delivery, apartment repairs, ticket sales for entertainment events, communication services, etc. Here, the same Sberbank can serve as a striking example. Its ecosystem includes such partners as "Domclick", "Yandex Money", "SberMobile", "Beru!", "Okko", "Citymobil", "Delivery Club", "DocDoc".

Portal "Domclick" provides services on search and purchase of apartments in the mortgage. The Service "Yandex.Money" provides electronic transfer services. The virtual operator "SberMobile" provides mobile communication services, thanks to the online marketplace "Beru!" "Sberbank" customers can order goods at a discount, online cinema services are provided by the service "Okko", and taxi services are provided by "Citymobil". With the help of the "Delivery Club" service, you can order food, and also medical services can be ordered through the "DocDoc" service.

Another giant of the banking sector, "VTB Bank", started creating its own ecosystem recently – in 2019. But there are already some developments in the real estate sector. In addition to mortgage programs, the VTB platform will provide services for selecting real estate, financing its purchase or construction, registering real estate, and repairing it. Following the example of Sberbank, "VTB Bank" launched its own Telecom operator VTB mobile.

"ROSBANK" uses its subsidiary bank "DeltaCredit" (after being renamed "Rosbankdom") as a platform for developing services around mortgages: it provides services for registration in Roseestr, real estate insurance, interior design, sale and delivery of furniture, and cleaning of premises.

"Tinkoff Bank", the most ambitious bank in the digital format, has been developing its ecosystem since 2013. Provides insurance services through the "Tinkoff Insurance" service, investment services through "Tinkoff Investments", clients of the small and medium-sized business segment can perform business registration, settlement and cash services, acquiring, and receive loans using the "Tinkoff Business" and "Tinkoff Payment" services. The Bank also has a virtual Telecom operator "Tinkoff mobile". The "Tinkoff Bank" ecosystem includes partners such as "Rambler.Checkout" and "Afisha-Restaurants".

Another option is to synchronize the banking infrastructure with key stages of the client's life. For example, in the case of an individual entrepreneur and a legal entity: financing the implementation of an idea (startups), supporting and supporting the business, expanding it, entering new markets, and so on. The system tracks the occurrence of such stages and provides information about this to partners within the framework of cooperation agreements. A striking example of such successful work can be the ecosystem built for small and medium-sized businesses by "Tinkoff Bank".

The use of such models, which take into account various customer requests, makes the market more mobile and more transparent. But the functioning of such a model is associated with a number of problems that are common to all open systems.

Along with competitive advantages, this can also be a serious reason for price discrimination. An example of a money switching barrier for consumers is the current practice of banks charging different amounts of commission for internal and external money transfers. According to a study conducted by the FAS of Russia together with the Bank of Russia, there is usually no commission for transfers within banks, but the average commission for external transfers is 1.5 %, and in some cases it can reach 3–4 % of the transfer amount. The current pricing practice restricts the consumer's ability to freely move their funds to a financial service provider that provides the best offer, and, thus, negatively affects competition in the financial market.

In maintaining and developing a competitive environment in the financial market, the role of the regulator that sets the rules and monitors their implementation is important. That is why the Bank of Russia, when developing its financial market development strategy, has identified competition as one of the four main areas of financial market development.

To ensure the dynamic development of financial services, increase competition, and protect the rights and interests of all financial market participants, including consumers, financial organizations, and FINTECH companies, the Bank of Russia opened a separate activity development of open API, in the financial technology development Association [17], established jointly with major Russian financial organizations in December 2016 [9]. This direction is engaged in studying and testing the possibilities of open banking in the financial market. This is about ensuring access for all market participants to open information from credit institutions: the location and opening hours of ATMs, branches, POS terminals, etc. It is also possible for third parties to access confidential information about the Bank's clients (their accounts) in order to provide new financial products and services.

The main issue is the development of approaches to regulation in this area. The development of open banking raises the regulator's concerns related to conflicts of interest, uneven distribution of influence and increased financial isolation. The convenience, speed, and simplicity of transactions can be worth losing control of your data.

That is why the Bank of Russia initiated a regulatory sandbox (a mechanism for piloting and modeling the processes of new financial services and technologies in an isolated environment that require changes in legal regulation) as part of the implementation of the main directions for the development of financial technologies for the period 2017–2020.

Moreover, in August 2019, the Bank of Russia coordinated a pilot project in the field of API for the integration of banking services within the Eurasian economic Union. 20 credit organizations were invited to this event.

The success of many major banks, which determine the vector of transformation of the financial sector, is impossible without the financial support of the state. The Bank of Russia is taking the necessary steps to do this. In addition to the FINTECH Association, which works with major banks to develop and test the latest technologies in the banking sector, there is also the State Corporation of development "VEB.RF". This Institute operates on the basis of the Federal law "About the State Corporation of development "VEB.Russia" and in partnership with Russian commercial banks are engaged in financing projects aimed at the development of infrastructure, innovation and high technologies.

In order to ensure that all individuals can access banking services remotely, the Bank of Russia has initiated a remote identification mechanism that allows individuals to receive financial services remotely by verifying their identity using biometric personal data at any Bank. After registering in the Unified identification and authentication system (USIA) and the Unified biometric system, a citizen can remotely contact a new Bank to receive financial services from any place at any time.

This procedure makes it possible to increase the availability of financial services and reduce the costs of citizens moving to other financial organizations.

In addition, the "Digital profile of a citizen" – infrastructure created with the participation of the Bank of Russia will allow for convenient, secure and fast data exchange between citizens, the state and business in a "one-stop shop" format. The creation of this platform will facilitate competitive access to data, as well as the transfer of services to digital format.

An important project implemented by the Bank of Russia in this context is the "Marketplace" project. It provides an opportunity for individuals to distantly receive a wide range of financial services from the largest possible range of financial organizations, as well as to record information about concluded transactions in a single financial transaction Registrar. The creation of this type of infrastructure will expand the opportunities for financial organizations that do not have a wide branch network and large-scale infrastructure for remote services to effectively compete with market leaders. At the same time, the marketplace will significantly reduce switching barriers for consumers.

7 Conclusion

As a result of the research, new factors that ensure the competitiveness of banks in the financial market were identified. The main one is "digital capital" – resources (tangible and intangible assets) that are necessary for the development and promotion of new products and services in the digital economy. Human capital becomes an advantage if Bank employees are endowed with initiative and creativity, and are guided by current trends, both in the field of financial services and in the field of new digital technologies. New created or purchased products or technological processes implemented should be able to adapt and modify so that the Bank can change its business model, adapt it to changing environmental conditions and new customer needs, without compromising the quality of existing services. Developing the Bank's ecosystem may be an option. Models for building ecosystems can be different: from combining banks, banks and financial and non-financial structures, to synchronizing your own business with your clients business.

To ensure the availability of digital capital and the development of ecosystems, banks need large financial investments in highly qualified personnel and powerful technical support in information technologies. Even large banks cannot afford such infusions. State support in this regard is very important. To ensure the dynamic development of the financial market, its transition to a digital format, there is State Corporation "VEB.RF", FINTECH Associations that actively participate in the development and testing of the latest technologies in the banking sector.

Another important direction for the development of the financial market in modern conditions is the development of approaches to regulation in this area. The widespread of open banking exacerbates the problems of information security and financial isolation. In this direction, the Bank of Russia is implementing a pilot project – "a regulatory sandbox". To ensure remote access of individuals to financial services, the Bank of Russia initiated the creation of a mechanism for remote identification, a "Digital profile of a citizen", and launched the "Marketplace" project. The development of this kind of infrastructure will provide equal opportunities for competition between financial market players of various scales of activity.

Acknowledgments

The reported study was funded by RFBR, project number No. 19-010-00100-A.

References

1. E.V. Ageeva, M.A. Afanasyeva et al. Digitalization of the financial and credit sphere in modern Russia (Direct-Media, Moscow; Berlin, 2019)
2. J. Bughin, J. Manyika. Measuring the full impact of digital capital (2018). Retrieved from: https://www.mckinsey.com/industries/high-tech/our-insights/measuring-thefull-impact-of-digital-capital#0
3. O. Ermolenko. Development of the banking sector on the basis of modern technologies of service. Sci. bull. of the South. Inst. of Manag., **2**, 26–30 (2018)
4. O. Konovalov. Organizational anatomy: five reasons why companies die (2017). Retrieved from: https://www.forbes.ru/biznes/340993-gazprom-na-neve-dve-treti-administracii-monopolii-uzhe-v-peterburge
5. V. Kosarev, G. Iarajuli. Ecosystem as a new model of the Bank's development. Finan. Mark. and banks, **1**, 58–62 (2020)
6. S. Matveyevskoye. FINTECH and development banks: impact on economic growth (for example, VEB of Russia). Finan. Mark. and banks, **1**, 105–109 (2020)
7. E. Okomina, M. Matveeva. The trends of «digitalization» of the banking sector. Syst. Anal. & Mathem. Model., **1(2)**, 47–53 (2020)
8. M. Ragnedda. Conceptualizing Digital Capital. Telemat. and Inform., **35**, 2366–2375 (2018)
9. Report for public consultation (2017). Development of open API (OpenAPI) on the financial market. Retrieved from: http://www.cbr.ru/content/document/file/50679/consultation_paper_171229.pdf
10. A. Rukavishnikova. Role and Importance of Artificial Intelligence in Russian Banking Industry. Glob. & reg. res., **1(1)**, 58–61 (2019)
11. M. Safronchuk. Influence of information economy and globalization on factors of economic growth. Bull. of the Acad., **4**, 31–34 (2014)
12. M. Safronchuk. Influence of digital transformation on business and the business environment. Econ. and manag.: probl., solut., **2(3)**, 38–44 (2018)
13. V. Samarukha, T. Krasnova, A. Dulesov. Integrating Digital Production Systems. Bull. of Baikal State Univ., **2(30)**, 309–317 (2020)
14. P. Vigna, M.J. Casey. The Age of Cryptocurrency: How Bitcoin and the Blockchain Are Challenging the Global Economic Order (Picador, USA, 2016)
15. G. Westerman, D. Bonnet, A. McAfee. Leading Digital: Turning Technology into Business Transformation. (Harvard Busin. Rev. Press, 2014)
16. M. Zhigas, S. Kuzmina. Blockchain and Decentralized Money System: Principles of Building and Ways of Its Development. Bull. of Baikal State Univ., **1(3)**, 79–88 (2020)
17. FINTECH services penetration index (2019). Retrieved from: https://www.ey.com/Publication/vwLUAssets/ey-fai-2019-rus/$FILE/ey-fai-2019-rus.pdf

The indigenous methodology as a method for studying social processes in the arctic

Ulyana Vinokurova [1,2,3,*], *Akulina* Mestnikova[1], and *Galina* Alekseeva[1]

[1]Churapcha State Institute of Physical Education and Sports, 2a, st. Sportivnaya, Churapcha, 678671, Russian Federation,
[2]Arctic State Institute of Culture and Arts, 4, st. Ordzhonikidze, Yakutsk, 677027, Russian Federation,
[3]North-Eastern Federal University named by M.K. Ammosov, 58, st. Belinsky, Yakutsk, 677000, Russian Federation

Abstract. In 2020, under the guidance of Doctor of Sociological Sciences Ulyana Vinokurova, the textbook "Social Processes in the Arctic" was published. It presents results of the studies on the sociology of the Arctic as a circumpolar macro-region conducted by sociologists of Yakutia. The textbook is the beginning of the series "Sociology in the Arctic" and the scientific and educational project of Arctic research, which forms the basis of Arctic education. This article presents the summary 4 modules containing the results of scientific research introduced into the educational disciplines of the social and humanitarian cycle. The first module presents the theoretical part; the second module is "The Arctic as a social phenomenon"; module 3 is "Social dynamics in the Arctic"; module 4 is "Social differentiation". Special attention is paid to the social processes in the Arctic investigated with the indigenous methodology, factors of stability and risks of geo-eco-socio-systems of the indigenous peoples of the Arctic. This manual creates an opportunity for an independent search for information, trends in socio-humanitarian knowledge in Arctic studies.

1 Introduction

The study of specific social processes in the Arctic through the Indigenous Methodology, identification of resilience and risk factors for geo-eco-socio-systems of indigenous peoples of the Arctic are aimed at enriching science through the interpretations of prospects and interests of indigenous peoples made by indigenous scholars in order to provide the best understanding of indigenous peoples as subjects of history. There is an indigenization of scientific knowledge both in the form of academic research and results of civil science.

Arctic education is an innovative model of continuous general and professional practice-oriented education aimed at training specialists with an Arctic identity, possessing complex competencies to improve the quality of life in the Arctic.

2 Materials and methods

The northern expanses became the Arctic after the collapse of the Soviet Union. Russia has lost its southern underside and has become a more northerly cold country than the Soviet Union. The expanses of the Arctic Ocean opened up thanks to the collapse of the "Iron Curtain". In Soviet times, we were not interested in other Arctic Ocean shores. Moreover, many Russians did not even know that it has an international name – the Arctic Ocean.

It was only in the 1990s that Russia began to recognize itself as a maritime power due to the growing importance of the Arctic Ocean, whose shores were perceived as alien at the beginning of perestroika. At that time the reformers mercilessly dealt with the naval, military, fishing, transport bases on Russian waters and the coastal land of the ocean. The population of the Far North was declared surplus. The collapse of the unified national economic complex in industry led to the liquidation of workers ' settlements and industrial sites, and the indigenous population was left without state protectionism.

Currently, about 2.4 million people live in the Arctic and subarctic zones of Russia. Arctic residents ask themselves "Do we need us?", "Do the Arctic need us?" Also, the question remains relevant: "To settle in the Arctic or to master it by shift methods?" This question shows that the social problems of the indigenous population of the Arctic, who inhabited the Arctic territories tens of thousands of years ago, are still not taken into account.

Indigenous peoples in a situation of scientific and educational inequality, lacking scientific competence may lose the ability to act as an independent subject of social progress in modern human history. Adherents of the Indigenous Methodology strive to develop strategies for social transformation for the benefit of these communities, to involve people who accept ideas and values of indigenous peoples and do not have the opportunity to enjoy the right of self-government and self-determination, to influence social and educational policies aimed to protect intellectual property and rights established by the UN Declaration of the Rights of the

* Corresponding author: uottaah1707@gmail.com

World's Indigenous Peoples (2007). New paradigms are required to find answers to modern challenges:

• the Earth's climate, ecological transformation: a response to global warming;

• the Limitrophe: a response to civilizational shifts;

• openness: a response to the growing multi-ethnicity of the social environment and active migration, global information society;

• body's reserves: response to changes in the pace, rhythm and content of adaptation to the new;

• the mentality paradigm: a response to the change in the materialistic worldview, the sixth technological order;

• destiny for indigenous peoples: a response to the influence of the overwhelming majority of communities, among which it is necessary to preserve own identity and integrity;

• power: a response to changing political systems.

The Indigenous Methodology expresses a general direction of development of this type of scientific rationality, which recognizes the subjectivity of the scientist, his social experience, social status, responsibility for the fate of the native people under study. The scientific systematization of knowledge does not go from the general to the particular. It goes from His own to the Other as the most important principle of selection and building of knowledge about the world. In this focus, the path of scientific knowledge is a process of understanding culture, entering it, discovering and protecting it.

Methodological peculiarity in the epistemology of indigenous scientists: Indigenous Knowledge for indigenous scientists does not have the prefix "traditional" knowledge, since it is presented in the temporal continuity of the past-present-future. Attention is paid to the education of the younger generation, the Indigenous Methodology is used as an educational paradigm [1].

3 Results

In 2020, under the guidance of Doctor of Sociological Sciences Ulyana Vinokurova, the textbook "Social Processes in the Arctic" was published. It presents results of the studies on the sociology of the Arctic as a circumpolar macro-region conducted by sociologists of Yakutia. This manual creates an opportunity for an independent search for information, trends in socio-humanitarian knowledge in Arctic studies. The textbook is the beginning of the series "Sociology in the Arctic" and the scientific and educational project of Arctic research, which forms the basis of Arctic education. The first issue consists of four modules containing the results of scientific research introduced into the educational disciplines of the social and humanitarian cycle.

4 Discussion

The first module «Theoretical frameworks» presents the theoretical part and consists of sections devoted to social institutions as regulators of the social structure (Vasiliy E. Okhlopkov, Doctor of Sociological Sciences); describes different approaches and identifies possibilities of using the Indigenous Methodology in the study of modern society (Uliana A. Vinokurova, Doctor of Sociological Sciences); describes methodological aspects of sociological research of historical memory and cultural trauma (Akulina E. Mestnikova, Candidate of Sociological Sciences) and social policies in the Arctic (on the example of activities of the First President of the Republic of Sakha (Yakutia) M.E. Nikolaev) (Natalia N. Kolodeznikova, Main Advisor, Permanent Office of the Republic of Sakha (Yakutia) to the President of the Russian Federation).

In sociology, there are various discursive practices, approaches to new cultural history, research of collective trauma, which make it possible to see the complexity and ambiguity of experience, to turn the perspective from the total to the particular, returning history to the human dimension in specific spatial-temporal and national-cultural coordinates.

Within the social theories, the past is viewed through the prism of social interactions, as a phenomenon generated by social reality and influencing it; the philosophical perspective implies the interpretation of the past as an abstract category, an intrinsically valuable phenomenon, studied outside the historical and cultural context; in semiotic studies, the past is perceived as a renewable "archive" of texts that invade the space of the present and form its value-normative appearance; from the standpoint of historical knowledge, the past is interpreted as an object and a result of scientific reconstruction.

In the 2008 study aimed to identify features of life strategies of the second or third generation of forced migrants to the North, the methodological foundation was the Indigenous Methodology; the theory of the "social framework of memory" by M. Halbwachs [2,3]; the theory of cultural trauma by P. Shtompka [4] and the theory of cultural memory by J. and A. Assman [5,6,7].

The research used the methods of questionnaires, in-depth interviews, media content analysis, historical and comparative analyzes.

One of the most important questions was "What helped the Churapchins to withstand the years of resettlement?". According to the settlers' descendants, these were strength of mind, vitality – 29,4%, faith that they will return home - 26%, diligence, inexhaustible optimism - 14.5%, mutual assistance - 13,1%, good health - 6.6%, strong family relations - 6.3%, traditional worldview - 2%.

Thus, collective memory remains one of the main sources of historical information for representatives of the Yakut ethnic group, and more modern methods for collecting information remain unclaimed. More than half of the respondents do not use scientific sources when receiving new information about the forced resettlement, which reduces the level of reliability of knowledge about the tragedy outside the family circle and restrains the process of the trauma of grief and loss in the narratives of the collective memory.

The second module "The Arctic as a social phenomenon" examines various approaches to the Arctic identity, its features and processes of formation; the results of the 2016-2019 research are presented. "Arctic Identity: Global, National and Local Processes of Construction and Transformation" describes Arctic identity as a continuous, socially structured process based on a sense of the Arctic homeland for both indigenous peoples and settlers (Daria V. Burnasheva, PhD (Cultural Studies)). It presents the concept of "Arctic" in terms of values of students of the North-Eastern Federal, Murmansk Arctic, and Northern (Arctic) Federal Universities (Daria N. Filippova, post-graduate student); concept, principles and functions of local self-government in the Arctic countries are described (Mikhail A. Pogodaev, Candidate of Economic Sciences, Oksana D. Romanova, Candidate of Sociological Sciences, Anna V. Mikhaylova, Candidate of Economic Sciences, Klavdiya D. Barashkova, Senior Lecturer).

Module 3 "Social dynamics in the Arctic" describes results of the studies on the development of volunteerism in the Republic of Sakha (Yakutia) (Galina G. Alekseeva, Candidate of Sociological Sciences); origins, development factors and a resource potential of Arctic tourism (Evdokiya P. Vinokurova, Candidate of Culturology); socio-cultural aspects of the Arctic sports games are described. The role of the Arctic in the formation of a coach's personality is identified (Vasiliy N. Alekseev, Associate Professor). Destructive social processes in the Arctic are presented Yuriy I. Zhegusov, Candidate of Sociological Sciences.

Sport as a socio-cultural institution has a huge potential; it can influence social problems, preserve and develop society, shape public opinion, create a positive image of the region, and contribute to the socialization of the younger generation. There are studies on the international sports games "Children of Asia" as a socio-cultural project that contribute to a healthy lifestyle in the republic; sports facilities performing socio-cultural functions; promotion of ethno and geo-cultural ideas and symbols of Yakutia; ways to strengthen the attractiveness of Yakutia in the Asian world [8].

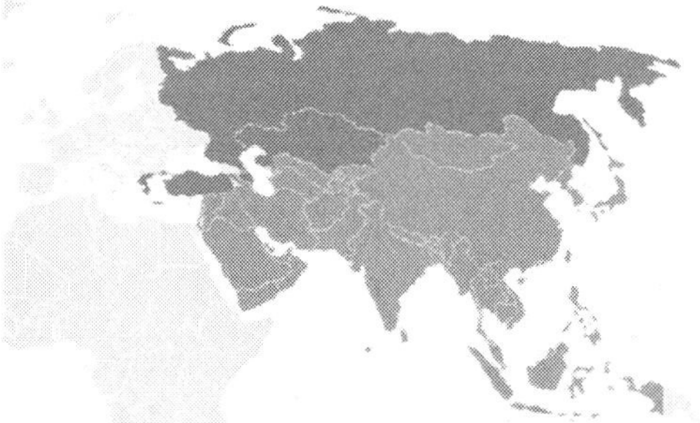

Fig. 1. Map of the geography of the «Children of Asia» (1996-2016)

The experience of holding international competitions of the countries of the circumpolar region — the Arctic Winter Games-influences the formation of the ethnic, regional and transnational identity of the peoples of the Arctic. Today, the traditional sports of the peoples of the Arctic attract the attention of the world sports community not only by their originality, but also by special techniques of exercises and training, honed over many centuries, which are of considerable interest to those who are engaged in professional sports. Sports and mass events are designed to introduce the Arctic peoples to each other's cultural traditions. Such a unique event, which harmoniously combines the sports and cultural components, is the Arctic Winter Games. The Arctic Winter Games are the largest sports and cultural event for the youth of the Arctic [9].

National sports in Yakutia were born in the process of life activity of the population of the northern territory in extreme climatic conditions. The formation of distinctive sports was greatly influenced by the natural, geographical, industrial and ethnic characteristics of the peoples inhabiting the vast territory of Yakutia.

The national exercises reflect the struggle for existence in the harsh, extreme conditions of the Far North. Traditions of physical education and spiritual culture should be considered as a harmoniously merged unity and as one of the most important and effective means of physical education of the modern younger generation [10].

High-performance sports go beyond the value consciousness of HomoLudens. The highest achievements of athletes have deep historical and personal prerequisites formed in the fates of their coaches. The life path and spiritual values of the coach are a kind of catalyst for the sports achievements of his students, who learn not only sports lessons, but also lessons of citizenship, spiritual and moral values of the coach. Identification of the role of the Arctic in the formation of the coach's personality (on the example of D. P. Korkin). The connection of the life path of the outstanding freestyle wrestling coach Dmitry Petrovich Korkin with the peculiarities of his coaching work and the achievements of his students [11].

Volunteerism as an object of sociological research is becoming increasingly important in connection with the current demands of the management of social practices of volunteerism. Such activity is connected, on the one hand, with positive events – the need to recruit volunteers for major sports competitions. On the other hand, these are social emergencies, which require the help of not only specialists, but also a large number of volunteers. In the third module, volunteerism is considered as a culture of mutual assistance and reveals the typology of approaches to the study of this phenomenon, regulatory and legal regulation, and the stages of the development of volunteerism in Yakutia.

Volunteering is the most active component of civil society, a practical manifestation of civic consciousness and civic responsibility for solving problems that are not subject to individual efforts. At present, the essential characteristic of its development is mobilization and self-organization of the population. In Russian sociology, volunteering is studied in terms of its economic efficiency and management aspects, which is associated with the need to assess the social usefulness of the phenomenon and the need to organize spontaneous activities of citizens [12,13].

Peculiarities of the volunteering in Yakutia must be considered within the socio-cultural, constructivist, and ecological approaches. The constructivist approach is used to study the mechanisms of constructing social problems as the basis for creating new areas for volunteer associations. The sociocultural approach makes it possible to study the role of volunteering as an institution of civil society, its specificity determined by ethnocultural traditions, peculiarities of motivation of volunteers in the context of the value-normative structure of society. The ecological approach can explain peculiarities of the volunteering in the Republic of Sakha (Yakutia) determined by harsh climatic conditions, a low population density and a vast territory [14].

The specificity of the regional aspect of volunteering is due to the peculiarities of worldviews of the inhabitants of the North, sustainable life of peoples living in permafrost and extreme climatic conditions. Volunteering in the regional aspect is seen as a method for preserving and strengthening spiritual and moral values, the continuity of culture, folk traditions.

Module 4 "Social differentiation" deals with the problems of northern cattle breeding (Elvira A. Mikhaylova, post-graduate student); "northern delivery" as an urgent social and professional problem in the Arctic (Maria E. Osipova, post-graduate student), social aspects of the mammoth fauna in the Arctic (Avgusta P. Dyakonova, post-graduate student).

Thus, the textbook reveals various approaches using the Indigenous Methodology in the study of modern society, identifies possibilities of the Indigenous Methodology in the study of modern processes in the Arctic, presents an analysis of the practice of using the Indigenous Methodology in sociological research conducted by sociologists of Yakutia over the past three years.

Using this textbook, students will be able to critically analyze modern social processes, social institutions and structures in the Arctic territories. They will know: social problems in the Arctic as a macro-region and in its circumpolar sectors;

be able to: identify causes, trends, challenges, risks and consequences in the Arctic; identify actors in social processes; formulate research and practical tasks, including in interdisciplinary fields and in geoeducational processes;

possess: the ability to carry out theoretical and applied research of social processes in local and regional territories; develop recommendations for eliminating practical problems in protecting social interests of the peoples of the Arctic, participate in the development of programs for the socio-economic development of local communities and the region.

The structure of the textbook is unified for perceiving key points and using electronic educational platforms. It meets the requirements of the federal state educational standard.

5 Conclusion

The book is the beginning of the series "Sociology in the Arctic" and the scientific and educational project of Arctic research which forms the basis of Arctic education. Since the current academic year, the discipline "Social Processes in the Arctic" has been introduced into the curricula of Churapcha State Institute of Physical Culture and Sports, the Arctic State Institute of Culture and Arts and the North-Eastern Federal University named after M.K. Ammosov.

References

1. U. Vinokurova *Sibirica*, **17(3)**, 3-15, (2018).
2. M. Khalbwachs *Emergency reserve*, **2-3 (40-41)**, 8-27, (2005).
3. M. Khalbwachs, *Social framework of memory*. (Moscow: New Publishing House, 2007).
4. P. Shtompka, *Sociological research*. 1, 6-16, (2001).
5. M.Ossowska, *Studia Socjologiczne*, **2**, 47-51, (1963).
6. A. Assman *Long shadow of the past: memorial culture and historical politics*. (Moscow: New Literary Review, 2014).
7. J. Assman, *Cultural memory. Letter, memory of the past and political identity in the high cultures of antiquity* (Moscow: YASK, 2004).
8. U. A. Vinokurova, Yu. I. Zhegusov, A. E. Mestnikova, G. G. Alekseeva, V. N. Alekseev *Theory and practice of physical culture* **3**, 94-96, (2017).
9. D. V. Burnasheva, G. G. Alekseeva, L. P. Neustroeva, *International sports games "Children of Asia" - a factor in promoting the ideas of Olympism and training sports reserve: materials of the International Scientific Conference*, 7-8 July, Yakutsk: RIO media-holding, pp. 166-168. (2016).
10. E. G. Vinokurov, V. N. Alekseev, *Socio-philosophical problems of the development of physical culture and sports: monograph* (Moscow: Lap Lambert Academic Publishing, 2017).
11. I. I. Gotovtsev, Scientific and pedagogical activity of D. P. Korkin: textbook. (Yakutsk: Churapchinsky State Institute of Physical Culture and Sports; 2016).
12. M. V. Pevnaya *Management of volunteerism: international experience and local practices*. (Yekaterinburg, 2015).
13. A.V. Ermilova, I. A. Isakova Bulletin of the Perm National Polytechnic University. Socio-economic sciences. **4**, 48–57, (2017) https://doi.org/10.15593/2224-9354/2017.4.4
14. P. A. Kropotkin, *Mutual aid as a factor of evolution*. (Moscow. Ed.: Self-education, 2007).

Transportation and subsoil use: integration for the Arctic economic development in Russia

Anatoliy Sleptsov[1], *Violetta* Gassiy[2,*], *Jean-Pierre* Desideri[3] and *Afanasiy* Postnikov[2]

[1]North-Eastern Federal University, 58 Belinsky str., Yakutsk, 677000 Russia
[2]Kuban State University, 149 Stavropolskaya st., Krasnodar 350040 Russia
[3]University of Versailles Saint-Quentin-en-Yvelines (UVSQ), Bâtiment Buffon, 55 Avenue de Paris, 78000, Versailles, France

Abstract. The issues of the Russian Arctic development are considered in the paper. The authors study the trends and factors in the formation of the Arctic transportation. The subsoil use is researched as an industry that stimulates the active development of the Arctic at the present stage. The analysis of the interdependence of the two sectors of the economy is carried out as well as the reasons for their integration are substantiated. The authors consider government incentives for investment in the Arctic. The article discusses the reasons why it is still not possible to use the resource potential of the Arctic to its full capacity. According to the authors, the transport infrastructure does not meet the needs of investment. They suppose that an obstacle to the further inclusion of the Russian Arctic in an active economic turnover is the delay in the construction of the infrastructure to support the Northern Sea Route: ports, icebreakers and rescue vessels. A breakthrough in transportation is required for further successful implementation of investment projects for the extraction of hydrocarbons and other minerals. Therefore, the development of the Arctic should be considered only in the context of the integration of subsoil use and transportation. In conclusion, the authors outline the main tasks that the state needs to solve for the Arctic policy implementation.

1 Introduction

In last decade, the vector of state policy in Russia has headed towards the Arctic. The government's attention to this important region is explained by the need to ensure the national security and economic interests of the country. Climate change is transforming the usual attitude towards the Arctic, expanding the potential for the development of the Arctic territories, including logistics and mining. Melting ice and increasing the availability of natural resources on the continent stimulate technological progress, which leads to the development of innovative solutions in the field of subsoil use, storage, processing and transportation of raw materials and goods. An example of such rapid technological change is the production of liquefied natural gas (LNG). The technologies applied at the LNG fields, related not only directly to hydrocarbon production, but also to environmental protection, organization of communications, and labor safety, have actually led to an innovative shift in the mining industry. According to the Brookings Institution research, the LNG was named as a key priority in Russian energy sector in the future to gain the leadership in world oil market along with Saudi Arabia [1]. In the future, it is planned to expand the share of liquefied gas in the total volume of produced hydrocarbons. At the federal level, a decision was made to implement at least ten LNG projects in the future. By 2035, the government plans to increase LNG production at least threefold. According to Rosstat in 2020 the production of LNG in Russia has increased by 3.5%, to 30.5 million tons [2]. The environmental friendliness of natural gas can be considered as a factor in the development of subsoil use in this direction. Since the EU countries have adopted a strategy of transition to zero consumption of hydrocarbons in 20-30 years, natural gas can be considered as a "transitional" type of fuel. The development of LNG and the need to create new capacities for its transportation is a separate example of significant shifts in the subsoil use industry. Thus, the transformation of the industrial sector for the extraction of natural resources means improved transportation, including modes of transport and logistics. In the article, the authors investigate the relationship between the transport industry and subsoil use, identify the problems of their development in the Arctic, and offer recommendations for their solution.

2 Materials and Methods

The research methodology includes the use of a wide range of data collection and analysis. Among the main

* Corresponding author: vgassiy@mail.ru

methods used in the article, the authors applied the following:
- method of retrospective and legal analysis;
- data analysis and assessment;
- deduction method to identify causal relationships.

In addition, during the research the system, essential and institutional approaches were used. In this study, the Arctic is considered as a global ecological and economic space, which is part of the world economic system. The Arctic space is permeated with institutional subsystems, and their elements function under the influence of local political, economic, social and environmental factors. For this reason, it is important to study the Russian Arctic as a part and its contribution to the world geopolitical and economic space, as well as the role of this macro region in the country's economy and social sphere.

The institutional approach assumes that a legal environment has been formed in the Russian Arctic, which determines the norms and foundations of regulation and the functioning of its subjects. These entities carry out social ties, have associations and conduct economic activities in the Arctic. The main subjects of the Russian Arctic include the following:
- government (federal, regional, local);
- population, represented both directly and through the system of non-profit organizations;
- business of any kind (multinational companies, medium and small businesses).

It should be noted that the specificity of the Arctic region is the diversification of the population by the nature of activities, lifestyle. Moreover, residents of urban and rural areas of the Arctic have a big difference in living standards. Arctic cities are more concentrated on the production and extraction of any natural resource (Norilsk, Vorkuta), the functioning of the port (Murmansk). The economies of such cities have been transformed by monopolies. In fact, the development of the territory directly depends on the success and profitability of the city-forming enterprise. The rural area of the Arctic is characterized by the problems of poverty, unemployment, orientation towards traditional types of farming. These are indigenous peoples who play a significant role in the ethno-cultural life of the Arctic and form the historical heritage of northern culture. They conduct traditional nature management, are engaged in reindeer husbandry, fishing, picking berries, mushrooms, medicinal herbs and other plants. Therefore, the management of the Arctic macro regions in Russia is due to the complexity: the need to take into account national interests and external influences with the need for social transformations without damage to the environment.

The complexity of management as well as modern trends has required a revision and update of the basic documents for the implementation of state policy in the Arctic. The Russian Arctic strategy was adopted in 2020 [3]. It outlines the importance of the Arctic for the realization of the national interests of the Russian Federation. The Arctic is a treasure trove of minerals. According to experts, about 80% of natural gas reserves are concentrated in the Arctic territories. The uneven development of the Arctic regions, their remoteness from industrialized centers and the need to transport raw materials are among the main factors in the transformation of state policy in the Arctic. An important aspect is the expansion of the capabilities of the Northern Sea Route as an international corridor for the transportation of products between Europe and Asia. In the last decade, non-Arctic countries such as South Korea, Japan and China have become more active in the Arctic. They have achieved observer status in a number of important Arctic international organizations, primarily in the Arctic Council. The reason for this attention to the Arctic is easy to explain. For many years there has been a discussion about the options for trans-Arctic shipping: the Northeast Passage between the Atlantic and the Pacific north of Russia, the Northwest Passage through Canada's Arctic Archipelago, and the Central Route across the North Pole [4]. For the largest Asian carriers, the NSR is the most attractive today. Moreover, we believe that Arctic shipments from Asia to Europe are the most sensitive area in the implementation of the political interests and economic policies of Asian countries. Therefore, such attention is paid to the problems of the NSR development.

Undoubtedly, the NSR is a tool for expanding Russia's presence in the Arctic and internationally as an important operator of the world economy. The total volume of traffic by the Northern Sea Route is constantly increasing. In 2020 it amounted to almost 31 million tons. There was also an increase in the volume of cargo transportation in the Arctic by rail, Tab. 1:

Table 1. The costs of transportation, storage and delivery of goods by main-line freight rail transport, US$ [5]

	2016	2017	2018	2019
Russian Arctic	74,2	42,2	45,8	73,9
Russian Federation	2842	3304	4479	6209

Russia's ambitious plans for the development of the Arctic and its resources require the solution of an important task. The transport infrastructure does not meet the goals of subsoil use and investment projects' implementation. In most of the territory, there is no port infrastructure, the necessary sites for light aircraft, which transports personnel, small cargo between fields and settlements. Also, an obstacle to further inclusion of the Russian Arctic in an active economic turnover is the delay in the construction of the infrastructure to support the Northern Sea Route: ports, icebreakers and rescue vessels.

It should be noted that the transportation framework of the Russian Arctic can be divided into two parts: western and eastern [6]. The western part is characterized by a greater degree of the development cause of located industrial facilities, railways, cities with a population of more than 100 thousand people (Murmansk, Arkhangelsk, Norilsk). Historically, the western part of the Russian Arctic is more developed. The eastern part

includes hot spots for industrial development, is less socially developed. It has also extensive economy and a high dispersion of the population. For example, the Arctic Yakutia, being the largest region of Russia, has a population density of 0.04 people per km^2.
To speed up the development of the Arctic territories, the government proposed to introduce a special economic regime. According to this approach, investors receive benefits for the implementation of innovative projects in the field of subsoil use, construction of engineering and transport infrastructure. Particularly interesting for investors is the instrument for providing a 15-year preferential tax regime for the projects of hydrocarbons production on the Arctic shelf. In recent years, a significant increase in the investment in the Russian Arctic is associated precisely with the implementation of subsoil use projects on the ocean shelf.
All economic entities in the Arctic can obtain resident status. It makes it possible to use tax incentives during the implementation of the investment project. To obtain the status of the Arctic zone resident, the investor must meet several conditions. For example, the volume of capital investments in a project should not exceed 25%, and their minimum volume should not be less than 1 million rubles. Also, to support investors, the government offered financial instruments that were not previously used in the practice of developing the Arctic territories. We are talking about subsidizing coupon bonds issued for the modernization of production and the introduction of green technologies, as well as leasing of the equipment intended in the Arctic.
The data of the Russian Federal Statistics Service show the growth of innovations applied in the Arctic, fig. 1:

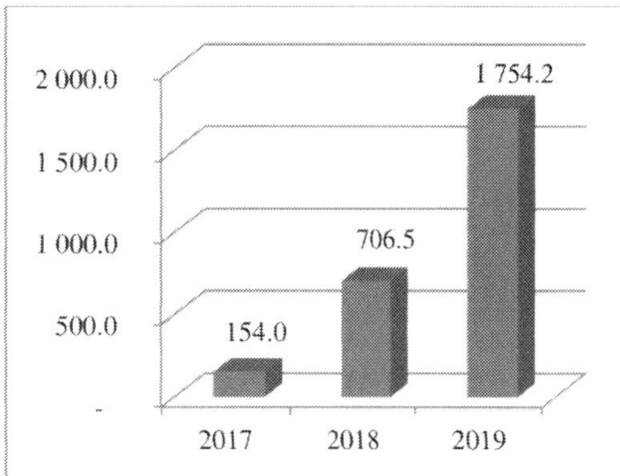

Fig.1. Volume of innovative goods and services in the Russian Arctic, $US [5]

In recent years, business costs for innovation have increased significantly in the Arctic zone. These include, first of all, the costs of the business for the ecological innovation of production. Each year, environmental requirements for subsoil use in the Arctic are tightened. Therefore, the state considers the stimulation of green activity as a basis for subsidizing investors.

3 Results

In 2021 a transport strategy for the Arctic is currently being developed in Russia. It should provide a number of measures to expand the network of railway, sea and airport hubs. To finance high-cost projects for the Arctic infrastructure development, this strategy will use public-private partnerships. For the Russian Arctic practice, projects implemented on the basis of PPP are a novelty, although in other Arctic countries there are successful examples of the implementation of infrastructure projects based on this mechanism. For example, in Canada over the past 20 years, 98 projects have been implemented on the basis of public-private partnerships. The cost of all investment projects is estimated at $ 61 billion. In 2017, the Iqaluit International Airport Improvement Project was implemented [7]. In 2013, the Government of Nunavut decided to modernize the airport in the capital Iqaluit. The land of Nunavut is characterized by the fact that the communities are located on the islands and on the mainland, and communication between them is maintained only by aviation. Therefore, the modernization of the airport was seen as an opportunity to improve the only way of transport links in the region. This project aims to improve the quality of life for Nunavut residents, businesses and tourists. Three actors took part in the project. Capital expenditures for the construction of the new airport are $ 288.5 million of which the government of Nunavut contributed $ 68.7 million, PPP Canada contributed $ 72.8 million, and the Arctic Infrastructure Partners contributed the shortfall from its own and borrowed sources. The experts estimate the net present value (NPC) of the 30-year project agreement at $ 418.9 million.
It is important to use in the Russian Arctic the experience of foreign countries in the implementation of projects for industrial development especially in transportation based on PPP [8]. The Arctic regions of Russia are experiencing a serious need for infrastructure projects, especially underdeveloped flight connections between remote settlements. Transport communication between them is irregular, and flights are carried out on the principle of the fullness of the aircraft cabin. At the end of 2020, the Ministry for the Development of the Far East and the Arctic decided to implement a new type of PPP for regional development - the Arctic concession. It will allow capitalizing tax revenues that the budgetary system will receive through the implementation of investment projects in the Arctic region.
The global trend towards more environmentally friendly fuels is reflected in the Arctic subsoil use. Oil production is gradually decreasing, while the implementation of investment projects for natural gas production is increasing, fig. 2:

https://doi.org/10.1051/shsconf/202111200023

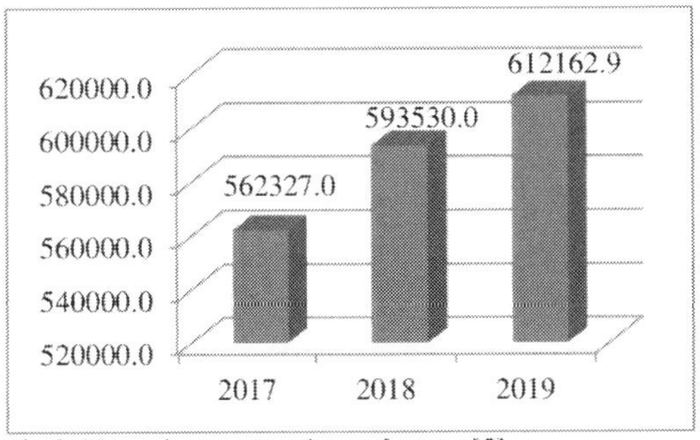

Fig.2. Natural gas extraction, mln tons [5]

Perhaps the decline in oil production in the Arctic zone is also associated with general global trends in the energy market. We are talking about the deals of the OPEC + countries to reduce the daily volumes of crude oil production. Since 2019, the price per barrel of oil has continued to fall actively. The 2020 pandemic has exacerbated this trend. The world economy entered a period of recession; the need for fuel has significantly decreased, as countries everywhere introduced restrictions on the movement of vehicles and citizens or even lockdowns. The tourism business has actually collapsed. The energy market is currently undergoing an ambiguous recovery phase, fig. 3.

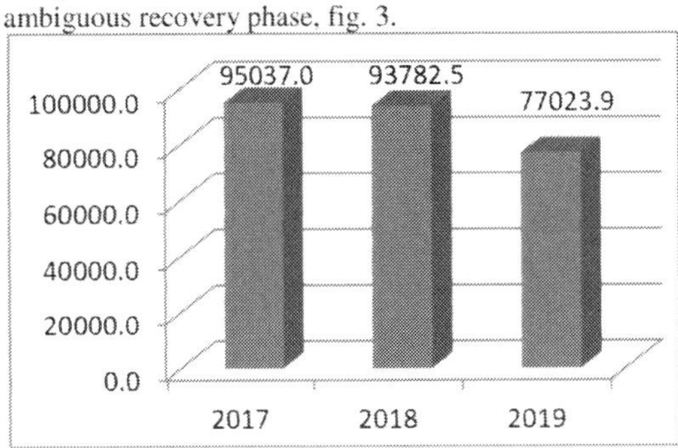

Fig.3. Oil extraction in the Russian Arctic, mln tons [5]

Investment Arctic projects are, first of all, the extraction of minerals, the volumes of which make this region a priority for economic development [9]. The subsoil of the Arctic territories of Russia contain reserves of hydrocarbons, non-ferrous and rare earth metals. According to official data, the Arctic has over 80 billion barrels of oil, one and a half trillion cubic meters of natural gas; about 800 billion tons of coal. These reserves are estimated at more than $ 30 trillion. The reserves of non-ferrous and rare-earth metals are represented by platinum, copper-nickel ore, iron, phosphorus, gold and diamonds, niobium, etc.

Comparison of data on the extracted minerals in the Arctic zone and in the country as a whole shows that for Russia, subsoil use in the Arctic is of great importance. For example, Table 2 shows the volume of extraction of the main types of minerals, according to which from 30% to 100% depending on the type of raw material is mined in the Russian Arctic. Moreover, there are deposits of rare earth metals, whose deposits were discovered relatively recently, but they are also located in the Arctic territories.

Table 2. Mining in Russia

Types of the raw materials	2019
Bituminous and brown coal, million tons	
Russian Arctic	11,3
Russian Federation	*439*
Crude oil, including gas condensate, million tons	
Russian Arctic	96.5
Russian Federation	*561*
Combustible natural gas (natural gas), billion m^3	
Russian Arctic	586
Russian Federation	*644*
Associated petroleum gas (combustible natural gas from oil fields), billion m^3	
Russian Arctic	27.9
Russian Federation	95,8
Apatite concentrate, million tons	
Russian Arctic	5,4
Russian Federation	*5,4*
Diamonds, million carats	
Russian Arctic	*38,5*
Russian Federation	*38,5*

Effective development of the Arctic opens up new opportunities for financing government programs and implementing the social goals of the state. There is an acute shortage of modern social and engineering infrastructure in Russia. Arctic subsoil use is actually a chance to increase budget expenditures by developing economic approaches to using the potential of the energy sector [9]. This approach presupposes intensive construction of infrastructure for the extraction and transportation of minerals. As the data analysis of the Arctic subsoil use has shown, currently the greatest interest for Russia is the hydrocarbons production, which is export oriented. However, taking into account the existing risks, it is necessary to diversify the transport infrastructure and partners.

A number of political decisions taken by the Russian government, as well as the concluded agreements on the construction of trunk pipelines, indicate that states are striving to insure themselves against possible disruptions in the transportation and sale of raw materials.

4. Conclusion

What are the prospects for Russia to wake up the Arctic region? Still this "sleeping" region with its colossal economic potential hasn't been fully developing due to natural limitations (economic activity is not always possible all year round), as well as significant financial costs due to harsh climatic conditions and complex logistics. Of course, these factors need to be taken into account when developing Arctic policy. The Russian

policy in the Arctic should be based, first of all, on the solution of the following tasks:
- Development of transportation and logistics corridors connecting industrial centers and transport hubs. This will give an impetus to territorial development, since the Arctic is characterized by a significant dispersion of the population, and settlements are actually cut off from each other. The construction of roads and other transport infrastructure for the purposes of the subsoil use industry will lead to the associated inclusion of existing settlements in a single Arctic transport framework.
- Reassessment of natural resources, which will provide updated data on the potentially available reserves of minerals. This applies not only to reserves of gas, oil and other popular resources. It is also necessary to refresh the data on the quality of the landscape, coastlines, geobotanical characteristics of the territory. For example, geobotanical studies of the Arctic regions of Yakutia were carried out in 1977-1978. Over 40 years there have been significant changes in the quality and composition of the vegetation cover and soil, which should be taken into account during investment. For example, geobotanical maps are used to calculate losses to indigenous peoples, whose lands are temporarily withdrawal during the implementation of an industrial subsoil use project.
- Selection of investment projects according to their priority. Here the following question must be resolved: what criterion will be used in the basis of priority? What is the purpose of its implementation? On the one hand, it is important for the investor to make a profit, on the other, it is important for the regional government to ensure a decent standard of living for the local population. In Russia, approaches have been developed for ranking and selecting projects that could take into account the interests of both [10].
- Development of the icebreaker fleet, transport based on safe innovations, as well as the introduction of green technologies in construction and facilities providing subsoil use. The implementation of the principles of a green economy will help preserve the fragile and sensitive nature of the Arctic [11].
These tasks cannot be solved without government assistance and incentives. It is necessary to develop mechanisms of administrative and financial support, among which the special tax regimes and public-private partnerships should be highlighted.

Acknowledgement

The paper is prepared by the financial support of RFBR project 21-510-22001.

References

1. F. Hill, F. Fee *Demokratizatsiya*, **10 (4)**, 462-487 (2002)
2. Novak announced the construction of more than ten LNG plants in Russia, retrived from: https://www.rbc.ru/business/29/01/2021/6013dc059a79473d601ea315
3. Decree of the President of the Russian Federation of N 645Strategy for the development of the Arctic zone of the Russian Federation and ensuring national security for the period up to 2035 (2020) retrieved from: http://www.kremlin.ru/acts/news/64274
4. Moe, O. S. Stokke, *Arctic Review on Law and Politics*, **10**, 24–52 (2019) http://dx.doi.org/10.23865/arctic.v10.1374
5. Publication calendar of official statistical information on the socio-economic development of the Arctic zone of the Russian Federation in 2020, retrieved from: *https://gks.ru/free_doc/new_site/region_stat/calendar1-2020.htm*
6. V.I. Denisov, V.N. Chernogradskii, I.M. Potravny, P.Y. Ivanova, *Studies on Russian Economic Development*, **31(4)**, 404–410 (2020)
7. Government of Nunavut Iqaluit International Airport Improvement Project 2014 Retrieved from: https://gov.nu.ca/sites/default/files/iqaluit_international_airport_improvement_project_report.pdf
8. V. Gassiy, V. Stoikov, *E3S Web of Conferences*, **177**, 04005 (2020)
9. K. Mohn, *The Energy Journal*, **40 (3)** 199-226 (2019) http://dx.doi.org/10.5547/01956574.40.3.kmoh
10. Novoselov, I. Potravny, I. Novoselova, V. Gassiy, *Polar Sci.*, **14**, 68–77 (2017)
11. Kyriaki Noussia, *European Energy and Environmental Law Review*, **3**, 98-109 (2020)

Priorities for the sustainable development of inland capture fishery

Anna Shishigina[1,*], *Vladimir* Sosin[1]

[1] North and Arctic Interdisciplinary Research Division of Arctic Scientific Research Centre, Academy of Sciences of Sakha Republic, 22 Kurashova st., Yakutsk, 677000, Russia

Abstract. Fishing in inland waters significantly differs in terms of production, seasonality, scale, methods of production and organization of fishing, equipment and technologies. In addition, fishing in inland waters is characterized by insignificant volumes of catch of aquatic bioresources and other socio-economic significance. As a rule, this type of activity is occupied by small organizations that use seasonal labor with temporary employment, which, at the same time, are of significant socio-economic importance for the local population. The article presents the characteristics of the situation in the field of industrial fishing in the Sakha Republic (Yakutia) and the main problematic trends of the industry. The purpose of the article is to show the main obstacles to the sustainable development of inland fisheries in the Sakha Republic (Yakutia) in fresh waters. As a result of the study, measures are proposed to solve the problems. The management should be focused on the rationalization of fishing to increase profits per unit of production assets used, as well as on attracting highly qualified specialists to fishing, that is, increasing the investment attractiveness of this industry. With this approach, inland fisheries can contribute substantially to increased food security, poverty alleviation, livelihoods, human well-being and ecosystem function.

1 Introduction

Managing the sustainable development of fisheries is an international challenge. In 1988, FAO agreed on the definition of sustainable development given by the World Commission on Environment and Development (1987) - "development that meets the needs of the present generation without compromising the ability of future generations to meet their own needs". All things considered that the management for the sustainable development of industrial fisheries has the object of the conservation and rational use of biological resources. That is enabling conditions when the reproduction of exploited populations remains at the highest possible level and produces maximum high-quality products with the minimum possible expenditure of effort and resources while preserving a balanced exploitation of stocks.

Inland fisheries in Yakutia is the only traditional economic activity in the North that produces real industrial products. Fish is not only the most important biotic resource for the population in the Arctic, but also a source of income and plays an important role in the social structure of the local population. At the same time, in Russia, the management of inland fisheries in freshwater is limited to the approved total allowable catch and the volume of possible catch of aquatic bioresources including a ban to fishing in the regions of the Russian Federation. But, as world practice has shown, inland fisheries have never been included in any major global fisheries assessment and, in particular, have not been included in the Sustainable Development Goals, their life support component is underestimated and largely ignored [10, 12]. This is related to the landing size. Global catches in inland waters are for 12.5 per cent of total fishing production. This capture varies wildly among the top catch producers, accounting for less than 1 per cent of total catches in the United States, Japan and Peru, compared to 44 per cent and 65 per cent of total catches in Myanmar and Bangladesh respectively. Inland fisheries are also important for food security in Africa, which accounts up to 25 percent of global inland fisheries, when combined catches in Europe and the Americas are about 9 percent [1].

It would be hard to overestimate the importance of inland fisheries for the food security of the Arctic peoples. The livelihoods of the Arctic peoples in Yakutia are becoming increasingly vulnerable due to recent changes in climatic and environmental conditions, alongside with the political and economic transformations [11].

2 Results and Discussion

Despite the relatively small size of the production of aquatic bioresources in the Sakha Republic (Yakutia), the life of the population of the Arctic region of the Sakha Republic (Yakutia), including the indigenous peoples of the North, critically depends on the functioning of the fishing industry.

* Corresponding author: schanan@yandex.ru

The analysis of fish catch on the main rivers for the specified period shows that the main share of industrial fish is caught in the basins of 4 main fishing rivers, in which 25,086. 6 tons or 94.8% of the total volume were caught, including:

- Lena River – 40.1% or 10,676. 6 tons,
- Kolyma River – 19.2% or 5,111. 4 tons,
- Yana River – 18.4% or 8,886. 7 tons,
- Indigirka River – 16.6% or 4,411. 9 tons.

As of January 1, 2020, in the Sakha Republic (Yakutia), a total of 952 fishing areas were formed, including 471 river and 481 lake areas. In view of the latest legislative changes, the situation with ensuring public access to the use of aquatic biological resources has improved. Nevertheless, in prospect it is necessary to optimize the regulations on the formation of fishing areas with the exception of all inconsistencies of various federal and republican departments, to provide open access to the List of fishing areas of the Sakha Republic (Yakutia) via the Internet, to facilitate and speed up the procedures for holding competitive tenders for fishing areas consolidation.

The implementation rate of allocated quotas on lakes is much lower than on rivers (59.5% against 77% of river ones) due to its assortment consisting of less valuable fish, such as perch, roach, char, minnow and others. At the same time, one of the main reasons for the partial implementation of quotas is also the extreme remoteness of the lake areas, which requires high costs for transporting the fish from the sites to the receiving points. For this reason alone, in the basins of the Khrom and Alazeya rivers more than 700 tons of quotas have not been reached, including chir and ryapushka (Table 1).

Table 1. Dynamics of the catch of aquatic bioresources and the development of allocated quotas for 2015-2019

Years	Allocated quotas, tons	Actual catch, tons	percent of development
2015	6 848.30	4 704.96	68.7
2016	7 265.52	4 921.34	67.7
2017	7 502.79	5 474.77	73
2018	7 932.16	5 757.5	72.6
2019	7 706.41	5 718.74	74.2
Total	**37 255.18**	**26 577.31**	**71.3**

Industrial fish catch sizes for the period of 2015-2020 in the Republic are about 26,577. 31 tons with the fishing quotas of 37,255. 18 tons. The average annual catch of fish is 5,315. 5 tons. Over the last years, the volume of catch of burbot and pike has increased, which is probably due to the introduction of technologies for their processing and the growing demand for these products. If the catch of burbot and pike in 2014 amounted to 75 and 184 tons per year, then in 2019 they were already 205 and 386 tons respectively. Also the percentage of the implementation of quotas for these fish species has also increased. The question is raised by the very low implementation percentage of quotas for catching carp, which is less than 50% annually. Even in the Kobyai ulus, the quotas implementation percentage for catching carp is only 43%.

The main share of industrial fish catch is provided by seven fishing areas that are part of the Arctic zone: Bulunsky, Ust-Yansky, Allaikhovsky, Nizhne-Kolymsky, Zhigansky, Sredne-Kolymsky and Abysky uluses, which produce the lion's share (88.1%) of the total volume of extracted fish. The largest fishing farms in the Republic are Artel (fishing collective farm) "Arctic", Municipal unitary enterprise "Bulunskoye" from Bulunsky, PC KRO "Russian Ustye", PC KRO "Allaikha" from Allaikhovsky, PC KRO "Turvaurgin", SPK KMNS "Pokhodsky", KRO KMNS-Yukagirov "Chobul" from Nizhnekolymsky, SPK KRO Ilin-Shar, LLC Ust-Yansky and other medium-sized farms from Ust-Yansky.Yansky ulus. These fishing farms are economic bases in their locations and the socio-economic situation of such settlements as Bykovsky, Kyusyur, Taimylyr, Russian Ustye, Pokhodsky, Kolymskoye, Andryushkino, Kazachye, Ust-Yana, etc. depends on the results of their economic activity.

The analysis of the catch structure of aquatic biological resources by the organizations of all forms of ownership shows that more than half (55%) of these aquatic biological recourses are fished by production cooperatives (25 large and medium-sized farms), 30% are fished by individual entrepreneurs and LLC, and indigenous communities of small-numbered peoples of the North account for no more than 5% of the total catch (Figure 1).

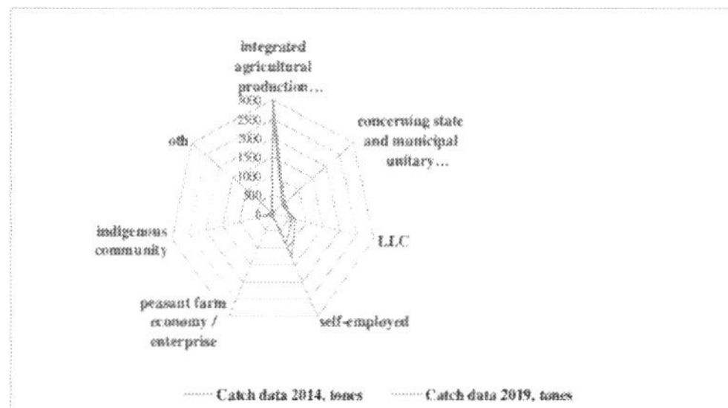

Figure 1. Catch data of organizations for 2014 and 2019.

At the same time, there is an increase of small farms in the industry, which is an undesirable factor. In today's market conditions the introduction of modern fishing technologies, freezing, storage and transportation of fish in order to preserve the high quality fish produced requires large financial investments for the purchase of appropriate equipment, equipment and fleet, which is beyond the power of small farms and individual entrepreneurs.

Fishing in the conditions of the Far North is a complex production process that carries large financial costs for its organization, and these costs occur before income is received, that is, before the period of fish production in connection with the preparations for the harvest - this is the purchase of fish, other materials, products, their delivery to the fish sites by the winter road, the input requirements should be met before harvesting products. Thus a huge cash gap is generated, which in previous years was partially covered by advancing state subsidies for industrial fishing. This made it possible to carry out a successful summer fishing season, of course, taking into account the natural and hydrological conditions. But starting in 2018, this type of state subsidy is given only to small and medium-sized fishing farms with quotas of up to 50-70 tons.

Fish in the Republic is mainly sold unprocessed through all sales channels via small enterprises (66%), peasant farms and individual entrepreneurs (19.7%), large and medium-sized organizations (13.4%). On average in the domestic market of the Republic 1 300 – 1 600 tons of fresh-frozen fish is purchased for processing per year.

Despite the fact that the Republic cannot export large volumes of products outside the territory of Yakutia, as there is a dependence on imports and a weak level of food self-sufficiency, the Yakut fish is supplied to the Amur, Tyumen, Kemerovo, Novosibirsk, Moscow and Leningrad regions, Altai and Primorsky Territories, as well as to Moscow. According to the Far Eastern Customs Administration, the export of Yakut fish is carried out to the People's Republic of China, the volume of which amounted to 10.9 tons in 2019. According to the report of the Federal Customs Service, in 2019, the export of fish products from the Sakha Republic (Yakutia) amounted to 60.3 thousand US dollars in monetary terms.

About 90% of especially valuable fish species, such as sturgeon, nelma and taimen, are sold mainly by direct deliveries, and in recent years the volume of omul, chir and muksun supplied by direct deliveries has also been growing, with fish often being shipped outside the Republic.

At the same time, according to the federal statistical observation (excluding microenterprises and individual entrepreneurs), in the Sakha Republic (Yakutia) since 2014, there has been almost four-time increase in the import of fish products and in 2018 it amounted to 144.1 tons. While the export of 1 500 – 2 000 tons of fresh frozen fish outside the Republic per year, the share of imported fish products is growing, there is clearly a dependence on imports and a weak level of food self-sufficiency for the residents of the Republic. Some authors [2] attribute this to the deficiency of opportunities for the majority of fish processing enterprises to produce high-quality and competitive products that meet the requirements of world standards (the capabilities of fish processing organizations in Yakutsk are severely limited due to lack of working capital, the small capacity of the Yakutsk market, and the low purchasing power of the population). This fact is contrary to the general trend of the Russian Federation, which is characterized by a significant change in the size and co-relation of export and import operations in terms of fish and fish products with a steady trend of reducing the latter [3].

In recent years, there have been situations when every second fishing enterprise has been operating on the verge of loss. According to the Federal State Statistics Service in the Sakha Republic (Yakutia) in 2016 and 2018, the share of unprofitable organizations in the total number of organizations by the type of economic activity "fishing" was 50% and 53.8%, respectively.

The fisheries sector of the Sakha Republic (Yakutia) is currently in a state of crisis and is facing a whole range of different problems related to the influence of such well-known and characteristic factors for the Arctic region as the deficiency of territorial connectivity, long distances, extreme climate, etc. However problems that directly depend on the effectiveness of the state policy in the field of fisheries. There are the underdevelopment of the basic fishing infrastructure, the obsolescence of the material and technical base of fishing farms, a significant increase in the number of small enterprises, and an acute shortage of qualified personnel, have no less impact on the state of the fisheries complex.

Planning and forecasting are among the most important functions of public administration in the fisheries sector. At the same time, at the current stage of development, such work actually consists in making forecasts of the biological productivity of fishing areas by industry institutions, as well as developing industry strategies and programs initiated by the Federal Agency for Fisheries and the regional government, which is clearly not enough.

Considering these problems in order to find ways to solve them, it is necessary to proceed, first of all, from the general principles of state regulation of industrial fishing in Russia, in the construction of which the existing economic relations in the field of industrial fishing are built.

When establishing the modern Russian system of Fishing Industry State Management, the so-called "historical principle" of granting the right to use aquatic biological resources was defined as the basis for its development, which is characteristic of the legislation of the most leading fishing countries. Their fishing is based on the exploitation of coastal and marine aquatic biological resources, which are characterized by significant scales and sizes of production.

In the Russian Federation, the historical principle is legally enshrined in Article 36 of Federal Act No. 52 of 24 April 24, 1995 "On the Animal World", according to which priority in providing the animal world for use in a particular territory or water area is given to Russian legal entities and citizens of the Russian Federation who previously carried out certain types of use of the animal world in this territory or water area and to land owners, landowners with appropriate funds and specialists.

This principle is connected with the need to preserve the objects of the animal world and their habitat, and its essence is to provide the subjects of nature management with a guarantee for the long-term and stable implementation of activities, as well as to ensure the continuity of rights to nature use. The reverse side of this principle is the artificial creation of a certain closeness of the fishing industry, which complicates the entry of new participants into the industry. However, with regard to marine fishing, this closeness does not particularly hinder its independent development, but rather contributes to the preservation of the potential and compliance with the environmental responsibility of nature users.

The historical principle in the field of fisheries is reflected mainly in the features of the distribution of quotas for the production (catch) of aquatic biological resources in marine and inland waters, established by Article 31 of Federal Act No. 166 of December 20, 2004 "On Fishing and the Conservation of Aquatic Biological Resources" (hereinafter – the Law on Fishing and the Conservation of Aquatic Biological Resources).

In accordance with parts 10 and 11 of Article 31 of the Act on Fisheries and Conservation of Aquatic Biological Resources, quotas for the extraction (Catch) of aquatic Bioresources in marine Waters, in Inland water bodies and international quotas granted to the Russian Federation are distributed in a proportionate ratio between persons whose contract on fixing the share of the quota for the extraction (catch) of aquatic bioresources expires by concluding a new contract with such persons on fixing the share of the quota for the extraction (catch) of aquatic bioresources for 15 years. The remaining quotas for the extraction (catch) of aquatic bioresources, for which industrial fishing and (or) coastal fishing were not previously carried out, or for aquatic bioresources in new areas of their extraction (catch), are distributed according to the results of auctions.

Thus, according to the basic principle of functioning, marine industrial fishing and industrial fishing in inland water bodies are practically equated.

Meanwhile, fishing in inland waters significantly differs in terms of production, seasonality, scale, methods of production and organization of fishing, equipment and technologies. In addition, fishing in inland waters is characterized by insignificant volumes of catch of aquatic bioresources and other socio-economic significance. As a rule, this type of activity is occupied by small organizations that use seasonal labor with temporary employment, which, at the same time, are of significant socio-economic importance for the local population.

This circumstance also affects the peculiarities of the administrative regulation of fishing. For example, according to the Fishing Rules for the East Siberian Fishery Basin, approved by the order of the Ministry of Agriculture of the Russian Federation of June 26, 2020, No. 347. All legal entities and individual entrepreneurs in the implementation of fishing are required to submit to the territorial bodies of the Federal Agency for Fisheries information on the extraction (catch) of aquatic bioresources on the 5th, 10th, 15th, 20th, 25th and last day of each month no later than one day after the specified date (with the exception of fishing carried out on ships filing daily ship reports). At the same time, fishing in inland water bodies is carried out without the use of vessels that submit daily ship reports. When fishing in lake fishing areas in the Arctic zone of the Sakha Republic (Yakutia), there is no communication with the outside world during the entire season.

However, the main problem of the lack of a clear distinction between types of fishing and differences in approaches to the regulation of marine and inland fisheries is the unjustified economic isolation of the latter, which results in the inability to create the necessary conditions for attracting investment in the fishing industry of the Sakha Republic (Yakutia). Fishing organizations of the Republic that have the right to extract (catch) aquatic biological resources, which is actually the most important asset, today can not fully implement it and are not able to independently get out of this situation without external support.

This is due to the fact that, in accordance with article 32 of the Act on Fisheries and on the Conservation of Aquatic Biological Resources, the transfer of the right to extract (Catch) aquatic biological resources from one person to another person is allowed only in cases defined by this Federal Act, as well as in the order of universal succession in accordance with Civil Law. However, certain cases of transfer of rights are not currently provided for by this Act, so there is only one way to transfer the right, it is the reorganization of the legal entity that has this right. This implies that there are limited ways to realize the economic potential of the right to extract (catch) aquatic bioresources.

The inability to attract long-term project financing on acceptable terms and low financial returns from users of aquatic bioresources do not cover the Republic's budget investments in the industry. On the one hand, due to the cancellation of subsidies from the state budget of the Sakha Republic (Yakutia) for industrial fishing to fishing farms with industrial quotas of more than 50 tons, the volume of fish harvesting in the Republic has decreased more than twofold (from 3,225.7 tons in 2016 to 1,271. 2 tons in 2019). On the other hand, the possibilities of the state budget are strictly limited and cannot serve as a donor as the only basis for the existence of the fisheries industry.

Thus, one of the most important tasks of the sustainable development of industrial fishing in the Sakha Republic (Yakutia) today is the creation of favorable conditions for attracting investment.

The crisis in the activities of organizations of the fisheries industry poses a threat to the sustainable development of the Arctic zone of the Sakha Republic (Yakutia), primarily, this concerns the Bulunsky, Allaikhovsky, Nizhnekolymsky, Ust-Yansky uluses of the Sakha Republic (Yakutia), whose population directly depends on the activities of organizations in this industry.

On the way to overcome this problem, the most promising solution is to soften the legal regime for the use of aquatic biological resources by expanding the methods provided for by law for transferring the right to extract (catch) aquatic biological resources from one person to another, provided for in article 32 of the Law on Fisheries and Conservation of Aquatic Biological Resources.

Since we are talking about the right to use natural resources that are in state ownership, we should consider expanding the methods of transferring the right to extract (catch) aquatic bioresources by analogy with the provisions on the transfer of the right to use subsurface areas and the reissue of licenses for the use of subsurface areas, established by Article 17.1 of the Law of the

Russian Federation of February 21, 1992 No. 2395-I "On Subsurface Resources", which, in addition to the transfer of the right to use in the order of universal succession, also allows the following methods:

1) when a legal entity-a user of subsurface resources acts as the founder of a new legal entity established to continue operations on the provided subsurface area in accordance with the license for the use of the subsurface area, subject to the conditions specified by law;

2) when transferring the right to use a subsurface area by a legal entity - a subsurface user who is the main company, to a legal entity that is its subsidiary, or when transferring the right to use a subsurface area by a legal entity-a subsurface user who is a subsidiary, to a legal entity that is its main company, subject to the conditions specified by law;

3) when a business entity acquires, in accordance with the procedure provided for by Federal Law No. 127-FZ of October 26, 2002 "On Insolvency (Bankruptcy)", the property (property complex) of a bankrupt enterprise (a subsurface user), subject to the conditions specified by law.

These methods of transferring the right of use do not contradict the general logic of natural resource law of the Russian Federation and the principles of rational nature management, since they do not exclude compliance with all the requirements and conditions for the implementation of the licensed type of activity to the new owner of the right.

The solution of this problem opens up new opportunities for the development of the fisheries industry and will ensure an increase in the economic value of aquatic biological resources in accordance with the opportunities and needs of the market. This will allow, for example, fishing organizations to cooperate with each other and other investors not only by reorganizing the legal entity, which implies benefits for one of the parties, to create joint subsidiaries with the preservation of management and the right to receive a share of income, as well as to use the new opportunities provided by Federal Act No. 193 of July 13, 2020 "On State Support for Business Activities in the Arctic Zone of the Russian Federation". In the current situation, the limits of the provision of state support measures provided for by this law for business activities in the Arctic zone of the Russian Federation are very limited, since, as a general rule, the types of business activities that a candidate for Arctic residency plans to carry out must be new to him, not previously carried out. Thus, existing fishing enterprises cannot apply for the status of a resident of the Arctic zone of the Russian Federation. However, if the Law on Fisheries and Conservation of Aquatic Biological Resources provides for the possibility of transferring the right to use aquatic biological resources to subsidiaries or parent organizations, the latter will already be able to apply for the benefits and preferences provided for Arctic residents.

3 Conclusion

The supply of fish in freshwater has a set of features related to the seasonality of fishing, the dynamics of aquatic bioresources, the complexity of predicting stocks of aquatic bioresources and determining the rational share of their withdrawal without compromising reproduction in natural conditions. Of course, the sustainable development of industrial fisheries should primarily be based on reducing the anthropogenic load on the ecological system to the maximum permissible level. At the same time, the management should be focused on the rationalization of fishing to increase profits per unit of production assets used, as well as on attracting highly qualified specialists to fishing, that is, increasing the investment attractiveness of this industry. With this approach, inland fisheries can contribute substantially to increased food security, poverty alleviation, livelihoods, human well-being and ecosystem function. A robust fishery cannot function sustainably against the background of inflexible and limited solutions.

References

1. FAO. *The State of World Fisheries and Aquaculture 2020. Sustainability in action. Rome.* (2020) Retrieved from: http://www.fao.org/publications/card/en/c/CA9229EN
2. V.A. Fedorova *International journal of applied and fundamental research* **5-3**, 478-482 (2015)
3. Stecyuk V.V. *Customs Policy of Russia in the Far East* **2(87)**, 65-76. 2019
4. Law of the Russian Federation № 2395-I *"On Mineral Resources"* (1992)
5. Federal Law № 52 *"On the Animal World"* (1995)
6. Federal Law№ 166 *"On Fisheries and Conservation of Aquatic Biological Resources"* (2004)
7. Federal Law № 93 *"On State Support for Business Activities in the Arctic Zone of the Russian Federation"* (2020)
8. Federal Law№ 127 *"On Insolvency (Bankruptcy)"* (2002)
9. Order of the Ministry of Agriculture of the Russian Federation № 347 *"On Approval of the Fishing Rules for the East Siberian Fisheries Basin"* (2020)
10. D. Islam, F. Berkes. *Maritime Studies* **15**, 1 (2016).
11. S. Ksenofontov, N. Backhaus & G. Schaepman-Strub *Polar Record*, **53(3)**, 289-303 (2017)
12. S.J. Cooke, E.H. Allison, T. Douglas Beard, R. Arlinghaus, A.H. Arthington, D.M. Bartley, I.G. Cowx, C. Fuentevilla, N. Leonard, K. Lorenzen, A.J. Lynch, V.M. Nguyen, S.-J. Youn, W.W. Taylor, R.L. *Ambio*, **7** (2016)

NSR and NSTC projects: environmental and economic efficiency assessment of the goods' overseas transportation

Sergey Nikonorov, *Alexander* Krivichev*, *Vladimir* Sidorenko

M.V. Lomonosov Moscow State University, Leninskie Gory, GSP-1, Moscow, 119991, Russia

Abstract. The authors attempt to assess the environmental and economic efficiency of investment projects for the goods transportation by the northern seas of Russia taking into account the relevance of the topic of cargo transportation by means of the "Northern Sea Route" (NSR) and new concept of the digital format of the "Northern Sea Transit Corridor" (NSTC) implemented by the "Rosatom" State Corporation. The structure of import-export and transit traffic by the NSR in 2020 is considered. The goal of the article is to evaluate the efficiency of investment project for the goods transportation forwards and backwards by the NSR from Murmansk city to Petropavlovsk-Kamchatsky city using the example of the "Andrey Osipov" shipping facility. Some indicators were assessed: planned revenue, payback period; NPV, environmental damage. The following results are obtained: the payback period is 10 years; by using the naval fuel oil over a ten-year period, 1101.6 tons of SO_x, 143.1 tons of CO_2, 288.9 tons of NO_x will be emitted. In addition, the authors propose an innovative approach to creation of ships from new materials and technologies based on the modern paradigm of "Accelerated Digital Certification".

1 Introduction

The Arctic is a unique ecosystem with an amazing fauna, a reserve source of drinking water and huge reserves of minerals. The polar possessions of Russia make up 18% of the entire territory of the Russian Federation, where less than 2% of the population of Russia lives. The Arctic produces 10% of the world's oil and 25% of natural gas, while 80% of all Arctic oil and gas is concentrated in the Russian part.

The sustainable development of the Arctic territories of Russia is one of the priority areas for the development of the Russian economy [1,2,3].

2 NSR, NSTC, SSR logistics

According to the Decree of the Russian President dated May 7, No. 204, 2018, the Federal project "Northern Sea Route" (2018-2024) and the "Infrastructure Development Plan" of the "Northern Sea Route" (NSR) until 2035 (borders of the NSR determined by international law from Novaya Zemlya to Chukotka) are established (Fig. 1).

According to the documents, the main tasks for the NSR development are: increasing the freight traffic by the NSR up to 80 million tons per year up to 2024 and up to 120 million tons per year up to 2030; navigation and hydrographic support of navigation on the NSR; ensuring year-round navigation of ships on the NSR; creation and reconstruction of transport and port infrastructure; construction of the fleet. The ports will be involved: Arkhangelsk, Murmansk, Sabetta, Dikson, Dudinka, Pevek, Khatanga, Petropavlovsk-Kamchatsky. The implementation of the tasks set within the Federal project of NSR is carried out at the expense of the federal budget.

"Northern Sea Route" is the shortest sea route between the European part of Russia and the Far East. The Russian legislation defined it as "the historically established national unified transport communication of Russia in the Arctic".

The use of cargo transportation by the NSR, in addition to reducing the travel time from 33 to 20 days, provides a number of other advantages:

- Saving on fuel almost 2 times compared to the southern sea route.
- No need to pay for the passage of the vessel (unlike the Suez Canal).
- No queues (as is the case with the Suez Canal).
- There is no risk of attack by pirates.

Import-export traffic by the NSR in 2020 was carried out mainly by river-sea ships, without class, Ice2, Arc4, Arc5, Arc7, and most of the traffic was performed by the Arc7 class of ships.

* Corresponding author: krivichev@live.ru

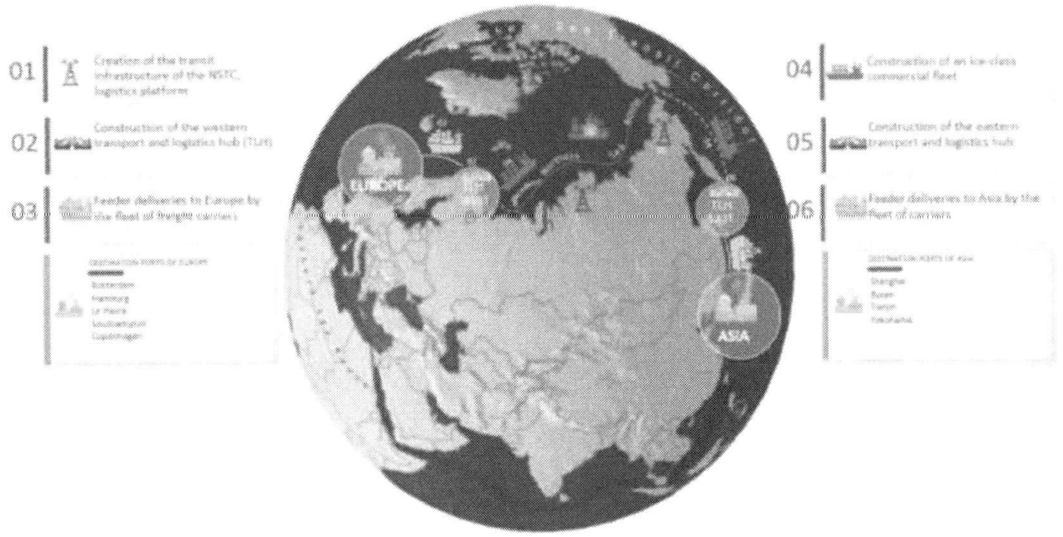

Figure 1. The logistic of the "Northern Sea Route" (NSR), "Northern Sea Transit Corridor" (NSTC), "Southern Sea Route" (SSR)

The structure of import-export traffic (Fig. 2) in 2020 was as follows: LNG – 59%; oil – 24%; general cargo – 11%; gas condensate (NGL) – 3%; oil products – 2%; coal – 1%; ore concentrate (Au) – 0.2%. Moreover, 3.2 million tons of cargo were transported.

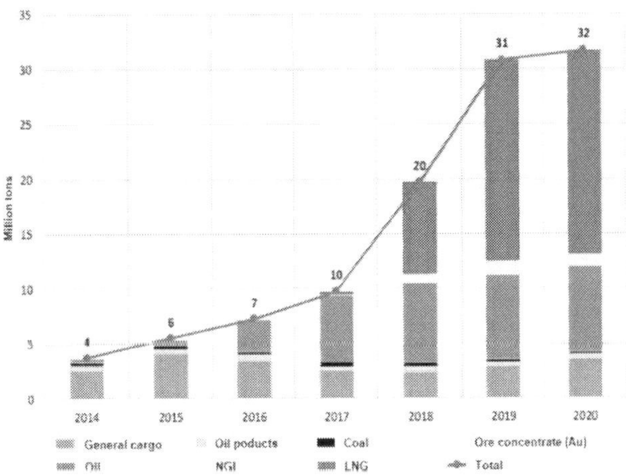

Figure 2. The structure of import-export traffic by NSR / NSTC

LNG exports from Russia have shown good results in recent years. Even in Europe, as stated in the report of the International Association of Liquefied Gas Importers, Moscow bypasses Washington in terms of LNG supplies (in 2018 – 4.4 million tons versus 2.7 million tons, in 2019 – 15.07 million tons versus 12.72 million tons).

Transit transportation by the NSR in 2020 was carried out, among other things, by ships of four ice classes: Ice1, Ice2, Arc4, Arc5, and more than 800 thousand tons of cargo were transported.

In 2020, transit traffic by the NSR in the structural section of freight traffic was as follows: iron ore – 78%; general cargo – 7%; cellulose – 5%; mineral fertilizers – 5%; bulk cargo – 2%; equipment – 2%; petroleum products - 1%; frozen fish – 0.5%. Moreover, more than 1 million tons of iron ore were transported, and the rest of the transported goods accounted for about 250 thousand tons.

To develop the transit potential and geographic advantages of the NSR, the "Rosatom" State Corporation in 2019 initiated a project to develop the logistics system of the "Northern Sea Transit Corridor" (NSTC). The concept of the project was developed within the framework of a strategic program for the creation of an international operator-logistics services.

The project includes:

- creation of the transit infrastructure of the NSTC and a logistic platform,
- construction of a western transport and logistic hub,
- feeder deliveries to Europe by the fleet of cargo carriers,
- construction of a commercial ice class fleet (30 units),
- construction of the eastern transport and logistics hub,
- feeder deliveries to Asia by the fleet of cargo carriers.

The core of the project is the creation of a commercial fleet with a design service life of the vessel of at least 40 years. The potential of the project is 4.5 million TEU per year (A twenty-foot equivalent unit (TEU or teu from English twenty-foot equivalent unit) is a conventional unit for measuring the capacity of freight vehicles. It is often used when describing the capacity of container ships and container terminals).

The Far East Development Fund, together with VEB.RF, the Ministry for the Development of the Russian Far East and the M.V. Lomonosov Moscow State University, are developing a digital platform Arctic Labs, which will be based on the digital twin of the "Northern Sea Route" (CD NSR): a dynamic mathematical model that will allow developing scenarios development of both the Arctic as a whole and its individual territories. With the help of the NSR CS platform, it will be possible to develop logistics routes, taking into account in real time more than 10 thousand parameters: Arctic resources, infrastructure, transport, shipbuilding, transportation options, economy, ecology, etc. This will allow a comprehensive assessment of: the feasibility of investment and infrastructure projects, the impact of projects on the dynamics of development of regions and individual territories, the impact of the Arctic economy on the GDP of the Russian Federation.

The authors assessed the efficiency of the investment project for the transportation of goods by the NSR from Murmansk to Petropavlovsk-Kamchatsky and vice versa, using the example of a shipping facility – "Andrey Osipov" (hereinafter referred to as the Arc5 class vessel). The goal was to transport cargo in containers from the port of Murmansk to the port of Petropavlovsk-Kamchatsky and back, taking into account downtime, calculating revenue, net present value, payback period and environmental damage.

In connection with this goal, the following tasks were solved: the container capacity of the ship, the volume of heavy fuel oil tanks, the ship's speed per hour, and the number of flights were determined. In addition, the calculation of the total investment costs for the project has been made. At the same time, the salary of the crew and services when renting containers were taken into account:
- forwarding to the port of departure,
- acceptance of cargo at the port,
- temporary storage on board,
- mount and fastening material,
- sea freight,
- icebreaking dues,
- unloading from the ship,
- temporary storage in the port,
- intraport movements.

3 Results

As a result of a study to assess the effectiveness of an investment project, the following results were obtained:
- the payback period of the project was 10 years, and the average amount per year for business development was 14 million rubles;
- with naval fuel oil using over a ten-year period the 1101.6 tons of sulfur oxides, 143.1 tons of CO2, 288.9 tons of nitrogen oxides will be emitted.

If we consider the simultaneous operation of a pool of ships of this type on the NSR, then the ecological footprint will be multiplied.

This is relevant, since the volume of emissions from ships of the world Merchant fleet is about 3% of greenhouse gas emissions in the world, since these ships operate on heavy fuel oil. It is known that the concentration of sulfur in heavy fuel oil is 1800 times higher than in diesel fuel for cars. Therefore, an ecological and economic assessment of the efficiency of the work of vessels of various classes is necessary.

At the same time, applying an innovative approach to creating a ship from new materials and technologies based on the modern paradigm of "Accelerated Digital Certification", it is possible to significantly increase the efficiency of using ships of a new generation. The use of carbon fiber materials in the process of creating a ship for NSTC can give the following advantages: reduction of the ship's weight by 30-40%; the possibility of building on the slipway. A decrease in the mass of the ship's hull will lead to an increase in the specific power of the gas-power plant (GPP) by 30%, as well as to a decrease in the draft to 10-12 m, which will allow the ship to pass through all the "depths" of the NSR. And the use of CFRP shipping containers, in addition to reducing weight, will increase the container service life from 5 to 30 years and reduce painting costs. The use of the power plant will many times reduce the ecological footprint and increase the ecological and economic characteristics of the pool of ships carrying cargo in the water area of the NSR.

In terms of environmental impact, ships with a nuclear power plant (NPP) have zero CO_2 emissions [2]. That is, it is a 100% ship with no carbon footprint. The use of ships with nuclear power plants on the NSTC routes will make it possible to make a real one hundred percent "green mile".

In addition, digital modeling was chosen as a fundamental basis for the implementation of the NSTC project, which includes:
- model-oriented system engineering,
- simulation modeling (Anylogic etc.),
- digital technologies and platform solutions,
- design and use of digital twins (Fig. 3) and digital shadows on the CML-BenchTM platform at all stages of the ship's life cycle,
- other digital technologies.

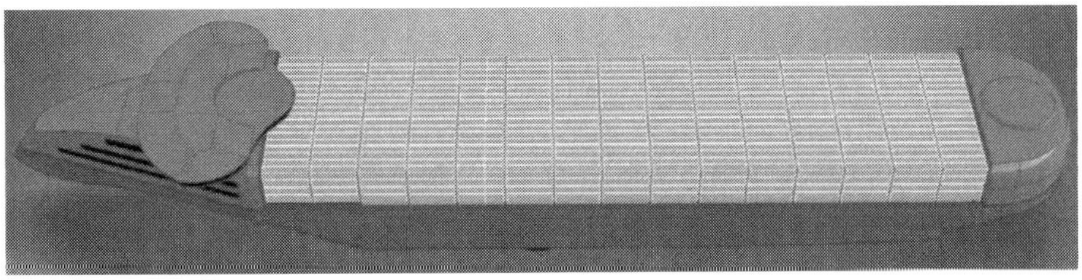

Figure 3. The example of digital twins for container ship "Ice Needle" (Marine Engineering Center, Moscow)

It is necessary to link the work on the creation of digital twins at various stages of the ship's life cycle, the use of information support systems, covering research, design, production, integrated logistics support, optimization of ship routes. This approach will create opportunities for more efficient use throughout the life cycle of ships for the carriage of goods by the NSR, as well as faster and more accurate economic calculations taking into account the environmental factor.

With global warming, the rate of ice melting in the Arctic has increased, and the importance of the Arctic strategy has become increasingly evident, which has caused competition among many countries. The Arctic Development Strategy 2035 is the first step towards sustainable development of the Arctic. To achieve a favorable situation in the Arctic, we need to formulate appropriate Arctic regional strategies in the Arctic Zone of the Russian Federation (AZRF) and in the Barents macro-region, so that Russia occupies an advantageous position in the future world order. Russia's 2021 chairmanship of the Arctic Council can partially ensure energy security and the security of sea transport channels, which is becoming a strategic direction for Russia's economic development. My country needs to assess the impact of the Arctic on the future international structure of the world and formulate an international Arctic strategy, with the involvement of 8 countries of the Arctic Council and with the involvement of observer countries (primarily China).

The opening of the Arctic waterway will change world trade and help create a super-Arctic economic circle dominated by Russia, North America and Northern Europe, which in turn will affect the economic and geopolitical structure of the entire world. If the new channel is opened, it will reduce the sea freight traffic, significantly reduce transport costs, save time and promote the development of foreign trade. The energy security of China, Korea, Japan and other countries, when importing and exporting goods, will be guaranteed. Russia's International Arctic Strategy can strengthen exchanges and cooperation with seven other Arctic countries, enhance its influence in the Arctic region and protect strategic interests. On the other hand, it attaches great importance to common interests with non-polar countries and strengthens exchanges and cooperation with non-polar countries such as China, Korea, Japan, India and Australia in Arctic affairs. By pursuing common interests, this will contribute to the internationalization of the management and development of the Arctic and will help guide the development of Arctic issues in a reasonable and fair direction.

4 Conclusion

Thus, the projects of the "Northern Sea Route" and the "Northern Sea Transit Corridor" are key vectors for the development of the Russian Arctic, and despite the fact that these projects are aimed at increasing the competitiveness of the Russian economy in the domestic and world markets, in solving the tasks set, they will allow the implementation of a number of other related projects (use of the best available technologies, research and protection of Arctic ecosystems, optimization of logistics solutions, use of digital twins and platforms, big data analysis [5,6,7], etc.).

References

1. Vitaly Sergeev, et al. *Transportation Research Procedia.* 54. (2021).
2. Alexander I. Krivichev, Eugene G. Smirnov *Management sciences in the modern world.* 2 (2). (2016).
3. V.M. Gruzinov, et al. *Arctic: Ecology and Economics* 1, 33 (2019).
4. I. Petrov, et al. *IOP Conference Series: Earth and Environmental Science* 377, 1 (2019).
5. Eric Van Der Horn, Sankaran Mahadevan *Decision Support Systems.* 145 (2021).
6. Vladimir N. Sidorenko, Pavel M. Krivenkov Dynamic scenario modeling. Optimization of the logistics solutions. Calculation of the economics of complex systems at some levels of detail at all stages of the life cycle on the example of small NPP projects in the "Rosatom" State Corporation. *About approaches to the creation of digital models in the project of the Northern Sea Transit Corridor formation. Digital ship as a*

basis of the transit infrastructure. “Rosatom” *State Corporation,* LLC “Rusatom Cargo”, PJSC “Inteltech”, St. Petersburg, Russia (2019).

7. Sergey M. Nikonorov *journal Scientific Works of the Free Economic Society of Russia.* **6**, 226 (2020).

Practical recommendations for implementing the project funding mechanism in the transport construction

Vera A. Kudryavtseva [1,*]

[1] Irkutsk State Transport University, 15, Chernyshevskogo str., Irkutsk, Russia

Abstract. Changes in the Russian and global economies have been affecting all sectors. Neither the transport industry nor the construction one is an exception. These industries are budget-forming, and all participants in business processes should be interested in their strategic development. In order for the federal projects developed by the Ministry of Transport to be effective, a new project funding approach is required. Industries and companies have limited financial resources. The Russian Railways company is no exception due to its limited budgetary funds and own financial capabilities. It is necessary to develop effective mechanisms for funding railway infrastructure projects. Project funding mechanisms can be innovative ways of funding large-scale infrastructure projects, since they are based on a combination of banking, commercial and public resources. Project funding can unite financial resources and competences of project participants. In addition, qualified financial institutions will be involved in project funding transactions, which ensures the timely project implantation. Project funding mechanisms allow for a comprehensive project assessment, since the lender confirms the economic feasibility of the project. Project effectiveness evaluation has become relevant after the adoption of a strategy for the development of railway transport in the Russian Federation until 2030. Polygon technologies are promising for Russian Railways and used by the Trans-Siberian Railways.

Keywords: infrastructure, funding, project, investment, budget

1 Introduction

The economic and social development of Russia is at the stage of deep transformations. It is important to develop a model that would strengthen the world position of the country. These measures were declared by the President of the Russian Federation. They are aimed at creating a "smart economy" that would contribute to the advanced development of science. Thus, the need for the transition of the Russian economy to the sixth "technological order" has been determined.

Digital technologies have been implemented in various industries and in the entire economic system, which determines the development and promotion of new science-intensive technologies.

It is important to note that in spite of the ongoing activities, some industries and sectors of the economy are characterized by poorly developed infrastructures.

Within the sixth technological order, it is planned to develop all sectors of the economy. The transport industry is no exception. The urbanization and business processes require changes in the concept and tools for the development of transport infrastructure.

The sixth technological order will contribute to the creation of a powerful scientific foundation and transform it into practical results.

Within the sixth technological order, the requirements for transport communications will become stricter. These requirements concern:

- environmentally friendly consumption;
- safety;
- speed;
- energy efficiency;
- efficiency;
- comfort.

In general, the world economic system has already entered the development cycle of the sixth technological order, which affects the global modernization of the world transport system.

2 Problem Statement

Recently, the development of the railway infrastructure has become a crucial issue. In 2014, the government approved the investment project "Modernization of the railway infrastructure of the Baikal-Amur and Trans-Siberian railways and development of carrying capacities" [1]. The project began to be implemented in 2013 by Russian Railways,

* Corresponding author: kudravera@mail.ru

the Ministry of Transport of the Russian Federation and the Federal Agency for Railway Transport. It involves two stages. The implementation of this project was supposed to be completed in 2017.

However, the final deadline for the completion of the first stage was postponed to the end of 2020 due to the fact that the funds were not fully used. At the beginning of 2020, only 67 % of funds were disbursed.

Undoubtedly, this project is one of the largest and most important infrastructure projects in the railway industry. Therefore, it is necessary to speed up the process of commissioning of strategically important facilities. Measures aimed at accelerated modernization of the Eastern polygon are being implemented by the government. On November 12, 2019, the Chairman of the Government of the Russian Federation D.A. Medvedev gave recommendations to "update the plan for the implementation of the investment project "Modernization of the railway infrastructure of the Baikal-Amur and Trans-Siberian railways and development of carrying capacities". A detailed plan for accelerated modernization has been developed [1].

V.V. Putin said, that Russia needs a breakthrough in the development of the national transport infrastructure. Among the priorities is large-scale modernization of the BAM and Transsib. By 2025, it is necessary to increase their throughput by over 200 million tons of cargo per year.

Moreover, in the mid-2020, Russian Railways launched a new large-scale project "East. Point of attraction". It is important to note that within this project a landfill for advanced development will be created to develop the landfill, increase its competitiveness and efficiency. A railway technopolis with housing and social infrastructures, additional opportunities for training and career promotion of employees and their families will be created [2].

To implement infrastructure projects, financial resources are required.

At the same time, within the limited state budget and own funds, Russian Railways is forced to apply new funding methods. One of these is a project funding mechanism.

3 Research Questions

As part of the above projects, attention should be paid to the issue of project funding. Figure 1 shows the structure of funding sources for the investment project "Modernization of the railway infrastructure of the Baikal-Amur and Trans-Siberian railways and development of carrying capacities". It is clear that co-financing is a guarantee of the execution of any project.

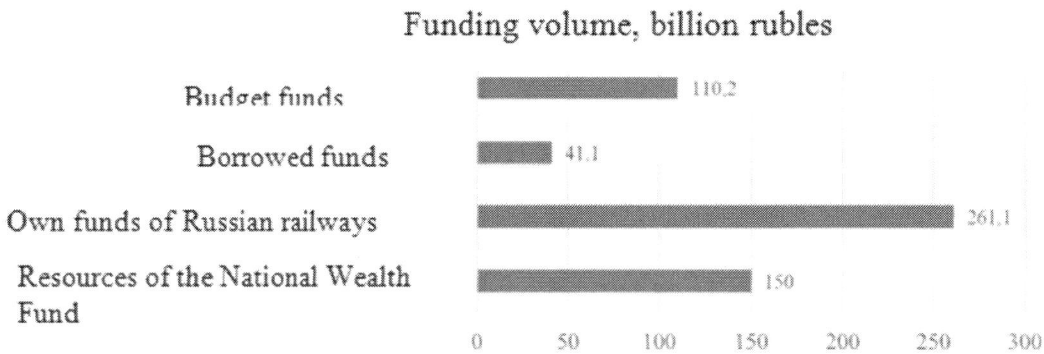

Figure 1. The structure of funding sources for the investment project "Modernization of the railway infrastructure of the Baikal-Amur and Trans-Siberian railways and development of carrying capacities"

In order for the federal projects to be implemented, a new funding method is required. That is, in order to achieve target results, it is necessary to revise the approaches to the allocation of budget funds. One of the ways to structure the project and distribute risks among all participants is to implement project funding or public-private partnership mechanisms. Project funding can be an alternative way to implement the project, and select an optimal combination of methods and mechanisms for attracting financial resources to the project [4]. Projects cannot be implemented using only own funds. Borrowed funds are required for funding any large project.

Since 2020, the project-funding scheme has been one of the main ways to finance the housing construction sector in Russia. At the same time, the concepts of project funding and "targeted loans" are not similar. For example, foreign project funding schemes are full-fledged investment products which are the basis for the Initial Public Offering mechanism. The project funding is effective within the investment project due to the advantageous position of fund recipients [4].

Figure 2 shows potential benefits of the project-funding scheme.

The Russian Railways-Invest company prepared an analytical review, according to which the project funding is an innovative investment mechanism for Russian Railways.

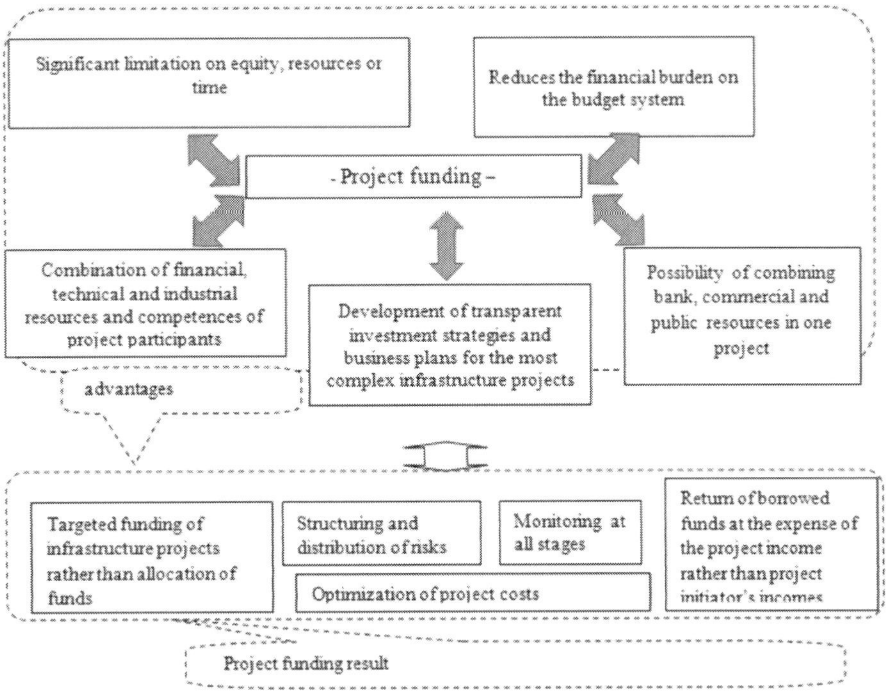

Figure 02. Project funding: advantages and result

4 Purpose of the Study

Currently, many industries and companies experience significant financial constraints. Russian Railways is no exception. Limited budgetary funds and own financial capabilities prevent them from choosing effective project funding mechanisms for the development of railway infrastructure.

It is necessary to create efficient high-speed transport systems. However, financial resources are required. They cannot be allocated by the government. Therefore, it is necessary to attract investment in the development of railway transport and transport infrastructures. The article aims to analyze the project-funding scheme as an alternative financial mechanism that contributes to the timely implementation of the project [3].

5 Research Methods

An analysis of the budget of national projects revealed that the most expensive national project is aimed at development infrastructure facilities. For its implementation, one fourth of the total amount of financial resources, equal to 25.7 trillion rubles, is required. Figure 3 shows the distribution of budget funds by industries and fields. As can be seen from Figure 3, an enormous burden falls on the budgetary system. Infrastructure projects should be funded from all types of budgets [5]. At the same time, it should be noted that budget funds cannot meet the financing needs of infrastructure projects, therefore, a mixed funding mechanism is required. Extra-budgetary resources should be used.

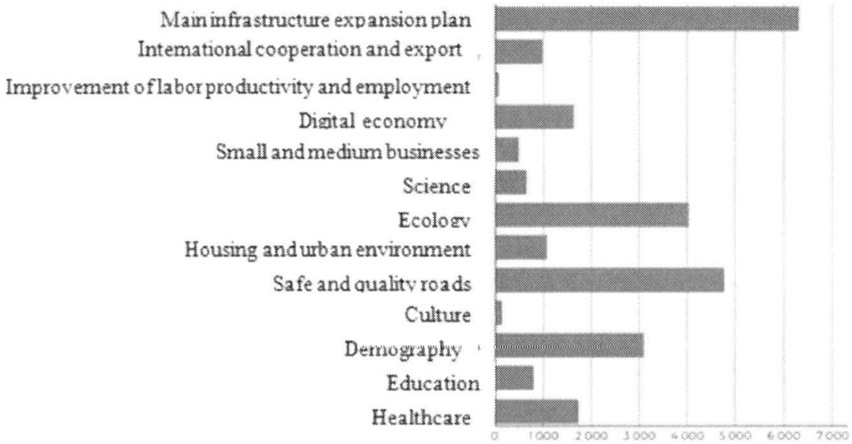

Figure 03. National project budget, billion rubles.

As can be seen from Figure 3, an enormous burden falls on the budgetary system. Infrastructure projects should be funded from all types of budgets. At the same time, it should be noted that budget funds cannot meet the financing needs of infrastructure projects; therefore, a mixed funding mechanism is required. Extra-budgetary resources should be used.

6 Findings

The project funding is part of business projects. It can be considered as an alternative funding method, especially when the company's investment policy is focused on raising funds from outside (Chelpanova, 2018). Sustainable development of the project funding consolidates financial resources for the implementation of infrastructure projects of Russian Railways with a significant reduction in project risks by increasing the number of project investment participants.

7 Conclusion

The project funding is part of business projects. It can be considered as an alternative funding method, especially when the company's investment policy is focused on raising funds from outside. Sustainable development of the project funding consolidates financial resources for the implementation of infrastructure projects of Russian Railways with a significant reduction in project risks by increasing the number of project investment participants.

Acknowledgments

The work has been published at the expense of the state task on the topic "Improving the mechanisms of interaction between polygon structures and regional corporate governance centers on the example of the Eastern railway polygon".

References

1. E.V. Bakharev. Substantiation of the effectiveness of investment project financing in railway transport. Transp. Busin. in Rus., **3**, 95–97 (2016)
2. N.A. Zhuravleva, A.Yu. Panychev. Financing projects for the development of railway infrastructure: topical solutions for Russia. Econ. of railway transp., **1**, 122 (2015)
3. M.A. Mova. Assessment of the economic efficiency of investment projects of JSC "Russian Railways". Socio-econ. probl., **1**, 175 (2013)
4. S.E. Prokofiev, V.I. Murar, I.V. Rashkeeva, M.V. Elesina. Project financing: nature and significance Sci., **6**, 1 (2014)
5. I. Grigorovich, M. Eliseev, P. Seregin. *Kilometers. Projects. Money. Analysis of the prospects for the development of project financing instruments* (JSC IC RZD Invest, Moscow, 2019)
6. M.B. Chelpanova. *Problems of efficiency of investment in railway infrastructure* (2018). Retrieved from: https://www.hse.ru/data/2018/05/31/1149339027 _final.pdf

Industrial symbiosis as an element of sustainable development of arctic companies

Sergei Nikonorov [1,*], and *Ekaterina* Utkina[1]

[1]Moscow State University, GSP-1, 1-46 Leninskiye Gory, Moscow, 119991, Russia

Abstract. Economic activity in the Arctic is playing an increasingly important role in the globalizing economy, given that such an economy is largely based on natural resources (oil, gas and minerals) as well as the long-distance transport of these resources and related goods. The Arctic seems to be one of the last opportunities to exploit vast resources, which, ironically, made it possible by the negative effects of this very exploitation in the form of global warming and the subsequent retreat of the ice cover of the circumpolar North. Thus, the Arctic has already become and will continue to be part of economic globalization. The article gives an analysis of the dynamics of the «Polar Index of the Barents region. Companies». The need to include the main elements of industrial safety in the sustainable development of the Arctic has been justified, as well as the establishment of industrial symbiotic cooperation rating based on it. Industrial symbiosis is an economically efficient and environmentally friendly solution for industry that reduces energy consumption and primary material consumption by deriving value from waste, by-products and heat losses.

1. Introduction

Despite significant advances in public understanding and awareness of sustainable development, the economic component of sustainability has traditionally prevailed, with states and individual firms working first. However, the traditional mode of economic development cannot guarantee the sustainable development of the Russian economy, and a scientific approach is needed to effectively correct the entire economic system. The development of the regional economy of the Russian Arctic is inseparable from the concept of sustainable development. The new economic system needs innovation from the top down, including the diversion of investment into new areas, the emergence of new markets for "green" industry and the acceleration of regional economic transformation. The management of environmental problems in our country has been ineffective for many years. The reason lies in the varying degree and dynamics of development of different regions. Economic development cannot ignore environmental costs and attempt to replace the necessary costs of environmental transformation with support for exceptionally high-speed economic development.

In 2020, a number of important strategic and long-term planning documents were adopted, defining the State policy for the development of the Arctic for the next 15 years. Among the most significant ones are the State Policy Framework for the Arctic Zone 2035 and the Strategy for the Development of the Arctic Zone of the Russian Federation and for National Security 2035. The Ambitious List of Infrastructure Projects was set out in the Strategy, On the one hand, it shows the leading role of Russia in the Arctic, and on the other hand it is intended to give a signal to business representatives, attracting them to participate actively in the investment of these facilities. At the same time, efforts to improve the activities of the federal authorities in the development of the Arctic were continued: the composition of the State Commission on Arctic Development was renewed, and its powers expanded; the newly formed Ministry of the Russian Federation for the Development of the Far East began its activities.

The State Policy Framework for the Arctic Zone 2035 and the Strategy for the Development of the Arctic Zone of the Russian Federation formulate and refine the Russian State policy in the Arctic in the context of medium- and long-term planning and could become the basis for the program of the Russian Federation's presidency of the Arctic Council for the period 2021-2023. The authors suggest that the future program should include themes of concern to the region, centred on its sustainable development with a balanced mix of environmental, economic and social components. These include improving the quality of life and living conditions of people in the Arctic, adapting to climate change; preserving fragile Arctic ecosystems and biodiversity while developing the region's economy, large-scale extractive and processing projects, Intensification of freight traffic on

* Corresponding author: nico.73@mail.ru

the Northern Sea Route. Addressing these challenges requires the collective efforts of all Arctic Council members and the development of new approaches to international cooperation.

The ambitious list of infrastructure projects set out in the Strategy shows, on the one hand, Russia's leading role in the Arctic, and, on the other hand, it is intended to send a signal to business representatives, encouraging them to participate actively in the investment of these facilities. The State Commission for the Development of the Arctic should constantly monitor the implementation of the established tasks.

2. Ranking methodology and dynamics

In order to correct the existing imbalance in the categories of stability of regions and companies, the Polar Index rating (hereinafter - the Rating) was created, which is a complex multi-level model for assessing sustainable development from the internal level, at which the companies of the region are located, to the external, regional level, where the sustainability of the region itself is assessed from the standpoint of the state, society and business.

The rating "Polar Index of the Barents Region", presented for the first time at the end of 2018, is the result of reaching the international level, developed earlier by the rating "Polar Index. Regions and Companies" aimed at assessing the sustainable development of regions and companies in the Russian Arctic. The indicators in the ratings taken in three key areas of sustainable development (ecology, economy and social) are updated annually.

The following indicators are used to calculate the Barents index:

- Economic. The basis of the indicator is the financial performance of companies according to international financial reporting standards.
- Social. The basis of the indicator is the quality of corporate responsibility (CSR) reporting.
- Environmental. The basis of the indicator is the qualitative characteristics of environmental management systems.

The Arctic companies reviewed in the Ranking are largely represented by the extractive sector, which has a strong impact on sustainability. Extractive industries are often characterized as energy-intensive and resource-intensive, with significant greenhouse gas emissions and high resource dependence [3] which makes it a good target for the implementation of new strategies and methods to improve resource and energy efficiency.

In the updated Rating of 2020, in terms of companies' research, the listed categories are represented as the vertices of a triangle (Fig. 1), in which companies are the core, while the vertices are mutually dependent on each other [1]. The outer vertices of the triangle are represented by government, business and society. Government, divided into federal, regional and local levels, is defined as measures taken by the authorities or administrations of the region to support, regulate and ensure the interaction of business and society. Businesses are large, medium and small operating enterprises operating in the study area. Accordingly, society is the sum total of all citizens of a given country, region or locality.

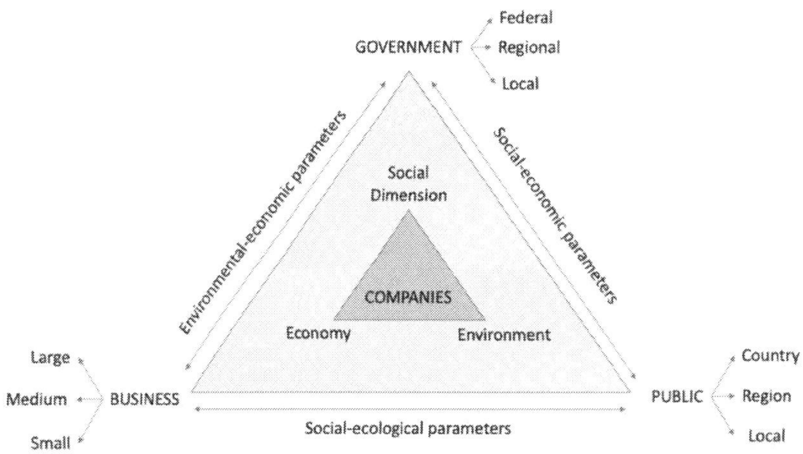

Fig. 1. Conceptual approach to the compilation of the Polar Index.

Table 1. Results of the Barents-Region Polar Index rating calculation. Companies

№	Company	Country	Polar Index 2018	Polar Index 2019	Changing Rating Position
1	Rosneft	Russia	8,51	9,00	↑1
2	Lukoil	Russia	8,01	8,33	↑2
3	Equinor	Norway	9,00	8,00	↓2
4	Norilsk Nickel	Russia	7,62	7,67	↑3
5	LKAB	Sweden	6,01	7,67	↑9
6	First Quantum Minerals	Finland	4,00	7,67	↑14
7	Agnico Eagle Mines	Finland	7,65	7,67	↓1
8	ALROSA	Russia	8,01	7,33	↓5
9	Gold Fields	Finland	-	7,00	New
10	Phos Agro	Russia	7,31	6,67	↓1
11	Petoro	Norway	-	6,67	New
12	Novatek	Russia	-	6,33	New
13	Gazprom	Russia	6,65	6,33	0
14	Boliden Group	Finland / Sweden	7,67	6,00	↓9
15	Nordic Mining	Norway	7,14	6,00	↓4
16	RUSAL	Russia	2,72	5,67	↑10
17	Magnus Minerals	Finland	-	5,67	New
18	Severstal	Russia	7,28	5,67	↓8
19	Sydvaranger Gruve	Norway	5,99	5,67	↓4
20	Concedo	Norway	5,95	5,00	↓4
21	Leonhard Nilsen & Sonner	Norway	5,66	5,00	↓4
22	Dragon Mining	Sweden	5,21	4,33	↓3
23	Norge Mineral Resources	Norway	5,33	4,00	↓5
24	Beowulf Mining	Finland / Sweden	3,37	3,00	0
25	Nussir	Norway	3,64	2,33	↓3
26	Aurion Resources Ltd.	Finland	3,66	2,00	↓5
27	Sunstone Metals	Sweden	3,03	2,00	↓2
28	Skaland Graphite	Norway	3,39	1,67	↓5
29	DEA Norge	Norway	7,50	1,00	↓21
30	Nenetskaya Neftyanaya	Russia	6,84	0,00	↓18

3 Results and discussion

The authors analyze the dynamics of the indicators calculated in the indices of previous years (Table 1).

The Polar Index used in this study, through its Integrated Corporate Sustainability Index, measures their success in key indicators: economy, social and environmental (including industrial safety). Consideration of the index in dynamics shows which companies are developing in the right direction and introducing, among other things, positive social and environmental technologies - this will be reflected in the growth of the index or in maintaining it at a high level. For example, the Integrated Corporate Sustainability Index of Lukoil varied from 0.909 in 2018, to 0.801 in 2019 and to 0.906 in 2020, and the Integrated Corporate Sustainability Index of Rosneft in these years was 0.893, 0.851 and 0.912, respectively [1, 2, 6]. From this example, it can be seen that these Russian companies reduced their financial, social and environmental performance in 2019 compared to 2018, but made up to some extent for the current year 2020 and are generally held at the top of the Polar Index. The report makes it clear that domestic companies are doing well not only through financial success but also through large-scale environmental programs, including the transition to "the best available technologies (BAT) within the framework of the Federal Project No. 11 of the national project "Ecology".

Given climate change and growing concerns about energy security, global markets need access to clean and reliable fuels. Natural gas is a net source of hydrogen, producing energy in fuel cells with minimal emissions, and is considered essential for the transition to a sustainable energy future. It can thus be seen as a link to a renewable energy economy [2]. The process of using natural gas for electricity generation results in

Most Russian large companies, due to Soviet traditions and lack of experience in competitive markets, demonstrate elements of monopolistic thinking. Nevertheless, state-owned enterprises are willing to collaborate in the Arctic to share high technological and economic risks, and to create broader partnership networks as part of their global business strategies. In general, the analysis of the performance of companies in the Barents Polar Index rating for the period from 2018 to 2020 shows the need for companies to pay more attention to environmental aspects of sustainability, due to the special value of Arctic nature and the importance of the Arctic zone to the world. On the contrary, as shown in the above studies, at the international level, environmental aspects of sustainability prevail over economic ones.

In order to increase sustainability, mining and processing companies are actively trying to minimize their waste and improve material efficiency by developing new products from their by-products, production residues and wastes.

In the case of the Russian Arctic, there is considerable scope for the formation of industrial-symbiotic networks. Thus, to date, the Russian Federation and the Commonwealth of Independent States, as well as other former socialist countries, have no examples of successful industrial-symbiotic interactions, and only a study based on counter-factual analysis of the Kola mining community is known from unsuccessful attempts. Industrial symbiosis model never implemented [5].

The long-term sustainability of the extractive sector may depend to a large extent on innovative solutions to energy intensity and carbon dioxide (CO2) management. The authors are invited to consider in more detail the possibility of setting up an industrial symbiotic network for extractive and processing industries.

According to the authors, there is a need for large-scale «carbon solutions" that can be used by several energy-intensive and carbon-intensive industries in relative proximity to each other. In particular, it is proposed to investigate the possibility of placing a mineral carbonization installation as a carbon sink at the centre of a mineral-energy complex in order to form an industrial symbiosis. Thus, a few resource-intensive industries could be synergistically integrated to create a complex of energy and mineral products with low net CO2 emissions.

significantly lower greenhouse gas emissions compared to using coal as a fuel on a megawatt hour basis [2]. A US study suggests that replacing coal-fired power plants with natural gas-fired power plants could reduce greenhouse gas emissions by 22% [3]. The largest LNG producer and supplier in Russia is Novatek with its three largest projects Yamal LNG, Arctic LNG-2, and Obskiy LNG.

4 Conclusion

The proposal and promotion of the concept of sustainable development are to change the economic development regime in the regions (especially in regions with complex environmental situations or fragile ecosystems, as in the Arctic zone of the Russian Federation), to raise awareness of available and produced resources from industrial cycles and environmental protection. It will also transform highly polluting enterprises in each region into environmentally friendly ones, promote the modernization of production methods in enterprises, realize the transformation and enhance sustainable development for the economy of each region, and, finally, implement the comprehensive and sustainable development of the Russian economy.

At present, while it is difficult to measure the real performance of companies, especially in terms of their compliance with environmental and social promises, the rating provides a comprehensive view of actual public commitments and institutional arrangements of companies. The more companies formally and publicly recognize the need for greater sustainability, the greater the pressure on other companies and, consequently, the stronger the basis for establishing business norms.

With regard to the analysis of industrial symbiosis, many different indicators have been proposed to date, while a clear map of the available tools is still lacking [8]. There was an obvious imbalance in the case studies analyzed, as environmental impacts of all aspects of sustainability were the most frequently analyzed and quantified, followed by economic and social indicators.

It is proposed to classify indicators in relation to industrially symbiotic interactions in order to answer three main questions: which indicators should be selected for analysis, in relation to what and how to measure them. This contributes to a clearer understanding of the available metrics in terms of purpose, context and methodology.

The present study provides a perspective for future research. 1) Indicators measuring the social benefits of industrial-symbiotic interactions should be developed. 2) New indicators or methods to identify potential, as well as an analysis of the current interactions in the concept of industrial symbiosis, which will allow the quantification of the impact of the three dimensions of sustainability: environmental, economic and social. Although the current Polar Index concept covers these three dimensions, it does not aim at industrial symbiosis. In addition, the coverage of environmental, economic and social components entails some

https://doi.org/10.1051/shsconf/202111200027

difficulties, such as the integration of qualitative and quantitative indicators into the same assessment framework [7].

Thus, new research is needed to overcome these barriers and to identify specific indicators for industrial symbiosis that will measure the overall impact of these practices on firms, the environment and society and compare industrial symbiosis in different realities, that is, different characteristics of the network, taking into account the characteristics of the region in which it develops.

References

1. S. M. Nikonorov, K. V. Papenov, *Ekonomika ustojchivogo razvitija*, 287-298 (2017)
2. B Shively and J Ferrare *Handbook on the Physics and Chemistry of Rare Earths* (Enerdynamics: Laporte, CO, USA, 2011)
3. MIT Energy Initiative 2011 *the Future of Natural Gas* Retrieved from: http://energy.mit.edu/research/future-natural-gas/
4. S. M. Nikonorov, K. V. Papenov, A. V. Lebedev et al 2019 *Polar index of the Barents region. Province and Company Sustainability Ratings* (Expert Center for Project Development of the Arctic "PORA», 2019)
5. E. E. Utkina *Vestnik of the Plekhanov Russian University of Economics* **17 (5)** 26–41, (2020)
6. T. Domenech, R. Bleischwitz, A. Doranova, D. Panayotopoulos and L. Roman *Resources, Conservation & Recycling*, **141**, 76-98, (2019)
7. S. Schoubroeck, M. Dael, S. Passel, R. Malina *Renewable and Sustainable Energy Reviews* **94**, 115–126, (2018)

The place of the concept of 'human security' in the strategies of the Arctic countries

Maria Lagutina*

Saint Petersburg State University, 7/9, Universitetskaya Emb., 199034, Saint Petersburg, Russia

Abstract. This paper is devoted to determination of the place of the concept of 'human security' in the modern strategies of the Arctic countries. The article examines the formation of the concept of 'human security', the features of human security in the Arctic, as well as the place of the idea of human security in the modern Arctic policy of Russia. Today, the Arctic, being a 'zone of peace and stability' in the categories of traditional security, faces a complex of new challenges and threats related to 'human security'. The author analyses the strategic documents of the eight Arctic countries with a view to reflect the ideas of human security in them. Special attention is paid to human security's issues in the AZRF. In conclusion, the author points out that none of the strategies of the eight Arctic countries explicitly mentions the term 'human security'. But at the same time, some elements of this concept are undoubtedly noted in strategic documents: the priority of most Arctic countries is economic security and environmental security, while the rest of the components of human security are found only in certain countries' strategies.

1 Introduction

The Arctic has always been imagined as a Terra Incognita, fraught with danger to humans: harsh climatic conditions, polar night, remoteness from developed territories and their inaccessibility, dietary habits, predatory animals, etc. From the very beginning, the population of the northern territories was engaged in adaptation to these harsh living conditions in the Arctic.

The modern Arctic is traditionally defined as a 'zone of peace and cooperation' [1], as a region where "traditional or state centred security elements, that is, those related to the survival of the state, present a low profile in the Arctic" [2]. At the beginning of 2021, despite the current crisis in relations between Russia and the Western countries on the world stage, the Arctic countries managed to avoid the transfer of geopolitical tension in their relations to the Arctic region. The crisis in Russia's bilateral relations with other Arctic countries caused by the events of 2014 lasted until about 2016-2017. Gradually, the media and official speeches of the leaders of the Arctic countries began to announce intentions to restore a full-fledged dialogue with Russia in the Arctic [3]. As a compromise, the most politically neutral, but at the same time very important for the region areas of cooperation were tacitly singled out: socio-economic, scientific, and educational, environmental protection and sustainable development. At the same time, in recent decades, it is in these areas the so-called new challenges and threats for the inhabitants and states of the Arctic have emerged, which are now commonly referred to as 'human security' and which today form the basis of threats to the national security of most Arctic states. In this regard, it seems important to determine the place of the concept of 'human security' in the modern strategies of the Arctic countries.

2 The concept of 'human security': basic provisions

Traditionally, the emergence of the concept of 'human security' is associated with the activities of the UN in the 1990s. But the idea itself was not new and many researchers point to the fact that it appeared back in the 1940s [4]. However, the term of 'human security' was first proposed in 1994 in the 'Human Development Report' of the United Nations Human Development Program (UNDP). There the phenomenon was defined as "safety from such chronic threats as hunger, disease and repression" and "protection from sudden and hurtful disruptions in the patterns of daily life-whether in homes, in jobs or in communities" [5]. Besides that, the main components of 'human security' were identified in the text of the Report, including economic security, food security, health security, environmental security, personal security, community security and political security. Among the threats were the consequences of climate change and environmental pollution, its degradation and depletion of resources, problems of food security and health, socio-economic development, various forms of violence, conflicts, as well as suicide and drug use as a form of individual violence against oneself, problems preservation of cultural heritage and traditional way of life, etc. A feature of this group of problems is that their solution requires the attention of not only individual governments, but the entire community, including residents and local communities.

* Corresponding author: m.lagutina@spbu.ru

At the same time, within the framework of the concept of 'human security' there has been an obvious shift in emphasis from the security of the nation state to security of individuals and their communities, which became a reflection of several world political transformations after the end of World War II. The concept of 'human security' in the UN interpretation is interconnected with the concepts of sustainable development and human rights.

However, according to Russian researcher Valery Konyshev, the UN's definition has a rather broad and abstract character and therefore serves only as a kind of general framework, but in practice there is a wide variety of approaches to the definition of 'human security', both at the level of international organizations and states [6]. It is because of the very broad interpretation of the term, as well as because of its 'Western' origin [7], many researchers assess the concept of 'human security' very critically [8]. Meanwhile, since the early 2000s human security's issues attracted the attention of the world community and expert circles, as well as stuck on the agenda of the UN and other international organizations.

3 Human Security in the Arctic

The case of the Arctic is unique. From the very beginning, the population of the northern territories was engaged in adaptation to the harsh living conditions in the Arctic. Despite the active role of states in the development of the Arctic territories, individuals and communities have always played a leading role in the Arctic, with unique experience of survival in difficult climatic conditions. The XX century made serious adjustments to the development of the Arctic territories, which was due to the active economic activity of several states in the region, the development of Arctic resources, etc. Against this background, new challenges to the security of Arctic communities have emerged, associated with the consequences of human economic activity, the impact of anthropogenic factors on the Arctic ecosystem, as well as the consequences of climate change. For most of the Arctic countries, the concept of 'human security' is not new. Moreover, countries such as Canada, Norway and Sweden are among those countries that in the 1990s actively promoted the theory of 'security of the individual and society' in their foreign policy [9]. Among these countries, Canada played a key role in including human security issues on the Arctic agenda: "Canada in particular has made stringent efforts to include human security issues on the circumpolar agenda, a fact related to the foreign policy priorities of the Canadian government when it led the establishment of the Arctic Council in 1996, and to the Canadian domestic situation, with its human security problems in the North…" [10]. But has this concept been reflected in the national Arctic strategies?

It is known that today all the Arctic countries have published their Arctic strategies, some of them have already been republished considering new trends in the development of international cooperation in the Arctic. It is in the texts of the strategies of the Arctic countries that one can find the goals, objectives, and main priorities of the Arctic countries in the implementation of their Arctic policy. However, none of the strategies of the eight Arctic countries explicitly mentions the term 'human security'. But at the same time, some elements of this concept (in the UN definition) are undoubtedly noted in strategic documents: "The policies and strategies cover not only economic or geopolitical strategic topics, but also social, environmental and human aspects of sustainable development in the region" [11]. Based on the analysis of these documents [12-19], it can be stated that the priority of most Arctic countries is economic security (sustainable use of natural resources, both renewable and non-renewable, economic development, business interests in the Arctic etc.), environmental security (climate changes, balance between environmental protection and economic development, protection of the Arctic ecosystem and biodiversity, etc.), community security (tackle societal and community-based problems, gender equality, promotion women and youth empowerment, create jobs, foster innovation, etc.), and political security (increasing participation of local inhabitants in decision-making processes, promotion wellbeing of indigenous people, etc.). The topic of economic development is the leading in all strategies because it corresponds to the national interests of all Arctic states. Besides that, it is important to note that mostly the development of the economic potential is regarded as a factor in improving the living standards of the local population and preventing depopulation. In many cases, the issue of economic security is considered in direct relationship with the sustainable economic development of the region, trade and investment opportunities, access to transportation network and infrastructure of the region. Special attention is also paid to the role and place of traditional economic activities, as reindeer husbandry, hunting and fishing, etc.

The rest of the components of human security are found only in certain countries. For example, references to food security issues can only be found in Canada's Arctic strategy, personal security - Canada and Finland, health security - Canada, Denmark, Russia, and Sweden. Obviously, attention is paid to various aspects of human security in national Arctic strategies to varying degrees, while in all cases there is no comprehensive approach to human security. As R. Martin notes, this is largely because Arctic strategies are mainly focused on determining the national interests of countries in the Arctic, they have truly state-centric character and "economic and environmental security, as elements of human security, can be better connected with the national priorities of the Arctic states" [20]. Thus, the strategic documents of the Arctic countries are mainly focused on the interests of the state, and the interests of the inhabitants of the Arctic take the second place. However, if we trace the evolution of the Arctic strategies of the countries that republished them, it can be noted that the 'human dimension' appears more and more in their content. One such example is Russia's Arctic strategy.

4 Human security in the Arctic zone of the Russian Federation (AZRF)?

For the Russia's Arctic, the concept of 'human security' is relatively new, which is because until recently the security of the Russia's Arctic was mainly defined through the state-centric approach that was reflected in Russia's strategic documents [21, 22]. This is due, in our opinion, firstly, to the leading role of the federal authorities in the development of the AZRF, and, secondly, to the definition of the AZRF as a resource base of the Russian Federation. However, in practice, it became clear that both approaches do not correspond to the modern realities of the development of the Russian Arctic. On the one hand, it became obvious that the state alone cannot adequately respond to all new challenges and threats, and the participation of other actors (business community and civil society, etc.) in solving urgent problems of the AZRF is necessary. On the other hand, a narrow 'resource' approach can make the AZRF a peripheral territory of Russia with a high level of population outflow to other regions and, as a result, create new threats to the security of Russia's northern borders. As a result, today the main task of the Arctic policy of the Russian Federation is the sustainable development of the region.

In 2020, two new strategic documents were adopted: in March, the President signed the 'The Basic Principles of The Russian Federation's State Policy in The Arctic Until 2035' ('The Basic Principles – 2020') and a decree approving the 'Strategy for Developing the Russian Arctic Zone and Ensuring National Security until 2035' ('Strategy – 2020'). An important innovation of these documents is the so-called 'human' component - the priority of improving the quality of life of people living in the Russian Arctic. So, in 'The Basic Principles – 2020' two of the five goals directly relate to the population of the Russian Arctic, indigenous peoples. One of the goals is related to improving the quality of their life, the second is to protect their original habitat and traditional way of life. The same can be found in the text of 'Strategy-2020'. The main problems of the development of the Russian Arctic and its national security include climate change, population decline, lagging indicators of the quality of life in the Russian Arctic from national indicators, lack of high-quality infrastructure, etc., all those questions that are included in the concept of 'human security'.

It seems possible to assert that today the problem of human security has acquired a special meaning for Russia in the context of the implementation of one of the most important tasks of the state policy of the Russian Federation in the Arctic: sustainable development of the region. The successful development of the Russian Arctic presupposes the creation of the foundation for the country's socio-economic development, and this requires the neutralization of the challenges and threats of human security.

5 Threats to Human Security in the AZRF

The successful development of the AZRF should create the foundation for an effective economic development of the country, which requires the neutralization of threats in the social sphere, with which the concept of human security is correlated.

The formation of threats to *economic security* in the AZRF is due to the inefficient social policy of the 1990s and early 2000s, outdated infrastructure and the lack of necessary investments that intensified the negative trends in the development of the Russia's Arctic. In the 1990s many industrial facilities were shut down, public utilities in several northern cities fell in disrepair, many ports, and other civil and military facilities of the AZRF were languishing in neglect. The region experienced a high level of out-migration to Russia's more prosperous regions, with the result that the region started experiencing a labor shortage and its population became concentrated in big cities mostly in the west of the AZRF (Murmansk, Arkhangelsk, etc.). At this stage, the specific feature of the AZRF is its heterogeneity in terms of the level of the economic and social spheres development: some regions near the areas of resource extraction and processing are very well developed and urbanized, but a big part of the territory has underdeveloped economy, poor infrastructure and living standards.

The AZRF is characterized by *food security* risks, which is a long-standing problem caused by several factors: "geographical location, food contamination with harmful substances, climate change and the economic vulnerability of the region" [23]. The extreme climate conditions of the Russia's Arctic create obstacles to the development of agriculture and, therefore, reduce the possibilities of food self-sufficiency in the Arctic regions [24]. As a result, a high degree of dependence on food imports from other regions of the Russian Federation is observed in the AZRF. The problem of the population's access to foodstuffs is also urgent for the AZRF, which is associated with the seasonal nature of the transport infrastructure in many regions and, consequently, the problems of delivering food to remote regions [25].

An important factor in ensuring *health security* of the population of the AZRF is again the climate factor: the extreme natural and climate conditions of most areas of the AZRF exerts serious pressure on the human body and lead to serious diseases and often premature death. Moreover, if the indigenous population is adapted to such extreme living conditions, then "people who come even for temporary work fall under the negative influence of the northern climate, which significantly affects their health, labor productivity ..." [26]. On the other hand, climate change creates additional health risks associated with melting ice and permafrost, namely: methane emissions, which, in turn, accelerate climate change; contamination with persistent organic compounds from nuclear and chemical waste disposal sites that appear as ice melts and finally ice melts also pose a risk to infrastructure that can be deformed [27].

The problems of the effectiveness of the health care system are also characteristic of the Russia's Arctic [28].

Environmental security, the threats of which are environmental pollution and degradation, resource depletion, anthropogenic disasters – mostly relate to the AZRF. Pollution of the territory of the AZRF is a legacy of the Soviet period of the development of the Arctic. Today, the scale of human activity in the Arctic continues to grow due to global climate change: the intensification of shipping through the Arctic sea routes, as well as strengthening the oil and gas development of the Arctic and the transportation of resources extracted on the Arctic shelf. As a result, the Russia's Arctic still has the largest number of so-called "hot spots" [29].

In the context of *personal security*, it is appropriate to mention the important problem of the AZRF - a struggle against nomadic lifestyles and separation of the children from their families (to socialize them) – these practices interrupt the centuries-old continuity and result in an acute shortage of labor in the traditional sectors, on the one hand, and most young people's inability to adapt to the modern living environment, on the other. Relocation to a city exposes members of the indigenous communities to certain social risks, first of all, the risks of alcoholism and drug addiction. 150,000 members of the AZRF's indigenous communities, meanwhile, practice nomadism [30], which, however, does not have any legal or institutional basis in the RF. As for the RF's Land Code, it does not provide for any forms of control over land other than leasing or privately owning it, and this is the main obstacle to recognizing the indigenous communities' land rights.

Finally, there are threats to *community security*. The Russian Arctic is home to more than 30 indigenous Arctic communities: counting no more than 200,000 overall, they maintain their centuries-old traditional way of life or practice a nomadic lifestyle. The living standards in these minorities' settlement areas, meanwhile, are lower than in Russia' other regions; unemployment levels are 1.5-2 times higher than elsewhere in the country, infant mortality rates are higher and life expectancy is lower than elsewhere. Shaped yet in the 1930s, the concept of developing the northern regions and managing their natural environment caused much harm to the traditional livelihoods of the Russian Arctic indigenous communities, ruining the environment in their traditional areas of settlement and generating highly polluted areas [31]. The issue at hand is the damage to deer pastures – reindeer husbandry is a staple of the Arctic indigenous communities' economy. Reindeer husbandry is a form of human beings' adaptation to extreme living conditions, and this form is so perfect, diverse, and all-encompassing that academics even talk about the 'reindeer civilization.' When the area was designated for industrial use and then became polluted with industrial effluents, its indigenous residents lost not only their rangelands and hunting grounds but also traditional fishing locations and sites where they used to harvest wild crops.

6 Conclusion

Summing up, it can be noted that a paradoxical situation has developed: on one hand, the Arctic is one of the regions where problems related to 'human security' are most acutely manifested, and on the other hand, at the national level, there has not been formed an integrated approach to solving this group of problems. The traditional state-centric approach and the traditional understanding of 'security' still prevail in national Arctic strategies. Nevertheless, the content of strategies contains elements of human security, mainly devoted to economic and environmental security. In the latest revisions of several strategies (for example, the Arctic strategy of the Russian Federation), there is an increasing manifestation of the 'human dimension' of the Arctic policy.

Today, the problem of human security has acquired a special meaning for Russia in the context of the implementation of one of the most important tasks of the state policy of the Russian Federation in the Arctic - the sustainable development of the region. The successful development of the AZRF presupposes the creation of the foundation for the effective socio-economic development, and this requires the neutralization of the above-mentioned challenges and threats in the field of human security. However, even though the main attention in the strategic documents of the Russian Federation has been drawn specifically to the socio-economic and environmental blocks of problems, there are still no direct references to the concept of human security and the mechanism for implementing the concept of sustainable development of the AZRF has not been spelled out.

Acknowledgment

This work was supported by the Russian Foundation for Basic Research and Foundation "Maison des Sciences de L'homme" (France) project no. 20 514 22001.

References

1. The speech of Mr. Gorbachev M.S. Murmansk, *Pravda* **275** 25262 (1987) Retrieved from: http://www.helion-ltd.ru/gorbachev-speech-part-1
2. J. M. R. Martín, *Human and Societal Security in the Circumpolar Arctic. Local and Indigenous Communities* (Leiden: Brill, Nijhoff, 2018)
3. N. Ermolaeva. Finnish Foreign Minister: We need to maintain a dialogue with Russia. *Rossiyskaya gazeta* (2021) Retrieved from: https://rg.ru/2021/03/25/glava-mid-finliandii-s-rossiej-nado-podderzhivat-dialog.html
4. R. F. Inglehart, P. Norris, *Scand. Pol. St.* **35** 1 71-96, (2012)
5. UNDP. Human Development Report (NY: Oxford University Press, 1994) Retrieved from:

http://hdr.undp.org/sites/default/files/reports/255/hdr_1994_en_complete_nostats.pdf

6. V.N. Konyshev, *Nat. int.: prior. i bezop.* **40** 277, 44-45, (2014)
7. G. Hoogensen, D. Bazely, J. Christensen, A. Tanentzap, E. Bojko, *J. of H. Sec.* **5** 2, 1-10, (2009)
8. D.R. Chandler, *Sec. Dial.* **39** 4, 427-438, (2008)
9. N.S. Evtikhevich, E.V. Israelian, *Puti k miru I bezop.* **1** 44, 38-51, (2013)
10. H. Exner-Pirot, *Human Security in the Circumpolar North: What Role for the Arctic Council?* (2008) Retrieved from: https://www.rha.is/static/files/NRF/OpenAssemblies/Anchorage2008/a/5th_nrf_anc_2008_exnerpirot_human_security_circumpolar_north.pdf
11. H. Exner-Pirot, *Human Security in the Circumpolar North: What Role for the Arctic Council? (2008)* Retrieved from: https://www.rha.is/static/files/NRF/OpenAssemblies/Anchorage2008/a/5th_nrf_anc_2008_exnerpirot_human_security_circumpolar_north.pdf
12. A Blue Arctic (2021) Retrieved from: https://media.defense.gov/2021/Jan/05/2002560338/-1/-1/0/ARCTIC%20BLUEPRINT%202021%20FINAL.PDF/ARCTIC%20BLUEPRINT%202021%20FINAL.PDF
13. A Parliamentary Resolution on Iceland's Arctic Policy. Approved by Althingi at the 139th legislative session March 28, 2011. Retrieved from: http://library.arcticportal.org/1889/1/A-Parliamentary-Resolution-on-ICE-Arctic-Policy-approved-by-Althingi.pdf
14. Canada's Arctic and Northern Policy Framework. 2019. Retrieved from: https://www.rcaanc-cirnac.gc.ca/eng/1562782976772/1562783551358
15. Finland's Strategy for the Arctic Region 2013. Government resolution on 23 August 2013. Retrieved from: https://vnk.fi/documents/10616/334509/Arktinen+strategia+2013+en.pdf/6b6fb723-40ec-4c17-b286-5b5910fbecf4
16. Kingdom of Denmark Strategy for the Arctic 2011–2020. Retrieved from: https://naalakkersuisut.gl/~/media/Nanoq/Images/Udenrigsdirektoratet/100295_Arktis_Rapport_UK_210x270_Final_Web.pdf
17. Order of the President of the Russian Federation №164 *The Basic Principles of The Russian Federation's State Policy in The Arctic Until 2035,* Retrieved from: http://static.kremlin.ru/media/events/files/ru/f8ZpjhpAaQ0WB1zjywN04OgKiI1mAvaM.pdf
18. Sweden's strategy for the Arctic region 2020. Retrieved from: https://www.government.se/information-material/2020/11/swedens-strategy-for-the-arctic-region-2020/#:~:text=Sweden's%20strategy%20for%20the%20Arctic%20region%20presents%20the%20Government's%20objectives,and%20the%20environment%3B%20polar%20research%3B
19. *The Norwegian Government's Arctic Policy. People, opportunities and Norwegian interests in the Arctic. 2021.* Retrieved from: https://www.regjeringen.no/en/dokumenter/arctic_policy/id2830120/
20. J. M. R. Martín, Human and Societal Security in the Circumpolar Arctic. Local and Indigenous Communities (Leiden: Brill, Nijhoff, 2018)
21. Order of the President of the Russian Federation №1969 The Basic Principals of The Russian Federation's State Policy in The Arctic Until 2020 And Beyond], Retrieved from: https://rg.ru/2009/03/30/arktika-osnovy-dok.html
22. Strategy for Developing the Russian Arctic Zone and Ensuring National Security until 2035, dated October 26, 2020. Retrieved from: http://en.kremlin.ru/acts/news/64274
23. M.E. Nikitenko, I.B. Trofimova, *Soc.: pol., ec., law,* (2016). Retrieved from: https://cyberleninka.ru/article/n/prodovolstvennaya-bezopasnost-v-arkticheskoy-zone-rossiyskoy-federatsii
24. V.A. Ivanov, E.V. Ivanova, *Arctic: ecol. and econ.* **2** 26, 117-130, (2017).
25. A.N. Degtyarev, R.S. Zheishev, Yu.A. Nikitin, *Bull. of the Altai Ac. of Ec. and Law* **5**, 120-127, (2018). Retrieved from: https://vaael.ru/ru/article/view?id=95
26. A. Yu. Popova, *Problems of maintaining health and ensuring sanitary and epidemiological well-being of the population in the Arctic: materials of a scientific and practical conference with international participation* (SPb: IPK Kosta LLC, 2017). Retrieved from: http://s-znc.ru/wp-content/uploads/2017/05/%D0%A1%D0%B1%D0%BE%D1%80 % D0% BD% D0% B8% D0% BA_% D0% BD% D0% BF% D0% BA_% D0% 90% D1% 80% D0% BA% D1% 82% D0% B8% D0% BA% D0 % B0.pdf
27. Yu. A. Bolsunovskaya, G. Yu. Boyarko, *Fund. res.* **9-12**, 2725-2728, (2014). Retrieved from: http://www.fundamental-research.ru/ru/article/view?id=35423
28. A. Yu. Popova, Problems of maintaining health and ensuring sanitary and epidemiological well-being of the population in the Arctic: materials of a scientific and practical conference with international participation (SPb: IPK Kosta LLC, 2017). Retrieved from: http://s-znc.ru/wp-content/uploads/2017/05/%D0%A1%D0%B1%D0%BE%D1%80 % D0% BD% D0% B8% D0% BA_% D0% BD% D0% BF% D0% BA_% D0% 90% D1% 80% D0% BA% D1% 82% D0% B8% D0% BA% D0 % B0.pdf

29. UNEP / GEF Project Russian Federation - Supporting the National Action Plan for the Protection of the Arctic Marine Environment (2008). Retrieved from: https://docplayer.ru/45130554-Proekt-rossiyskaya-federaciya-podderzhka-nacionalnogo-plana-deystviy-po- zashchite-arkticheskoy-morskoy-sredy.html
30. U. Vinokurova, Yu. Yakovets, *Arctic Circumpolar Civilization* (Novosibirsk: Nauka, 2016)
31. M.L. Lagutina, *Russia's Arctic Policy in the Twenty-First Century. National and International Dimensions* (Lexington Books, 2019)

Modern problems of Evenk language functioning and its development prospects

Tamara Andreeva*,[1], and *Kirill* Struchkov[2]

[1]Institute for Humanities Research and Problems of Indigenous Peoples of the North SB RAS 1, Petrovsky str. Yakutsk,677027, Russia

[2]M. K. Ammosov North-Eastern Federal University 42, Belinsky str. Yakutsk, 677018, Russia

Abstract. The article is dedicated to the modern problems of Evenk language functioning and its development prospects. The preservation and development of the language remain an urgent sociolinguistic and sociocultural problem to this day. The research conducted in Yakutia shows that the functioning of Evenk language is reduced to a minimum. The linguistic situation with the native language is assessed as definitely critical and can be one of the indicators of socio-economic ill-being of not only Evenks, but also other indigenous peoples. The main problems of linguistic life in the republic include the reduction of Evenk language functions and inadequate formation of a positive motivation for learning the native language. In the society, there is a complex system of objective and subjective factors that do not contribute tointensive development of positive processes in the preservation and development of Evenk language. In general, subjective factors include poor organization of the educational process in educational institutions, ineffectiveness of training sessions, insufficient attention to the development of speech culture and enrichment of the native language's vocabulary, low level of educational and methodological support of education and upbringing. The social environment for Evenk language functioning is continuing to narrow, which makes an objective factor.

1 Introduction

The article discusses modern problems of functioning and prospects for the functioning of the Evenk language. Referring to specific linguistic situations in various places of Evenks' compact residence in the republic – Iengravillage of NeryungriDistrict, Khatystyrvillage of Aldan District, Tyanyavillage of Olyokma District – allows one to obtain reliable assessments of the conditions and prospects for the development and revival of the natural intergenerational transmission of the native language. This takes into account 1) status, demographic and communicative characteristics; 2) assessing their vitality in the existing sociolinguistic classifications; 3) the conditions for turning the language shift [1] and the prospects for studyingthe language development.

Modern language policy strategies aimed at strengthening the prospects for the vitality of indigenous languages stem from the relevant articles of the UN Declaration on the Rights of Indigenous Peoples (2007) [2]. The main aspects include the differentiation of the goals of reviving or preserving the status of the minority native languages - functionally the first language or functionally the second language; the importance of teaching in native indigenous languages using modern teaching technologies. An equally important factor is the potential for the indigenous peoples' activity.

2 Results and discussion

According to the 2010 census, 21,008 Evenks live in the Sakha Republic (Yakutia); the share of people who indicated their native language corresponding to their ethnicity remains low. As compared with the previous census, thesharehas been steadily decreasing [3].

The functioning of the Evenk language is reduced to a minimum, the linguistic situation with the native language is assessed as definitely critical and can serve as one of the indicators of socio-economic ill-being of not only Evenks, but also other indigenous peoples of the North. According to the Law of the Sakha Republic (Yakutia) "On the Status of the Languages of the Indigenous Small-numbered Peoples of the North of the Sakha Republic (Yakutia)", the languages of the peoples of the North, including Evenk, are official atthe places of these peoples' compact residence.

* Corresponding author:taan2001@mail.ru

Table 1. Number of indigenous peoples of the North of the Sakha Republic (Yakutia) according to the census of 1989, 2002 and 2010, thousand people

Representatives of the indigenous peoples of the North of the Sakha Republic (Yakutia)	1989	2002	2010	Change in the number by 2010 as compared with 1989, %
Evenk	14,4	18,2	21,0	145,8
Even	8,7	11,7	15,0	172,4
Yukaghir	0,8	1,1	1,3	162,5
Chukchi	0,5	0,6	0,7	140,0
Dolgan	0,4	1,3	1,9	475,0
TOTAL	**24,8**	**32,9**	**39,9**	**160,9**

By analogy with the international practice, the official status ofa language presupposes the use of the language in the public life of the society. Unfortunately, this is not the case. The Evenk language is not used: on the official signs, in the street names, attributes and information, there are only a few examples; in none of the settlementsthe Evenk language is used in the work of the authorities and paperwork, etc. Thus, in fact, the Evenk language does not fully fulfill the role of the official language.

Changes in the modern linguistic situation among the indigenous small-numbered peoples of the North, including Evenks, in the Sakha Republic (Yakutia) deserve special attention [4].

In the Sakha Republic (Yakutia), the reports of the government, public organizations, educational and scientific institutions dealing with this issue contain a lot ofdifferent information on the measures taken to improve this situation.

The republic has adopted its own language policies. These are the Laws "On the legal status of the indigenous small-numbered peoples of the North", "On the languages in the Sakha Republic (Yakutia)" (October 16, 1992), "On the status of the languages of the indigenous small-numbered peoples of the North of the Sakha Republic(Yakutia)" (1992), "Oneducation" and otherregulatory documents of the Ministry of Education of the Russian Federation and the Sakha Republic (Yakutia).

To implement the Law "On the languages in the Sakha Republic(Yakutia)", theNationalProgram on the revival, preservation and development of the languages of the indigenous peoples of the Sakha Republic (Yakutia) was developed and adopted for implementation in 1993. The main developers of this program were the Institute of Humanities of the Academy of Sciences of the Sakha Republic (Yakutia) (Part 1) and the Institute of Indigenous Peoples of the North SD RAS (Part 2) that existed in those years. The goals of the program are to create favorable conditions for the development and support of the state and official languages of the Sakha Republic (Yakutia) [5].

Educational sovereignty has become one of the ways to manifest the republic's state sovereignty.Experts recognize that a systematic impact on the Evenks' linguistic situation, as well as purposeful activity of teaching the language in the educational institutions can stop the assimilation processes resulting in the loss of the language. Unfortunately, over the past 4 years, there has been a negative trend of reducing the number of schools where native languages are taught. To overcome this negative trend, the Ministry of Education and Science of the Sakha Republic (Yakutia) needs to tighten control and responsibility of the district education authorities for the indices; in particular, it shouldinclude these indices in the performance indicators of the senior education executives.

At present, the Native Language subject is taught in Evenk schools; however, the scope of Evenk language is shrinking, and the introduction of new subjects cannot really change this situation. The fact is that children barely hear Evenk speech at home and outside school in general, i.e., there is no language environment; they do not really need to use it anywhere [5].

Staffing the preservation of languages remains a topical issue.

Sociolinguistic approach considers the literary language as one of the forms of the language existence in general. "According to the experts in Tungus languages, modern Evenk languagestill does not have an established literary and written form" [6].

In this regard, the issues of identifying the prospects for further functioning of Evenk language arise. The results of surveying the degree of preservation and the prospects for the development of Evenk language among the Evenks in the areas of dispersed residence of the Evenks via surveys and interviews made it possible to reveal the ratio of those who have good knowledge of Evenk language,poor knowledge of Evenk language or no knowledge of Evenk language at all. Let us give an example of the linguistic situation of Evenks living in the village of Khatystyr of Aldan District. The analysis of the statistical data presented by a student of the Northern Philology Department of AmmmosovNEFU is presented as follows: total 59 people were interviewed, 35 of them being men and 24 women. Of these, 12 people have higher education, 25 secondary vocational education, 12 secondary education, 12 schoolchildren, and 10 students. 43 people recognized Yakut language as the most functional, 23 people recognized the need for Evenk language, 16 people considered Evenk language unnecessary. 59 respondents see the importance of Evenk language in preserving the traditional culture. Only 1 person of all the respondents spoke Evenk. As you can see, in the village of Khatystyr, mainly the Yakut-Russian type of bilingualism is observed, and it can be said that almost all residents have switched to the Yakut language.

The survey that we conducted during the expedition made it possible to define the level of distribution and use of Evenk language in the village of Tyanya of Olyokma District. 120 people were surveyed, which is about 1/3 of the number of the Evenks living in the village. The survey included a group of questions about the informants' language competence. Based on the results of the survey, we were able to divide the residents of Tyanya villageinto the following linguistic communities: 1) Evenk-Yakut bilinguals (3.6% of the total number of respondents); 2) Evenk-Yakut-Russian (72.9%); 3) Yakut-Russian bilinguals (23.5%). Most Evenks are Yakut-speaking. Only 44.7% of the respondents havepres erved their majority language.

Table 2. Number of indigenous peoples of the North of the Sakha Republic (Yakutia) (according to the All-Russian census of the Russian Federation in 2010)

№	Indigenous people of the North	Total number in RF (people)	Including in SR (Y) (people)	in % from the total number	consider the language native (people)	In % from the total number
1	Evenk	38396	21008	54,7	4802	12,5
2	Even	21830	15071	69	5656	25,9
3	Yukaghir	1603	1281	80	370	23,1
4	Chukchi	15908	670	4,2	5095	32
5	Dolgan	7885	1906	24,2	1054	13,4/55

The situation among the Evenks of Iengra village is different – the Evenk-Russian type of bilingualism is observed here. Almost all Evenks speak their native language to varying degrees. All Evenks in the village speak Russian. In mixed families, Russian is the main language of everyday communication. The Evenk language is spoken only with the older relatives. In purely Evenk families, parents most often speak Russian with children and among themselves. The Evenk language is mainly used to discuss the topics they want to hide from the children or strangers.

According to the survey, all Iengra adult Evenks would like their children to be able to speak their native language. In addition, they believe that it would be good if Evenk language were taught at kindergartens and schools. Most Evenk children also expressed a desire to know their native language. Almost all the respondents (except 3 people) aged 26 to 60 indicated that they learned their native language in the family; they have been fluent in their native language since childhood and can write and read Evenk.

Thus, the position of Evenk language has been changing in two opposite directions simultaneously. On the one hand, after receiving the Cyrillic alphabet, Evenk languagebecame a written language; the norms of the so-called literary Evenk language, based on thePoligusdialect, were formed; the measures to spread this language standard to all spheres of use of Evenk language have been taken, largely in vain. On the other hand, the number of people speaking Evenk fluently has been reducing steadily; Yakut and Russian language are becoming a universal communication standard in both public and private life.

At present, the residents of Yakutia note that the native language is very poorly spread among the Evenks, with forecasts regarding its future being quite pessimistic. However, a simple statement of this situation seems to be too general an assessment. The level of native language proficiency is not the same among different age groups of Yakutia's population and in different regions. Conducting a more detailed analysis of the modern linguistic situation in the future seems interesting.

The process of replacing Evenk language with Yakut and Russian in everyday communication, turning it into a kind of ethnic rarity, fits into the timeframe of upbringing about 3 generations. In turn, the fieldwork materialmakes it possible to clearly distinguish three age groups matching them among the residents of the surveyed settlements of Tyanya(Olyokma District), Iengra (Neryungri District), and Khatystyr (Aldan District).

For the people of the older generation, born in the 1930s-1940s, Evenk language is native in the full sense of that word;as remembered by them, it was the language of their communication with their parents, they still freely communicate in Evenk with their families. Quite often, their children, and even more so grandchildren, are unable to adequately respond in Evenk, but there is no information about the cases when old people force their families to communicate in the native language. In general, the people of this generation slightly disfavor the ignorance of the native language by anyone in the family, but they do not make any attempts to change this situation.

Most adults, born in the 1950s-1960s, use Yakut and Russian languages in most life situations. Nevertheless, as a rule,the representatives of this generation consider they know theEvenk language at a simple level. Fieldwork observations show that a long conversation, even a completelyeveryday one, in the Evenk language is quite rare and is usually limited to a few short remarks.

Youth speaks Evenk even less;the language has become a school subject rather than a living means of communication.

We see that at present, the spoken Evenk linguistic environment does not really exist or is extremely deformed. It is obvious that the fragmentary information possessedby modern youthcannot be considered a language. That is why, it is no accident so many of our middle-aged informants are confident enough to say that they know their native language and are fluent in it. The mass media, the education system – all these mainstream information resources, should, in one way or another, be aimed at formingthe ethnic identity. However, this is also a very difficult problem.

The results of monitoring the functional distribution (linguistic distribution, the natural combination of languages with each other) revealed the respondents' dissatisfaction with the volume of broadcasts in the languages of the indigenous peoples of the North, including Evenk. At present, the linguistic ratio of television programs in the republic does not differ from the traditional one: 62% of broadcasting is in Russian, 38% in Yakut, 1% in the languages of the indigenous peoples of the North, including Evenk.

4 Conclusion

In this regard, in the context of international experience and the experience of language legislation, the task is to develop new guidelines and priorities to preserve and develop the linguistic and cultural diversity in the republic.

Acknowledgement

The work was conducted within the framework of the project "Verbal traditions of the indigenous small-numbered peoples of the North: documentation, description, systematization" of the Federal Research CenterYakutsk Scientific Center RAS # 121033100192-9 dated March 21, 2021.

References

1. Fichman J. *Reversing language Shift: Theoretical and Empirical Foundations of Assistance to Threatened Languages.* (Multilingual Matters Ltd, (1991).
2. *UN Declaration on the Rights of Indigenous Peoples* Retrieved from: https://www.un.org.
3. Federal State Statistics Service. *All-Russian Population Census* (2010) Retrieved from: http:www.gks.ru
4. A.A. Burykin, The dynamics of the language situation, functioning of writing and stages of evolution of the ethnic cultures of indigenous peoples of the North: on the problem of the relationship and interdependence of linguistic and socio-cultural processes. *Small-numbered peoples of Siberia and the Far East. Problems of preservation and development of languages.* (Saint-Petersburg, 1997)
5. K.N. Struchkov, *Functioning of Evenk language in the field of education in the Sakha Republic (Yakutia).* (Novosibirsk, Nauka, 2009).
6. N.Ya. Bulatova, *Evenk dialects of Amur Oblast.* (Leningrad, Nauka, 1987)
7. N.Ya. Bulatova, Evenk language and its regional variants in the sociolinguistic aspect. *Small-numbered peoples of Siberia and the Far East. Problems of preservation and development of languages.* (Saint-Petersburg, 1997)

Efficiency indicators of the eastern polygon and its development prospects

Natalya N. Grigoryeva[1,*], and *Tatyana A.* Bulokhova[1]

[1] Irkutsk State Transport University, 15, Chernyshevsky Str., Irkutsk, Russia

Abstract. The development strategy of "Russian Railways" OAO provides for increasing competitiveness and strengthening the company's market position. Due to the increase in Russia's foreign trade turnover with the countries of the Asia-Pacific region, the growth of transit freight traffic between Europe and Asia, and the need to implement the country's transit potential in the context of the globalization of the transport service market the railway industry faces the task of handling the growing volumes of traffic in the direction of the Far Eastern ports. One of the essential goals until 2030 is to increase the country's transit capacity. Its implementation is a priority in the field of international standing of our country. Along with the Baikal-Amur Mainline, the Eastern polygon is pivotal in solving this problem. Its development will ensure the connection of transport flows from West to East, will make changes in the routes of world trade, thereby increasing the share of Russia in Eurasian transit. The degree of efficient railway transport operation depends on the creation of favorable conditions for modernization, transition to the innovative development path and sustainable growth of the national economy. This generally contributes to the conditioning of Russia's leadership in the world economic system, as there is a new geopolitical landscape globally evolving at present, whose contours should be considered in the foreseeable future. In today's conditions of railway management, it is impossible to head over to an innovative path without proper infrastructure development, affecting the performance of structural units and requiring investments.

Keywords: reformation, transit potential, polygon technologies

1 Introduction

In the Russian transport system, railways have a leading role characterized by long transportation distances, the lack of communication of other types of transport in the direction of East-West, primarily in the regions of Siberia and the Far East, by spatial disintegration of production and consumption of basic raw materials. A significant increase in the volume of traffic from the West Siberian Railway [8] to the ports of the Far East resulted in the need for structural changes in the field of traffic management of the Krasnoyarsk, East Siberian, Transbaikal and Far Eastern railways. They have common technological features (origination (dissipation) of freight traffic volumes, provision of traction service, logistic management of the delivery to seaports and interstate division points, etc.) and infrastructure parameters (in particular, the Eastern polygon).

The Eastern polygon, located within the boundaries of 4 railways – Krasnoyarsk, East Siberian, Transbaikal, and Far East – serves the transport needs of 14 constituent entities of the Russian Federation and the transit needs of the country (Grigoryeva, 2020).

2 Problem Statement

Today, as with the origination of Russian railways, the main remaining problem is the chronic shortage or complete lack of technological infrastructure reserves due to the extremely high network load and its negative impact on the quality of customer service. The targeted work, carried out at "Russian Railways" OAO to organize and plan train traffic on the network polygons (instead of regional management principles), is aimed at ensuring the growth of freight turnover, passenger turnover and indicators of quality operation performance [7]. At the same time, the basis of the polygon technology, consisting of uniform principles of car and locomotive fleet and infrastructure management, requires the development of the entire production complex, for the needs of which investments are required.

3 Research Questions

At the current stage of development, the polygon technologies have proved to be positive in the organization and direct management of the transportation process of rail transport.

* Corresponding author: zolotkina@mail.ru

Today the Eastern polygon, formerly the first pilot project, clearly proves its effective functioning.

Implementation of the transport management model on four roads allows one to organize and control car and train traffic, based on the feasibility not only of the economic criteria set by the interests of the Holding as a whole. This creates the possibility of moving to a technology-linked framework in the rationalization of train traffic management and makes it possible:

- to increase the operation efficiency by improving the handling of the transportation process;
- to improve the quality of planning and organization of "track possessions" and long-term shutting-downs for infrastructure repairs;
- to increase the length of runs and sections within which the technically safe passage of cars between train service points is guaranteed;
- to improve the quality of service for rail transport users [9];
- to ensure a rational organization of working hours and rest period of locomotive crews [6].

4 Purpose of the Study

The purpose of this article is to analyse the performance of the Eastern polygon and its future development through investment in its infrastructure.

5 Research Methods

Since 2012, the effectiveness of polygon technologies has been proven not only by real-time work indicators, but also by locomotive crew organization indicators, which are steadily improving.

During the study period, the operating volume increased by 27 % (Figure 1). With a relatively small increase in the work force of locomotive crews (2.5 %) (Figure 2), overtime hours are reduced by 46 % (Figure 3) and hours of non-productive losses have been reduced by 15 % (Figure 4).

Since 2012, the average daily capacity of the working fleet locomotives at the polygon has increased by 15.3 % (Figure 5). This result was achieved by driving full-weight and full-length trains.

The growth of the average train weight (Figure 06) by 3.2 % and the average daily kilometerage of the locomotive (Figure 7) by 6.1 % also had a positive effect on the locomotive performance.

The reasons for the average train weight growth were the increase in the average train length and the decrease in the percentage of empty kilometerage.

The service speed depends on the throughput capacity of separate sections, the amount of freight and passenger train traffic, the technical condition of the track, block service and rolling stock, train schedule and centralized traffic control. This, in turn, had a positive impact on its tendency (Figure 8).

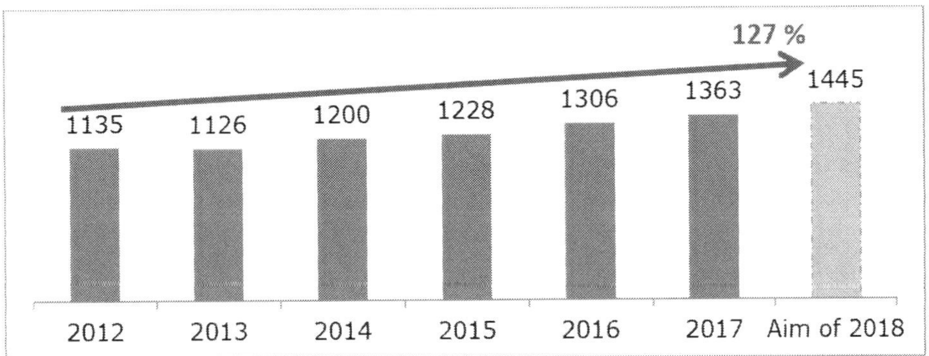

Figure 01. The amount of work in the billions of km, gross

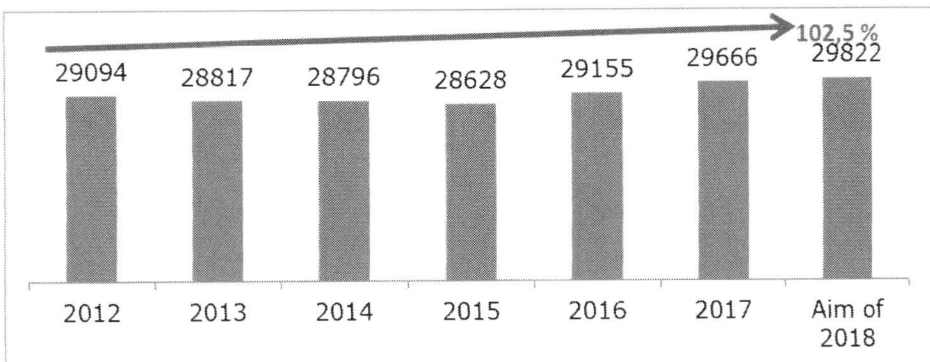

Figure 02. Average listed work force, pers

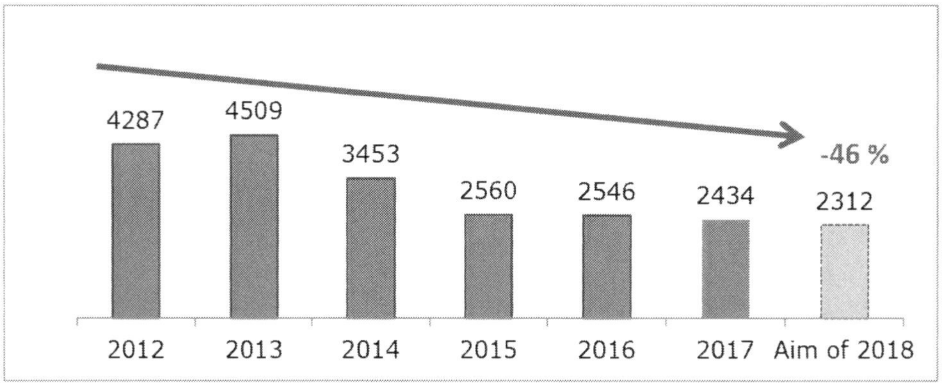

Figure 03. Overtime work, hrs

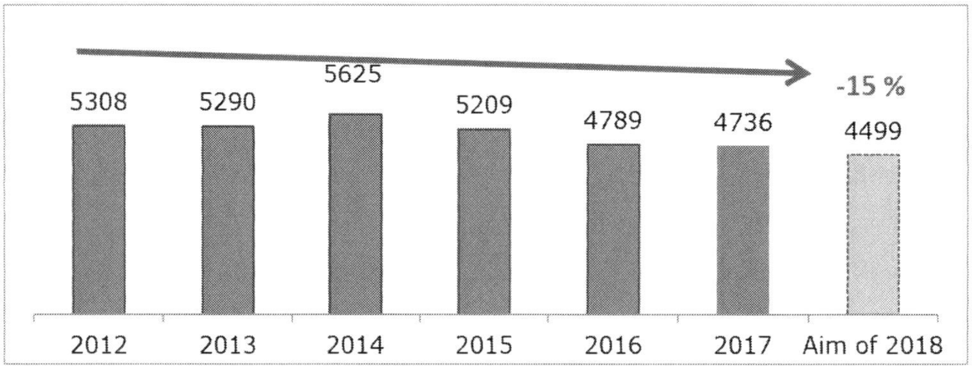

Figure 04. Non-productive losses, thousands of hours

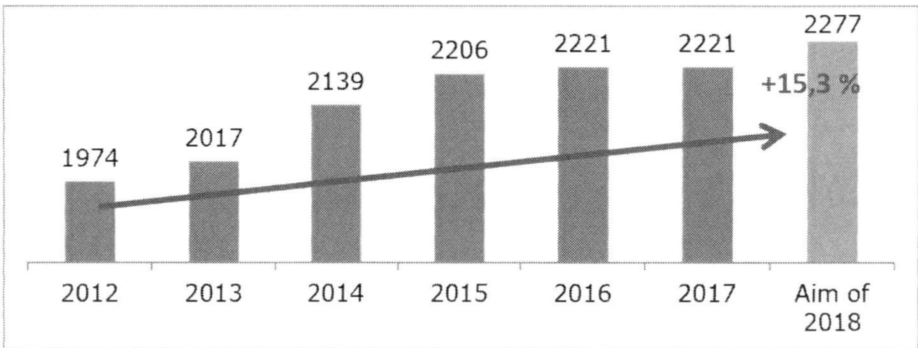

Figure 05. Average daily performance, thous. km, gross

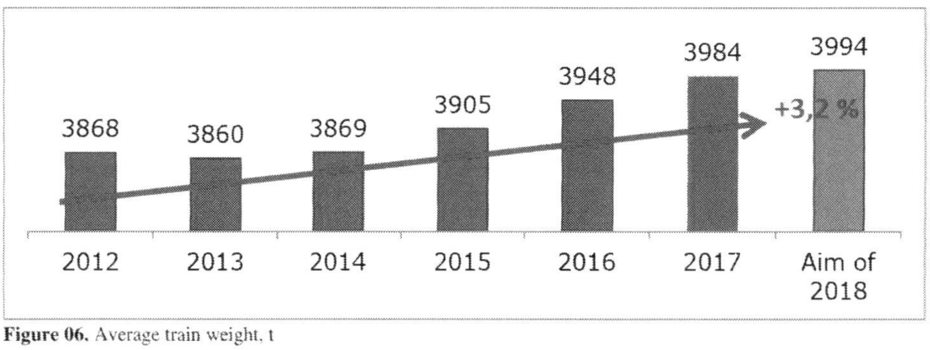

Figure 06. Average train weight, t

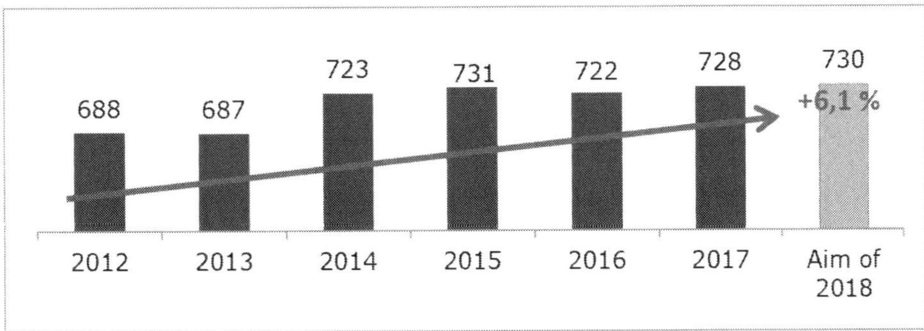

Figure 07. Average daily kilometerage, km

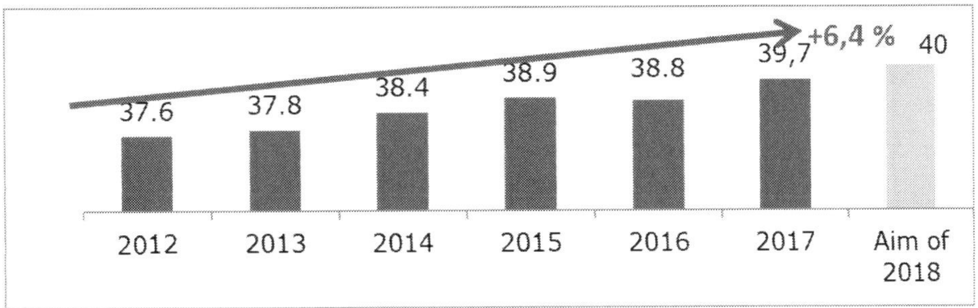

Figure 08. Service speed, km/h

The introduction of uniform parameters of train work at the Eastern polygon and its traction maintenance, carried out under the management of the Eastern Polygon Freight Control Center, allows one to implement the principles of technological interaction of all participants in the transportation process on the problems of:

- long-term and real-time planning of transportation process (in accordance with the railway production activities (regional corporate management centres), directorates, railway rolling stock operators and owners of non-public tracks);
- standardizing of transportation process;
- real-time control of the technological standard implementation, including the speed of car traffic volume and train traffic movement;
- dispatching real-time management of the transportation process,
- information support of interaction.

One of the main functions of the Eastern Polygon Freight Control Center is to ensure the logistics of transporting foreign trade goods by appointment to the ports and border crossings of the Far Eastern Railway. The largest Far Eastern ports account for a quarter of the cargo turnover of Russian ports [5].

Logistic support of transportation within the boundaries of the Eastern polygon is one of the most important functions of polygon technology. The main task is to monitor the car traffic volumes to the Eastern polygon at inter-road division points and to create proposals for the standardization of the presence of cars on the polygon roads and the amount of their handover at the Arkhara and Khan junctions to port stations and oil depots.

The concentration of this work in the Eastern Polygon Freight Control Center will not only make it possible to arrange the rhythmic supply of trains to ports, but also to significantly increase the train traffic progress prediction depth on the Eastern polygon transit roads, starting from the freight traffic origination.

Currently, the Trans-Siberian railway and the Baikal-Amur mainline's throughput capacity is more than 100 billion tons and will be additionally increased by 65 million tons in the coming years. However, such results can only be achieved by increasing the investment attractiveness of the transport and logistics sectors [4].

For example, the volume of the investment program of the East Siberian Railway, which is part of the Eastern polygon, amounted to 42.8 billion rubles in 2017 (this is 7.8 % more than in 2016). At the end of the two-year period of 2016–2017, 91 facilities were commissioned on the Eastern polygon development project [2].

In 2018, about 57 billion rubles were provided to implement the investment program of Russian Railways OAO within the borders of the Eastern Siberian Railway. Most of the money is used to modernize the railway infrastructure of the Baikal-Amur and Trans-Siberian Mainlines to increase the carrying and throughput capacity of the road. Investments in the development of the East Siberian Railway in 2019 amounted to 54 billion rubles. In 2019, the total volume of repair and track work on the East Siberian Railway was 406.8 km. Within the framework of the campaign, the track was reconstructed for more than 232 km: on the sections of the Trans-Siberian (130.4 km) and the Baikal Amur Mainline (102.3 km), 170 km of full and intensive medium repairs were carried out. The continuous or "velvet smooth" track has been laid on 38 sections of the mainline with a total length of 340.9 km. Thus, over the past three years, the amount of investment in infrastructure upgrades on the East Siberian Railway has increased [1].

As a result of the renewal of infrastructure facilities, it was possible to reduce the duration of sections with existing speed limits by 21 %, to reduce the length of barrier locations by 6.5 %.

6 Conclusion

Thus, the development and renewal of the railway infrastructure is a prerequisite for ensuring the balanced economic development of the Eastern polygon and the country as a whole, forming an internal production base, implementing industrial potential, expanding foreign economic ties and ensuring the integrity and security of the state.

Acknowledgments

The research "Improving interaction mechanisms between polygon structures and regional corporate governance centers as illustrated by the Eastern Railway Polygon of Russian Railways OAO" was publicly funded.

References

1. T.A. Bulokhova, A.S. Peretolchina. The level of development of the railway infrastructure as a factor in ensuring the industry customer focus and the sustainable development of the regions. In: *The development of economic science in transport: the economic basis for the future of transport systems* (pp. 147–154) (Inst. of Independent Soc. and Econ. Stud. OOO, St. Petersburg, 2019)
2. T.A. Bulokhova. Modeling the quality of public transport services. Modern Technol. Syst. anal. Model., **1(65)**, 128–136 (2020)
3. N.N. Grigoryeva. Railway transport management system transformation in passing to polygon technologies. IOP Conf. Ser. Mat. Sci. and Eng. **918**, 012194 (2020). DOI: 10.1088/1757-899X/918/1/012194
4. N.N. Grigoryeva. The Problems and the Future of Implementing Innovations in Railway Transport. *Transport infrastructure of the Siberian region. The mat. of 9th Int. sci. and pract. conf.*, vol. 2 (ISTU, Irkutsk, 2018), pp. 97–101
5. V.A. Olentsevich, V.E. Gozbenko, S.K. Kargapoltsev. Automated system for placing and securing cargo on open rolling stock of railway transport. Bull. of Irkutsk State Techn. Univ., **21**, 157–165 (2017)
6. V.A. Olentsevich, N.P. Astashkov, Yu.I. Belogolov. Improvement of Transport Processes Operational Management on Railway Transport. *Proceedings of 6th International Symposium on Innovation and Sustainability of Modern Railway ISMR 2018* (ISTU, Irkutsk, 2018), pp. 602–609
7. V.E. Kautz, E.V. Nefedieva. Innovations in the transportation process by the example of the East Siberian Railway and the Eastern Polygon of the Trans-Siberian Railway. In: *The Transport Infrastructure of the Siberian Region*, vol. 2 (pp. 120–124) (IrGUPS, Irkutsk, 2019)
8. O.I. Mokretsova, Yu.N. Malanina, N.V. Kretova, A.A. Zaikovskaya, I.R. Libenson. Governmental Support and Efficiency Management of JSC "Russian Railways" Investment Activity. In: *Advances in Economics, Business and Management Research*. Vol. 113. *Factors of Regional Extensive Development (FRED 2019)* (2019). Retrieved from http://creativecommons.org/licenses/by-nc/4.0/
9. G.V. Silicheva, T.A. Bulokhova, E.S. Podgornova. *Methodological approaches to quality management in railway transport in the context of industry reform* (IrGUPS, Irkutsk, 2016)

Economic connectivity of international transport corridor projects and the trans-Siberian railway

Mariia A. Khazheeva[1,*], and *Ekaterina A.* Bondarchuk[1]

[1] Irkutsk State Transport University, 15, Chernyshevskogo str., Irkutsk, 664074, Russia

Abstract. The development of Russia's foreign economic potential in the current conditions of globalization is inextricably linked with the assessment of the prospects for implementing the country's transport and infrastructure potential. The importance of transport in the conditions of increasing instability of the world economy is a factor in the sustainability of the national economy. Implementing a cluster approach to the development of an efficient and modern transport and logistics infrastructure will help to accelerate the movement of goods and reduce costs in international trade, as well as the consistent integration of the country's transport system into the global network of international transport corridors. Hence, the Trans-Siberian Railway occupies a leading place in the Russian state policy on the formation of international transport corridors of latitudinal striking. The Trans-Siberian Railway's ability to provide international transit container transportation as an alternative to the traditional sea route is being considered. In this regard, consideration of the prospects for integrating the projects of international transport corridors with the participation of the Trans-Siberian Railway is most relevant. This article discusses both economic, technological and political factors that influence the potential effectiveness of this project. The prospect for the development and integration of the Trans-Siberian Railway with the Trans-Korean Railway leads to the creation of an effective transport corridor for the foreign trade of the countries of Southeast Asia and Europe, primarily the Republic of Korea with the EU countries, as well as with Russia. Analysis of the foreign trade turnover of these participants of economic cooperation shows the perspective of dynamics and the huge potential of foreign trade. Attention is focused, on the one hand, on the competitive advantages of the Trans-Siberian Railway, including an increase in throughput capacity, which is largely due to the use of the polygon management model.

Keywords: international transport corridors, Trans-Siberian Railway, transport and logistics infrastructure, container transportation

1 Introduction

International transport corridors (ITCs) are becoming particularly relevant in the conditions of increasing interconnectedness and interdependence. Existing as a single integrated complex, they have different transport systems of the states participating in the corridors, representing several countries with different national regulations, customs regimes and different widths of the railway track.

The Trans-Siberian Railway is the longest railway in the Russian Federation, guaranteeing direct and safe access to both Russian and transit goods from the ports of the Far East to the European Union. The modernization and development of the Trans-Siberian Railway are carried out with the active support of the state.

However, the attractiveness of rail transport is increased by the performance and capacity development capabilities of continental freight flows generated by international trade. The dynamic development of trade and economic cooperation between countries and the world's trade and economic blocs requires appropriate transport support, which can improve the increase in traffic to service regions with the global transport system and the dynamic growth of the cargo base characteristic of the countries of the Asia-Pacific region (APR) with regard to the East-West trade [5].

2 Problem Statement

Currently, as the trend of environmentalization increases, most coal terminals in the Baltic Sea ports are reducing transshipment due to falling demand for this type of fuel in Europe. This trend is clearly visible already as early as 2019. This adverse trend will continue in 2020. Due to the changing structure in the Baltic, a major oversupply of transshipment capacity is formed, which cannot be said about east direction. In east direction, on the contrary, there is an increase in demand for coal [7].

* Corresponding author: marya77@list.ru

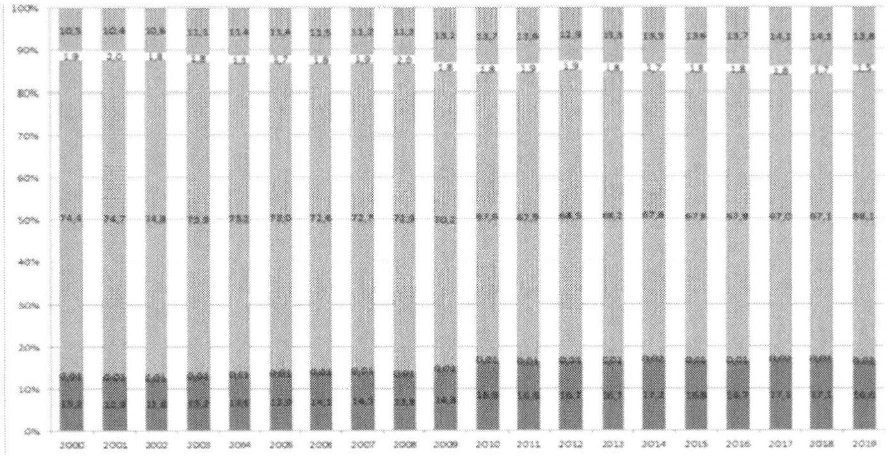

*Blue – railway; red – air; yellow – water transport; green – automobile; orange – pipeline

Figure 01.The structure of transportation by mode of transport in the Russian Federation in dynamics from 2000 to 2019 (Federal State Statistic Service, 2019)

In the "East-West" project, the basis of the Eurasian direction is the Trans-Siberian Railway, the vector of which is aimed at transport support of foreign trade relations in both Siberia and remote regions of the East and Russia as a whole.

Exploring the current economic state, we can see a slowdown in GDP growth in 2019 compared to 2018 from 2.5 % to 1.3 % [3]. The main reasons behind the decline in growth were: the slowdown in economic activity in the world as a whole, the maintaining and strengthening of sanctions pressure, the increase in the VAT rate, the reduction in export volumes, etc. Accordingly, we are seeing a decline in traffic across all modes of transport (Figure 1).

On the basis of the presented figure, it can be stated that over the past 19 years there has been a trend of growth in rail transport and it has been observed since 2010. But despite the growth, the leading mode of transport is still automobile transport, which occupies 68.1 %. However, when comparing 2019 with 2018, it can be stated that the use of railway transport has decreased by 0.5 %.

3 Research Questions

The advantages of railway transportation while implementing transit transportation are the following [12].

1. High overload of particular sections of the Trans-Siberian Railway.

2. The movement of main tracker trains is carried out strictly according to the schedule, which allows manufacturers and owners of goods to track and calculate the time of movement of these goods using the Internet.

3. The cargo capacity of the train is twice as high as in the EU.

The main cargo flows in this direction on the Russian railway are coal, oil, iron, ore, black metals, grain, and machines. The cargo turnover by mode of transport is represented in Table 1.

Table 01. Cargo turnover by mode of transport, 0669 (billion-t.km) [3]

Modes of transport	**2000**	**2005**	**2010**	**2015**	**2016**	**2017**	**2018**	**2019**
Transport – total	**3 638**	**4 676**	**4 752**	**5 108**	**5 198**	**5 488**	**5 636**	**5 674**
including railway transport	1 373	1 858	2 011	2 306	2 344	2 493	2598	2602
automobile	153	194	199	247	248	255	259	275
pipeline – total	1 916	2 474	2 382	2 444	2 489	2 615	2668	2686
including gas pipeline	1 171	1 317	1 259	1 176	1 181	1 300	1336	1318
oil pipeline	718	1 123	1 084	1 226	1 262	1 265	1276	1318
oil product pipelines	27	33	39	42	46	50	55	51
marine	122	60	100	42	43	50	37	37
inland water-borne	71	87	54	64	67	67	66	66
air	2.5	2.8	4.7	5.6	6.6	7.9	7.8	7.4

Based on the data presented in the table, we can infer that there is an increase in freight traffic by railway transport. The largest growth was such cargo as: loading of iron ore (by 3.5 million t, or 3.0 %), of chemical and mineral fertilizers (by 1.3 million tonnes, or 2.2 %), cement (by 1.2 million tonnes, or 4.7 %), cargo in containers (by 4.1 million tonnes, or 13.1 %) [6]. In the study of the share of domestic and export communications, the loading structure was not transformed in 2019: 61.7 and 37.4 %, respectively. In loading, the main cargoes were coal, oil and construction cargoes, and ore [8].

4 Purpose of the Study

The aim of the research is to study and substantiate the possibilities of integration of projects of international transport corridors with the participation of the Trans-Siberian Railway and the use of polygon management technologies [4].

The integration of projects involves coordination between different economic entities on the same territorial plane. If there are contradictions between the entities in solving the unified issues of the connection or there are all sorts of obstacles, it all interferes with the spatial integration. Among the theoretical concepts, one of the most interesting ones, in our opinion, is the concept of the "pole of growth" by F. Perroux, according to which the economy begins to grow only in certain points, which are the "poles of growth". According to the concept of F. Perroux, the initial growth of the economy, having begun at the poles of growth, spreads and creates an "increase effect" on a national scale [13].

Another researcher, P. Pottier, considered his vision of the theory of "poles of growth", basing it on the fact that the territories among themselves are in constant cooperation, so some act as "locomotives" of growth and drag along less developed ones. This is possible due to increased cargo flows, the spread of innovation, the development of infrastructure. Therefore, the adjacent territories, located between the poles of growth, increase their level of well-being of the population through or with the aid of corridors of development, "by poles of growth – the spatial framework of economic growth of the country" [14].

The implementation of infrastructure modernization includes several stages; the initial cost of the project was 562.4 billion rubles, as a result of the audit by the Accounts Chamber of the Russian Federation was reduced to 520.5 billion rubles. The Russian Railways has already spent 30 billion rubles on the development of the Eastern Railway Polygon, but there is an "accumulate delay" [3]. However, as of 2020, the issue of financing has been practically resolved by increasing investment by 2023. The volume of investments was significantly increased by 137.5 billion by 2023.

Intensive modernization of the Baikal-Amur Mainline and the Trans-Siberian Railway is possible only if resources are available to finance it. Therefore, the investment of the Russian Railways until 2023 will have the following indicators (Figure 2).

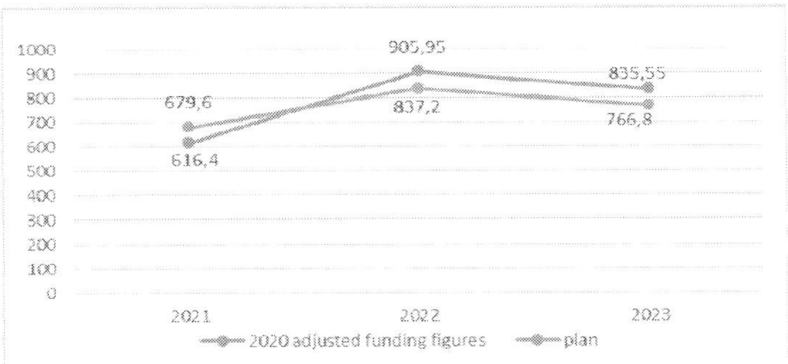

Figure 02. Adjusted financial indicators (JSC Russian railways, 2019)

To create optimal conditions for foreign trade, Russia should actively participate in the formation of the ITC. The Trans-Siberian Railway has a competitive advantage, acting as an economic bridge between the continents, both to the East and to the West. The Trans-Siberian Railway modernization program is expected to increase throughput capacity to more than 200 million tons of cargo per year by 2025 [11]. To achieve this result, it is necessary not only to build additional track railway lines, tunnels and junctions, but also to increase the speed of traffic, to increase the volume of transit container transportation.

5 Research methods

In the study, the authors used comparative analysis techniques to solve strategic management problems that require an analytical rationale [10].

It should be pointed out that the reconstruction of the Baikal-Amur Mainline and the Trans-Siberian Railway will have a positive impact not only on the growth of Russian export, but also on foreign trade traffic. The Trans-Siberian railway traffic is increasing every year, and China-Europe directions have the fastest growth dynamics. Japan also regards the Trans-Siberian Railway as the main channel for shipping goods to Europe. It is very expensive to perform transportation by plane, and shipping carries high time expenditures. Based on the proposed comprehensive infrastructure modernization plan, one can observe the dynamics of the growth of coal transportation in East direction. It is based on the Plan of export transportation of coal by railway transport in East direction (Figure 3) [11].

According to the Plan, there are two scenarios, one of which provides for the possibility of increasing coal exports to the Asia-Pacific region, both through the ports of the Far East and through the ports of South and

Northwest directions, as well as the Arctic basin. The increased export of coal from the new coal-mining centers of the eastern regions of the country will improve the competitiveness of Russian coal in the markets of the Asia-Pacific region, by reducing transport expenditures as a result of the reduction of export shipment range [12].

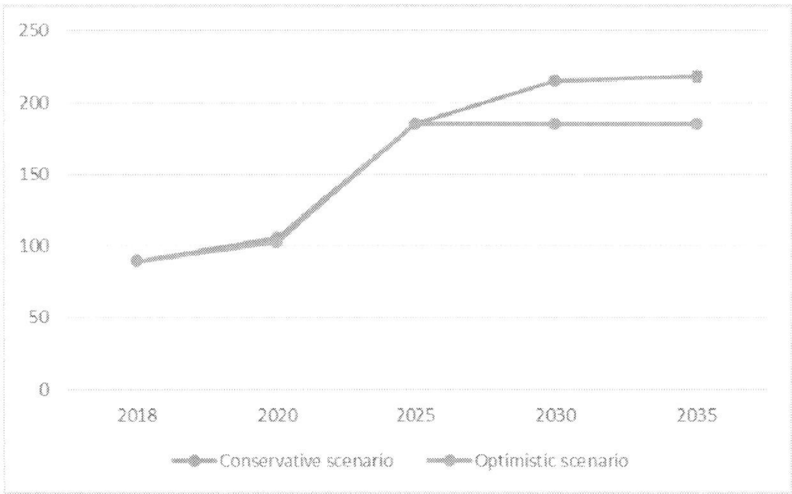

Figure 03. Coal transportation growth forecast in East direction [11]

6 Findings

Multi-vector development will cause high investment attractiveness of the mainline near the Polar coast, the development of raw materials of Siberia and the Far East and the development of regions. The project of connecting the Trans-Siberian and Trans-Korean mainlines (Figure 4).

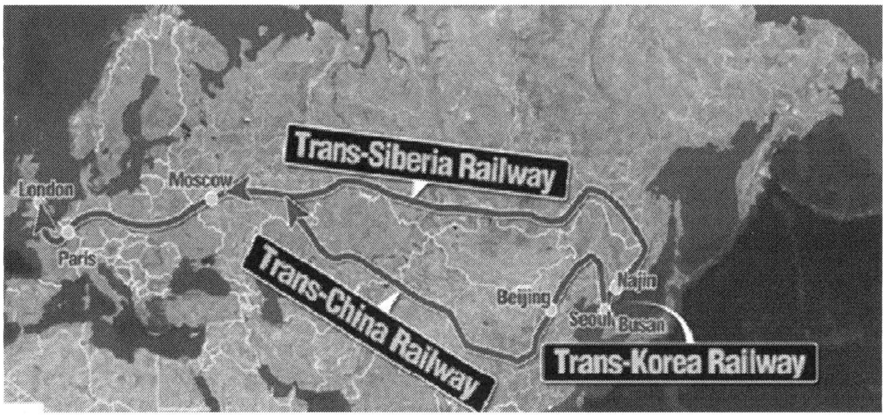

Figure 04. The variant of integration of the Trans-Siberian and Trans-Korean Railways

The junctions of the Trans-Siberian and Trans-Korean highways increase the load on the Baranovsky-Khasan section [15]. This leads to a border crossing between Russian and Korean railways. Through it, the Trans-Siberian Railway is connected to the international transport corridor "Primorye-2", and it is one of the priority areas of reconstruction. The implementation of the Trans-Korean Railway project involves 4 accesses to the Trans-Siberian Railway. Only one of them passes through Khasan to the Trans-Siberian Railway to cross China. The other two options also represent leaving the Trans-Siberian Railway but provide transit through China and Mongolia, with West-Chita and Naushki. The fourth option includes leaving Xinjiang Uygur Autonomous Region and Kazakhstan and accessing the Trans-Siberian Railway; the Southern branch goes along the route of the Great Silk Road. This option in the railway version is called TRAKEKA (the Central Asian – Caucasian-European road) [1].

Russia's interest is outlined in Korean transit through a number of factors affecting all variants of the Trans-Korean rail connection to the Trans-Siberian Railway:

Technological:

- State railways and energy of the DPRK (need to restructure financing, electrical materials, etc.) The modernization of transport infrastructure involves the construction of a high-speed railway connecting the capitals of the two states Seoul and Pyongyang. Rajhin's advantageous position at the junction of the borders of the three states (RF, DPRK and China), being in close proximity to other dynamically developing countries (South Korea, Japan, etc.), the growth of trade and economic cooperation in the region, increasing the demand for logistics services allow one to think about the more global prospects of the project [9].

• Discrepancy between the width of the railway tracks in different countries. A modern automatic train control system has been introduced, integrated with the system of the Russian Railways OAO [10].

• Cargo base assessment. The entry of the Trans-Siberian Railway to the international level of transport contributes to the increase of the domestic cargo base, the development of competition in the transport sector, the implementation of a goal-oriented state policy aimed at increasing the level of quality acceptable to foreign cargo owners, as well as the speed of their implementation, the formation of long-term, profitable, stable tariff policy, understandable for counterparties, and thus, as a result, achieving stability in terms of cargo delivery.

Economic:

• Economic competitiveness of the route, especially the level of tariffs. A considerable part of goods is economically profitable to carry with the help of sea transport, as the level of tariffs is much lower. At the moment, the distribution is the following: 15 % is for railway transport, the rest is for sea transport. Another reason is the technical inability to transport expensive goods by rail.

• Study of freight rail tariffs shows that there is an increase by 3.5 % since January 1, 2020. Indexation of tariffs on the principle of "inflation minus." It is used from 2018 and will be in effect between now and 2025. The cumulative effect on the economy from the adjustment of tariffs for reduced inflation for the period up to 2025 will be 2.7 % indexation or 41 billion rubles [12].

Political:

• Border, customs and other formalities (at the moment, there is no single automated workflow system, allowing to reduce the time for the registration of relevant transport documents).

• U.S. sanctions policy towards Russia and North Korea. The difficulties of doing business in the DPRK, due to the country's reluctance to cooperate on the basis of generally accepted economic practices and the high degree of political risks associated with the remaining military and political tensions of inter-Korean relations, deter potential investors willing to pour cash in the development of the project.

The above-mentioned groups of factors influence the effectiveness of the planned project to a different extent. Taken together, they will provide a comprehensive look at the project and its implementation prospects, which in turn can be viewed from both positive and negative sides [15].

The main drawbacks are the extremely high cost of the project and the difficulty in finding sources of financing. As noted, for the time being a Russian company has invested heavily in the reconstruction of the railway infrastructure of the DPRK. However, the modernization of North Korea's railways, the resumption of direct trans-Korean railway communication, with access to the mainline, can be a huge burden for any country [16].

Among the prospective consequences of the project, it is advisable to highlight:

• **access to the non-freezing port.**

One of the huge advantages in the future is the obtaining of another non-freezing port of Rajhin for the Russian Federation at the current operating capacity 5 million tons annually. Currently, the mediator is the People's Republic of China. Coal from the Far East goes to China by rail through the border crossings of Makhalino-Hunchun and Grodekovo-Suifenhe, the volume of transshipment is comparable to the volumes going through Rajhin [3].

• **increased economic security.**

The development of mutually beneficial cooperation will return dividends not only to both Koreas, but will help to reduce tension in this part of the world, which has a positive effect for Russia as well. Among other things, this project retains Russia as an influential player on the Korean peninsula. In the case of the DPRK, the North Korean side has received a ready-made infrastructure project on its territory. The cooperation has created job opportunities and also allows North Korea to make profits with port fees [13].

Some of the negative effects are expected to be leveled by increasing throughput capacity, including the Trans-Siberian line. Improving the speed of transportation of raw materials through it, domestic and foreign, following in transit from the superpowers of the Asia-Pacific region to Europe is one of the priorities for today. To a great extent, this is facilitated by the application of the polygon management model.

In 2019, a new, unified traffic management directorate was created. A single technological space in the entire direction of the cargo flow from Kuzbass to the seaports of the Far East will contribute to the transportation process efficiency on the basis of the polygon management model. So far, the new structure has included two directorates – East Siberian and Trans-Baikalian. In the long run, the functions and staff numbers of the Krasnoyarsk and Far Eastern directorates of the traffic management will be transferred under its authority. The new centre allows the management of operational and technological communication systems within the borders of the Eastern Polygon site. The centralization of the management of communication systems and their smooth operation will have a positive impact on the East Siberian mainline's task performance on the increase of the cargo flows through the Trans-Siberian Mainline and the Baikal-Amur Mainline. Work in the conditions of the polygon management model allowed the East Siberian Railway to increase the service speed from 44 km/h in 2016 to 45.1 km/h in 2018, the journey speed of delivery of all shipments by 44.8 km/day; the overall reliability of delivery increased by 3.1 %.

7 Conclusion

Summarizing the above, we can say with confidence that this integration is very beneficial in a long term for both Russia and the DPRK. The development of transport corridors will contribute to the growth of the level and quality of the population's well-being, strengthening geopolitical, infrastructure, economic positions on the entire route of the transport corridor [14]. Therefore, the formation of the transport complex of the world economic system is manifested in its dynamic development. Gradual development and modernization of the existing infrastructure subsequently open huge sales markets for domestic companies abroad.

Acknowledgments

Published at the expense of the state task on the topic "Improving the mechanisms of interaction between polygon structures and regional corporate governance centers by the example of the Eastern railway polygon".

References

1. V.A. Batyr. The Hasan-Rajin Russian and North Korean Project in the System of International Economic Constraints (Impregnable Russian-Chinese Bastion in the DPRK). Lex Russica. **4(137)**, 53–69 (2018)
2. Y.Y. Eglit, O.Y. Ogaltcova, A.V. Andorskaya, M.A. Shapovalova. Digitalization of container transportation and its impact on logistics. Syst. Anal. and logist., **4(22)**, 40–46 (2019)
3. *Cargo turnover by mode of transport in the Russian Federation* (2019) Fed. State Stat. Service. Retrieved from: https://rosstat.gov.ru/folder/23455?print=1
4. *Modernization of the Baikal-Amur Mainline and the Trans-Siberian Railway* (2018) Interfax. Retrieved from: https://rzdfuture.interfax.ru/transsib/
5. *Cargo and rail tariffs rose by 3.5 % from 01 01* (2020) Interfax. Retrieved from: https://company.rzd.ru/ru/9401/page/78314?id=183368
6. *The Annual Report of the Russian Railways OAO over 2019* (2019) JSC Russian railways. Retrieved from: https://ar2019.rzd.ru/ru/performance-overview/market-overview
7. M.A. Khazeeva, E.A. Belobrova. Current state of railway transport in Russia. Econ. and Entrepr., **6(107)**, 101–106 (2019)
8. N. Kucherov. *The Baikal-Amur Mainline and the Trans-Siberian Railway upgrades will pay off in spades*, ***4*** (2019). Retrieved from: https://www.ritmeurasia.org/news
9. *The plan of the export transportation of coal by rail in East direction* (2020) Ministry of energy. Retrieved from: https://minenergo.gov.ru/node/18169
10. *"Eternal fix income securities" have been found for the Baikal-Amur Mainline and the Trans-Siberian Railway* (2020) Nezavisimaya Gazeta. Retrieved from: https://company.rzd.ru/ru/9401/page/78314?id=191616
11. *Strategy for the development of service exports until 2025* (2019) Order of the Government of the Rus. Fed., no. 1797-r. Retrieved from: https://www.economy.gov.ru/material/file/6a532b0291a562e597a55aa491061df7/strategiya.pdf
12. O. Perepelitsa. *From the Trans-Siberian Railway to the Eurasian Corridor* (2013). Retrieved from: http://magazines. russ.ru/oz/2013/3/22p.html
13. F.L'. Perroux. *Economie du XXe siècle* (Paris, 1961)
14. P. Pottier. Axes de communication et développement économique. Rev. écon. **14**, 58–132 (1963)
15. *Russian-North Korean logistics project "Khasan – Rajhin": Current state and prospects for development in new conditions* (2018) RASONCONTRANS. Retrieved from: http://rasoncontrans.com/images/files/Lproject-ru.pdf
16. *Russian Railways in numbers* (2019). Retrieved from: https://old-www.rzd.ru/static/public/ru?STRUCTURE_ID=5232&layer_id=3290&refererLayerId=162&id=4085

Long-term economic growth and investment: features of the relationship between the North and the Arctic of the Russian Federation

Alexander Kugaevsky*, and Anastasiia Dolgunova

M.K. Ammosov North-Eastern Federal University, 58 Belinsky Street, Yakutsk, 677891, Russia

Abstract. For the North and the Arctic constituent entities of the Russian Federation, the mechanism for the impact of investments on economic growth is quite specific due to special conditions of investment activities in the North. Investment processes in the northern territories are characterized by "extension" or increased duration of investment cycles, instability, but at the same time inertia. The results of the study reveal that "northerness", as well as the raw material orientation of the economy, make it difficult to determine the relationship between such indicators as "fixed investment" and "gross regional product", therefore, for these territories there is often no clear and statistically significant connection. The investment growth models in the strategic planning of the socio-economic development of these regions should be applied with due diligence and broad understanding of all the features and factors of economic growth in the North.

1. Introduction

Targeting the level and dynamics of investment is a trend of modern state economic policy in the Russian Federation: strategic planning documents consistently indicate the leading dynamics of fixed investments or a significant level of investment as the most important factor of the economic growth.

The Decree of the President of the Russian Federation of May 7, 2018 No. 204 "On National Goals and Strategic Objectives for the Development of the Russian Federation for the Period until 2024" [1] considers the entry of the Russian Federation into the five largest economies of the world, thus ensuring economic growth rates above the world levels, as one of the national development goals. According to the explanations of the Government of the Russian Federation, its achievement implies slightly higher economic growth rates (more than 3%) [2]. The achievement of the desired growth rate is associated with the "investment growth model". This wording was given in 2018 by the Chairman of the Government of the Russian Federation D.A. Medvedev: "It is with the investment growth model that we associate plans to reach the desired growth rates. The share of investments in GDP is planned to increase from 21-22% to 25% in 2024". In the previous document, which determined the strategic goals and priorities of the socio-economic development of the Russian Federation (Decree of the President of Russia of May 7, 2012 No. 596 "On Long-term State Economic Policy" [3]), an increase in investment was established to at least 25% of gross domestic product by 2015 and to 27% by 2018. According to Rosstat, the share of fixed investments in the country's GDP in 2015 was 20%, in 2019 – 20.7% [4].

The relationship between the economic growth and investment dynamics is a fact that characterizes economic fluctuations, the existence of which is well known, but the correct application in practice is accompanied by certain difficulties.

The basic economic theory states that investment contributes to economic growth. The theory of modern economic growth considers investment in physical and human capital as correlates of economic growth. Empirical studies of economic growth and capital formation in the 1990s have largely led to the conclusion that investment has a significant impact on the economic growth. As an example, this was stated by the following authors: DeLong and Summers (1991, 1992) [5], Levin and Renelt (1992) [6]. Later, these studies have been criticized more than once, including by economist Robert Joseph Barro, as containing inflated estimates of this influence: the relationship between economic growth and investment may ultimately be characterized as direct but weak. The scientist's conclusions are based on a significant number of research works devoted to the determinants of the economic growth over the past thirty years (including the studies conducted in 1991, 1993, 1996, 1999, 2003, 2008 [7-12]). In his recent research *Double Investment Account* (2019) [13] Barro says that there is a double account of investment in the national income accounts.

The description of empirical studies of Russian authors suggests that at present the fixed investments

*Corresponding author: nastya_batozh@list.ru

are accepted as the most important factor in the economic growth of regional economies. At the same time, the specifics of working with this indicator are not always taken into account, for example, the need to consider its impact on economic growth with some lag, the bi-directional or inverse relationship between indicators is not supposed, the work mainly covers short time series.

2. Methodology

To study the regional features of the relationship between such indicators as "fixed investments" and "gross regional product", the series of dynamics were compiled according to official statistics published on the Rosstat website, in the context of all constituent entities of the Russian Federation from 2000 to 2017, i.e. for 18 years (the earlier period is not considered since mainly the economies of the regions and their investment activities were in a state of deep recession, and the financial and economic crisis in Russia in 1998 only aggravated this situation).

When phenomena are related not in the same period of time, but with some delay, it is necessary to measure the relationship of the factorial feature with the effective feature not by simultaneous levels, but by taking into account some lag. In the study, the lag is understood as the value that characterizes the time interval (number of years) of the relationship (joint variability) of "fixed investments" and "gross regional product", where the first is the cause, while the second is the consequence. The absolute values of these indicators are reduced to a single base year (2017). The lag value is determined by the correlation analysis. In view of the previously performed detrending, the correlation analysis was applied to deviations of individual levels from trends (the study assumes the hypothesis that the time series are stationary in relation to the trend).

Traditionally, the scientific research include the following constituent entities of the Russian Federation in the northern regions: Arkhangelsk, Murmansk, Sakhalin, Magadan regions, Nenets, Khanty-Mansi, Yamalo-Nenets, Chukotka Autonomous districts, Republic of Karelia, Komi, Sakha (Yakutia), Kamchatka Territory.

3. Results and Discussion

Investment projects are implemented in the regions. At the same time, the economy of each constituent entity of the Russian Federation always has objective development features, which imposes a certain imprint on the nature of the relationship between investment and growth of the regional economy. The economic dynamics of the constituent entity is also determined by other factors that, in terms of the effect size, may significantly exceed investments in fixed assets. Therefore, in order to effectively manage the region's economy, it is necessary to understand precisely the specifics of the relationship of indicators.

Projecting the national development goal to the regional level, i.e. to achieve an investment level of at least 25% of the gross regional product, it must be recognized that this sets the goal for half of the regions to do what they have only one step closer to (in the "best" years for the Russian economy) over the past more than 15 years.

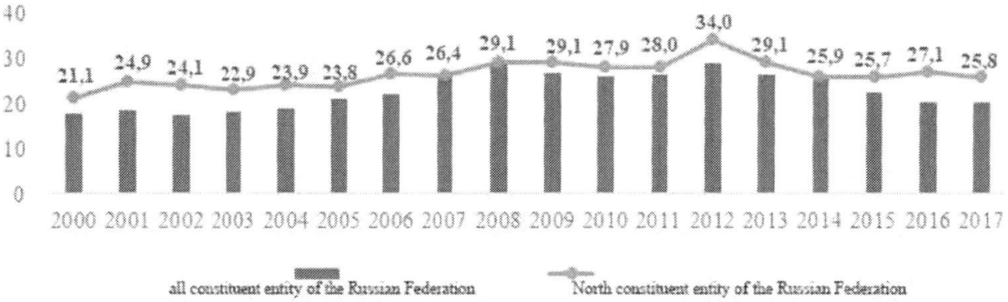

Fig. 1. Median values of the share of investments in fixed assets in the gross regional product for the constituent entities of the Russian Federation over 2000-2017

Then let us implement a retrospective analysis of the ratios between the values of the indicators "fixed investments" and "gross regional product" in the regional context.

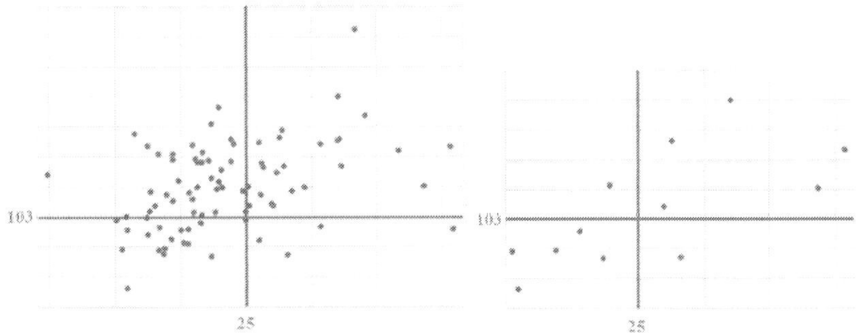

a b

Fig. 2. Ratio of the average share of investments in fixed assets in the gross regional product (horizontal axis; %) and the average growth of gross regional product (vertical axis; %) in 2000-2017: a – for all constituent entities of the Russian Federation, b – for northern constituent entities of the Russian Federation

The graphical analysis suggests that higher levels of investment usually correspond to higher economic growth. At the same time, more than 40% of constituent entities and with an investment level below 25% were able to ensure a 3% average annual growth of their economies. If we remove more successful years from the analysis, when the existing economic conditions could simultaneously predefine significant financial investments, i.e. high level of investment, and the growth of regional economies, then there is a change in the visualization of possible dependencies.

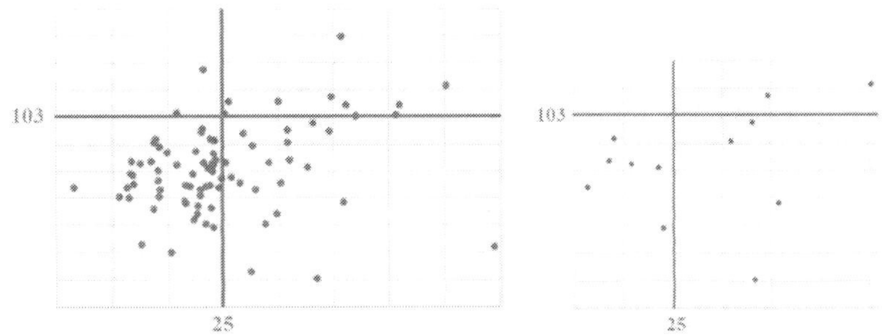

a b

Fig. 3. Ratio of the average share of investments in fixed capital in the gross regional product (horizontal axis) and the average growth of gross regional product (vertical axis) in 2013-2017: a – for all constituent entities of the Russian Federation, b – for northern constituent entities of the Russian Federation

At this period, there are already much fewer regions that were able to achieve high growth rates of their economies at low level of investment. Most regions have an investment rate of less than 25% and an economic growth rate of less than 3%. There are also significantly more regions with a high level of investment, but economic growth of less than 3% per year. For the northern regions, it may be stated even more unequivocally that the investment rate of 25% is not the key to high economic growth. Nevertheless, in the general conclusion that there may be a weak direct link between the studied indicators, for the northern regions it should be assumed that in order to achieve the accepted "high" economic growth, the level of investment should not just be "high", but "notably high".

Investments have a distinctive feature – a focus on future result from actions in the present and past, so it is quite natural to expect a temporary lag in the dynamics of the studied indicators. This process, as a rule, is extended over time: the period of design, project approval and decision making, the construction period or the purchase and installation of equipment, the period of reaching the design capacity and, finally, the first result of the economic (financial) return on investments.

The presence of a significant relationship of indicators, as well as the length of the lag, is determined in 82.9% of the regions of the Russian Federation (68 regions out of 82 analyzed). There is no lag, i.e. the relationship of indicators is manifested only "year-on-year", in 23 regions (28%). The largest number of regions – 34 regions (41.5%) – the relationship of indicators is significant over 2 years, namely, in the absence of a lag ("year-on-year") and with a lag of 1 year. A very small number of regions have a longer relationship: for 3 years – 3 regions, 4 years – 1 region. It is logical for the constituent entities of the northern macroregion to assume the predominance of prolonged investment cycles, however, according to the obtained results, it is impossible to unequivocally formulate this conclusion.

The determination of group specifics for the Arctic zone of the Russian Federation was not carried out due to the impossibility of currently conducting a correct comparison of indicators for its territories [14]. At the same time, it is assumed that for the Arctic territories the northern specifics will be manifested to a much greater extent.

Table 1. Assessment of peculiarities of the relationship between "gross regional product" and "fixed investments" in time for the northern regions of the Russian Federation over 2000-2017.

n/n	Constituent entity of the Russian Federation	Time lag in the relationship
1	Republic of Karelia	No time lag in the relationship of the studied indicators (0 lag)
2	Komi Republic	No time lag in the relationship of the studied indicators (0 lag)
3	Nenets Autonomous District	Specifics in the relationship in time (from 1 to 2)
4	Arkhangelsk Region (without Nenets Autonomous District)	Short lag (from 0 to 1)
5	Murmansk Region	No statistically significant relationship
6	Khanty-Mansi Autonomous District-Ugra	No statistically significant relationship
7	Yamal-Nenets Autonomous District	No statistically significant relationship
8	Republic of Sakha (Yakutia)	No statistically significant relationship (zero-lag correlation coefficient value 0.395)
9	Kamchatka Territory	No statistically significant relationship
10	Magadan Region	Specifics in the relationship in time (2-years lag)
11	Sakhalin Region	No statistically significant relationship
12	Chukotka Autonomous District	Specifics in the relationship in time

In most cases, there is no statistically significant correlation of indicators for the northern regions with a resource-based economy. Besides, occasionally there are certain features in the relationship over time (Magadan Region, Chukotka Autonomous District). In the vast majority of the northern regions of the North-Western Federal District it is possible to find a statistically significant relationship between the indicators "fixed investments" and "gross regional product", and in some of them (Nenets Autonomous District and Arkhangelsk Region (without the Nenets Autonomous District)) there is some time lag in this relationship.

The absence of a statistically significant clear relationship between "fixed investments" and "gross regional product" for the northern regions with a resource-based economy may be explained by significant features of investment activities in the North, as well as the distinctive features of investment processes formed from them, including their influence on the duration of investment cycles and the stability of investment trends over time.

Special conditions for investment activities in the North may be formulated as follows:

- careful and long-term planning of all operations;
- limited transport accessibility;
- predominantly raw material focus;
- predominance of the corporate sector;
- special role of the state;
- complexity and specificity of construction in the North;
- special requirements for material and technical resources and technologies;
- shortage and high cost of manpower;
- low sustainability of environmental systems and, accordingly, special requirements for environmental protection [15].

These features predispose to the following characteristics in the dynamics of investment processes in the northern territories: "extension" or increased duration of investment cycles, instability (significant investment decline after the completion or sudden shutdown of the project), but at the same time inertia (investments may be continued contrary to crisis phenomena, for example, in case of accumulated safety margin, due to the scale of projects). As a result, these conditions significantly hamper economic and statistical modeling of these regions, which may be more effective if the time period of the study is increased (as confirmed for the Republic of Sakha (Yakutia) [16]) before the analysis according to the type of economic activity.

4. Conclusion

Despite the fact that macroeconomic dynamics is not indifferent to the ongoing investment processes, the quantitative increase in investment is not the main factor of the economic growth. The mechanism for the impact of investment on the economic growth has significant features for the northern regions of the Russian Federation due to special conditions for investment activities in the North. In the general conclusion that there may be a weak direct link between the studied indicators, for the northern regions it should be assumed that in order to achieve the accepted "high" economic growth, the level of investment should not just be "high" (25%), but "notably high" (more than 35%). Nevertheless, this factor of the "northern rise in price" as a feature of investment activities in the North is far from sufficient to assess the impact and success of investments, and the question of the quality of investments is still open.

References

1. Decree of the President of the Russian Federation No. 204 *"On national goals and strategic objectives for the development of the Russian Federation for the period up to 2024"* (with amendments and additions) Retrieved from: http://base.garant.ru/71937200/#ixzz5WkF4IWj8.
2. *Main directions of activity of the Government of the Russian Federation for the period up to 2024* Retrieved from: http://www.consultant.ru/document/cons_doc_LAW_307872/.
3. Decree of the President of the Russian Federation of No. 596 *"On Long-Term State Economic Policy"* etrieved from: http://base.garant.ru/70170954/#ixzz5WkDBYmFo.
4. *Official website of the Federal State Statistics Service (Rosstat)* Retrieved from: http://www.gks.ru.
5. J. B.De Long, L. H. Summers, *QuarterlyJournal of Economics* **106** 445-502 (1991)
6. R. Levine, D. Renelt, *American Economic Review* **82** 942-63 (1992)
7. R. J. Barro, L. Jong-Wha, *NBER Working Papers* **4341**, (1993)
8. R. J. Barro, T*he Quarterly Journal of Economics*, **106(2)**, 407-443 (1991).
9. R. J. Barro, *NBER Working Papers* **5698**, (1996)
10. R. J. Barro, *NBER Working Papers* **7038** (1999)
11. R. J. Barro, *Annals of Economics and Finance, Society for AEF*, **4(2)**, 231-274, (2003)
12. R. J. Barro, Asian Development Bank. **11,** (2008)
13. R. J. Barro, *NBER Working Papers* **25826,** (2019).
14. R. R. Nogovitsyin, V. A. Grigoryev, A. S. Lukovtseva, V. N. Markova, N. V. Prokhorova, J. A. Filippova, *International Journal of Supply Chain Management*, **8(5),** October 2019, (2019).
15. A.T. Dolgunova, *Regional economy: theory and practice*, **17(12)**, (2019).
16. A.A. Kugaevsky, A.T. Dolgunova B*ulletin of M.K. Ammosov Northeast Federal University. Series: Economics. Sociology. Cultural studies*, **1 (17)**, (2020).

SHS Web of Conferences **112**, 00033 (2021) https://doi.org/10.1051/shsconf/202111200033
Northern Sustainable Development Forum 2020

Ways to optimize process of modernization of trans-Siberian railway using digital economy technologies

Maria Samarina[1], and *Ekaterina* Bondarchuk [1,*]

[1] Irkutsk State Transport University, 15, Chernyshevskogo str., Irkutsk, 664074, Russia

Abstract. The article discusses the issues of improving measures to optimize the process of modernizing the Trans-Siberian Railway using digital economy technologies. The digital economy provides for the implementation of a number of measures that stimulate the introduction of information and telecommunication technologies in most of the economic sectors of Russia. In the new interpretation, three components of the "Digital Economy" methodology are noted: information infrastructure of e-business, network e-business, e-commerce. All of these components are also part of the Digital Railroad Bratskoe. The railway is the backbone of the entire transport complex of the Russian Federation and is of great socio-economic, political and defense importance. Compared to other modes of transport, railway transport is determined not only by the large value of the main economic indicators, but also by the volume of transportation work performed, both in freight and passenger traffic. The technological and economic effects from the development and implementation of digitalization projects for both freight and passenger transportation have been determined. A comparative analysis of the used services based on digital technologies was carried out for the predicted economic effect. The project "Digital Railway" was analyzed, in particular, one of the pilot projects "Digital Depot", implemented on the basis of the service locomotive depot SLD Bratskoye. The effectiveness of this measure is difficult to fully assess due to the relatively short implementation period. The first data for calculating the technical and economic efficiency are analyzed: a forecast for the future is presented.

Keywords: Digitalization, railway transport, CRM system, customer focus, efficiency

1 Introduction

In modern society, issues related to the prospects for the development of the digital economy and affecting the interests of both economic entities and ordinary people are becoming more and more relevant. Today, we can talk about the active development of digital technologies with the continuation of this process in the near future, as a result of which these trends cannot be ignored for many areas. The transport sector will not be an exception, as a result of which this issue should be studied in more detail [8].

It is difficult to imagine a person's life without digital technologies that contribute to increasing the competitiveness of the transport industry, improving interaction with customers, taking into account their needs in order to understand what proposal should be developed, and approve new services based on them for demand.

2 Problem Statement

The fundamental factor for mastering the achievements of digitalization in accordance with the Long-term development program of the Russian Railways company is considered as the possibility of creating a truly unified information space for freight transportation and logistics in order to increase the profitability of the entire freight market, logistics business in particular, without unforeseen costs arise [1].

In the modern world, the Digital Railroad Model of the Russian Railways Company is a special matrix of service blocks in which certain categories of automated solutions are placed in conjunction with aspects of the digital model of a business project.

The structure of the digital railway is presented in the following areas:

1. Passenger transportation: mobile application, ticket sales, Wi-Fi, information services for travelers, analytical reporting CRM.
2. Freight transport: digital exchange of wagons, 4 PL logistics, multimodal transport, paperless technologies, CRM analytical reporting.
3. IT services: catalog of IT services and SLA.
4. ACS: complex automation.
5. IT Infrastructure: Continuity Services.
6. Motion control: intelligent motion control system, paperless technology.
7. Management of railway infrastructure: maintenance of infrastructure as per condition.

* Corresponding author: samarina-mashenka@mail.ru

8. Customer services: mobile services (CRM, BigData), etc. [2].

3 Research Questions

It will be possible to achieve the efficiency of the production activity of each of the participants in the transportation market if we ensure a truly high-quality fulfillment of the obligations assumed. These are fully agreed with the owner of the infrastructure of applications for the transportation of goods, the movement of wagons that belong to operators, as well as other owners of railway rolling stock, seeking minimum costs. This in turn directly depends on the volume of existing traffic in accordance with a single technology for organizing this process, avoiding serious violations [4].

As you can find out from the analysis of global technological trends, digital technologies contribute to the creation of special conditions in order to make adjustments to the classical models of doing business, to help with the discovery of new opportunities in absolutely all directions, in particular:

1) quick and easy access to the services provided for the transportation of goods;

2) predictive analytics of clients, development of proposals taking into account the actual needs of the target audience, which will appreciate the new services, services;

3) creation of new programs to attract organizations of the Russian Railways company, other enterprises, other types of transport infrastructure, in order to achieve a complete catalog of services;

4) introduction of Russian Railways directly into the customers' logistics system, based on joint planning of shipment, subsequent transportation of finished products, careful and fast delivery of goods, taking into account the actual need of the entire production process of customers who are interested in long-term cooperation [7].

4 Purpose of the Study

The purpose of this study is to assess, both from a technological and economic point of view, the possibilities of applying digital achievements at Russian Railways, in particular, on the Trans-Siberian Railway.

The program provides for a fairly large list of activities, each in its own way reflecting a different degree of automation.

In order to competently implement end-to-end coordinated automated technology, it is important to organize a single information space in an industry such as transport, so that it is possible to establish effective interaction of all interested participants in the transportation process. These are the automated control system of owners and customers, the operator, and the carrier, which is directly Russian Railways. As a result, the aspects associated with artificial intelligence and Big data technology will be really widespread for the automation of these processes.

5 Research Methods

Based on the purpose of this work, theoretical and quantitative methods were chosen as research methods. Due to the fact that the digitalization of the company's activities is taking place relatively recently and in completely different directions at the same time, it is quite difficult to quantify many indicators [3]. The main research method was theoretical, which made it possible to analyze the Strategy of Russian Railways within the framework of the digitalization section. In addition, the concept for the implementation of a complex scientific and technical project "digital railway" was chosen as the basic document of the study. The application of the quantitative method within the framework of the analysis of the dynamics of changes in indicators characterizing the operation of the depot became possible on the basis of the study of the Bratskoye digital depot project. This method was applied to assess the first results obtained since its commissioning, data for a period of 10 months.

6 Findings

The analyzed data on the directions of digital transformation in Russian Railways are presented in Table 01.

According to the data presented in the table above, one of the projects within the Digital Railway is the creation of the Digital Depot by the LocoTech holding. Having analyzed the Digital Depot project, which really corresponds to the concept of the Russian Railways company, as well as Transmashholding JSC regarding digital initiatives, in particular, Digital Railway, Smart Locomotive, which made it possible to make a choice regarding the Vikhorevka Vostochno- Siberian Traction Directorate, where AC electric locomotives "Ermak" were delivered [5].

An important aspect of monitoring the Digital Depot project is considered as the urgent need to increase the objectivity of data on the existing state of the locomotive fleet at absolutely all levels of management of maintenance processes, repair work, the ability to organize technological support for employees when making important decisions, automate processes associated with training, restoration, operations for the assembly and disassembly of parts, related equipment that is used.

A similar principle should be followed when the maintenance of the linear infrastructure takes place after a preliminary analysis of the current technical condition, which allows you to subsequently achieve acceleration, subsequent reduction in the cost of design, construction work, reduction in operating costs, and implementation of planned projects.

It is supposed to carry out competent automation of the terminal and warehouse complex in order to introduce innovations, ensure the presence of robots in the warehouse, various sensors. That make it possible to draw up a plan, and subsequently take into account the use of available resources that will be useful in certain circumstances on a long-term or short-term basis.

Table 01. List of services organized on the basis of digital technologies (up to 2025)

Direction of digital transformation	Services	Projected economic effect
Implemented CRM system to ensure full interaction with every interested client	A personal approach to each client, taking into account the previously gained experience of cooperation, based on current needs, including the sale of complex services in the transport sector, receiving confirmation of applications on a convenient Internet portal with a clear interface	Increasing the share of rail transportation, as well as generating income from the use of additional services
Multimodal electronic system that functions directly in the passenger segment	complex transportation of clients "from door to door", using third-party carriers from other types of the transport sector on the market	Increase in commission income, as well as the ability to ensure the consolidation of traffic in one place, in order to facilitate future monetization, increase loyalty indicators, and trust of each passenger, regardless of their age, gender, social status
Intelligent systems that allow you to establish competent management of the transportation process	digitalization of the work of all represented stations and service of clients interested in cooperation at the points of unloading and loading	Increasing the speed of forecasting traffic, reducing the influence of the human factor, increasing the productivity of traction rolling stock, reducing losses from non-schedule delays and train stops
"Smart locomotive" and "digital depot"	An open service, which provides information about the manufacturer, user, directly to the service organization, reporting accurate and accessible information about the current state of the rolling stock, taking into account automated and remote diagnostics, high-quality repair work, if necessary, without waiting for the deadline of the schedule drawn up in advance	The ability to reduce the cost of service, taking into account the long period between repairs, as well as making efforts to prevent unplanned downtime

A separate block of the entire program is the aspect associated with the optimization of corporate processes, taking into account the introduction of electronic circulation of documents, tax monitoring with access to information from the tax service, digital personnel service, services. They allow you to quickly find qualified and experienced specialists, minimize the time spent on execution routine operations, which will help to increase labor productivity, reduce the time needed to search for candidates for vacancies, and reduce administrative costs directly [6].

For example, the Digital Depot project fully contributes to the achievement of the criteria of the target KGE (readiness for operation) when specifying economic indicators, downtime for planned types of service. As a result, it can be called a truly unique project of the entire organization, the implementation of which will make it possible to correct the existing nuances during the operation of locomotives, to obtain a visible effect from not only reducing penalties for the CGE, but also reducing direct and overhead costs on the network.

The principle of organizing repair work by the example of the Digital Depot project consists of the following elements:

- diagnostics;
- controlled repair process;
- control over personnel;
- transparent ITC management system;
- automated release to the line.

At the same time, each of the stages, in turn, is subdivided into smaller operations, thereby allowing carrying out a whole range of measures aimed at more efficient organization of the depot operation. Thanks to this project, it was planned to solve a number of problems encountered in practice and which led to economic losses.

Among such problem areas are:

- inappropriate use of the time factor in repair activities;
- low labor productivity;
- irrational use of labor;
- high rate of fines;
- low indicators of KGE, etc.

The introduction of any technology involves investment. To assess the effectiveness of an investment project for the implementation of a CRM system, the fundamental principles are modeling flows, resources and cash over periods, a comprehensive accounting of external and internal factors in the implementation of the project, determining the effect by comparing the expected investment results and costs.

The standard payback period for this project is considered less than 3 years, which makes it quite attractive.

A fairly short payback period is possible if the mandatory list of requirements is fulfilled according to the organization of the process. In this case, the economic effect will manifest itself in reducing the time for types of repairs, changing the form of workers' wages, and eliminating unproductive time.

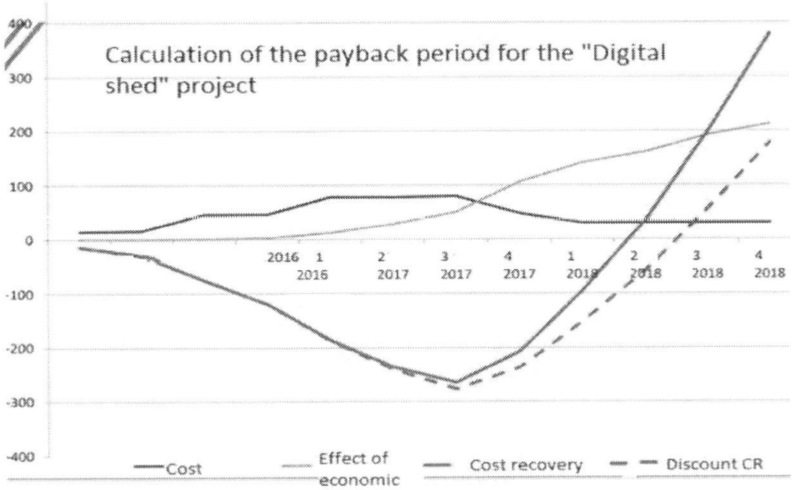

Figure 01. Predicted calculation of the payback period of the Digital Depot project

The main requirements for the personnel are the uninterrupted delivery of business travelers to the place of work. With regard to equipment, there is budget approval for a one-time supply and constant replenishment of an irreducible stock of linear equipment, as well as its installation on a locomotive at TR-1, TR-2, and HP.

Indicators of goods and materials are introduction of a simplified procedure for the purchase and reconciliation of invoices – directly through the TD, as well as the approval of new norms for writing off goods and materials.

At the moment, it is not entirely advisable to talk about the attainability of this indicator. During the first months of operation of the pilot project, the following results were obtained.

The effect expected from the implementation of this project is most pronounced in the time spent on repair work and the number of sections. This relationship is shown in the figure below.

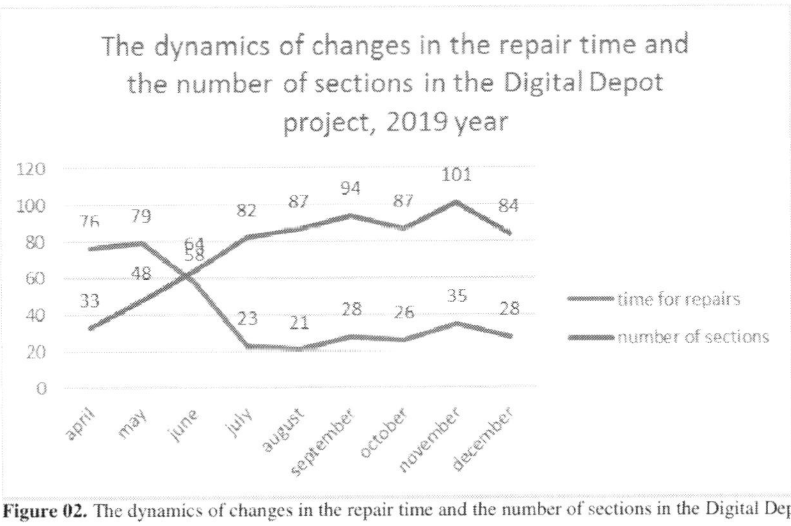

Figure 02. The dynamics of changes in the repair time and the number of sections in the Digital Depot project 2019 year

As you can see from the figure above, the amount of time for repair work is reduced, while the number of sections increases. A record value was reached in November with 101 sections. The relationship between the indicators shown in the figure above is also confirmed by the correlation analysis carried out earlier. According to him, the R-squared value is 0.87, which indicates a close relationship between these indicators. In the future, the time indicators for repairs are also planned to be reduced to possible values, while the quality of work does not suffer, and labor productivity increases. The performance indicator for this period, according to the ratio of the time spent to the volume of work performed, varies from 0.4 to 3.4. According to the planned indicators, labor productivity should grow by 30 %. In the future, this manifests itself in the number and amount of fines associated with the AGE indicator.

Table 02. Comparison of data on the time of work and the number of work performed before and after the implementation of the Digital Depot project

Indicator name	Types of repair work carried out at st. Bratskoe (Digital Depot project)					
	maintenance work – 1		line maintenance – 2		unscheduled maintenance	
	actual value	target value	actual value	target value	actual value	target value
Repair time, hours	64	18	1.5	0.3	21	4
Number of sections, pcs	40	130	-	-	-	-

At the same time, the availability factor also tends to decrease from 0.950 to 0.914. The amount of fines for the AEG is therefore reduced. In the most optimistic scenario, fines for some locomotives can be avoided.

The pilot includes over 30 different technological and organizational solutions combined into one system. They will improve the efficiency of production processes without increasing the workload of employees.

The technical condition of the locomotive is the main resource with which the project works. In the short term, the implemented systems will make it possible to assess and predict the technical condition of components and assemblies, as well as to offer the optimal solution in terms of the time and volume of repairs.

The predicted result from the competent implementation of the entire project is an increase in the readiness factor for the operation of the fleet of existing locomotives that are under maintenance, repair work, which is up to 97 % with the ability to talk about:

- an increase in the quality indicators of the direct maintenance and subsequent repair of the presented locomotives, contributing to a decrease in the time spent in the workshops by at least 50 %;
- guarantees of indicators of objectivity of results, allowing to establish control over the performance of maintenance operations, repair work in accordance with the technology that has been approved;
- the results of competently carried out diagnostics, which makes it possible to assess the technical condition of the existing locomotive, as well as those components that are part of it, allowing for operation, subsequent repairs and further maintenance, observing all necessary measures, safety measures;
- optimization of the system of well-functioning management of high-quality inventory items;
- introduction of automation of operations, allowing for restoration and repair work;
- reduction to the minimum idle time of wagons that must be sent for the carriage of goods.

7 Conclusion

The integration of all digital solutions into a single system is planned to be completed by the end of 2022. Successful digital innovations will be replicated and implemented in other service depots, which will lead to a complete and perfect technological modernization of locomotive service maintenance throughout the country. Improving the operating model of information technology management is aimed at ensuring effective information technology management, ensuring transparency of costs and increasing labour productivity, which will optimize staff and achieve savings in procurement.

It is safe to say that digitalization is a truly global process that is actively taking place in many countries, making it possible to improve many areas, in particular, rail transport, which makes it possible to optimize transportation, the maintenance of the entire infrastructure, and the provision of comfortable conditions for interested passengers and owners' cargo for various purposes.

Acknowledgments

The work is printed at the expense of the state assignment on the topic "Improving the mechanism of interaction between polygon structures and regional centers of corporate governance by the example of the Eastern polygon of railways".

References

1. O.A. Alekseenko, I.V. Ilyin. Digitalization of the global world and the role of the state in the digital economy. Inform. Society, **2**, 25–28 (2018)
2. V.S. Dubov. Indicators for assessing the development of the digital economy. Sci. through the prism of time, **7(16)** (2018). Retrieved from: http://www.naupri.ru/journal/1116
3. A. Kitaev, I. Mironova, A. Pogodaeva, D. Sokolov, E. Guseva. Railway station 2.0: a new pattern for the development of digital railway. Int. J. of Open Inform. Technol., **5(2)**, 85–96 (2017)
4. A.A. Kudryashov, A.I. Sholina. Infrastructure of the digital economy. Actual probl. of the modern econ., **5**, 25–32 (2018)
5. A.N. Menovshchikova, I.D. Belaya, O.V. Selina. Development of information economy in Russia. Actual probl. of modern econ., **3**, 177–181 (2018)
6. A.S. Sagynbekova. Digital economy: concept, prospects, development trends in Russia. Theory. Pract. Innovat., **4** (2018). Retrieved from: http://www.tpinauka.ru/2018/04/Sagynbekova.pdf
7. I. Sokolov, A. Misharin, V. Kupriyanovsky, O. Pokusaev, Y. Kupriyanovsky. Robots, autonomous robotic systems, artificial intelligence and the transformation of the market of transport and logistics services in the digitalization of the economy. Int. J. of Open Inform. Technol., **6(4)** (2018)
8. N.A. Stefanova. The concept and general principles of the formation of clusters of the digital economy in Russia. Azimuth of sci. res.: econ. and manag., **1**, 237–241 (2018)

Knowledge in the model of dynamics and stability of an industrial enterprise

Sergey Chuprov [1,*]

[1] Baikal State University, 11, Lenin Str., Irkutsk, Russia

Abstract. The formation and development of the knowledge economy logically lead to the need for an in-depth study of the nature and properties of information in economic systems. The priority value of the information resource and knowledge in the activities of industrial enterprises is determined both by the prospects for the development of science-intensive products and their competitiveness in the world and domestic markets, and by the desire to ensure the stable operation of the enterprise in a highly disturbed business environment. With the search and analysis of the dependences of the production, financial and economic processes taking place in them on the control information, the systemic interaction of the enterprise resources and the dynamics of key indicators of the activity of industrial enterprises are revealed.From the standpoint of thermodynamics, statistical physics and cybernetics, it is proposed to spread the information concept of V. Trapeznikov to substantiate and modelling the dependence of a number of economic and financial indicators on the amount of accumulated information in the controlled complex of an industrial enterprise.Modeling and interpretation of the relationship between the volume of control information with the uncertainty of the functioning of an industrial enterprise and the achieved level of knowledge about it is carried out. The condition for the stability of the indicator of the effect of activity and the financial stability of the enterprise (autonomy ratio) is formulated. The mathematical condition for the range of changes in the level of knowledge of industrial enterprises has been substantiated.

Keywords: entropy, knowledge, industry, information, model, stability

1 Introduction

The harmonization of financial and economic processes in industry is under the increasing influence of the flow of information, the resource of which becomes dominant to ensure favorable conditions for their stability and strengthen competitive positions. The painful institutional transformations experienced by the Russian industry under the pressure of a powerful stream of indignations prompt analysts to investigate not only the nature and dominant trends of the business environment, but also to form the mechanism of adaptive management of domestic enterprises. At the same time, the high dynamism of the outraged environment and the mastery of the tools for managing the development of enterprises in the globalizing world reasonably concentrate the efforts of enterprise managers on balancing financial proportions and increasing the economic efficiency of industrial enterprises as part of their intensive innovative modernization.

The relevance of understanding information and financial and economic interaction in the activities of an enterprise increases even more if we take into account the fact that the dependence of the effective and solvent activities of enterprises on the volume of accumulated information has not yet found deep theoretical and methodological study. The significance of the scientific results published by analysts, both in conceptual and applied aspects, is higher for understanding the nature and dynamics of communication of information and financial and economic processes at industrial enterprises in the context of innovative development of enterprises [1, 9, 11, 14].

2 Problem Statement

The modern knowledge economy is being formed in the space of world challenges and threats, saturated with a stormy stream of technological innovations that are fundamentally transforming the face of industrial production. Aggravating and motivating indignations excite the business environment and dictate the need for a radical modernization of economic structures and the widespread development of science-intensive means of manufacturing products and providing services. Organizational, managerial, design and technological innovations now not only determine the competitiveness of manufactured products, but also, with the increase in information and innovation resources of enterprises, strengthen the stability of their activities in an outraged competitive environment.

* Corresponding author: ChuprovSV@yandex.ru

The full use and involvement of useful information in management loops and production processes was objective and did not meet any conceptual objections, since they served to revolutionize the development of the technical base of the enterprise, optimize product output and achieve targets due to improved coordination of actions of all its departments. After all, the progressiveness of technological methods of manufacturing products and rendering services, and the quality of management decisions are ultimately determined by the valuable information contained in them. At the same time, the postulate of an innovative economy, recognized in principle, needs to deepen theoretical concepts, search and systematic understanding of the relationship of a complex of resources from the point of view of the impact of information on the structural and behavioral attributes of an industrial enterprise.

3 Research Questions

Within the research task set, the following questions are subject to studying:

1. Justification of the application of theoretical approaches to solving the problem of analyzing the influence of information on the performance of an enterprise.
2. Formalization and analysis of the model of the dependence of the indicators of the effect of the enterprise on the size of the information resource available to it.
3. Determination of the dependence and interpretation of the model of the dynamics of financial and economic indicators and the stability of the enterprise on its level of knowledge.

4 Purpose of the Study

The purpose of the study is to clarify the nature of the information impact on the activities of an industrial enterprise, to assess and analyze the impact of the level of knowledge on economic dynamics and the stability of its functioning.

5 Research Methods

For the research, the following methods were used: thermodynamics and statistical physics, cybernetics and information theory.

6 Findings

Information theory draws its postulates from the basic concepts of thermodynamics and statistical physics about the behavior of macroscopic physical systems that are in a state of equilibrium and pass from it to other states. The fundamental studies of R. Clausius [7], L. Boltzmann [5], J. Gibbs [8], M. Smoluchowski [7] created and developed the science of the nature and laws of thermodynamic processes, which are quite universal for a wide class of systems. Their dynamic picture is characterized by a measure of the probability of a certain state of a macroscopic system, or its entropy. In particular, a closed system left to itself over time irreversibly comes to an equilibrium state with maximum entropy.

The organic relationship of entropy, organization and the amount of information in the system was analyzed by the classics of cybernetics. The organization of a system, N. Wiener believed, is the amount of information in it, while the entropy of a system is a measure of its disorganization [16]. Such an interpretation of the attributes of a dynamical system was found logical and was successfully developed in the works by L. von Bertalanffy [4], S. Beer [3], L. Brillouin [6], C. Shannon [13], W. Ashby [2] et al.

Statistical physics not only linked the entropy of a system with the probability of its macroscopic state, but also takes into account fluctuations in which systems move from more probable states to less probable ones when the entropy of the system decreases [12]. Such an interpretation of entropy served to introduce the concept of information entropy as an estimate of the uncertainty of a message described by a set of quantities with the corresponding probabilities of their appearance in it. In C. Shannon's information theory [13], communication channels are stochastic and can also be characterized by the uncertainty of behavior, and its decrease is perceived by obtaining information about the functioning of the system.

If the initial entropy H_* of the system in the process of observing it decreases to the value H, the amount of information extracted in this case is equal to the difference between them:

$$H_* - H = I.$$

For an economic system with its inherent disorder of behavior to one degree or another (disruptions in the supply or movement of resources, equipment failures, mismatch of personnel actions, violation of executive discipline, downtime, technological defects, etc.), saturation with control information improves the neutralization or localization of interference. It also extinguishes abnormal disturbances and thereby increases the coordination of personnel activities and leads to an increase in the effect of the system functioning. In the statistical interpretation, the model of the dependence of the effect E of the functioning of the system on the amount of the information I entered into it is expressed by the formula obtained by V. Trapeznikov [15]:

$$\frac{E}{E_{max}} = 1 - e^{-\frac{I}{I_0}}, \qquad (1)$$

where E_{max} – effect of a perfectly functioning system (maximum possible effect),

I_0 – amount of information typical for a given control object.

The exponential dependence of the effect of the functioning of the system (productivity, income of the enterprise, etc.) on the amount of control information received into it states the fundamental feature of the economic dynamics of the activity of an industrial enterprise with different rates of increase in the effect. At first, it is almost proportional to the amount of information entered, but after the region of the greatest curvature of the exponent, it rapidly decreases, and therefore the level of the effect in the region of its small values requires less increment of control information, in contrast to the field of large values. Moreover, as the effect level approaches the maximum possible ($E \to E_{\max}$ or $\frac{E}{E_{\max}} \to 1$)the amount of additional input control information required at the enterprise is rapidly increasing, as a result of which a balanced approach to reducing entropy in the activity of an industrial enterprise and investments to maximize its effect is justified.

In the development of this conclusion, we will find the condition for its stability, bearing in mind the change in the magnitude of the effect in the noncritical range when the effect of the disturbances destabilizing its functioning on the enterprise. Let us assume that the relative indicator $\frac{E}{E_{\max}}$ stability will be inherent if the value of this fraction, despite the influence of harmful disturbances, remains in the permissible range of values from the lower K_e^l to the upper K_e^u boundary.

With the transformation of formula (1), we find the required condition for the stability of the indicator of the effect of the enterprise:

$$\ln(1 - K_e^l)^{-1} \le \frac{I}{I_0} \le \ln(1 - K_e^u)^{-1}.$$

This equality implies the expected economic dynamics: ensuring the stability of the effect size in the area of its small values is less resource-intensive than in the area of large values. For example, comparing ranges of the same length, but at different points on the number axis, we make sure that when moving from the allowable range 0.4…0.5 ($0{,}4 \le \frac{E}{E_{\max}} \le 0{,}5$) to the "far" range 0,8…0,9, it is required to ensure that the amount of control information is entered into the enterprise by more than 3 times (for lower limits of 3.16 and upper limits of 3.33 times). And this entails the attraction of additional resources for the collection, transmission, processing and storage of information arrays, the complication of the enterprise management system with equipping it with innovative computer technologies.

In terms of the relative value of the economic effect $\frac{E}{E_{\max}}$, we summarize that it can be used to judge the degree of suppression of disturbances, increase in orderliness (decrease in entropy) in the functioning of the enterprise, and, therefore, the achieved quality of its management. In the concept of V. Trapeznikov [15], ratio $\frac{E}{E_{\max}}$ is determined by the amount of information entered and its usefulness and assesses the level of perfection in managing the economic system L_m, reflecting the effectiveness of the activities of its management personnel.

Let us formalize the dependence of the amount of control information I in the economic system on the level of perfection of its control L_m. Omitting the mathematical calculations, let us give the final form of the formula:

$$I = I_0 \ln\left(\frac{B_0}{1 - L_m}\right), \tag{2}$$

where B_0 – disorder of the system in the initial state.

From the point of view of the analysis of economic dynamics, it is confirmed that the amount of information entered into the system I is logarithmically related to its disorder B_0 and the level of perfection of system management L_m. In an effort to ensure an increase in the latter, it is easier to achieve an increase I if the disorder in the system B_0 is sufficiently low. Conversely, when B_0 it is quite large, it is more difficult to increase the level L_m, since to decrease the entropy it is necessary to increase the amount of information I.

According to formula (2), if the amount of control information I in the economic system is equal to the initial I_0, then with the initial disorder B_0= 0.50 level of perfection of system management L_m = 0.82, and at higher disorder B_0= 0.75 value L_m below (L_m = 0.72). With increasing control information I the discrepancy between the values L_m at different B_0 decreases. So, if in the economic system the amount of control information I is 5 times more than it was before its receipt ($I = 5I_0$), then with the same B_0 (0.50 and 0.75) the values L_m match up to hundredths (0.99).

Meanwhile, it is obvious that the successful financial and economic activity of an industrial enterprise is based not only on circulating control information, but also on the competent use of the knowledge accumulated by society, materialized in the technical base of the enterprise (designs of machines, units, equipment, their complexes, etc.), which determine the level of perfection of the system L_s. The more modern science-intensive design and technological solutions are embodied in the production equipment of an enterprise, the higher its capabilities in manufacturing competitive products, adaptability and stability in the business environment. Aggregate indicator combining levels L_m and L_s, according to V. Trapeznikov [15], characterizes the level of knowledge and skills used, or in short, the level of knowledge L_k, thus mathematically there is a work L_k=L_mL_s.

The level of knowledge L_k is affected by the influence of labor (qualifications and skills of workers), social and psychological (dedication, persistence of workers, discipline, etc.) factors. In formal terms, the value L_k is proportional to the multiplication of labor productivity P_l, measured by the generated net product, and capital productivity P_c. Since the coefficient introduced into the definition formula L_k, taking into account industry conditions, for the country's economy and most industries is taken equal to 1.0, we can write an expression for calculating the level of knowledge: $L_k=P_lP_c$ (ruble/man).

Now we will focus on the formalization and analysis of the relationship between the level of knowledge accumulated in the economic system of an industrial enterprise, with its characteristics of work efficiency and financial stability, which allows us to find a remarkable economic dynamics of the enterprise.

To do this, we will use the author's proposed model of dependence of one of the main indicators of the financial stability of the enterprise– its autonomy ratio on a number of indicators of the enterprise functioning:

$$K_{a.r} = L_k \frac{K_{f.w.c}}{P_l E_{o.f}},$$

where $K_{a.r}$ – autonomy ratio,

L_k – the level of knowledge at the enterprise,

$K_{f.w.c}$ – the share of fixed assets and standardized working capital in the value of the property of the enterprise,

P_l – labor productivity, measured by the generated net product,

$E_{o.f}$ – efficiency (turnover) of own funds (the ratio of proceeds from product sales to the cost of the company's own funds).

Despite the direct dependence of the autonomy ratio $K_{a.r}$ on the level of knowledge L_k, in reality, the tendency to change their values was predominantly opposite (only occasionally did they have a unidirectional change) due to the fact that the increase in labor productivity P_l was very noticeable and was accompanied by a relatively low unevenness of indicators efficiency (turnover) of own funds $E_{o.f}$ and the share of fixed assets and standardized working capital $K_{f.w.c}$. At the same time, the indicators P_l and $E_{o.f}$ experienced the positive impact of increasing the level of knowledge L_k (saturation with control information, accompanied by the buildup of hardware and software products for management purposes) of the enterprise management system, and $K_{a.r}$ the restraining effect of the accumulation of short-term liabilities.

Along with this, in order to clarify the dynamics of the magnitude of the level of knowledge L_k, it is pertinent to ask about the extent to which it varies. Taking into account that, by definition, the values of the coefficients are $K_{a.r}\leq 1.0$ and $K_{f.w.c}\leq 1.0$, simple mathematical transformations lead to an inequality that sets the limits of change in the level of knowledge L_k:

$$K_{a.r} P_l E_{o.f} \leq L_k \leq \frac{P_l E_{o.f}}{K_{f.w.c}}.$$

It shows that at the achieved values P_l and $E_{o.f}$, the lower the limit of the level of knowledge L_k at an industrial enterprise is determined by the coefficient $K_{a.r}$, and the upper limit – by the coefficient $K_{f.w.c}$. Thus, the characteristic of the financial stability of the enterprise $K_{a.r}$, which summarizes the effectiveness of management activities, the structure and return of the enterprise's resources, sets the minimum level of knowledge L_k. At the same time, its maximum is fixed by the coefficient $K_{f.w.c}$, while the smaller it is, the further the upper limit of the value L_k is pushed back and the range for increasing the level of knowledge expands. Thus, thanks to skillful management and filling with useful information, an industrial enterprise manages to increase the effect of its activities with a decreasing share of fixed assets and normalized working capital, expressed by the coefficient $K_{f.w.c}$.

7 Conclusion

The phenomena of global challenges and threats, accelerated by the rapid development of high-tech production, give rise to extraordinary dynamic processes in the space of national economies, prompting industrial enterprises to adapt to the stunning disturbances of the business environment. In view of this, embedding them in an innovative economy and gaining stability demanded both the search and implementation of modern science-intensive technologies and widespread use in scientific research of the symbiosis of natural and social branches of knowledge.

The growing influence of an information resource on increasing the competitiveness and efficiency of industrial enterprises logically leads to the need to understand and apply universal laws of behavior of macroscopic systems. Revealing the features of the economic dynamics of enterprises, they enrich the theoretical concepts, methods and models of analysis and information support of the effect of enterprise functioning. Drawing on the views of thermodynamics, statistical physics, cybernetics and information theory, it becomes possible to give an entropic interpretation to the information resource of an enterprise and to develop the approach of V. Trapeznikov to the relationship of entropy, information and the effect of the enterprise.

The analytical value of the approach and model proposed in the article lies in the fact that they allow one to assess the measure and understand the nature of the influence of the information resource and the level of knowledge on the economic dynamics and stability of the efficient operation of an industrial enterprise.

References

1. B.Y. Akcali, E. Sismanoglu. Innovation and the Effect of Research and Development (R&D) Expenditure on Growth in Some Developing and Developed Countries. Proc. - Soc. and Behav. Sci., **195**, 768–775 (2015)
2. W.R. Ashby. *An Introduction to Cybernetics* (Foreign liter. Publ., Moscow, 1959)

3. S. Beer. *Cybernetics and Management* (Sci. Publ., Moscow, 1965)
4. L. von Bertalanffy. General system theory – a critical review. *Studies on the General Systems Theory* (Progress Publ., Moscow, 1969)
5. L. Boltzmann. *Articles and Speeches* (Sci. Publ., Moscow, 1970)
6. L. Brillouin. *Scientific Uncertainty and Information* (Mir Publ., Moscow, 1966)
7. S. Carnot, W. Thomson (Kelvin), R. Clausius, L. Boltzmann, M. Smoluchowski, A. Timiryazev (Ed.). *The Second Law of Thermodynamics* (LKI Publ., Moscow, 2007)
8. J. Gibbs. *Thermodynamics. Statistical Mechanics* (Sci. Publ., Moscow, 1982)
9. M.J. Hashim, I. Osman, S.M. Alhabshi. Effect of Intellectual Capital on Organizational Performance. Proc. – Soc. and Behav. Sci., **211**, 207–214 (2015)
10. R. Kamasa. Determinants of innovation performance: A resource-based study. Proc. – Soc. and Behav. Sci., **195**, 1330–1337 (2015)
11. L. Márquez-Ramos, I. Martínez-Zarzoso. The Effect of Technological Innovation on International Trade. A Nonlinear Approach. Open-Assessm. E-J., **4**, 1–37 (2010)
12. A.M. Prokhorov (Ed.). *Physical encyclopedic dictionary* (Sov. Encycl., Moscow, 1983)
13. C. Shannon. *The Mathematical theory of Communication* (The Univ. of Illinois Press, Urbana, Illinois, 1949)
14. G.L. Sipos, A. Ionescu. The Innovation – Corporate Performance Relationship in Emergent Countries. Proc. – Soc. and Behav. Sci., **238**, 124–132 (2018)
15. V.A. Trapeznikov. *Management and scientific and technological progress* (Sci. Publ., Moscow, 1983)
16. N. Wiener. *Cybernetics or Control and Communication in the Animal and the Machine* (Sci. Publ., Moscow, 1983)

Language rights and education of the small numbered indigenous peoples of the European North of Russia: from the Soviet period to modern time

Vadim Atnashev[1], *Art* Leete[2], *Natalya* Komelina[3], *Arina* Bildyug[3], and *Elena* Eltc[1*]

[1]St. Petersburg State University, 7/9, Universitetskaya Emb, St.-Petersburg, 199034, Russia
[2]University of Tartu, 18, Ülikooli, Tartu, 50090, Estonia
[3]Institute of Russian Literature (Pushkin House), Russian Academy of Sciences, 4, Makarova Emb., St.-Petersburg, 199034, Russia

Abstract. The article examines the experience of Russian constituent entities (Murmansk region, Nenets Autonomous Okrug, Archangelsk Oblast) in the support of indigenous ethnocultural rights in the context of current models for protection of minority languages and initiatives in the field of education. The contemporary scene of interaction between the state and indigenous minorities in the region is considered from a historical perspective. Special attention is paid to the early decades of the Soviet regime when key discussions concerning the status, way of integration with the state and direction of cultural development were established. The cultural bases and boarding schools, collectivization, forced relocations contributed to the construction of social reality of indigenous peoples which we can observe in the post-Soviet period. The specificity of modern transformation in the sphere of language and life of the reindeer herders are discussed in the context of the everyday practices of the interethnic relations of the Nenets and the Pomors of the Kanin peninsula and the northern parts of Mezen Region.

1 Introduction

Culture and language are the main elements of the identity preservation, therefore the support of the culture of minorities is necessary to protect them from assimilation and is vital for their survival in general. The UN Declaration on the Rights of Indigenous Peoples recognizes the right of these peoples to self-determination, autonomy and self-government, to complete freedom in the development of their economic, social and cultural institutions.

The definition of the concept of an indigenous people is contained in the International Indigenous and Tribal Peoples Convention, 1989, which obliges states to use special means to protect representatives of these peoples, their institutions, property, labor, tradition, culture and environment, as well as to provide conditions for the preservation of indigenous languages.

The UN indigenous right bodies highlight cultural loss and assimilation as the main reasons for the disappearance of indigenous languages. The Arctic peoples also face with such challenges, more or less acutely, depending on a country's situation. The practice of placing indigenous children in boarding schools, applied by governments in many regions of the world, has led to the children's assimilation, since their ties with their culture, language, and traditions were destroyed. It largely concerns the Russian North. All four dialects of the Sami language spoken in Russia are represented in the UNESCO list of languages in danger. Among the most pressing problems of Saami and Nenets of European North of the Russia are the low level of development of traditional activities, decrease in motivation for learning the native language, weak media coverage of the traditions, living conditions and life (which, like the projects "Voice of the tundra" and «Kola Saami Radio» is largely initiated within the framework of international cooperation in the north of Europe).

As a report on Russia presented at the UNPFII indicated in 2010, in many regions of Russia rural schools were closed. "Learning native languages is not only limited, but practically abandoned altogether.... Powers in the culture sphere have been delegated to the municipal level and the culture sector is financed in the last turn" [1]. Steps implemented to support the traditional culture and languages of small indigenous peoples in the regions, which are under discussion in this study, don't stop the threat of their assimilation and loss of languages.

2 Cultural development of indigenous people in the north-west of Russia in the early Soviet era

The contemporary scene of interaction between the state and indigenous

* Corresponding author: elenaelts@mail.ru

minorities in the region was very much shaped by specific reforms, employed since the early decades of the Soviet regime. It was the epoch when key discussions concerning the status, way of integration with the state and direction of cultural development were established. Central topics of these discussions focused on the degree of autonomous development that ended by the 1930s with the official decision to integrate the northern peoples tightly to the general progress of the state and a possibility of special ways of evolution was denied. Establishment of cultural bases and boarding schools, collectivization and forced relocations changed people's life in indigenous communities fundamentally and produced social reality for them in the way we can observe it also in the post-Soviet period.

2.1. Missionaries of Socialism and Culture bases

The Committee of the North was established in 1924 and it implemented two strategies of modernising the indigenous peoples of the North. Following the first approach, proposed by Vladimir Bogoraz-Tan [2: 48-50], 'missionaries of the new culture' were sent to taiga and tundra for facilitating socialist change among the indigenous groups. These 'missionaries of socialism' were young professional ethnographers, educated after the tsarist time, willing to explore indigenous culture in deep and having new, socialist ideals. They were supposed to provide practical help in turning indigenous peoples into socially self-conscious and economically self-sufficient social agents, able to follow the socialist path of development on their own.

Soon it became clear that a small number of inexperienced experts could not fulfil these gigantic plans of reorganising the life of the indigenous minorities. A more structural strategy was needed, and the Committee of the North introduced a more comprehensive plan of reforming the North by constructing culture bases. This was the most complex and intriguing socialist experiment conducted by the Soviets in the Arctic. The construction of culture bases was decided in 1925 [3: 111]. Nineteen culture bases were established in the remotest regions of the Russian North, Siberia and Far East by the end of the 1930s. Culture bases were exemplary settlements, built for indigenous groups in different regions of the North and consisting culture houses and many objects of infrastructure.

Culture bases were assumed to provide essential benefits and services typical of a modern society. These new settlements were meant to convince indigenous peoples in superiority of the Soviet life [4: 107]. Experts of the Committee of the North outlined the idea that culture bases must be founded in the areas almost entirely inhabited by natives, surrounded by native regions and characterised by a traditional way of life (as "the main hearth of the indigenous culture"). A culture base had to be established in order to influence the strongest indigenous groups and through them the surrounding communities. Through a chain mechanism of "self-influence," culture bases should execute Soviet control over indigenous communities in rather large territories. Special attention was supposed to be paid to preparation of specialists of indigenous origin. [5, 6, 7: 77-79].

Despite ambitious plans, actual impact of the culture bases on indigenous population remained somewhat limited. Culture bases' budget was insufficient; they lacked means of transportation, and even food and heating [8: 46]. Besides, there was a general scarcity of suitable employees [9: 41, 66]. The problem of lacking competent staff concerned also the challenge of developing literacy in indigenous languages. Schoolbooks in languages of the northern peoples were published but it is complicated to estimate how many teachers were capable to use these textbooks in practice [10: 208].

Potential effect of culture bases was significantly restricted by resistance from local administrators and indigenous groups who mostly just evaded invasion of the Soviets to the tundra, until it was still possible. However, violent episodes of resistance and even indigenous uprisings took also place during the 1930-40s, especially in the Western Siberia [4: 103-117, 11: 167-176, 12, 13, 14]. This lack of enthusiasm by the indigenous population slowed considerably down reforms, induced through culture bases in tundra. After the Committee of the North was liquidated in 1935, the culture bases were handed over to the *Glavsevmorput* and later officially dismissed but these settlements exist also nowadays as administrative centres or economic meeting points for indigenous people.

2.2 Soviet reforms in the Russian North

'The missionaries of socialism' managed to facilitate organising cooperative production units among the Nenets and Pomors in the Russian North [2: 50]. By 1927, all Nenets of Small Land tundra belonged to these small cooperative units [15: 83, 16: 135]. However, these new 'missionaries' were not able to initiate principal innovation in the local communities. Therefore, culture bases became the next step of socialist reforms also in the Russian North. In the region, two culture bases were built – in Khoseda-Khard (Big Land tundra) among the very first culture bases [5: 36, 17: 84] and Kolokolkovo (Small Land tundra) in the later phase of establishing culture bases.

As there were no medical facilities in the tundra of the Russian North (as well as among the indigenous groups of Siberia) until the 1920s [2:45], 'flying squads' of medical workers were arranged for the Nenets of the Big and Small Land tundra (and for some other peoples of the Soviet Arctic) [15: 82, 16: 135-136, 17: 82, 18: 102]. In the late 1930s, construction of the hospital in Kolokolkovo culture base was not finished yet and most of equipment and medical staff was still missing. Despite this, the small existing staff of the hospital worked hard in order to convince the Nenets to visit hospital regularly. The staff travelled around the tundra, providing medical help. Doctors had a mobile sauna with them and, they managed to encourage hundreds of Nenets to visit that tent-sauna. Later, the mobile sauna was attached to the red tent as an attribute of Soviet propaganda [19].

There were other conditions or achievements of the Committee of the North that distinguished the Russian North from the rest of the Soviet Arctic. Here, schools were more common even from tsarist time, teachers with experience of working with indigenous children were available and it was easier to arrange educational work for indigenous groups [15: 79, 16: 135, 20: 20]. The 10th extended plenary meeting of the Committee of the North issued the statement that nomadic councils, as newly established bodies of indigenous self-administration, functioned better in the Nenets National Region in the Russian North than in Siberia [16: 135, 21: 140]. Situation in the Russian North was so promising that even utopic economic experiments were introduced here. In 1926, Arkhangelsk regional Committee of the North initiated experiments of reconditioning a growth of moss in tundra and an experimental reindeer station was established on Kolguyev Island [22: 89].

On Kola Peninsula, no culture bases were established but similar trends of development can be traced. The indigenous minority policy in the region was rather weak during the 1920s and 1930s, when the Committee of the North enforced reforms in other parts of the Soviet Arctic [23: 46]. It has been argued that the Kola Sami suffered more than Nenets people did during the Stalin era because of "the border-transcending aspect" of the Sami population. However, repressions and devastative effect of collectivisation and boarding schools was rather common among the indigenous groups in the USSR [4: 110-115, 11: 176-178, 23: 47-48].

Decades later, the Kola Sami elders remembered collectivisation and the first wave of resettlement during the 1930s and 1940s as relatively smooth social challenge. The Sami population was relocated from tundra to small native villages. They lost their nomadic way of life but continued to dominate their social and cultural environment. But the second resettlement reform (from the 1950s to 1970s) had devastative consequences. Almost all the Kola Sami were removed to Lovozero village where they were amalgamated with the Russians, Izva Komi and Nenets, remaining socially and economically marginalised minority [23: 46-48, 24].

The Committee of the North found consistently positive examples of development (indigenous education, self-administration, health care, economic cooperation) from Big and Small Land tundra. Overall estimation of progress in socialist reforms among indigenous and non-indigenous population of the Russian North was also one of the highest among regions of the Arctic [25: 111, 116, 120].

On the basis on presented assessments, it can be argued that reforms of the Committee of the North were more successful in the Russian North than in Siberia. However, from the second half of the 1930s, after the Committee of the North was dismissed in 1935, prospects of successful integration of indigenous population into a larger society were lost. The potential of special development of indigenous groups in the Russian North did not materialise (local indigenous councils and courts were abolished, kolkhozes and boarding schools sustained). From the end of the 1930s, no distinctive indigenous policy was implemented in the Soviet Union.

3 The current experience of Russian constituent entities in the support of indigenous ethnocultural rights

Arctic settlements preserve the memories of 'missionaries of the new culture', the enthusiasts of the 1930s, mobilized and called up by the revolution, who worked and roamed in the tundra, who treated the indigenous population with respect and investigated the life of reindeer herders. Widely known ones are the names of N. Karpov, the founder of a nomadic school for children and youth on the Kaninsky Peninsula; doctor A. Koroliov, who made weekly visits from the hospital to the reindeer herders' camps in the Nes' village. Since the 2000s several projects have been implemented in the Nenets Autonomous Okrug, using the models of the late 1920s-1930s.

A social project "Red Tent" to ensure the availability of medical care for the nomadic population in remote regions was launched at the initiative of the "Association of the Nenets People "Yasavey", and with the support of the district administration, as well as the oil business. The project received a long-term planning character.

At the heart of the debate is the need to revive nomadic schools, an example of which was the school in the village Shoyna back in 1930-1935. Restoring the nomadic educational model in the late 1990s was a joint initiative of the administration of the Nenets Okrug, along with the Norwegian Barents Secretariat. It was associated with the urgent need to adapt a group of Nenets reindeer breeders, most of whom were illiterate and did not know Russian, to the modern socio-economic conditions of the region. In view of current urgent problems of preserving the Nenets language, the nomadic schools are considered as the practice of strengthening the language of the indigenous people and hereditary transmission from generation to generation, the model of a boarding school for children of reindeer herders is associated with economic advantages, and therefore, it often seems to be more preferable.

The pilot project of the nomadic «Kaninsky (tundra) kindergarten», launched in 2016 and financed under the regional state program is often assessed as an instrument of adaptation to studies in boarding schools in Nes' and Shoyna rural settlements. Due to the factual lack of children's skills in speaking their native language, only the elements of the Nenets language were used in joint games, in the classroom and in communication.

The opportunity to learn native language and to study in the Nenets language is enshrined in the regional law on the Nenets language. There are schools where the native (Nenets) language is being studied in municipal districts, where the share of the Nenets population exceeds the half, as well as in the city of Narian-Mare. The mother tongue is taught optional or as a national regional component of the curriculum, in grades 1-9 (with parental consent) and takes 2 hours a week. However, under the conditions of development of ethnic identity among the indigenous peoples, associated with the revival of the original culture, economy and language, since the 1990s, situations arise, when the children of reindeer herders (Nenets, Izhma-Komi) drop out of school after four years of study [26].

The current interest in wider dissemination of the nomadic school model, which will require changes to the regional law on education, teachers' training and additional funding, arose against the background of the approval of this model within the framework of the Euro-Arctic Barents Cooperation, Council of Europe as well as the successful experience of neighboring regions – Yamalo-Nenets Autonomous Okrug and the Republic of Sakha.

The end point of the winter routes of the Kaninsky reindeer herders is the Mezensky and Pinezhsky districts in the Archangelsk Oblast, where the Pomor identity is most clearly expressed. The districts' pasturelands were given in long-term lease to Kanin reindeer herders' teams. The inhabitants of the Mezen villages (Ruchyi and Dolgoshchelye) have abandoned reindeer husbandry in the absence of state support and due to a high land tax, of which the Nenets of the Nenets Autonomous Okrug are exempted.

Currently, 175 Nenets (1%) live a sedentary life in the Mezensky region. However, in 2018, the Council of Civil Society Development and Human Rights under the President of the Russian Federation recommended adding the municipal district to the list of places of traditional residence of the Nenets. Nevertheless, the proposal has been subject to widespread criticism, as potentially contributing to increased conflict between the Nenets and Pomors.

The mistakes of the second half of the XX century, related to the radical breakdown of the economic and everyday life of the Kola Sami, violation of the system of traditional upbringing of children, the transformation of individual farms into collective and state farms, the resettlement of indigenous peoples to larger settlements have been officially recognized in the Murmansk region. Since the 1970s the Kola Sami are concentrated in four settlements, while they are adjacent to other ethnic groups: Lovozero – the cultural capital of the Sami (Sami, Komi, Nenets, Russians); Krasnoschelye (Sami, Komi), Tuloma (Sami, Russians), Ena (Sami, Finns, Karelians).

The emphasis on the danger of losing the language of the Kola Sami, in the face of a decrease in the number of speakers, and the predominant role of enthusiasts in the transfer of knowledge, was put forward by the working group on the indigenous peoples of the Barents region. However, the Sami language is currently only taught as a one-hour-a-week elementary class at the only boarding school in Lovozero.

4 The Nenets and the Pomors of the Russian North: "Now all are confused..."

Our case study aims to add some touches to the scene of modern changes in the reindeer herders' everyday life through the narratives of the Nenets and the Pomors around this subject. Recently, the process of sedentarization of nomadic population is under discussion along with the problem of adaptation of the reindeer herders to economic transformations, gender shift in tundra communities, and the role of village in the nomadic identity [27, 28, 29: 80-83]. The same problems of Nenets community on Kanin peninsula (Nenets Autonomous Okrug) are also discussed in a number of works [30, 31, 32: 16-22, 33].

Our work is based on the results of fieldwork of Pushkin House (seasons 2014 and 2016) in villages Koyda and Dolgoshchelye of the northern part of Mezen' Region of Archangelsk Oblast. The materials of fieldwork 2006-2007 on Kanin peninsula are included, as well. It was conducted in terms of joint project of University of Aberdeen and Faculty of History, Saint-Petersburg State University. The Kanin peninsula and the northern part of Mezen' Region are the territories of the reindeer-herding by the Nenets and the Izhma-Komi 30: 6]. The Pomor villages started to appear on these territories from the 16th-18th centuries. Village Shoyna was founded on Kanin peninsula as the centre for fishing at the Soviet times.

During the last decades, the intensifying process of sedentarisation of reindeer herders has formed stronger ties between nomads and sedentary groups of the Nenets people (for the same on Yamal see Stammler [29: 80-83]). At the same time it has led to the closer inter-ethnic relations between the Nenets and the Russians in the field of everyday life in the villages: horticulture, leisure and so on [33]. Our informants mentioned the following reasons of sedentarisation: the inaptitude of someone for reindeer herding (problems with orientation in tundra, difficulties in the reindeers earmarks distinguishing, etc. (the Archive of the Department of Ethnography and Anthropology, the Institute of History, Saint-Petersburg State University, Northern expedition – 2006, Field Notes, Bildyug A. (further in the text - FN 2006): 19)); difficult living conditions in tundra; reluctance of young people to nomadic life (FN 2006, 20; Digital Fund of the Phonogram Archive of the Institute of Russian Literature (Pushkin House) RAS (further in the text - FA DF) 905-A039-04; for comparison see Liarskaia [34: 55-58]); a lack of money; the system of boarding schools (FN 2006: 19-20).

The ongoing changes are considered as rather complicated by the Nenets and the Pomors both: "If they settle, then they are not in their own surroundings," "people feel themselves confused" (FN 2006: 18). The reindeer herders "are being pressed". However, very often our informants related the reasons of the conflicts to bad tempers of its parties, regardless of their ethnicity. The stories about the alcohol addiction of the Nenets form the special group in the narratives about conflicts (on stereotypes of this kind see Leete [35]; Dudeck [36]). But still in these cases the boarder of the collision lay to a greater extent between settled group of people (the Nenets and the Pomors both) and the nomads (*toundroviki*) (FN 2006: 11, 13-14). Conflicts, which reasons can be traced to the ethnicity, are often provoked by the difficulties in the legal status of sedentary and nomadic peoples, still living in the same conditions of the North: "The Mezenians were not allowed to fish for flounder at the sea, so the Mezenians do not allow reindeer herders to stay at the forests, they shoot their reindeers. They say, Mezen' is not NAO" (the Archive of the Department of Ethnography and Anthropology, the Institute of History, Saint-Petersburg State University, Northern expedition – 2007, Field Notes, Bildyug A. (further in the text - FN 2007): 14 (see also: FN 2007: 6)).

The process of sedentarisation leads to gradual fading of the Nenets language on Kanin peninsula and Mezen' Region. People there can use three languages (including Komi) in their everyday practices (Ushakov [31:7]; FN 2006: 7). Our informants noticed that the Nenets language is used mostly in the tundra, since a lot of special phenomena related to the everyday nomadic life can't be properly reflected in another way. But on the whole, nowadays the Russian language is the main one almost for all the Nenets of Kanin peninsula and Winter coast (FN 2006: 7): "That's it, now that's it. And why did they forget what is theirs? Young people do not speak at all, well, there are some girls who already do < …> and the rest, they now do not speak it at all…" (FA DF 908-A004-01). The problem of language is closely related to the interethnic marriages, which were not common before. More often the wives learn the language of their husbands (FN 2006: 7). Usually the Nenets girls marry the Pomors and stay in the villages, but sometimes the Pomor girls move to tundra and start the nomadic life, which is more surprising for their neighbors: "They told her, where you are going, what all this is for?" (FA DF 905-A039-04) (on gender shift in tundra see Liarskaya [34]).

In the narratives related to the everyday life and the Pomor–Nenets interethnic relations the ongoing changes are reflected in the opposition of the well-ordered past and somewhat chaotic present: "They used to respect/follow (*sobljudali*) everything before. <…> Now all are confused/mixed up (*zapoutalis'*)" (FA DF 905-A037-01).

5 Conclusion

The territory of indigenous peoples, which are vital for their subsistence, cultural and spiritual life and for achieving and implementing autonomy and self-government, should be guaranteed to them. Though the constitutions of the Russian Federation and its subjects (e.g. the Charter of the Nenets Autonomous Okrug) recognize the right to preserve and develop the languages and cultures of indigenous peoples, the effectiveness of various programs is limited.

The only way of language revitalization is associated with a change in the status of the language through the formation of language policy and the expansion of the spheres of language functioning and training of civil servants in indigenous languages. In particular, it is necessary to provide intercultural and multilingual education for the entire population of the Russian Arctic. The governments, both Soviet and Russian ones, more often have just worried about teaching the official, state language. At the local level, measures to preserve practice of the indigenous languages are very weak let alone to promote their development.

Acknowledgements. The work was supported by the University of Tartu (grant PHVKU19913, 2019-2021) and the Russian Science Foundation (grant 17-78-20194, 2020-2021).

References

1. *Indigenous peoples: development with culture and identity: articles 3 and 32 of the United Nations Declaration on the Rights of Indigenous Peoples Report of the international expert group meeting, 12 - 14 January 2010,* Retreived from: https://undocs.org/E/C.19/2010/14 (12 June 2020)
2. V. Bogoraz-Tan, *Northern Asia* [In Russian] **3**, 40–50 (1925)
3. The protocol of the extended plenary meeting of the Committee of the North with the delegates of the 12th Congress of the Soviets of the RSFSR *Northern Asia* [In Russian] **3**, 111 (1925)
4. M.M. Balzer, *The Tenacity of Ethnicity. A Siberian Saga in Global Perspective.* (Princeton University Press, Princeton, 1999).
5. A.K. Lvov, *Northern Asia* [In Russian] **3**, 28–37 (1926)
6. V.M. Novitskii, *Ural Regional Studies* [In Russian] **2**, 117–124 (1928a)
7. V.M. Novitskii, *Northern Asia* [In Russian], **5–6**, 68–73 (1928b)
8. P. Terletskii, *Soviet North* [In Russian] **1**, 36–48. (1935)
9. E.D. Kantor, *Soviet North* [In Russian] **6**, 66–68 (1933)
10. E. Toulouze, L. Vallikivi, A. Leete., *Ethnic and Religious Minorities in Stalin's Soviet Union: New Dimensions of Research*, 199-223 (Södertörn University, Huddinge, 2017)
11. A. Golovnev, *Talking Cultures: Samoyed and Ugrian* [in Russian] (Ural Branch, Russian Academy of Science, Ekaterinburg, 1995)

12. O.D. Ernykhova, *The Kazym Rebellion: About the History of Kazym Uprising 1933–34* [in Russian] (Siberian chronograph, Novosibirsk, 2003)
13. A. Leete, *Temenos - Nordic Journal of Comparative Religion* **39-40**, 235–245 (2004)
14. A. Leete, *Asian Folklore Studies* **64(2)**, 233–246 (2005)
15. Extended plenum of the Committee of the North at the Board of The All-Russian Central Executive Committee (from 28/II to 5/III 1927) *Northern Asia* [In Russian] **3**, 76–90 (1927)
16. K.S. Korelin, in *The 1926/27 Soviet Polar Census in the Russian North (Arkhangelsk province and autonomous region of Komi)* [In Russian], 127-139 (EthnoExpert St-Petersburg, 2010)
17. Work Results of the Committee for Assistance to the Peoples of the Northern Borderlands of the All-Union Central Executive Committee. *Northern Asia* [In Russian] **3**, 80–85 (1926)
18. P. Khatanzeiskii, in. *The 1926/27 Soviet Polar Census in the Russian North (Arkhangelsk province and autonomous region of Komi)* [In Russian], 99-106 (EthnoExpert, St-Petersburg, 2010)
19. G. Khabarov, *Soviet Arctic* [In Russian] **5**, 62–63 (1938)
20. A.V. Lunacharskii, *Northern Asia* [In Russian] **3**, 18-22 (1927)
21. Presentation of Reports. *Soviet North* [In Russian] **3**, 139–143 (1934)
22. Plenary Meeting of the Committee of the North . *Northern Asia* [In Russian] **3**, 85–94 (1926)
23. M. Berg-Nordlie, *Acta Borealia* **32 (1)**, 40–67. (2015)
24. A. Afanasyeva, *Forced relocations of the Kola Sámi people: background and consequences. (Thesis submitted for the degree: Master of Philosophy in Indigenous Studies.).*(University of Tromsø. 2013)
25. The protocol of meeting of the extended plenary meeting of the Committee for advancement of peoples of the northern peripheries at the Board of The All-Russian Central Executive Committee, held from the 7th to the 13th of March 1928. Northern Asia [In Russian] **4**, 109–123 (1928)
26. M.Kh Belianskaia, *Journal of Siberian Federal University* **10**, 2342-2350 (2016)
27. A.V. Golovnev, D.A. Kukanov, E.V. Perevalova, *Arctic: Atlas of Nomadic Technologies* [in Russian] (MAE RAS Publ, St. Petersburg, 2018)
28. E.V. Liarskaia, *Anthropology. Folklore Studies. Linguistics* [In Russian] **1**, 36–55 (2001)
29. F. Stammler, *Reindeer Nomads Meet the Market: Culture, Property and Globalisation at the End of the Land* (Lit-Verlag, Muenster, 2005)
30. S.B. Kiselev, *Traditional Economy of the Nomadic Population of the Kanin Peninsula in the First Third of the 20th Century and at the Beginning of the 21st Century. Thesis for a PhD in History* [In Russian]. (St. Petersburg, 2018)
31. N.V. Ushakov, *Field Study Materials MAE RAS* [In Russian], **15**, 6-22 (MAE RAS, Saint-Petersburg, 2015)
32. S. Usenyuk, S. Dudeck, N. Garin, *The Polar Journal* **5(1)**, 170-202 (2015)
33. O.V. Sokolova, O.A.Subareva *Field Study Materials MAE RAS* [In Russian] **8**. Saint-Petersburg, 24 – 38 (2008)
34. E.V. Liarskaia *Anthropology of East Europe Review* **28**, 51-84 (2010)
35. A. Leete, *Acta Ethnographica Hungarica*. **50 (1)**, 241–258 (2005)
36. S. Dudeck, *Folklore* **61 (August)**, 89-116. (2015)

Features of the adaptation of the indigenous people of the Arctic

Evgeniya. Nikolaeva[1,*], *Olga* Kolosova[2], and *Elena* Belchusova[2]

[1] M.K. Ammosov North-Eastern Federal University, 27, Oyunsky street, Yakutsk, 677013 Russia.
[2] The Yakut Scientific Centre of the SBRAS, Institution of the Russian Academy of Sciences, 41, Lenin avenue, Yakutsk, 677007, Russia

Abstract. To determine age-specific differences in adaptation of indigenous residents, nonspecific adaptive body reactions (NARs) have been studied among the Evenki men aged 13 to 86, permanently living in the Arctic. The results indicated the presence of age-specific differences in NARs. The decline of nonspecific body resistance in every fourth teenager is correlated with adaptive reactions developing in parallel with underlying low-level body reactivity. Adaptive reactions in men aged 75-90 were characterized by a medium to a high reactivity level, which enabled preservation of body reserves even under extreme climate conditions. Two age groups (22-35 and 61-74) included the highest percentages of people with adverse NARs. Therefore, men in these age groups should be considered as a major risk group for developing both psychosomatic and somatic diseases.

1 Introduction

Development of some specific patterns of the physiological function in individuals arriving to live in the Arctic is defined by the region's extreme climate, geography, and heliogeophysical conditions. At the bottom-line of an adaptive response to changing environment conditions (including extreme climate factors), there are nonspecific adaptive body reactions (NARs). The study of strategies for adapting to the specific ecological and climatic factors of the body of the indigenous people of the Arctic, who in the process of evolution developed specific mechanisms for biochemical and physiological adaptation, is extremely important and necessary. Previous studies have led to the conclusion that there are adaptive types in the population of people that are adapted to well-defined climatic and geographical conditions of existence. Adaptive types of human organisms determine survival and can also be constraints in possible adaptive processes in new environmental conditions [1]. The effectiveness of adaptation to extreme conditions depends not only on genetic factors, but also on the morpho-physiological parameters of a person, which are formed in ontogenesis under certain climatic and geographical conditions [2-7].

In the literature, views on NARs are closely related to the study of adaption mechanisms. According to the theory of Hans Selye about a 'general adaptation syndrome', a state of stress (strain) can be induced by any strong stimulus (stressor). Because stimuli affecting the body may be of various intensity, theory of NARs was further developed by L.KH. Garkavi and colleagues (1968). The body, reacting to stimuli of varying strength and quality, causing fluctuations in the parameters of homeostasis, tries to keep them within optimal boundaries, using a common periodic system of nonspecific, anti-stressor reactions of the body. With long-term exposure to a complex of extreme factors in the body, in the process of evolution, the most optimal adaptive reactions are fixed, which allow a person not only to survive in these conditions, but to fully live, work, develop, etc. Under the influence of factors of varying strength and intensity, regardless of their nature, physiological and biochemical changes take place in the body, leading to nonspecific adaptive reactions of the organism. Following this recent theory, the system of human adaptive reactions includes, alongside with the stress reaction (SR) and over-activation (OA) reaction, the following types of reaction: training reaction (TR) developing under weak stimuli; activation reaction (AR) subdivided into quiet activation reaction (QAR) and advanced activation reaction (AAR) developing under moderate stimuli. Each type of adaptive reaction corresponds to a specific set of indicators describing neuro-endocrine and immune body subsystems, processes of metabolism and energy metabolism, and mental status. Good adaptation to changing environmental conditions can be achieved if there is a sufficient adaptation potential of a body. This capability is conditioned not just by the presence of a functional body reserve, but also by body reaction adequacy and economy, as well as efficiency of control over functional reserve expenditure and recovery [7-10]. In addition, adaptive reactions can be correlated with different body

* Corresponding author: en.nikolaeva@mail.ru

reactivity levels (RL), depending on the magnitude of the stimulus. The hallmark of the high reactivity level is low sensitivity threshold in response to factors of minor magnitude, while a lower reactivity level is associated with growing sensitivity threshold developing under factors of large magnitude. Apparently, the tendency is such that a higher body reactivity level correlates with more effective synchronization of body's protective systems due to multiplicity of minor interactions. A decline in the reactivity level is associated with the onset and escalation of the signs of tension, which signalize desynchronized functioning of body systems. Thus, reactions of training, quiet activation, and advanced activation alongside with underlying a high reactivity level are considered beneficial body reactions. Unfavourable NARs, which pave the way for pre-disease, include an over-activation (OA) reaction, stress reaction, and reactions of training, quiet, and advanced activation developing in parallel with the low body reactivity level.

In the course of evolution, under certain climatic and geographic factors, native northern minorities, living in the harsh Arctic environment from generation to generation, phenotypically developed specific adaptation mechanisms, which formed the basis for optimal functioning and regulation of body systems [2,6,8]. Since genetically fixed adaptive mechanisms in the process of ontogenesis are manifested depending on the living conditions, it is very important to understand their phenotypic manifestation in different age groups of the same ethnic group living for a long time in the Arctic region.

In light of the above said, it was deemed relevant to study the age-related features of the ecological and physiological adaptation mechanisms developing in residents of the Arctic in response to extreme environmental factors.

The aim of this study was to detect patterns in the development of nonspecific adaptive reactions in indigenous males (Evenki) residing permanently in the Arctic.

2 Material and methods

259 men, permanently residing in Olenek Evenki National Region, were observed. They volunteered to be included to the study. All patients were Evenki by ethnicity, aged 13 to 86. All were divided to 6 age groups: 13-16 (A1); 17-20 (A2); 21-34 (A3); 36-60 (A4); 61-74 (A5); 75 or more (A6).

Types of NARs and reactivity levels (RL) were determined based on L.KH. Garkavi et al. Fasting peripheral blood samples were obtained in the morning; blood smears were prepared according to the standard technique.

To assess the vegetative regulation, the method of heart rate variability analysis (HRV) according to R.M. Baevsky was used. The study was conducted on the hardware-software complex "VNS-Micro" company Neurosoft (Ivanovo) for 5 minutes in the prone position in the morning. Using a heart rate variability analysis based on processing data on the distribution of cardio intervals, a stress index of regulatory systems (Stress Index - SI) was determined, which reflects the degree of centralization of heart rhythm control. The method of identifying the hidden periodicity of the dynamic range of cardiointervals determined the power spectrum in the ranges:

1. High frequencies (High Frequency - HF) - 0.15–0.40 Hz.
2. Low frequencies (Low Frequency - LF) - 0.04–0.15 Hz.
3. Very low frequencies (Very Low Frequency - VLF) - 0.003–0.04 Hz.

The study is based on a simple random sample. Storage of research results and primary processing of the material was carried out in the original database Microsoft Excel 2007. Statistical processing was made using conventional methods with the package "SPSS 10". Mean (M), mean square deviation (SD) and mean (m) error were determined. The nature of the distribution was assessed using the Kolmogorov-Smirnov criterion. In the normal distribution of variables, the student's unpaired t-test was used to determine the differences between the two independent groups, and in the nonparametric one, the Wilcoxon – Mann–Whitney test was used. Correlation analysis methods for parametric and nonparametric types of distribution – Pearson and Spearman criteria, respectively - were used to identify the relationship between the studied parameters. Differences and correlations were considered significant at $p < 0.05$. The study is complied with ethical guidelines of World Medical Association Declaration of Helsinki and 'The fundamentals of the legislation of the Russian Federation on the protection of citizens' health' (issued in 1993).

3 Results and discussion

The results of studying the state of adaptation processes of the organism of indigenous small ethnic groups of the Arctic region, regardless of age, indicate that more than half of men (54%) show a high level of reactivity in the body, ensuring the possibility of rapid mobilization of functions and an adequate response of the body to external influences (Fig. 1).

Here is the structure of the types of nonspecific adaptive reactions (A) and (B) reactivity level (UR) of the body of Evenk men (p=259;%):

Against this background, a rather high percentage of people (30%) is found in a state of low reactivity characterized by a predominance of inhibitory processes in the body, both at the level of the central nervous system and at the periphery, whose body is unable to develop an adequate, rapid response of the adaptive reaction. The adverse types of NARs (RS and OA) of the body of Evenki men are 14%. Despite the high percentage of men with favourable NARs (TR, QAP and AAR) (86%), taking into account the fact that 30% of them occur at low RL, it can be said that 35% of other people in the body have a mismatch functioning of the subsystems of the body, which is the cause of the intensity of adaptive systems

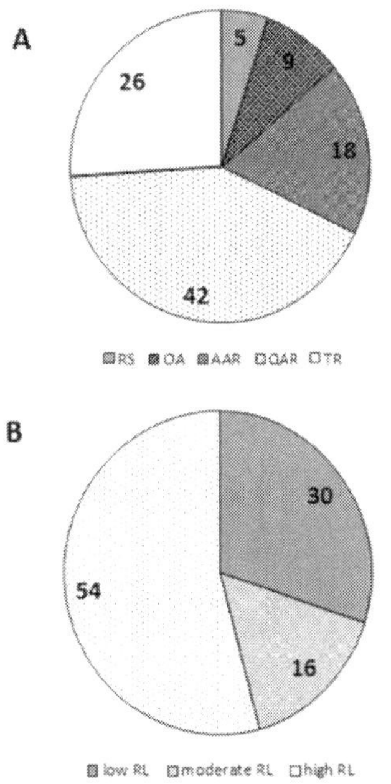

Fig. 1. Structure of types of non-specific adaptive reactions (A) and the level of reactivity (B) of the body of Evenki men (n=159, %): RS - stress reaction, RT-training reaction, RSA-quiet activation reaction, RPA-increased activation reaction, OA - over-activation reaction.

A more detailed study of the structure of adaptive reactions of the organism was carried out in 6 age groups (Fig. 2). Statistically meaningful differences ($p<0.05$) were demonstrated in the adaptive response patterns between study groups.

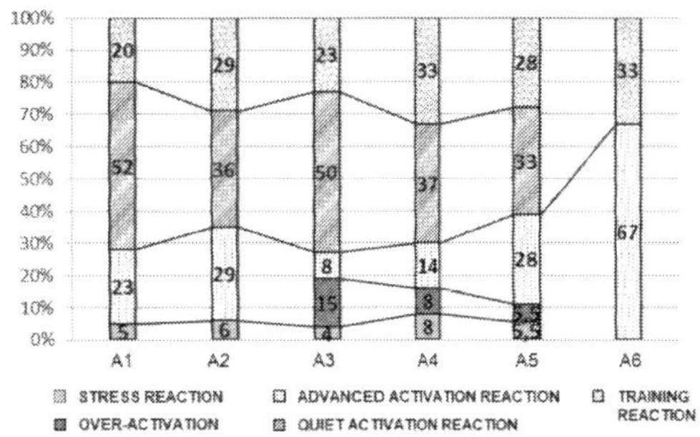

Fig. 2. Prevalence of nonspecific adaptive body reaction indicators in males in relation to age (%) (A1 – 13-16 years, A2 – 17-20 years, A3 – 21-34 years, A4 – 35-60 years, A5 – 61-74 years, A6 – 75 – 90 years)

In age group A1, activation reactions prevailed in the majority: QAR and AAR (52% and 23%, respectfully), suggesting high level of health, increased activity of regulatory and protective body subsystems, predomination of anabolism with moderate energy expenditure, indicating that adaptive responses are developing in a favourable way. TR was observed in 20% of adolescents, associated with the presence of protective inhibition in central nervous system (CNS), decreased catabolism, accumulation and preservation of plasticity and energy reserves of the body. TR in adult age would be labelled as an anti-stressor recovery reaction, but in adolescent age, with its intensive physical growth and mental development, extensive changes affecting hormonal and nervous control processes, this signals a worsened health level. Moreover, 5% of adolescents were assessed to be in a stressed state.

The study of body reactivity levels showed that in 29% of Evenki adolescents, adaptive reactions were associated with the low or moderate reactivity level, suggesting a decline in sensitivity of CNS, desyncronization in regulatory systems, and consequently, tensed adaptive reactions (Fig. 3).

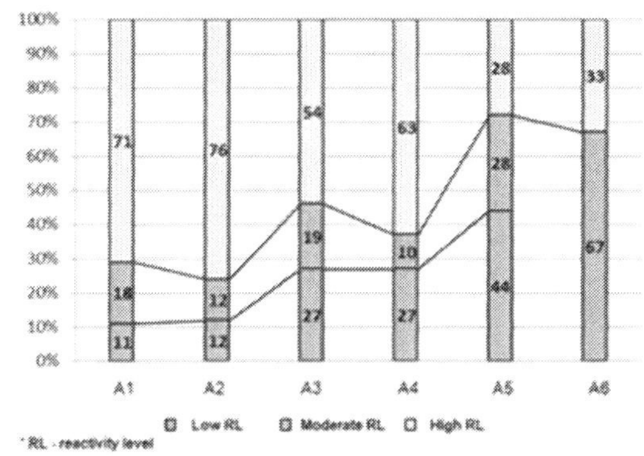

Fig. 3. Prevalence of reactivity levels of nonspecific adaptive body reaction indicators in males in relation to age (%) (A1 – 13-16 years, A2 – 17-20 years, A3 – 21-34 years, A4 – 35-60 years, A5 – 61-74 years, A6 – 75 – 90 years).

In adolescence (A2), the proportion of individuals with QAR was significantly (by 16%) smaller ($p<0.05$) and proportion of those with TR was 9% larger compared to A1 group. Aside from experiencing physiological changes, life of adolescents is filled with social events of utmost importance such as graduating from school, entering university, starting an independent life, etc. A compound effect of climate, social, psychological factors may, probably, be overwhelming and excessive for Evenki teenagers, leading to development of protective inhibition in CNS and training reaction.

Among men in the early middle age (A3), AAR was observed in 21% less than it was in A2 group. Over-activation reaction was demonstrated in 15%, which may indicate that adaptation mechanisms were tensed, and AAR was transitioning to OA reaction. Compared to A2 group, there were twice more ($p<0.01$) men with a low body reactivity level. At the low reactivity level, adaptive reactions develop following stimuli of strong magnitude, which is a sign of diminished CNS sensitivity, leading eventually to desynchronization of humoral and neural control mechanisms, increased energy expenditures, and reduced functional body reserve.

In groups A4, A5, and A6, the number of men demonstrating QAR gradually decreased with aging, while people in this group increasingly show AAR, which speaks for favourable development of adaptation mechanisms. The maximum number of persons with unfavourable NAR is observed in age group A3. In subsequent groups, the number of people with unfavourable NAR is gradually decreasing (Fig. 4).

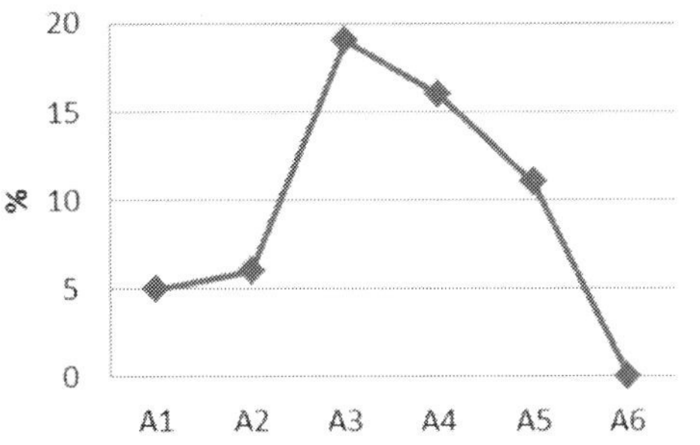

Fig. 4. Dynamics of unfavourable nonspecific adaptive body reaction indicators in males in relation to age (%) (A1 – 13-16 years, A2 – 17-20 years, A3 – 21-34 years, A4 – 35-60 years, A5 – 61-74 years, A6 – 75 – 90 years).

But at the elderly age (A5), the proportion of individuals with high reactivity levels declined dramatically ($p<0.05$), alongside with the increase (more than 50% higher) in the number of men with low level reactivity. As far as an average life span in male population of northern regions is less than 64 years, it could be deduced that Evenki men making it to this age would have exhausted functional body reserve, energy and protective body capacities, so that the body would fail to adequately respond to any stimuli. And this finally could be leading to irreversible desynchronizing deleterious processes.

In senile men, AAR and TR develop in association with moderate or high reactivity levels, which is why individuals with high adaptive body reserve are more likely to reach this age.

Thus, the largest percentage of people with unfavorable NARs are men aged (22–35) years and (61–74) years.

Since, as it is supposed, mobilization of energy reserves and development of the corresponding NARs in the body occurs as a result of a change in the level of activity of regulatory systems, a study was conducted concerning the stress index (SI) of the regulatory systems of men in adult age groups (Fig. 5). In a healthy person in central Russia in a calm state, the magnitude of the SI is 60-120 cu. In our study, in three age groups (A3, A4, and A5), the SI value was significantly ($p<0.05$) higher than the standard indicators. The high voltage of the state of regulatory systems in the men of the Evenks forms unfavorable NARs.

The acrophase SI falls on the age group A3, and the level of tension of the regulatory systems in comparison with the A2 group increases significantly ($p<0.01$) by more than 7.7 times.

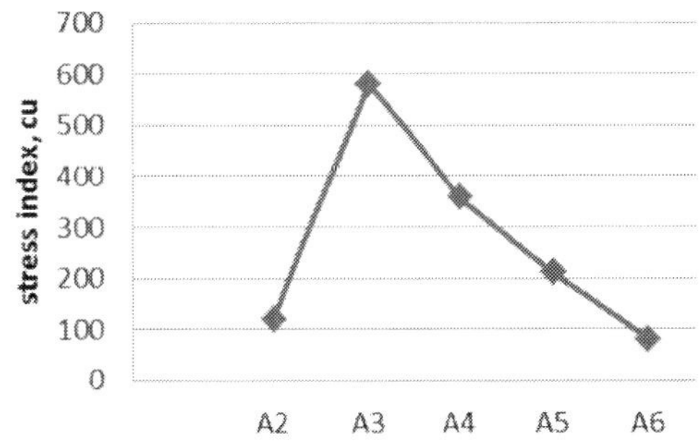

Fig. 5. The magnitude of the stress index of the body's regulatory systems in males in relation to age (cu), (A2 – 17-20 years, A3 – 21-34 years, A4 – 35-60 years, A5 – 61-74 years, A6 – 75 – 90 years).

The physiological mechanisms of the formation of unfavorable NAR in the age category from 21 to 34 years (A3) are associated with increased duplication of adaptation processes by the central regulation. This is noted in cases where the body, when a voltage condition of regulatory systems occurs, constantly suffers a shortage of functional reserves. The results obtained to identify the spectrum of regulation of functions in men indicate that in the age group A3 almost 47% occupy a very low frequency range (VLF%). The power of this range is associated with the central neuro-humoral regulation, which corresponds to the third, highest, uneconomic level of activity - the level of central, intersystem control, which causes the highest level of SI.

The formation of unfavorable NAR in the A5 age group is based on somewhat different mechanisms; in particular, this is associated with an increase in the tone of the sympathetic division of the autonomic nervous system. The power of the low frequency range (LF%) rises, which is a marker of the sympathetic nervous system and corresponds to the second level of activity - the level of intrasystem regulation.

4 Conclusion

Study results indicate the presence of age-specific differences in NARs observed among Evenki men permanently residing in the Arctic. Decline of nonspecific body resistance in every fourth teenager correlated with adaptive reactions developing in parallel with underlying low-level body reactivity. Adaptive reactions in men of the senile age group were characterized by medium to the high reactivity level, which enabled preservation of body reserves, even under extreme climate conditions. It was established that in two age groups - early adulthood ((21-35) years) and old age ((61-74) years), the highest level of tension of adaptive reactions occurs due to low recovery and mobilizing potential, immunodepression, desynchronization of functional and stress-realizing systems of the organism, the predominance of catabolism processes on the background of transcendental arousal (leading to transcendent inhibition) of the central nervous system. The physiological mechanisms of NAR formation in these groups differ in different levels of regulatory systems and, therefore, in the magnitude of the body's functioning tension. In group A5, the formation of NAR is carried out by intrasystemic rearrangements, by increasing the tone of the sympathetic division of the autonomic nervous system. In the age group A3, the mechanism of the formation of adaptation reactions is based on energetically extremely costly processes of intersystem rearrangements, due to the increased duplication of adaptation processes by the central regulation.

Acknowledgement

The study was carried out within the framework of the projects: "Study of chronophysiological, molecular mechanisms of adaptation and psychoemotional stress of a person under the conditions of the combined action of extreme climatic, technogenic factors in the Arctic and Subarctic and a pandemic, post-pandemic, development of ways to increase the adaptive potential" 121021600269-0), Federal Research Center YSC SB RAS and "Physiological and biochemical mechanisms of adaptation of plants, animals, humans to the conditions of the Arctic / Subarctic and the development of biological products based on natural northern raw materials that increase the efficiency of the adaptation process and the level of human health in extreme environmental conditions" (No. 0297 - 2021-0025 registration number AAAA-A21-121012190035-9) of the Institute for Biological Problems of Permafrost, Federal Research Center of YSC SB RAS.

References

1. J. Launay, G. Savourey, *Ind Health* **47** (2009)
2. T.M. Mäkinen, *Am J Hum Biol* **19** (2007)
3. T.M. Mäkinen, *Front Biosci - Scholar* **2** (2010)
4. J. Bittel, *Int J Sports Med* **13** (1992)
5. A.T. Steegmann, *Jr. Human, Am J Hum Biol* **19** (2007)
6. A.J. Young, M.J. Fregly, C.M. Blatteis. *Handbook of physiology. Section 4: Environmental physiology*. Oxford University Press (Oxford, UK, 1996)
7. N.I. Douglas, E.N. Nikolaeva, N.S. Baisheva, O.N. Kolosova, E.A. Belchusova, L.E. Vasilieva, *Int J of Biomed* **9** (2019)
8. T. Adams, B.G. Covino. *J Appl Physiol* **12** (1958)
9. W.D. van Marken Lichtenbelt, H.A. Daanen, *Curr Opin Clin Nutr Metab Care*. **6** (2003)
10. J.W. Castellani, A.J. Young, *Auton Neurosci*. 196 (2016)

Construction of the Irkutsk-Baikal section of the trans-Baikal railway: state of research

Valery V. Tretyakov[1,*], *Yuri A.* Petrushin[1], *Anastasia V.* Nekludova[1], and *Elena A.* Yakovleva[1]

[1] Irkutsk State Transport University, 15, Chernyshevsky Str., Irkutsk, Russia

Abstract. The article considers the problems of railway construction in Russia in the 19th and early 20th centuries and finds out the available information on one of the fragments of this grandiose construction in the historical literature. Attention is focused on the construction history of one of the important sections of the Trans-Siberian railway – the Irkutsk-Baikal branch railway, which became the foundation of the Circumbaikal railway. This part of the Great Siberian Railway proved to be a very technically complex construction facility and required significant investments from the state. However, it was not easy to approach to its construction. Its creation was preceded by the work of numerous research teams, selfless labour of designers, engineers, technicians and workers. Thanks to their efforts, our country ultimately managed to establish a continuous rail transportation service and unite the western and eastern parts of Russia, which made it possible to maintain the country's defense capability during the sudden outbreak of the Russian-Japanese War of 1904–1905. The authors specify the main problems that arise when studying the history of the construction and operation of the Irkutsk-Baikal line. They define the range of sources put into circulation by researchers of the designing and construction processes of this road. It is emphasized that the largest number of sources relate to the creation of the line, while the sources which deeply reflect the line operation are few and fragmented.

Keywords: Railway construction, Trans-Siberian Railway, Irkutsk-Baikal Railway Line, Baikal Railway Crossing, Trans-Baikal Railway, Great Siberian Railway.

1 Introduction

The construction of the Great Siberian Railway is an outstanding achievement of the Russian state. This road provided a reliable communication between the vast territories of Russia and helped to increase the economic potential of the country. Concurrently, it became a significant geopolitical factor. It caused radical changes in all spheres of life of the Siberian society, contributed to the development of modernization processes on the outskirts of the country.

The main chapters of the history of the Trans-Siberian Railway are generally known. But each of the sections of this railway is unique in its own way, and in each case, engineers had to creatively solve the problems that they faced before the construction. Perhaps the most difficult problem was the connection of the Mid-Siberian and Baikal sections of the Trans-Siberian Railway by continuous track line. Here the foothills of Lake Baikal – an area poorly studied in geological and hydrological terms – stood in the way of the construction workers. The designers developed several variants of the line bypassing the Baikal, and each of the projects of the Circumbaikal Railway had its advantages and disadvantages. These discussions continued for a long time. Meanwhile, to solve the problem, a temporary solution to the problem was implemented – the construction of a lake ferry railway crossing. This is the reason why it was originally intended to build a railway linking the Irkutsk station with the Baikal. It wasn't easy to construct this section, but its importance for the Trans-Siberian railway can hardly be overestimated. The "sticking point", the "bottleneck" of the Trans-Siberian railway – these sharp characteristics are given to this area in the book prepared for Russian readers by C. Wolmar, an English journalist [23].

2 Problem Statement

In the 1990s, researchers actively studied the process of constructing the Irkutsk-Baikal railway and its Irkutsk-Baikal section. The results of this work were presented in several monographs [18, 20]. At that time, interest in the topic was partly fueled by the desire to recreate the realities of the Trans-Siberian railway construction process on the eve of the upcoming anniversary – the 100th anniversary of the arrival of the first train in Irkutsk. However, in the succeeding years, new publications appeared that address this topic. In addition, new aspects of the study of the topic were identified; new materials were used [5, 9, 19]. Nevertheless, the new knowledge has not yet been analyzed.

* Corresponding author: tretvv@yandex.ru

3 Research Questions

In recent decades, there has been an increase in researchers' interest in the topic. The document collections of the Irkutsk (GAIO, The State Archive of the Irkutsk Region) and the St. Petersburg (RGIA, The Russian State Historical Archive) archives and materials of local and central libraries were studied, which allowed expanding and deepening knowledge about the Trans-Siberian railway construction process. This resulted in the emergence of new publications and preparation of dissertation studies. Previously unknown stories of the history of the construction of the Irkutsk-Baikal line have been revealed. Many of the hypotheses made earlier have been substantiated. At the same time, the range of issues that attracted the attention of researchers on this topic was clearly defined.

4 Purpose of the Study

The aim of this work is to analyze the state of study of the processes of the Irkutsk-Baikal section of the Trans-Siberian railway. To do this, the following tasks are solved:

- summing up the results of the previous studies;
- defining a range of works emerged in the last two decades that describe the construction of the Irkutsk-Baikal line at the end of the 19th century;
- defining the source base of the conducted studies, considering the sources previously used;
- determining the prospects for developing the subject.

5 Research Methods

The authors used several groups of methods in the study. General theoretical methods of induction and deduction, abstraction and generalization, systematization and interpretation of facts were used as the study basis. In the preparation of the conclusions, the authors used sociological methods of working with empirical data (content analysis, measurement, comparison). But the main method for working with sources and literature was a historical method, and the guiding principles of the study were the principles of objectivity and historicism.

The empirical basis for the study was the data of various nature and meaning, collected in archives and libraries in the course of work.

In interpreting the data and preparing the results of the study, the authors relied on special historical methods. The historical and genetic method made it possible to systematize the data, to establish the main milestones of the process of preparation for the construction of the Irkutsk-Baikal mainline. The application of the historical and comparative method helped to clarify the various states of the process of preparation for the construction of the railway line. At the same time, it helped to establish the specific features of individual states of the process, to determine the meaning of certain solutions in the general case of finding the best way to solve the problems that arose in the design and construction of this line. The use of cause-and-effect analysis allowed the authors to establish links between major and secondary events, which was aimed at building a logically sound sequence of the states of project development processes and their implementation. The application of the historical reconstruction method allowed one to consider the processes of organizing the works of designers, construction workers and railway workers in the unity of subjective and objective aspects. When working with literary sources, the authors relied on retrospective techniques.

6 Findings

Previous studies have revealed that the start of construction of the Siberian Railway in 1891 made many problems come to light, and research on the future route line in Eastern Siberia has temporarily receded into the background. But in February 1893 it was decided that the railway would go towards Irkutsk, and it was necessary to accelerate the research. By this time, the projects to bypass Baikal from the north had already been rejected, but the issue of the direction of the route from Irkutsk on the western shore of the lake to the village of Mysovoi (the modern city of Babushkin) on the eastern shore was not solved. Engineers developed different variants of the direction of the Circumbaikal railway through the valleys of the Irkut, Olkha, and Kitoi rivers (Figure 1).

The plan for the construction of this road along the shore of Lake Baikal from the source of the Angara to the village of Kultuk and on to Mysovaya – the project of the line was the shortest in length, but at the same time very costly and complicated in construction – was also considered. It was this option that was eventually implemented. This choice was largely encouraged by the factors of reducing time costs and improving the safety of the railway line: the shortest way was supposed to be arranged on a smooth and geologically solid foundation, which was provided by the rocky shores of the lake. In November 1893, while the decision on the Circumbaikal bypass was not approved, the Committee of Siberian Railways decided to temporarily connect the railway between Irkutsk and the Trans-Baikal ferry crossing. For this purpose, it was planned to specially build icebreakers capable of transporting rail cars, cargo and passengers across the lake. But it was also necessary to construct the railway from Irkutsk to the shore of the Baikal. In February 1894 the Ministry of Railways obtained permission to produce research on the line of Irkutsk – the village of Listvenichnoe, located on the shore to the right of the source of the Angara. This project involved significant construction difficulties, for example, it was necessary to build a float bridge over the Angara in Irkutsk, as the Mid-Siberian Road approached Irkutsk on its left bank. The construction of the line was supposed to begin in 1895. However, more detailed studies have shown significant inconveniences of the Irkutsk-Listvenichnoye direction: the water level in the Angara fluctuated considerably, and during the spring high waters and icefall the float crossing could be interrupted for a month and a half; the construction of a permanent railway bridge was very costly.

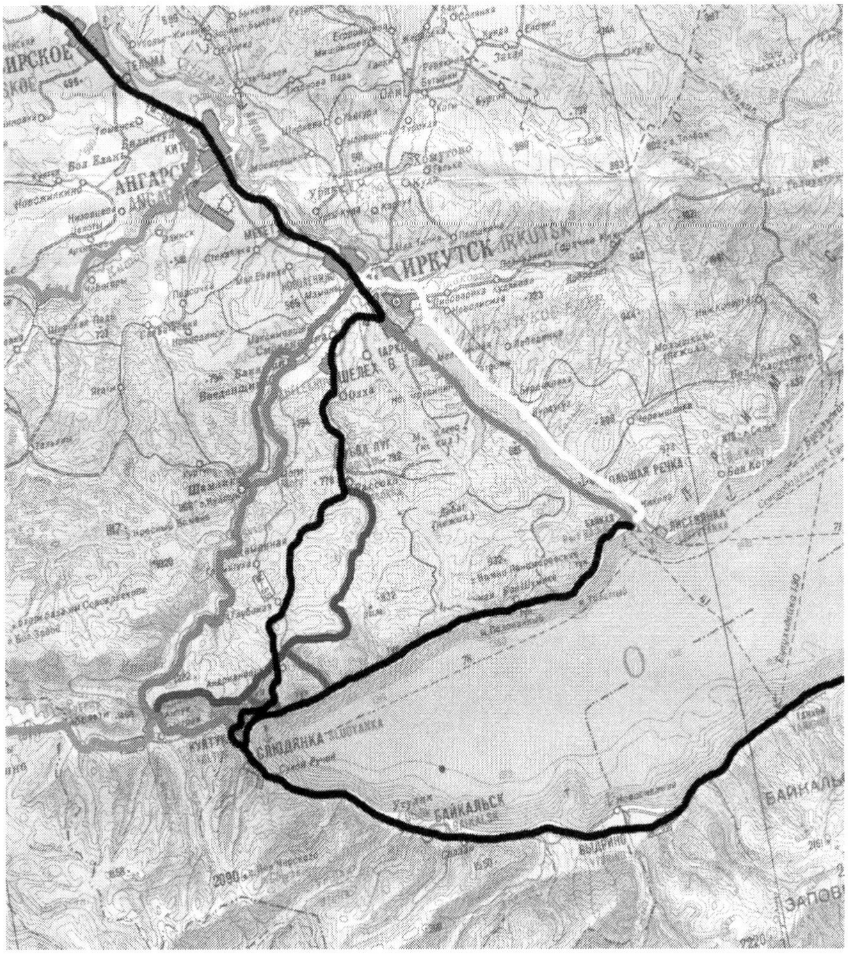

Figure 1. Projects of the Circumbaikal railway. The modern Trans-Siberian railway route is marked in black; the Irkutsk-Listvenichnaya line project – in yellow; the Irkutsk-Baikal line – in gray. (The illustration is from the collection of A.V. Hobta)

It was decided to carry out additional surveys on the left bank of the Angara, which showed that laying a branch there is quite possible. The left-bank project reduced the overall length of the route and did not require the construction of additional bridges, although in comparison with the right-bank version it assumed significant amount of work on the mountain excavation and track construction. At the same time, the project of the Circumbaikal railway construction along the lake shore began to attract special attention, and in this case the Irkutsk-Baikal line would become not a branch, but a section of the Trans-Siberian railway. After numerous discussions, the project was approved and the line was scheduled to be built in 1896. According to the project, it started from the final point of the Mid-Siberian Railway at he Innokentyevskaya station on the left bank of Irkut, crossed it and approached the Irkutsk station on the second verst. The station of Mikhalevo was located on the 26th verst, and the final station of Baikal – on the 64th verst, where the track ended with the wharf exit for the delivery of trains to the icebreaker ferry. To ensure the traffic on the single-track line, passing loops were arranged. In July 1896, the Highest consent for the line construction followed. In October 1898 the laying of rails from Irkutsk to Cape Baranchik on the shore of Baikal was completed, and the preliminary operation of the line was started. In January 1899, a temporary traffic was open on the section. It was put into permanent operation in July 1900 and became part of the officially opened railway [18]. This line was flooded in the 1950s due to the construction of the Irkutsk hydroelectric power plant. Most of it was immersed in the waters of the Irkutsk reservoir. But by that time it was only a siding line, as the main track of the Trans-Siberian railway was laid in a different direction [2, 4, 5].

The main milestones in the history of the Irkutsk-Baikal line were found out in the studies of the 1990s. But interest in the subject has not subsequently faded. We have studied the main publications in which it has been touched upon to varying degrees of completeness. The characteristic features of these studies were the expansion of the source base and the addressing to new aspects of the subject.

The range of archival sources was expanded through deep development of the funds of the Irkutsk and St. Petersburg archives. An important depository is the funds of the Russian State Historical Archive: in 1906, after the construction of the Circumbaikal railway, collections of documents for the construction of this site were taken to St. Petersburg. The main documents are concentrated in the fund No. 1273 (documents of the Committee of the Siberian Railway), but modern researchers turned to the documents of the Fund of the Office for the Construction of the Circumbaikal railway (F. 326), the Fund for Personal Affairs of Communications Engineers (F. 229) and the Fund of Plans and Drawings for the Construction of Railways and Artificial Facilities (F. 350) [8, 9]. This allowed, in particular, finding out the issues of the formation of estimated costs of working expenses, as well as clarifying many details of projects. It was possible to find the documents confirming the existence of plans for the Irkutsk-Baikal line project, which envisaged the construction of a kilometer bridge crossing from the village of Listvennichniy on the right bank of the Angara to Cape Baranchik on the left bank, and then the line was to continue along the shore of the Baikal [9].

The use of the new documents of the Irkutsk Archive made it possible to expand the understanding of the processes of construction work and the emergence of new problems brought by the Trans-Siberian railway: the road disturbed the usual way of life of people, made conflicts escalate. Thus, the Fund of the Office of Construction and Road Units under the Irkutsk Governor-General (F. 31) maintains records of disputes between residents of the areas adjacent to the line under construction and railway workers. For example, residents of the village of Mikhalevo insisted upon crossing the railway tracks for unhindered access to the Angara, as well as asked for a doubling of the bridge spans over the Kurma River (from 3.8 m to 7.5 m), as the narrow span did not allow handling rafts. The documents show that in all cases the administration of the route under construction had to look for compromise solutions [19].

"Journal of the Commission that carried out the survey of the Trans-Baikal railway and the Irkutsk-Baikal railway line on the occasion of their transfer by the construction department to the operation department" [7], "Album of the as-built drawings for the construction of the Trans-Baikal railway, the Irkutsk-Baikal railway line..." [1], "Report with applications for the construction of a railway line from the city of Irkutsk to Lake Baikal" [16] still remain the base technical documents for the researchers of the subject. But modern researchers began to use them more widely to pay more attention to the technical characteristics of the route. They use longitudinal profiles of the Irkutsk-Baikal line to recreate images of stations and arranged engineering structures [5, 9].

Literature sources have become more widely used in modern research. The historical essay "Siberian Railway in its past and present" based on extracts from official sources [17] remains an important source. For the first time, the book gave a general overview of the process of construction of the Irkutsk-Baikal line, but also provided summary statistics on the total cost of construction of separate sections of the Trans-Siberian railway. This allows comparing the Irkutsk-Baikal line with other sections: for example, if the length of the Tomsk branch was 28.912 rubles, then for the Irkutsk-Baikal branch it was 49.477 rubles. The authors have compiled diagrams of the Trans-Siberian railway performance indicators, which make us think about the importance of the Irkutsk-Baikal road in the transportation of goods and passengers. These data have not been used critically for a long time, but modern studies have shown that they are largely based on project documents and are not always factually accurate [10].

New data were also clarified from the memoirs of the participants of the construction of the Trans-Siberian railway. Thus, from the memoirs of A.V. Liverovskii [11] we became aware of the contents of the minutes of the commission meeting chaired by N.P. Petrov, Deputy Minister of Communications [14], according to which the construction of the Irkutsk-Baikal branch in 1895 was considered as appropriate, but this construction still depended on the development of a common project of the Circumbaikal railway [8]. Another interesting fact was found in the memoirs of A.N. Pushechnikov [15], he appointed the head of construction of the Irkutsk-Baikal branch and the Trans-Baikal railway in the spring of 1895. Preparing for construction and studying the project of the right-bank branch from Irkutsk to the Baikal, it was prepared by engineer G.V. Adrianov. A.N. Pushechnikov decided to discuss it with another prominent engineer and a well-known public figure. As a result of this meeting, the left-bank project of the Irkutsk-Baikal branch was introduced and then implemented [9].

Starting from the 2000s, researchers began to use photographic documents extensively, which allowed them to illustrate the volume of work performed, to help readers forming vivid ideas of the bridges and other engineering structures built on the line, to recreate the paintings of the line in its natural environment [5, 21]. The jubilee editions dedicated to the 100th anniversary of the Circumbaikal railway are particularly noteworthy here [8, 9].

Among the works based on foreign sources, special mention should go to the book by C. Wolmar, the aforenamed British journalist, describing the process of construction of the Trans-Siberian railway in journalistic terms [23]. The merit of the book is the author's brief analysis of the sources from which it turns out that in the review of the history of the Irkutsk-Baikal line, the author rests upon the works of H. Tupper [22] and the book by W. Greener written in 1902 [6], on descriptions of American travelers who were on the Trans-Siberian railway in the early period of its operation [13]. Thus, C. Wolmar notices that H. Tapper emphasizes the special role of M.I. Hilkov, the Minister of Communications, in the choice of direction of the Irkutsk-Baikal line; the conclusion is interesting but it requires confirmation. The question of using books prepared by travelers as sources should be critically and deeply worked out, especially since their authors traveled along the Trans-Siberian railway after the construction of the Circumbaikal railway.

The article by A.I. Makurin deals with the analysis of the views towards the Trans-Siberian railway of L. Bates, the American engineer on the Trans-Siberian railway, who visited the Trans-Siberian railway in 1908 and published a special book [12]. Although the book by L. Bates [3] does not have a broad historical description, this edition presents the author's general impressions of a trip along the Trans-Baikal railway line, reflects his view of Irkutsk, and presents a set of illustrations, among which there are views of Listvenichny, the Baikal station and other places of the Irkutsk-Baikalskaya line. It may happen that future researchers of this subject will make it possible to use this book more widely as a source.

Finally, one should mention a source such as materials from local newspapers. However, there were few periodicals in those years, and today many of them are in a worn-out state. Carefully flipping through the pages, researchers collect materials on the subject bit by bit. There are also few of them too, since the construction of the Irkutsk-Baikal line was still a small fragment of the grandiose construction of the Trans-Siberian railway. Nevertheless, in modern research on the subject, materials from the "Irkutsk Gubernskie Vedomosti" and "Eastern review" newspapers are actively used – their collections are relatively complete and include valuable information about the period of the construction of the Irkutsk-Baikalskaya line.

7 Conclusion

The construction of the Irkutsk-Baikal line of the Trans-Baikal railway is one of the unique pages in the history of the Trans-Siberian railway. The general plan for this construction is already well known, but many important details are still to be clarified. For example, further development of the topic may be related to determining the performance indicators of this line, as existing studies often provide target figures. This will require a deep study of the archives. Important aspects of the history of the construction of the railway in the Baikal region can be revealed when assessing the personal relationships between the supervisors of the railway construction. At this point one should study the archival materials of the personal funds of M.I. Khilkov, A.N. Pushechnikov, G.V. Adrianov, and others in detail.

New aspects of the topic can be discovered when working with foreign literature on the history of the Trans-Siberian railway, with the memories of travelers. Although there are few books of this kind, they will make it possible to reveal estimates of the state of the Siberian railway at the turn of the 19th and 20th centuries from foreign observers. Here the researchers will be granted assistance with the materials on the Transsib bibliography, posted on the website Transsib.ru (Transsib.ru. Bibliography).

References

1. *Album of the as-built drawings for the construction of the of the Trans-Baikal railway, the Irkutsk-Baikal railway line, the railway line to the Chinese border, and the berths of the icebreaker ferry on lake Baikal: 1895–1901* (St. Petersburg, 1902)
2. A.S. Aseeva, A.V. Khobta. Design and exploration work on the new direction of the railway line Irkutsk – Slyudyanka in 1938–1940. J. of the Labor. of Ancient Technol., **1(10)**, 100–107 (2014)
3. L.W. Bates. *The Russian Road to China* (Houghton Miffin Comp., Boston; New York, 1910)
4. V. Bryanskij. Second bypass of Baikal (Trans-Siberian line Irkutsk – Andrianovskaya – Sludyanka). Zemlya Irkutskaya, **3(26)**, 39–43 (2004)
5. A.K. Chertilov. Line Irkutsk – Baikal – the beginning of the railway bypass of the Sacred Sea (Historical and architectural essay). J. of Ancient Technol. Labor., **15(2)**, 139–157 (2019)
6. W.O. Greener. *Greater Russia; the continental empire of the Old world* (The Macmillan Co, New York; Macmillan & Co, Ltd, London, 1903)
7. *Journal of the Commission that carried out the survey of the Trans-Baikal railway and the Irkutsk-Baikal railway line on the occasion of their transfer by the construction Department to the operation Department* (Printing of P.I. Makushin & V.M. Posokhin, Irkutsk, 1901)
8. A.V. Khobta, V.G. Tretyakov. *History of the Circum-Baikal Railway 1905–2005 (To the 100th anniversary of the commissioning of the Circum-Baikal Railway in permanent operation)* (Izdatel A.N. Garashchenko Publ., Irkutsk, 2005)
9. A.V. Khobta. *The road to the century: Guidebook album on the Circum-Baikal Railway* (Ottisk Publ. House, Irkutsk, 2004)
10. A.V. Khobta *History of the construction of the Circum-Baikal railway 1887–1915*, Cand. Dissertation thesis (Publ. house of Irkutsk state Univ., Irkutsk, 2005)
11. A.V. Liverovskij. *50 years of railway work* (Far Eastern Railway Publ. House, Khabarovsk, 1999)
12. A.I. Makurin. Trans-Siberian Railway in the early 20th century through the eyes of American engineer Lyndon W. Bates. St. Peter. Hist. J., **1**, 240–252 (2018)
13. D. Manley (Ed.). *The Trans-Siberian railway: A traveler's anthology*. (London etc. Century, 1988)
14. N.P. Petrov. *Report of the Chairman of the Commission for the study of the construction of the Siberian railway in the area of the Deputy Minister of Railways engineer General-Lieutenant N. Petrova* (St. Petersburg, 1896)
15. A.N. Pushechnikov. The case of my rail construction practice. Intellig. of the meet. of railway eng., **25**, 431–435 (1913)
16. *Report with applications for the construction of a railway line from the city of Irkutsk to Lake Baikal. 1897–1901* (St. Petersburg, 1904)
17. S.V. Sabler, M.V. Sosnovskii. *Siberian Railway in its past and present: To the tenth anniversary of the*

Siberian Railway Committee (1893–1903). Historical essay (St. Petersburg, 1903)

18. *Transsib.ru. Bibliography.* Retrieved from: https://transsib.ru/biblio.htm
19. V.G. Tretyakov (Ed.). *The East Siberian railway: a journey through a century (1898–1998)* (Publ. house of Irkutsk state Univ., Irkutsk, 1998)
20. V.G. Tretyakov (Ed.). *Railway transport in Eastern Siberia from XIX to XXI century*, Vol. 1. *1837–1941* (Oblmashinform Publ. house, Irkutsk, 2001)
21. V.G. Tretyakov (Ed). *Baikal ferry railway crossing: To the 100th anniversary of construction and operation* (Publ. house of Irkutsk state Univ., Irkutsk, 2000)
22. V.G. Tretyakov, V.V. Tretyakov. Irkutsk-Baikal Railway Line. Zemlya Irkutskaya, **3(26)**, 33–38 (2004)
23. H. Tupper. *To the great ocean: Siberia and the Trans-Siberian Railway* (Little, Brown, Boston, 1965)
24. C. Wolmar. *Trans-Siberian Railway. The history of the creation of the railway skeleton of the empire* (Publ. House "Kuchkovo pole", Moscow, 2016)

Human capital of Arctic Yakutia as a resource of ethnic group resilience

Sargylana Baisheva[1,*]

[1] Institute for Humanities Research and Indigenous Studies of the North SB RAS, Center for Ethno-sociological Research, 1, Petrovskogo str., Yakutsk, 677000, Russia

Abstract. In the Strategy of the Arctic zone of the Republic of Sakha (Yakutia) for the period up to 2035, an important role is given to human resources in the industrial development of the region. It was found that a whole range of factors influences the formation and use of human resources. These are an unfavorable demographic situation, migration processes associated with the outflow of the able-bodied population, a low standard of living, an unsatisfactory state of social infrastructure, transport inaccessibility, poor efficiency of public administration and impact on the development of territories inhabited by small people ethnic groups. As a result of our research, we identified indicators of a sociological assessment of labor adaptation of the economically active population and the main trends in employment against the background of problems of stability and vulnerability of indigenous communities. In general, the absence of a systematic, complex interaction between government bodies, local governments, the business community and the population itself complicates the sustainable development of the Arctic regions in the future. For the further development of the Arctic, special attention should be paid to the adaptation strategies of the aboriginal community, taking into account the personal resource.

1 Introduction

A focus should be shifted to integrated studies of indigenous peoples of the North that experience a complex period of transformation under conditions of new social challenges put forward by globalization.

Current ethno-social processes in the region characterized by dynamic social environment, imperfection of the legal regulation in traditional economic sectors of the North, the interrelation between political socio-economic, environmental factors complicate the everyday life of people in the Arctic region.

Large-scale capital projects aimed at improving quality of life of indigenous peoples in the Arctic have attracted genuine interest of the scientific community. Regional issues and perspectives of human development in the Arctic have been the focus of studies in Yakutia, Russia, and abroad [1, 2, 3, 4].

In general, employment and human resources is a multifaceted problem addressed by the Russian Academy of Sciences as well as its Siberian Branch within the context of ongoing social changes, including the demographic aspect, resilience of Arctic communities, the level of education, human health, gender relations, etc.

A number of collaborative studies and individual papers analyse the overall dynamics of modern Russian society and transformation processes under market conditions affecting employment patterns, all aspects of employer–worker relationship, the state of social classes, social groups, and strata both at a national and regional scale [5, 6, 7].

The labour market as a basic concept for any region has been paid great attention in sociology, political science, and regional studies. Interesting methods are suggested by the Institute of Regional Consulting led by A. N. Pilyasov that develops documentation for strategic policy in socio-economic development of federal subjects of the Russian Federation as well as municipalities; provides scientific, analytical, and methodological support for strategic socio-economic development of the Arctic zone of the Russian Federation and the Far North regions [8, 9].

The Arctic is viewed as a constant instable, fragile and mobile challenge to Russia that motivates to find new adjusted institutions, forms, techniques, relations and encourages innovative search.

By and large, the literature review shows that Arctic Yakutia has no unified paradigm for innovative search to develop local communities as the development of labour migration, local identity, infrastructure mega-projects and business structures, institutional factors of regional development, the design of strategies and programs for socio-economic development of regions and municipalities are poorly understood.

Most experts agree that traditional practices on ancestral lands is not only the basis of indigenous livelihood but a lifestyle integral to traditional culture

* Corresponding author: baisargy09@yandex.ru

and worldview of Northern ethnicities as a whole [10, 11].

2 Results

Table 1. Population dynamics in the areas traditionally inhabited by minor indigenous peoples of the North in the Republic of Sakha (Yakutia) for 1989-2018, persons.

Municipal districts	Census 1989*		average for 2013 - 2017 **	for 01.01. 2018**
	Total population	Indigenous population		
Total in the republic	**1 094 065**	**24 674**	**957 961**	**964 330**
In the areas traditionally inhabited by minor indigenous peoples of the North in RS(Y)	**48 416**	**18 682**	**42 469**	**42 121**
Share in RS(Y), %	***4.4***	***75.7***	***4.4***	***4.4***
1. Abyysky	572	88	489	479
2. Aldansky	2 323	1 551	2 296	2 268
3. Allaykhovsky	1 435	401	618	631
4. Anabarsky	2 776	805	3 425	3 567
5. Bulunsky	5 326	2 417	3 833	3 792
6. Verkhne-Kolymsky	811	266	1 124	1 097
7. Verkhoyansky	***-	***-	979	984
8. Zhigansky	6 005	1 977	4 255	4 222
9. Kobyaysky	1 267	587	1 135	1 109
10. Mirninsky	344	170	286	271
11. Momsky	1 496	479	4 078	3 949
12. Neryungrinsky	1 479	796	1 033	971
13. Nizhne-Kolymsky	2 160	1 160	1 771	1 731
14. Oymyakonsky	2 265	311	1 736	1 739
15. Olekminsky	2 375	733	2 345	2 317
16. Oleneksky	4 140	2 310	3 994	4 072
17. Sredne-Kolymasky	344	290	316	310
18. Tomponsky	1 279	707	890	909
19. Ust-Maysky	5 239	1 829	1 719	1 665
20. Ust-Yansky	4 198	863	3 357	3 225
21. Eveno-Bytantaysky	2 582	942	2 790	2 813

**Minor indigenous peoples of the North in RS(Y)*, 9-10 (Poligraphist, 1994). **Socio-economic status of the areas of traditional habitat and economy of minor indigenous peoples of RF on the territory of RS(AY) for 2012-2017, *Statistics* 17/466-dsp, 6-8 (2018). ***Verkhoyansky District was not listed as an area of habitat of indigenous peoples of the North in 1989.

More than 65 % of indigenous peoples of the North in Russia live in rural areas where ancestral traditions and lifestyle are very carefully preserved. 82.5 thousand indigenous people were recorded in the Arctic zone of the Russian Federation by the 2010 Russian population and housing census.

It is noteworthy that the share of indigenous peoples of the North in the population of the Sakha Republic (Yakutia) by ethnic group increased from 2.2 % in 1989 to 4.2 % in 2010.

In administrative composition 71 rural settlements in 21 districts (*uluses*) are recorded as areas of traditional habitat and traditional economic activities of indigenous peoples of the North of the Sakha Republic (Yakutia). They include 48 ethnic (nomadic) rural administrative divisions (*naslegs*).

As of January 1, 2018 the population of 42.1 thousand (4.4 % of the population of Yakutia), mostly persons of indigenous ancestry (minor indigenous peoples of Russia), was recorded there (Table 1).

The area inhabited by indigenous peoples of the North extends over a vast territory that is a home to 75.7 % of Republic's indigenous ethnic groups [12].

The Arctic zone conditionally includes 13 districts with 30.9 thousand people in 49 localities of traditional settlement of indigenous peoples, among them 5 districts of the Arctic coast with 12.9 thousand in 23 settlements, and northern territories not included into the Arctic zone as they are located at less high latitudes but still established the areas of the Far North (8 districts with 11.2 thousand people in 22 settlements).

Of particular interest for researchers is the Arctic zone including the Arctic coast as the most remote and poorly studied subregions of Yakutia with the most extreme conditions inhabited by indigenous peoples of the North.

The Arctic community has exhibited resilience in the face of changes with profound shifts in natural (climate), cultural (including language), and economic (emergence of mixed economy) processes. Most attention is paid to the analysis of institutional aspects of employment, including legislation, interaction between communities of descent and industrial business structures. Monitoring of indices to track changes in a society in the center of which is the man in the Arctic is a concrete step towards identifying priorities in further development of Arctic territories.

The key feature of the area under discussion where localities of settlement of indigenous peoples are concentrated that determines specific approaches to its socio-economic development and national security is high sensitivity of traditional lifestyle of persons of indigenous ancestry to external effects.

Major risks of vulnerability for economic activities of minor indigenous peoples of the North and the Arctic are a complex transportation scheme and a lack of year-round land transport; poor equipment and low staff capacity of educational institutions in the areas inhabited by minor indigenous peoples; underdeveloped social and productive infrastructure, high rate of thermal and electric energy losses, deterioration of infrastructure facilities and the increasing rate of dilapidated housing; low quality and slow mobile phone and internet service in majority of areas where minor indigenous peoples live. All risks mentioned above complicate everyday life of minor indigenous peoples of the North, being a cost-increasing factor in end-product value that is multiple times higher than the average in the Republic, and lead to lower quality of education.

Despite state support for education the problems of human (professional) resources in sectors of industry still remain unsolved in areas traditionally inhabited by minor indigenous peoples of the North. 50 % of schools in the Arctic are small with lower capacity, poor infrastructure and lack of qualified staff. 88 % of school graduates from Arctic districts in 2018/2019 academic year were admitted to institutions of vocational and higher education which shows their motivation to continue education. 90 % of the admitted students were provided with dormitory by the educational institution. From 2014 to 2019, 5 secondary schools with 965 student capacity, 16 pre-school institutions with 901 places were built.

At beginning of year 2018, the number of people employed in major industrial sectors in the areas inhabited by minor indigenous peoples of the North of the Sakha Republic including small businesses was as low as 13.4 thousand people including 5.0 thousand in the productive and 8.4 thousand in non-productive sectors [12].

Thus, every fifth inhabitant in places of residence of minor indigenous peoples of the North is employed in the non-productive (public) sector, with only every eighth being employed in the productive sector including mining (every 20th), agriculture (every 25th), and energy (every 43rd).

Surveys that we conducted show that indigenous people living in ancestral lands hope for improvement of life in the Arctic and in the North following not only traditional lifestyle but mostly industrial development and future infrastructure projects.

However, despite the measures taken, the situation of minor indigenous peoples of the North has been compounded by inadaptability of their traditional lifestyle to the modern economic situation in recent decades. The number of people employed in reindeer herding has been declining in the Republic of Sakha (Yakutia) since 2010 (Table 2).

Low competitiveness of traditional economic activities is related to low production, higher transportation cost, lack of modern factories and technologies for complex processing of raw materials and biological resources.

In this way, the steepest decline in employment rates in 2010 – 2017 of 8.7 % is found in the Arctic districts (Ust-Yansky, Anabarsky, Verkhne-Kolymsky, Srednekolymsky). Generally, the highest unemployment rate was in 2016 – 2017 in the Eveno-Bytantaysky and Kobyaysky districts 13.2 % as compared to 7.2 – 6.9 % in the Sakha Republic. The highest stress index in the labour market recorded in the Oleneksky district with 61.7 – 35.8 person per one vacancy as compared to 1.5 – 3.0 for the Sakha Republic.

For a number of years we have conducted sociological (field) surveys in areas inhabited by minor indigenous peoples of the North, including the Neryungrinsky and Aldansky districts (towns Aldan, Neryungri, villages Khatystyr, Iengra in 2010 – 2014 and 2017 – 2019), the Allaykhovsky district (Olenogorsk, Russkoe Ustye in 2011 and 2014), the Oleneksky district (2006).

Table 2. Population engaged in reindeer herding in the areas of traditional habitat of minor indigenous peoples of the North in the Republic of Sakha(Yakutia) for 2010-2018*.

	2010	2015	2016	2017	2018
Total in RS(Y)	**2157**	**2009**	**1842**	**1792**	**1490**
RS(Y), %	**100,0**	**100,0**	**100,0**	**100,0**	**100,0**
Arctic zone	**1314**	**1145**	**1053**	**1107**	**951**
share in RS(Y), %	***60.9***	***57.0***	***57.2***	***61.8***	***63.8***
Abyysky	8	7	6	6	0
Allaykhovsky	20	6	0	0	0
Anabarsky	98	98	97	109	112
Bulunsky	133	115	129	163	150
Verkhne-Kolymsky	19	13	13	13	14
Verkhoyansky	74	40	48	48	22
Zhigansky	111	76	75	76	75
Momsky	215	118	142	140	96
Nizhne-Kolymsky	172	216	162	162	112
Oleneksky	52	54	65	65	65
Srednekolymsky	61	40	37	28	28
Ust-Yansky	143	205	170	169	142
Eveno-Bytantaysky	208	157	109	128	135
Non-Arctic districts	**843**	**864**	**789**	**685**	**539**
share in RS(Y), %	***39.1***	***43.0***	***42.8***	***38.2***	***36.2***
Aldansky	267	260	160	150	140
Kobyaysky	148	118	130	93	83
Neryungrinsky	165	156	156	143	94
Oymyakonsky	63	65	57	69	56
Olekminsky	14	117	111	116	71
Tomponsky	157	132	108	109	65
Ust-Maysky	17	6	5	9	9

* Ministry of Agriculture of the Republic of Sakha (Yakutia)

We used questionnaire, interview of experts, non-standardized observation of changes in social reality (transformation of labour market and employment). Interviewed experts included representatives of the legislature and the executive, large and medium business, entrepreneurs, managers of industrial enterprises, educational institutions, representatives of non-governmental structures and native population including indigenous leaders.

We used snowball sampling to study labour issues at the new stage of industrial development of the territory (30 informants) as well as employment and vocational needs (50 informants). The first survey included informants under 30 (20 %), 30 – 40 (40 %), 40 – 50 (7 %), 50 – 60 (13 %), above 60 (20 %) years of age. Among them state employees made up 80 %, private companies 20 %; 67 % are married; 93 % have higher education; top managers comprise 13 %, managers 20 %, professional workers 20 %, managerial workers 47 %; 73 % have average income, 27 % below the average.

We used convenience sampling to interview informants on problems of development of small business (34 informants), opinion of entrepreneurs about the role of local authorities in their area of activity (52 entrepreneurs).

The quality of labour resources is affected by complications in the supply of industrial enterprises by highly qualified professionals from local population (low qualification, poor work skills of indigenous people in industrial sectors), the level of social and productive infrastructure, less opportunities for self-fulfilment, constant migration, long-term unemployment.

Interviewing administrators of educational institutions showed challenges in the system of search and selection of young professionals among graduates of educational institutions of higher education (branch of the Northeastern Federal University, Yakutsk) whose tuition fees were covered from the Republic's budget; complications in building local workforce and further recruiting local professionals in large industrial enterprises.

Lately, the M. K. Ammosov Northeastern Federal University (Yakutsk) and certain industrial enterprises have agreed on federal government funded quotas in technical disciplines. The issues of recruitment and long-term employment of local workforce at industrial enterprises operating in Yakutia have also been addressed. There is a strong need to build a vocational education cluster involving experts from large mining companies active in the region such as PJSC Gazprom, OJSC Surgutneftegas, PJSC Rosneft, PLC Transneft-Vostok.

Monitoring of major indices of the labour market in Yakutia based on survey data of the regional labour market, state statistical survey, and employment services revealed inadequate regulating effect of employment policy on the state of labour market.

Major factors impeding the job placement are partial disparity between educational institutions and the labour market; poor awareness of young people about their rights; the gap between expectation and demands of population and the reality of the labour market, persistence of old stereotypes among employers about young people not being able to prove oneself, build working relationships in a team, being too emotional.

The most important quality for employment is considered to be professionalism, commitment, and leadership. The survey showed that better quality of work life depends on job satisfaction, salary increments, moral or material rewards for work, the quality of social and labour relations (first of all, social and psychological environment in the workplace).

The informants of the survey answered affirmatively on the statements that work is "a means of personal gain for company owners" (53.3 %); "an important way to find and develop one's abilities, their personal potential" (33.3 %); "an activity respected by society", "foundation for team building than becomes valuable for its members", "a means to satisfy a man's need to be a useful member of society", "the main way to meet the needs of society for certain goods and services" (26.7 % each); "a basis for one's credibility among family and friends" (21.4 %); "a basis for professional development and promotion", "a means of self-determination and self-affirmation in society", "a basis for secure position and confidence in the future", "a means to satisfy one's need for the meaning of life" (20.0 % each); "a reliable source of material well-being" (13.3 %); "a source of joy and moral satisfaction for employees" (6.7 %). 40.0 % of informants provided a negative response to the statements, 33.3 to 78.6 % partially agreed.

3 Conclusions

Despite the region's remoteness from centers of civilization, minor indigenous peoples of the North and the Arctic remain in the focus of world community and science and are involved in globalization processes.

Everyday life of indigenous ethnic groups is a great example of resilience, engagement in traditional economic activity on vast territories, and the ability to adapt to severe natural reality and survive under impact of industry on their native habitat.

Rapid development of technologies, their integration into many spheres of social life affected the aboriginal community of Yakutia as well. In its progress the aboriginal community goes through a complicated process of adaptation to current transformations through breaking stereotypes and changing the socio-cultural paradigm. It is characterized by differentiation of new models of adaptation strategies depending on factors of sustainable development and reducing vulnerability based on terminal value orientations.

Implementation of national and municipal programs maintains major trends in employment of local population: use of labour resources from outside the region for employment at large industrial enterprises; employment at the "grey" labour market that concentrates around economically developed social sectors including services; increasing number of requirements from regional and municipal authorities resulted in transition of business to the informal employment sector to optimize expenditure and survive in the market.

The identified criteria for social evaluation of labour adaptation of region's economically active population fall into two groups. The quantitative indicators are as follows:

1. General demographic indices (total population, persons; urban population, %; population of working age, persons; economically active population, persons).

2. The state of the recorded labour market (recorded unemployment, %; average length of unemployment, month).

The qualitative indicators are:

1. Education (the number of people with higher and postgraduate education, persons; with vocational education, persons; with secondary education, persons).

2. Morbidity and mortality (morbidity, per 1,000 individuals; first diagnosed occupational diseases, per 1,000 individuals of working age; mortality at working age, per 1,000 individuals of working age).

Problems of employment in the Arctic region of Yakutia are investigated in the context of sustainable development of the Arctic. Sociological surveys were conducted in the areas inhabited by minor indigenous people of the North. They revealed main challenges faced by young people in the market-driven labour relations, their view of priorities, standards, and rules of working life.

The studies on young people's challenges in society implemented through grants, involving the author of this paper, have been used to adjust youth policy and to study how younger generation of Yakutia's aboriginal community constructs social reality in employment.

The interviews with representatives of industrial companies demonstrate that new technologies are used to communicate with potential employees; investments are shifted towards creating databases of potential recruits (external personnel reserve) and social network communities.

Acknowledgement

The study was done with a support of the state assignment of the Yakut Scientific Center SB RAS "The Republic of Sakha (Yakutia) and great challenges: social well-being, mobility, and adaptation strategies" 0297-2021-0029, reference number 121031300008-7.

References

1. K. Grint, *The sociology of work. An introduction* (The Polity Press & Blackwell Publishing, 2003)
2. T. Hiyama, H. Takakura, *Global warming and human: nature dimension in Northern Eurasia* (Springer, 2016)
3. H. Takakura, *Arctic pastoralist Sakha. Ethnography of evolution and microadaptation in Siberia* (Trans Pacific Press, 2015)
4. J. Larsen, G. Fondahl, eds., *Arctic human development report. Regional processes and global linkages* (TemaNord, 2014)
5. Z.T. Golenkova et al., *Social inequality of Russian regions* (Institute of Sociology RAS, 2017)
6. Y.V. Popkov et al., *Ethno-social processes and ethnic policy in Siberian regions* (SB RAS Publishing, 2015)
7. Y.V. Popkov, V.G. Kostyk, eds., *Socio-cultural approach to regulating interaction of ethnic groups* (Novosibirsk, Manuscript, 2013)
8. N.Y. Zamyatina, A.N. Pilyasov, *Russian Arctic: towards new understanding of development* (URSS, 2018)
9. A.N. Pilyasov, *And the last will become the first: Northern rimland on the way to knowledge economy* (URSS, 2015)
10. S.M. Baisheva, *H. Transb. St. Univ.* 12, 11 (2017)
11. S.M. Baisheva, T.N. Konstantinova, L.Y. Pisareva, *Theor. Probl. Polit. Stud.* 5A, 286 (2018)
12. Stat. Coll. *Socio-economic conditions of areas of traditional habitat and traditional economy of indigenous peoples of RF in RS(Y) for 2012-2017*, 14/466, 68 (2018)

Evaluation of the return of investments in transport route development

Ulyana I. Malakhova[1,*], and *Victoriya A.* Vinichenko[1]

[1] Siberian State University of Water Transport, 33, Shchetinkin Str., Novosibirsk, Russia

Abstract. The network of Russian railways provides the movement of goods and passengers both within the country and in the context of international transport, through which goods from different countries go in transit. In Russia the railway lines are double-track and fully electrified on the main directions of the East-West transport corridor. The modernization of the branches from the Trans-Siberian railway to the state border with China and Mongolia continues. The paper describes the business model of the project, the corresponding calculations for managerial decisions on the launch of a direct train route Novosibirsk – Urumqi and return. During the study, the monitoring of existing routes of passenger traffic was carried out: the capacity of railway routes, frequency, duration and capacity of alternative crossings. The result of the study is a developed model for launching a project of a direct connection between Novosibirsk and Urumqi and return. The paper describes the stations that are of particular importance for the route – the place of customs inspection, rearrangement of wheel sets of cars due to the different wheel gauge of neighboring countries. The result of the study is a multifunctional model of a transport route that allows analyzing the economic parameters of any route of a railway communication. According to the model, it is possible to analyze various parameters of the revenue side of the project, calculate the amount of investments and indicators for the assessment of the effectiveness of real investments.

Keywords: Transport, logistics, investment efficiency, management solution

1 Introduction

Nowadays the promotion of competitiveness of a business is inextricably connected with the maximization of the benefits of its owners, in particular, by the search for ways to expand the business [4]. The development of Russia today makes it necessary to establish and strengthen ties between domestic companies and Asian regions. Today the projects are being implemented in the world that has been united under the general name "One Zone – One Road" (Concept of the pan-Eurasian transport system "New Silk Road"). This determines the spatial structure of economic development – the increase in the turnover of transit cargo and flow of tourists. These indicators determine the relevance of the development of additional direct routes connecting Russia with Asian countries, in particular, with China. Thus, the relevance of the research topic is reasoned by the intensification of the movement of tourists and goods in this direction, reflected by the lack of capacity of railway routes, as well as the presented characteristics of the frequency, duration and capacity of alternative crossings.

In order to establish links in this direction, it is advisable to use railway service: the necessary infrastructure has been created. The network of Russian railways ensures the movement of goods and passengers both within the country and in the context of international transport, through which goods from different countries go in transit. The paper proposes the launch of a direct train Novosibirsk – Urumqi (and return), its economic justification. It is important to note that Novosibirsk and Urumqi are the largest transit points in their region, which are provided with a developed transport infrastructure.

2 Problem Statement

In Russia the railway lines are double-track and fully electrified on the main directions of the East-West transport corridor. The modernization of the branches from the Trans-Siberian railway to the state border with China and Mongolia continues. It becomes difficult to determine an investment-attractive route for the new transport infrastructure for a number of reasons; the key ones are urban sprawl and possible environmental consequences [2].

Since the creation of additional direct routes involves passing through the border territories of Russia and China, the railway network of the Republic of Kazakhstan will be an integral part of the train movement. The current state of the railway infrastructure in the areas of international transit is characterized by the lack of sufficient reserves of crossing capacity. In the

* Corresponding author: malakhova997@gmail.com

future, the development of directions for unified network normative indicators and systems for automatic control of train traffic is planned [3].

Nowadays the initiative is put forward to create a direct passenger train on the route from Novosibirsk to the north-west of China with the inclusion of postal and baggage cars in the scheme. The city of Urumqi was chosen to launch the project (it is the largest transit point of the Lanxin Railway in PRC). The implementation of the project became possible as a part of the creation of the concept of a new pan-Eurasian transport system "New Silk Road".

3 Research Questions

The research subject is determined by the object and subject of the study.

3.1 The object of research in this paper is the route Novosibirsk – Urumqi (and the way back).

Novosibirsk and Urumqi are the largest transit points in their region, which are provided with a developed transport infrastructure. This infrastructure includes the international airports Tolmachevo (Novosibirsk) and Divopu (Urumqi), large railway stations providing passenger and cargo traffic: Trans-Siberian Railway – Novosibirsk-Glavny, Lanxin Railway – Urumqi-Yuzhnaya. Automobile transportation is represented by an extensive network of roads. Novosibirsk has the following highways.

- Federal highways R-254 "Irtysh" (Chelyabinsk – Novosibirsk), R-255 "Siberia" (Novosibirsk – Irkutsk) and R-256 "Chuisky tract" (access to the Central Asian countries, Mongolia, China).
- Regional highways with access to the Northern latitudinal highway (Tomsk), North-East Kazakhstan, South Kuzbass.

In order to determine the key parameters for the organization of the route, it is necessary to monitor all types of alternatives for various types of transport: air travel, rail and bus crossings and mixed (road and rail) crossing [1].

In the market of passenger transportation services, direct flights are offered by one airline – S7 Airlines. The description of the route from Tolmachevo airport to Divopu airport: route number S7 5739, flight frequency is twice a week (every Wednesday and Sunday), flight time is 2 hours 35 minutes, the number of seats for passengers is 78. Including transfer, total travel time on average is 5 hours 15 minutes (to Tolmachevo airport (40 minutes), check-in (1 hour 30 minutes), time for baggage claim (30 minutes). Minimum price as of 12.01.2020 is 8606 rubles.

During the study, the monitoring of existing routes of passenger transportation by railway service was carried out. According to the website of the Russian Railways, there are six alternatives that are included in the train routes:

- 301N Novosibirsk – Karagandy;
- 054Ts Astana-Nurly Zhol – Urumqi;
- 013H Almaty 2 – Urumqi.

Passengers are transferred along the route at such stations as Akadyr, Sayak, Zharyk, Karagandy Pass., Karagandy Sort., Sary-Ozek. All transfer points are located on the territory of the Republic of Kazakhstan.

3.2 The subject of the research is the logistics and financial characteristics of Novosibirsk – Urumqi route (and the way back)

Mixed travel is determined by rail and road transport. The route starts at Novosibirsk-Glavny. In Almaty at the Sairan railway station, a transfer to a vehicle – a bus is carried out. Transport to Urumqi runs daily from Monday to Saturday. The total travel time is 3 days 1 hour 29 minutes, including the waiting time for a bus in Almaty of 23 hours 19 minutes. The cost of a train ticket varies from 3803 to 5384 rubles (depending on the class of service), transfer by bus in conversion from tenge to rubles at the rate of the Central Bank of the Russian Federation is 2744 rubles. The minimum fare is 6547 rubles. Prices are as of January 12, 2020.

A bus journey with a transfer to railway transport can be considered as an alternative to transfer in mixed traffic from Novosibirsk to Urumqi. The route starts at the Novosibirsk bus station. In Karaganda, there is a transfer to a train to the station Urumqi – Yuzhnaya. The total travel time along this route is 2 days 11 hours 24 minutes, taking into account the transfer time (1 hour 29 minutes). The total cost of transportation is 6190 rubles. Prices are as of January 12, 2020. It is necessary to note that the monitoring of this route revealed that there are no vacant places for sale by railway service for the next three months. This indicates that it is not possible to use the services of the above mentioned carriers.

The bus route from Novosibirsk to Urumqi is presented with transfers in cities such as Nur-Sultan (Saparzhai bus station) and Almaty (Sairan bus station). The total travel time is 2 days 4 hours 20 minutes. The total cost of the fare from Novosibirsk to Urumqi is 6563 rubles. The cost is indicated as of January 12, 2020.

Thus, monitoring of existing routes in the direction of Novosibirsk – Urumqi and the way back identified five possible options for moving passengers from Novosibirsk to Urumqi. They are: air flight (S7 company), six railway routes with transfers, a bus route (including Avtoservice – Novosibirsk bus station and two carriers of the Republic of Kazakhstan), mixed routes: railway and bus passages.

Only air transport has a direct route among the alternatives described above, but it does not provide the possibility of transporting a large amount of cargo. Motor transport does not have a direct non-stop connection and also does not provide transportation of sufficient cargo by passengers. In the future of the development of the route, only railway communication can provide both direct travel for passengers and transportation of goods in specialized cars.

Thus, the analysis of key parameters, namely the time of transportation, cost, frequency of flights, determined the insufficient capacity of the route on the market. It is advisable to develop an investment project to launch a direct route from Novosibirsk to Urumqi and back with the inclusion of postal and baggage cars.

4 Purpose of the Study

The purpose of this research is to substantiate the feasibility of the organization of a transport route in the direction of Novosibirsk – Urumqi (and the way back) by railway service.

5 Research Methods

During the course of the research we carried out:

1. The description of the logistics business model of railway transportation;

2. The description of the key parameters of the organization of the Novosibirsk – Urumqi railway transport route (and the way back) – the route, key stations, the composition of the train, especially the passage of border zones;

3. The monitoring of existing alternatives in the described route: the analysis of prices and transportation times, frequency of routes;

4. The evaluation of performance indicators and risk analysis of the transport route between Novosibirsk and Urumqi (and the way back).

The basis of the study was determined by the interpretation of scientific, methodological and educational literature, regulatory documents. In the analytical part of the paper, the results of monitoring alternative routes and those described by the author are substantiated by the prospects for the development of the route, describing problem areas that require analysis. The research result is the definition of the economic feasibility of the logistics project.

6 Findings

Railway transportation is a fundamental component of the economic development of both an individual transport company and the country as a whole. Nowadays the management of the transportation process in railway transport determines the factors that affect the composition of costs and they are represented by the volume of traffic, route range, technical equipment of railways, train composition, train load, average daily mileage etc. In the study, the costs are grouped by tariff components: the carriage part, the infrastructure component (including station one), locomotive part. Thus, the prime cost of railway transportation determines the composition of costs as a separate block in the formation of transport routes. These conclusions were drawn based on the result of a study of aspects of the business model of railway transportation and the composition of costs [5].

The justification of the design part: the feasibility of the organization of the Novosibirsk – Urumqi route (and the way back), monitoring of existing alternatives, description of the key parameters of the organization of transport route. According to the research results, the cooperation between Russia and China in transport and logistics sector can be considered mutually beneficial. Novosibirsk is a connecting link between the West and the East: there is an active process of the inclusion of the regions of Western Siberia in global logistics supply chains. Moreover it is one of the largest transport and logistics hubs. Urumqi, in its turn, provides promising volumes of transportation of both cargo and passengers.

The analysis of alternative ways of transportation Novosibirsk – Urumqi determined five possible options for transportation of passengers:

1) air travel (company S7);

2) six railway routes with transfers (KTZh);

3) bus route (including "Autoservice" – Novosibirsk bus station and two carriers of the Republic of

4) mixed routes: rail and bus passages.

At the same time, only railway service can provide a direct transfer to passengers and heavy cargo transportation. The analysis of logistics characteristics (transportation time, cost, frequency of flights) confirmed the demand for the Novosibirsk – Urumqi railway route (and the way back).

Key parameters for the organization of a direct train Novosibirsk – Urumqi (and the way back). It is planned to lay the route through the already existing infrastructure facilities of the Russian Federation, the Republic of Kazakhstan and the People's Republic of China. There are 38 railway stations along the route of the passenger train. The length is 1930 km. The estimated composition of the train is 14 cars (4 SV, 4 compartments, 5 postal-baggages and a restaurant car). Thus, the total number of seats for passengers will be 216: compartment carriage – 144 seats, SV carriage – 72 seats. The volume of the transported cargo per one postal and baggage car is 100 cubic meters; the weight varies from 20 to 26 tons. It is necessary to note that the described option is basic. In the future, in accordance with the demand, the composition of the train will be regulated and thereby the economic component of the project will change.

Financial characteristics of the transport route. The forecast of the revenue of the project determined the prices of passenger transportation services as: compartment car – 7,136 rubles, SV car – 11,390 rubles. Thus, the returns from ticket sales to passengers and returns from a restaurant car will amount to 4,619.03 thousand rubles per one flight. The returns from the transportation of freight units in postal-baggage will amount to 4,307.28 thousand rubles per one flight. Thus, the total revenues for the provision of services for the transportation of passengers and luggage will amount to 8,926.31 thousand rubles per one flight. Forecasting the annual income, the planned train load (80 %) and its planned frequency (four runs per month) were established in the work. Thus, the income will amount to 342,770.33 thousand rubles in a year.

In order to forecast the expenses of the project, they were grouped by tariff components. The cost of the carriage component will amount to 1,353.78 thousand rubles per one flight. The cost of the infrastructure component will be 2,837.55 thousand rubles per a flight. The cost of the locomotive component will be 1,069.79 thousand rubles per a flight. The extra expenses in the project were set at 3,000 thousand rubles per a month. Thus, the forecast costs for a direct train Novosibirsk – Urumqi (and the way back) will be 288,533.75 thousand rubles in a year.

For the assessment of the efficiency, the project implementation period was set (25 years – the service life of the cars). At the same time, the first target year is represented by nine months and the design load (85 %). Thus, according to these calculations, we found:

- net present value of 153,773 thousand rubles;
- internal rate of return of the project 37 %;
- profitability index of 2.73 rubles/rubles;
- payback period of 37 months;
- discounted payback period of 49 months.

The above mentioned indicators characterize the project for the creation of the Novosibirsk – Urumqi railway route as an attractive one for investment [6]. The present study substantiates its financial success. To get payback, it is necessary to achieve return of at least 222.09 million rubles per year, which in practice is presented by 32 flights.

During the development of the route, both national and supranational risks were determined, since the route is international. Among the most significant are:

- the rise in the cost of the project;
- underfunding;
- currency risk;
- geopolitical risk.

7 Conclusion

During the research, a multifunctional model of the transport route was developed, which allowed analyzing the economic parameters of any route of the railway transportation. According to the model, it is possible to analyze various parameters of the return of the project, calculate the amount of investments and indicators for the assessment of the effectiveness of real investments, thereby providing an economic justification for management decisions when implementing various transport projects. This model was used when performing the financial part of the project.

Acknowledgments

This research was performed with the support of the Ministry of Education of Novosibirsk Region within the framework of the state program of Novosibirsk Region "Development of Education, Creation of Conditions for the Socialization of Children and Students in the Novosibirsk Region". http://docs.cntd.ru/document/5499965.

References

1. R. Berežný, V. Konečný. The Impact of the Quality of Transport Services on Passenger Demand in the Suburban Bus Transport. Proc. Eng., **192**, 40–45 (2017). DOI: https://doi.org/10.1016/j.proeng.2017.06.007
2. A. Lukin, V. Yakunin. Eurasian integration and the development of Asiatic Russia. J. of Euras. Stud., **9**, 100–113 (2018). Retrieved from: https://doi.org/10.1016/j.euras.2018.07.003
3. S. Tischler. Finding the right way – a new approach for route selection proceures? Transport. Res. Proc., **25**, 2809–2823 (2017). Retrieved from: https://doi.org/10.1016/j.trpro.2017.05.247
4. I.C. Schutte, A. Brits. Prioritising transport infrastructureprojects: towards a multi-criterion analysis. South. Afric. Busin. Rev., **16(3)**, 97–117 (2012)
5. V.A. Vinichenko, V.P. Nekhoroshkov. *Evaluation of the efficiency of transport services: theory and practice*, Monograph (Publ. House Creat. Econ., Moscow, 2019), 244 p. DOI: 10.18334/9785912923029
6. N.A. Ivanova. The choice of methods for assessing the effectiveness of investment projects. Actual probl. of econ. and manag., **1(17)**, 41–44 (2018)

Application of the corridor management principle to the organization of the trans-Siberian railway

Daria A. Dinets [1,*]

[1] Irkutsk State Transport University, Irkutsk, Russia

Abstract. The article is focused on assessing the possibility of adapting the world experience of managing the transport corridor, including international ones, to the process of organizing and managing the Eastern railway polygon. Particular attention is paid to the analysis of the possibilities of implementing the corridor management principle on the basis of the Trans-Siberian railway in terms of competitive interactability with the railways of Europe, East and Central Asia in the framework of international transport construction projects. Conditions and prerequisites for achieving commercial return of transport construction projects and integration in terms of priority areas, types of cargo, types of rolling stock and the role of international transport integration in achieving the goals of socio-economic development of the regions have been revealed. It has been determined that the orientation to long-distance transport of bulk cargo hinders the implementation of the principle of corridor management on the Trans-Siberian Railway. It has also been proven that a vertically integrated structure with a network of unequal functional and territorial branches with fundamentally different goals hinders the possibility of ensuring non-discriminatory access to the railway transport infrastructure and specifically for intermodal transport operators. As a summary, the paper presents the key directions of improving the Trans-Siberian railway management and organization system to ensure competitiveness, obtain positive effects from the implementation of projects for organizing railway polygons, as well as effective integration into projects for the development of international transport corridors.

Keywords: transport economy, polygon technologies, international transport corridors

1 Introduction

Effective management of transport infrastructure means solving two main tasks in parallel: optimal resource allocation in terms of global value chains, as well as productivity growth that can be translated into all related sectors of the economy. To solve these problems, reformation pressure on the organization of transport operations and the principles of transport infrastructure management increases from time to time. Unfortunately, the reforms are mainly aimed at finding the best strategies for managing a vertical integrated holding in isolation from the tasks of horizontal integration [5].

With that, intermodal cooperation projects in transport industry are also being implemented in order to reduce the "economic distances" and to find ways to organize this interaction between the participants of the transport process, which will reduce the loss of added value in the transfer of goods between them. The lack of competitive advantages in interconnected types of transport is also a prerequisite for the non-profitability of transport construction and integration projects, the mass involvement of players in exchange trading in the carrying and thoughput capacity of integrated carriers.

The transport labor productivity growth is theoretically multiplied in related industries, which creates prerequisites for state interference in large transport infrastructure projects. In fact, however, we believe that the optimization mechanisms laid down in the structural reform in the railway transport industry have a limited impact on the multiplier effect and the growth of multifactorial productivity in the economy as a whole due to the wrong priorities between ensuring production and labor intensity in the task of productivity growth.

2 Problem Statement

The main problem of the modern stage of the development of the Trans-Siberian railway of the Russian railways is the lack of competitiveness of its proposed transport construction projects due to the impossibility of achieving their payback against the background of the expansion of international transport corridors of the global West and East. Geographical location has ceased to play an important role in the organization of logistics on the continent, and, therefore, natural advantages do not allow achieving efficiency. At the same time, it is impossible to achieve positive results

* Corresponding author: dinets@irgups.ru

by copying foreign models of transport development due to the huge differences in the principles of its organization and the achievement of commercial success.

3 Research Questions

The subject of this study is the social and economic relations that arise when applying the principle of corridor management to the organization of the involvement of the Trans-Siberian railway in the international transport corridors of Eurasia. To achieve the study goal, it is required to do the following interconnected tasks:

1. Critical analysis of the practices of applying the principle of corridor management by the example of European and Asian routes.
2. Identifying the conditions for the successful implementation of the best practices of transport corridor management to the activities of the Trans-Siberian railway.

4 Purpose of the Study

The main purpose of this study is to identify the conditions for successful implementation of foreign projects to the experience of managing international transport corridors involving the Trans-Siberian railway.

5 Research methods

The author used historical and systemic analysis methods, comparison and grouping methods, comparison methods, and dialectical methods of cognition.

6 Findings

It is believed that the degree of influence of the development of transport infrastructure on multifactor performance is higher when the railway network under study contains more "bottlenecks". In other words, the development of transport infrastructure can make various contributions to the rationality of value chains depending on the actual needs for the infrastructure complex development [15]. This position can be accepted in the context of road development, but in the case of the Trans-Siberian railway, the connectivity of the territories along the railway network can be provided only when there are no bottlenecks throughout the mainline.

In particular, if we compare the scheme of the Russian railways with the scheme of railway communication in Europe, North America or the planned scheme of railways within the framework of the CAREC project, it is obvious that horizontal connections prevail in the first case and there are a lot of circular routes in all other directions. Therefore, debottlenecking of the Trans-Siberian railway, especially the part of it that runs to the east of the Ural Mountains, is a single technological task, which will be projected into a united growth of production and efficiency across the entire railway network [4].

At the same time, it is believed that the organization of integrated logistics makes it possible to optimize value chains by reducing the transport component in the cost of traded goods and partly services. For example, in the European Union the optimization of logistics interaction and the application of the corridor management principle to the organization of deliveries on integrated logistics networks, according to experts, leads to a reduction in the prices of transported goods by 13–15 % [3].

Figure 1. The diagram of the European railways [13]

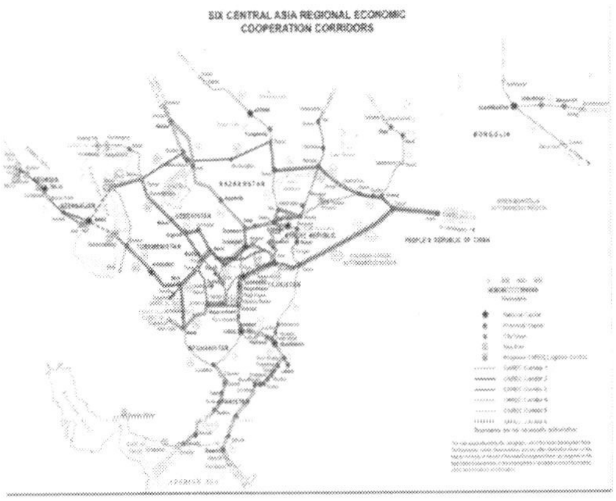

Figure 2. CAREC transport corridors [14]

We believe that such a high estimate can be obtained only if the transport service is optimized as a whole, and not only at the expense of the logistics organization in the "one stop principle". The availability of flexible infrastructure and high intermodality, stimulating the use of several modes of transport to build the most optimal route allow transport corridors to be effective at both long and short distances, when driving north-south and west-east without significant difference in the resulting positive economic impact. This is where we see the ability to have a positive impact on value chains and multifactor performance. And this, rather than the implementation of the technical process of management of the railway polygon, is where we see a kernel in the organization of transportation by the "corridor management" principle [6].

The matrix diagram of railways involves a matrix scheme of the value chain, while the vertically integrated holding company "Russian Railways" manages a horizontal integrated network of roads, and the formation of railway polygons is aimed at linking in a single process of vertically integrated functional branches.

If the national rail network is focused on increasing transportation range, it is not possible to compete on an equal footing in the problems of transport logistics and freight transshipment between different types of transport with matrix structures, and should never be the objective of structural reform in transport industry [7].

The inefficiency of the railway polygon management system is increasing due to the economically unreasonable outsourcing of logistics services in a united transport space in the process of forming an integrated transport service. Outsourcing is beneficial only if there is a positive economies of scale when transferring some of the fixed costs into variables or saving directly by a number of cost elements (most often, on labor costs and social contributions). The effectiveness of the latter measure is not proven if it is assessed not by the formal level of labor productivity, without taking into account the contribution of production and labor-intensive work. And the economy of scale, as it is known, acts both in growth and in reducing of indicators, as it is an objective economic phenomenon. In other words, if rail transport is competitive only for long-distance transportation of low-margin cargo, the expectation of a positive economies of scale from the outsourcing of logistics looks economically naive in the face of an increase in the cost of shipments and a decrease in average distance. At the same time, more than 70 % of freight flows have been outsourced to logistics operators, and they carry out a near-complete cycle of cargo processing [8].

In this context, the rationality of the use of rolling stock is also an important problem: the number of suspended freight railcars is constantly increasing, station tracks possession rates are increasing [12], which invariably has a negative impact on the station speed and, of course, leads to a decrease in the competitiveness of railway transport.

Political, country and global risks also offer little room for optimism when forming a forecast of realistic development of the level of transport and logistics service on the Trans-Siberian railway with the involvement of logistics outsourcers. The transfer of goods to maritime transport is illustrated by the decline in the share of transport services in the total volume of service export. In the context of the coronavirus pandemic, the situation is somewhat changing, especially in recent times – from the moment of the permission of the transit of sanctioned goods on the railways of the EAEU. However, we believe that this is a temporary phenomenon, and there is no reason to expect that a mass transfer of cargo in containers from sea to rail. This is evidenced by the location geography of the main container terminals of the leaders of commodity platforms – most of it is located in places adjacent to the port capacities [1].

According to Professor H. Krampe's theory, all participants in the transport and logistics hub, which is the point of intermodality of the transport network, are economically equal and carry out their activities on the basis of their own benefit. At the same time, they should have non-discriminatory access to transport and logistics infrastructure [10].

In this context, this poses the question on the applicability of this model to Russian reality in achieving the goal of integrating the Trans-Siberian railway into international transport corridors. We suppose that the implementation of the "one stop principle" in the use of rail transport infrastructure is not achieved at this stage, even at the level of coordination of activity of the functional and territorial branches of the Russian Railways JSC, and not only at the level of united transport and logistics services in international transport corridors, even at the level of the EAEU. The situation is aggravated by the life cycle stage of the main part of objects of the infrastructure complex of the Trans-Siberian railway. In the context of systematic non-compliance with stop-hour plans, it is difficult to imagine the possibility of additional economy of scale from the growth of the service's competitiveness [9].

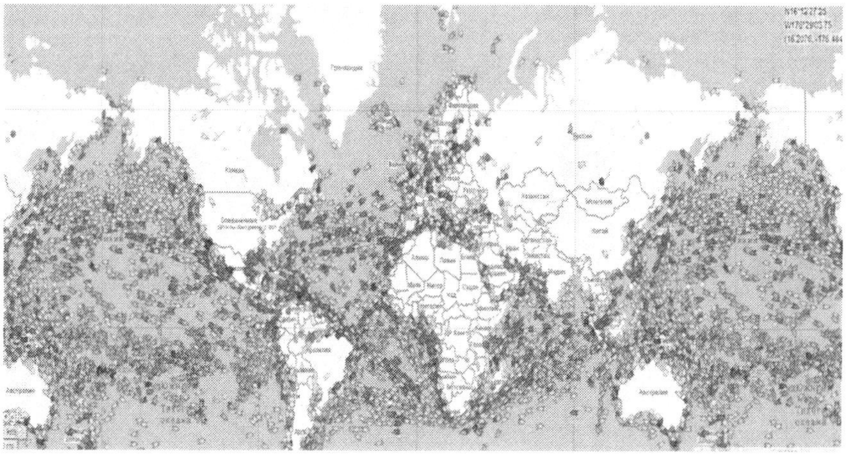

Figure 3. Sea ships traffic map [11]

If we pay attention to the map of the movement of sea cargoes, we can draw an unequivocal conclusion about the irrationality of the existing service and logistics strategy of involving the Trans-Siberian railway in international transport corridors. Directions for the use of railway transport infrastructure and the use of transport infrastructure of seaports providing transit will not allow achieving the required synergies from transport integration projects. Comparison of the data presented with data on the structure of cargo turnover allows us to draw the following conclusions about the prospects of the Trans-Siberian railway involvement in international transport corridors as an equal partner.

The railcar clearing systems and the relative autonomy of the regional polygon control center bypassing the Regional Center of Corporate Management, operating by the rolling stock sidings, reserves of sidings for non-public use for private rolling stock.

7 Conclusion

The key findings that allow achieving this study goal are presented in Table 1.

Table 1. Main directions for improving the efficiency of implementing the corridor management principle to the Trans-Siberian railway

Subject of management	Current state	Target model
Management and coordination	Vertical integration	Horizontal and matrix structures
Bottlenecks	Assessment without reference to the transport corridor	Impact on the throughput and carrying capacity of the entire transport corridor
Performance management	Production	Labour intensity
Operator's activity	Focusing on range and large-scale involvement	Focus on intermodality
Suspended railcars	Reduction of the service speed	There should be a system of clearing of one's own cars and stock trading the possibility of possessing tracks of non-public use
An infrastructure lifecycle stage	Systematic failure to fulfil the stop-hour plan	Changing the technological process in the transition to a matrix system of management and construction of third tracks on the heavy-traffic areas of the eastern part of the Trans-Siberian railway and the Baikal-Amur Railway

Acknowledgments

The work was carried out using the funds of the State task for the implementation of the project "Improving the mechanisms of interaction between the polygon structures and regional corporate governance centers by the example of the Eastern Railway polygon of the Russian Railways".

References

1. *Financial settlements*. Annual reports of JSC "UTLC ERA" (2019). Retrieved from: https://www.utlc.com/events/
2. B. Barisitz, A. Radzyner. *The New Silk Road*. Part II: *A stocktaking and economic assessment*. (2017). Retrieved from: https://www.wu.ac.at/fileadmin/wu/d/cc/cee/1__Startseite/Barisitz_feei_4_17_OF.pdf
3. I. Bond. *The EU, the Eurasian Economic Union and One Belt, One Road. Can they work together?* (2017). Retrieved from: https://www.cer.eu/sites/default/files/pb_eurasian_IB_16.3.17_0.pdf
4. B. Cosentino, D. Dunmore, S. Ellis, A. Pretti, D. Ranghetti, C. Routaboul, M. Garratt, A. Teodoro. *Research for TRAN Committee: The new Silk Route – opportunities and challenges for EU transport* (2018). Retrieved from: https://www.europarl.europa.eu/RegData/etudes/STUD/2018/585907/IPOL_STU(2018)585907_EN.pdf
5. D.A. Dinets, A.G. Dzhavrsuan. International transport corridors and transborder movement of fictitious capital. A. MATEC Web of Conf., 03013 (2018). DOI: 10.1051/matecconf/201823903013
6. D.A. Dinets, M.V. Konotopov. Transregionalism and fictive capital. E3S Web of Conf. Innov. Technol. in Environ. Sci. and Ed., 04033 (2019). DOI: 10.1051/e3sconf/201913504033
7. *Sixth report on monitoring development of the rail market pursuant to Article 15(4) of Directive 2012/34/EU of the European Parliament and of the Council*. Europarlament reports (2012). Retrieved from: https://ec.europa.eu/transport/sites/transport/files/staff_working_document_- 6th_rmms_report.pdf
8. *International trade of EU, the euro area and the Member States by SITC product group*. Eurostat (2020). Retrieved from: www.eurasiancommission.org
9. I.D. Grachev, M.M. Sharapov. *Evaluating the transition to gas engine fuel for the Western Europe – China international transit corridor* (2016) Retrieved from: https://www.elibrary.ru/item.asp?id=25314488
10. L. Grob, N. Craven. *Analysis of Regional Differences in Global Rail Projects by Cost, Length and Project stag* (2017). Retrieved from: https://uic.org/IMG/pdf/analysis_of_global_rail_infrastructure_investment.pdf
11. *Global ship tracking intelligence*. Marine live map (2020). Retrieved from: https://www.marinetraffic.com/en/ais/home/centerx:-12.0/centery:25.0/zoom:4
12. *Strategic Transport Infrastructure Needs to 2030*. OECD reports. Transport (2019). Retrieved from: https://read.oecd-ilibrary.org/economics/strategic-

transport-infrastructure-needs-to-2030_9789264114425-en

13. *Signaling systems*. Open Railway Map (2020). Retrieved from: https://www.openrailwaymap.org/?lang=ru&lat=48.019324184801185&lon=14.78759765625&zoom=5&style=signals
14. C. Secchi, S. Riela. *Infrastructure for growth: how to finance, develop and protect it* (2019). Retrieved from: https://www.ispionline.it/sites/default/files/media/pdf/infrastructure_study.pdf
15. A. Tunnicliffe. *Horizon 2020: a look at the EU's flagship rail projects* (2020). Retrieved from: https://www.railway-technology.com/features/horizon-2020-rail-projects/

https://doi.org/10.1051/shsconf/202111200041

International and national practice for estimating cash expenditures of road replacement, repair and maintenance

V Gogolev[1], and *M* Barysheva[2]

[1]Road Department, North-Eastern Federal University, 13 Krasilnikova Str., Yakutsk, 677013, Russia
[2]Law College, North-Eastern Federal University, Yakutsk, 677013, Russia

Abstract. The paper considers the importance of the uninterrupted functioning of the transport network, which is the most important condition for eliminating the transport inaccessibility of Arctic territories of the regions, thus excluding transport discrimination of the population living in the Arctic zone of the Russian Federation. Besides, the paper provides a brief overview of international road repair and maintenance experience on the example of Norway, analysis of the standard for repair and maintenance of roads in the Russian Federation, proposals on the method of calculating the cash expenditures of replacement, repair and maintenance of roads in the Russian Federation. The Decree of the Government of the Russian Federation No. 658 dated May 30, 2017 approved the standards of financial expenditures and the rules for calculating the size of federal budget allocations for replacement, repair and maintenance of federal roads. However, at the moment, not all regions have updated the regulatory framework on the norms of financial expenditures and the rules for calculating budget allocations for replacement, repair and maintenance of regional and inter-municipal roads. The existing methodology for calculating the cost of repair and maintenance of roads in the Arctic territory of Russia does not fully provide the need to bring the existing roads into normal transport and operational condition for their safe operation. The analysis shows that there is no systematic work to ensure sustainable financing of the road economy in the republic and the road industry is financed at a level much lower than the actual needs.

1 Introduction

The development of the Arctic territories of the regions of the Russian Federation is largely determined by the intensification of the transport industry, which in the new market conditions is turning into a cross-industry system. Smooth operation of the transport network, including all modes of transport, is an essential condition for eliminating the transport inaccessibility of the Arctic territories of the regions, which excludes transport discrimination of the population living in the Arctic zone of the Russian Federation [5].

Efficient transport infrastructure should combine the characteristics of branching, technical infrastructure and the length of a single network of roads. This requires a combination of costs for the construction of new roads, as well as replacement, reconstruction, repair and maintenance of existing roads.

2 Materials and Methods

The existing method of calculating the cost of repair and maintenance of roads in the Arctic territory of Russia does not fully meet the requirements for bringing the existing roads into normal transport and operational condition for safe operation. Due to insufficient funding for the repair and maintenance of roads, the term "under-repair" has appeared and this "under-repair" is dramatically increasing every year.

In the current conditions, the most important task is to maintain the road network in an adequate transport and operational condition, taking into account the efficiency of measures for road users and for the national economy.

International practice in the field of estimating the financial expenditures of replacement, repair and maintenance of roads shows that the state authorities of foreign countries pay more attention to the quality of the execution of state contracts starting from design and survey work to the end of road construction. The quality of construction guarantees compliance with the standards of time between repairs, which is ultimately reflected in the planning of costs for repair and maintenance of roads. At the same time, sufficient state funds are allocated to maintain the transport and operational condition of state and regional roads. Funds are also allocated for municipal roads taking into account funding from the local budget.

*Corresponding author: vega-54@mail.ru

3 Analysis of international and national practices

Let us consider Norway as an example, since this country is part of the Arctic countries.

The natural and climatic conditions of Norway are most severe among all countries of Europe with excessive rainfall and quite long winters with low temperatures. In conditions of overwetting and long periods with negative temperatures, deformations of the roadway due to frost heaving of the foundation are one of the main problems during the operation of the country's roads.

In accordance with the road legislation, the Norwegian Ministry of Transport provides guidance on the maintenance of national roads and finances it. Norwegian Public Roads Administration (NPRA) acts on behalf of the Ministry of Transport. NPRA also manages provincial road maintenance activities implemented by the Provincial Councils. Municipalities finance and control the maintenance of municipal roads.

The Norwegian Public Roads Agency (Statens vegvesen) is the public authority responsible for national and circular public roads in Norway. The agency is subordinate to the Norwegian Ministry of Transport and Communications, manages the Public Roads Administration (Vegdirektoratet), and is one of the largest government agencies in Norway in terms of its budget. The agency is the distributor of funds for the repair and maintenance of public roads accumulated in the trust road fund.

The Norwegian Public Roads Administration takes precedence over the views of municipalities, since only locally they know everything about the region's different conditions and needs of a particular territory. Therefore, all road construction and repair projects pass through public hearings with the participation of municipalities, governmental agencies, landowners, road users and other persons interested in the development of the territory. It is they who make decisions based on the assessments and recommendations of the Norwegian State Road Administration and public hearings.

What is the situation with our national practice in this field?

According to the report of the Accounts Chamber of the Russian Federation, as part of the control activity "Checking the use of budget allocations aimed at maintaining public federal, regional and inter-municipal roads in 2018 and the past period of 2019", on average, the maintenance of 1 km of the federal road network in 2019 amounted to 2.2 million rubles, while the maintenance of regional roads was almost 8 times less – 288 thousand rubles per 1 km. In general, on average, 29.6% of the normative need was allocated to regional or inter-municipal roads [1].

The Decree of the Government of the Russian Federation No. 658 of May 30, 2017 approved financial cost standards and rules for calculating the size of federal budget allocations for replacement, repair and maintenance of federal roads. However, at the moment, not all regions have updated the regulatory framework on the norms of financial expenditures and the rules for calculating budget allocations for replacement, repair and maintenance of regional and inter-municipal roads [3].

The length of the regional road network in the Republic of Sakha (Yakutia) is 13,079 km.

A characteristic feature of the Republic of Sakha (Yakutia) is that 73% of the regional road network is occupied by seasonal roads (ice – 35%, land – 38%) (Fig. 1.1) and these are mainly Arctic territories.

The estimated normative need for budgetary allocations for the maintenance of regional and inter-municipal roads in the Republic of Sakha (Yakutia) in 2019 amounted to 6017.19 million rubles. But it should be noted that the investment deflator index by year does not fully take into account the dynamics of the increase in the cost of goods, services (work) in the region, therefore it is proposed to use the consumer price deflator index in the region in a way that other constituent entities of the Russian Federation use it in similar calculation rules.

According to the Decree of the Government of the Russian Federation [3], the standards of financial expenditures for replacement, repair and maintenance of federal roads of category V are established in the amount of:

- maintenance – 806 thousand rub/km;
- repair – 4738 thousand rub/km;
- replacement – 12008 thousand rub/km.

The network of roads owned by the Republic of Sakha (Yakutia) consists of regional, inter-municipal and local roads. Annual surveys and diagnostics of existing roads carried out within the framework of legal regulations are designed to detect defects and violations in the transport and operational condition of road structures. Based on the acts of examination and diagnostics of the main regional and inter-municipal roads, the analysis and assessment of the public road network was carried out. The results of the assessment of the state of the network of regional roads in the main areas are satisfactory, and the assessment of inter-municipal and local roads is unsatisfactory. Next, let us consider the analysis of costs for replacement, repair and maintenance of roads.

According to the Decree of the Government of the Republic of Sakha (Yakutia) No. 352 dated August 27, 2009 [4], the standards of cash expenditures for the maintenance and repair of regional and inter-municipal roads of category IV with a transitional type of sand-gravel coating were established in the following amounts:

- maintenance – 319 thousand rub/km;
- repair – 4,576 thousand rub/km;
- replacement – 8,614 thousand rub/km.

https://doi.org/10.1051/shsconf/202111200041

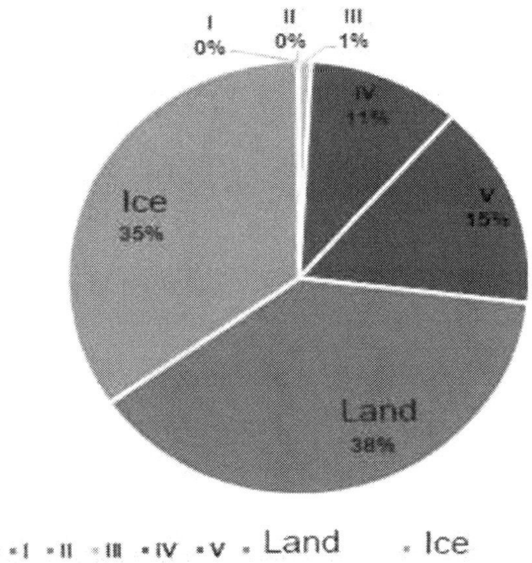

Fig. 1. Structure of regional and inter-municipal roads in the Republic of Sakha (Yakutia) in 2019 by categories

The difference between the norms of financial expenditures for replacement, repair and maintenance of federal roads of category V and for maintenance and repair of regional and inter-municipal roads of category IV with a transitional type of sand-gravel coating in the following amounts is as follows (excluding coefficients):

• maintenance – 2.51 times;
• repair – 1.03 times;
• replacement – 1.39 times.

As a result, for the maintenance of 1 km of roads of regional and inter-municipal significance from the Road Fund of the Republic of Sakha (Yakutia) in 2019, the provision of regulatory requirements was 27% or lower than the level in the whole of the Russian Federation (29.6%). That does not allow to maintain the proper technical condition of highways. Taking into account the additional funds allocated for the implementation of the national project "Safe and High-quality Highways" (hereinafter referred to SHH), the provision of regulatory requirements has increased to 48%.

Consequently, the full scope of work on the maintenance of the regional road network is not carried out. In fact, in 2019, 61% of the regional road network of the RS (Yakutia) funded by their road fund was covered, where maintenance work was carried out with an average cost of 157.7 thousand rubles per 1 km, which is 45% lower than the all-Russian level (288.1).

The insufficient level of funding for the repair and overhaul of the regional road network of the Republic of Sakha (Yakutia) causes the following negative consequences:

- does not allow to maintain the proper technical condition of roads and leads to a decrease in the level of maintenance of roads;
- the low level of maintenance of highways leads to an increase in the number of accidents, including due to the shortcomings of the transport and operational condition of sections of the road network;
- the limited financial security of the work leads to non-compliance with the inter-repair deadlines with the required increases, taking into account Decree of the Government of the Russian Federation No. 658.

The analysis showed that at the moment there is a situation in which the financial independence of the constituent entities of the Russian Federation in the road economy is quite limited. To adequately finance the replacement, repair and maintenance of roads at the regional level, each constituent entity needs to bring the regulatory framework in line with existing state programs, since the federal budget is spent 70% through the program method. This approach is caused by clearer targeted budget spending within the normative, estimated pricing in the road industry, which has developed over a quite long period of time.

To date, there are no legally established criteria for assigning roads against a certain level of their maintenance, as well as the mechanism, according to which Rosavtodor distributes budget allocations for the maintenance of roads to its subordinate institutions.

The law only stipulates the general standard for the size of budget allocations, which is used in the formation of the federal budget for the next year and the planning period. As a result, in different regions the volume of funds is quite different from the standard. The powers of regional authorities and local self-governmental bodies in the field of the use of regional, inter-municipal and local motor roads and the implementation of road activities provided for by the Federal Law of 08.11.2007 No. 257-FZ "On Roads and Road Activities in the Russian Federation and on Amendments to Certain Legislative Acts of the Russian Federation", are not fully regulated [2, 3].

So it turns out that in disregard of paragraphs 11 of article 12 and 13, paragraphs 2 and 3 of article 33 and 34 of the Federal Law No. 257-FZ and the Decree of the Government of the Russian Federation No. 658 of May 30 2017 the approved funding standards for replacement, repair, and maintenance of regional, inter-municipal and local roads are greatly underestimated. As a result, we get "under-repair", which is increasing from year to year.

The analysis shows that there is no systematic work to ensure sustainable financing of the road economy in the republic and the road industry is financed at a level much lower than the actual needs.

Insufficient funding for the repair and replacement of the regional road network in the Republic of Sakha (Yakutia) does not allow maintaining the proper technical condition of roads and leads to a decrease in the level of maintenance of roads and to non-compliance with the period between repairs with the need to increase it. Thus, to form a system of budgetary financing for road replacement, repair and maintenance at the regional level, there is a need to establish a road planning system for the reporting period, to adopt regulatory acts governing the procedure for assessing the technical condition of roads and artificial structures on them, to specify the procedure for diagnosing the state of roads, to update the standards of financial expenditures for maintenance and repair of regional and inter-municipal roads from the budget of the the Republic of Sakha (Yakutia), to modernize the methodology for their calculation in accordance with regional features.

Within the framework of the Safe and Quality Automobile Roads (SQAR) national project, the amount of funding for the region's road fund in terms of inter-budget transfers from the federal budget directly depends on the competent, systematic approach of executive and legislative authorities of the constituent entities of the Russian Federation to the preparation of regional road projects, the availability of regulatory acts governing replacement, repair and maintenance of roads.

4 Conclusion

The discrepancy of standards of financial expenditures for replacement, repair and maintenance of regional, intermunicipal and local roads, rules of calculation of the amount of regional and local budget allocations for specified purposes, techniques of their calculation does not allow the regional and local authorities properly controlling their functions:

• ensuring safety of regional, inter-municipal and local roads;

• conducting road activities on regional, inter-municipal and local roads;

• timely planning and calculating of expenses for the next fiscal year necessary for replacement and repair of regional, inter-municipal and local roads.

In this regard, it is necessary to develop and approve new financial cost standards for replacement, repair and maintenance of regional, inter-municipal and local roads and rules for calculating the amount of regional and local budget allocations for these purposes, as well as a methodology for calculating them.

5 Acknowledgements

According to the research project No. AAAA-A20-120121490014-1, the study is funded by the Automobile Roads Department of the Republic of Sakha (Yakutia).

References

1. Chamber of the Russian Federation, *Report on the results of the control activity "Verification of the use of budget allocations aimed at the maintenance of roads of federal, regional and inter-municipal public in 2018 and in the to date period of 2019" (with control and accounting bodies of the constituent entities of the Russian Federation)*, Retrieved from: https://ach.gov.ru/upload/iblock/c8e/c8e8afe3c8e348dc67f36bd6e24bf5c8.pdf. (2020)
2. Federal Law No. 257-FZ *"On Roads and Road Activities in the Russian Federation and on Amending Certain Legislative Acts of the Russian Federation"* Retrieved from: http://www.consultant.ru/document/cons_doc_LAW_72386/. (2007)
3. Decree of the Government of the Russian Federation No. 658 *"On Financial Cost Standards and Rules for Calculating the Size of Budget Allocations of the Federal Budget for Replacement, Repair and Maintenance of Federal Roads"*. Retrieved from: http://base.garant.ru/71689744/.
4. Resolution of the Government of the Republic of Sakha (Yakutia) No. 352 *"On the norms of cash expenditures for the maintenance and repair of regional and inter-municipal roads and the rules for their calculation"*. Retrieved from: http://base.garant No .t.ru/26715703/#xzz6aEz0Yf RC
5. V.N. Bugromenko, *Izvestia RAS. Series: Geography*. **4**, 7-16, (2010).

Federal and regional standards for highways maintenance and repair costs

Dmitry Filippov* and *Elena* Grigoryeva

North Eastern Federal University, 58, Belinsky street, Yakutsk, 677000, Russia

Abstract. The current conditions of Russian territories road infrastructure, regional budgetary funding, differences in economic potential and the economy structure of the constituent entities of the Russian Federation determine the relevance of the development and improvement of the regional regulatory base. This is important in the field of tariff policy, one of the elements of which is the approval of basic standards and rules for calculating the amount of budgetary funds for capital repairs, repairs and maintenance of regional road network. This article is devoted to the study of the regional practice that has developed in the Russian Federation for the establishment of standard costs for capital repairs, repairs and maintenance of regional and intermunicipal roads in comparison with federal ones. The work consists of two logically related structural elements: an overview and an analytical part. The review revealed the differentiation of the size of budgetary allocations in the context of subjects and federal districts. In the analytical part, an attempt is made to determine the closeness between federal and regional standards and, as a result, to identify the range of regions that practice linking regional standards to federal ones.

1 Introduction

According to the results of control measures of the Accounts Chamber of the Russian Federation [1] at the beginning of 2020, 85% of the federal road network meets the regulatory requirements for the quality of content. With regard to the regional road network, this indicator improved compared to 2017 by only 1.1% and amounted to only 44.2%. If the volume of budget allocations for the maintenance of the federal road network in 2019 amounted to 2,242 thousand roubles / km, then 288 thousand roubles / km, or 7.7 times less, were allocated for the regional road network.

At present, a "high" level of maintenance of the federal road network has been established; the values of the regional road network must be brought to the same level, taking into account their transport and operational status. As a result, it was proposed to revise and justify the change in the standards for the maintenance and maintenance costs of the regional road network in the constituent entities of the Russian Federation.

This paper presents the results of a study of the regional practice that has developed in the Russian Federation to establish the standard costs for capital repairs, repairs and maintenance of regional and intermunicipal roads in comparison with federal ones.

2 Research methods

The work uses statistical and comparative methods for analysing quantitative data. Thus, by calculating the standard statistical indicators of the standard deviation and the coefficient of variation, the numerical values of the scatter were calculated. Then we compared with the standard values of the deviations in accordance with the methodology proposed in Federal Law No. 44.

When studying the degree of influence of external and internal factors affecting the amount of cash costs for the maintenance, repair and capital repair of regional and intermunicipal roads, the authors relied on a number of scientific works on the research problem. Issues of regional de-differentiation of the cost of work (services) and goods, including road construction, are considered in the following works [2-7]. The methodological basis of the study was based on scientific works [8-10], justifying the influence of the factors of cost increasing in capital repairs of the regional road network, taking into account transport accessibility and remoteness from the local centres of the transport system. There are various techniques and methodological approaches to calculating the standards of monetary costs for the maintenance of highways [11-13], including taking into account the turnaround time [14], the cost of maintaining special road equipment (fuel, spare parts, repairs) [15] and others [16].

Due to the presence of variability in the implementation of repair and maintenance work, it is impossible to achieve strict adherence to real costs. In order to achieve the quality indicators of the work carried out, it is necessary to establish their standard values. In this regard, one of the most important tasks of quality management is to reduce the variation of processes and their results, the solution of which is impossible without the use of various statistical methods. The use of statistical methods makes it possible to

* Corresponding author: fil_dv@mail.ru

collect, process and analyze data in order to turn them into objective and reliable information necessary for making informed decisions.

3 Analysis

In accordance with adopted hypothesis, the target benchmark of the standards for financing federal highways was defined as a reference. However, "blind copying" of federal norms and standards in the corresponding documents of the regional level is unfounded. In this connection, regulatory documents, similar rules and practices for calculating the monetary costs of the Russian Federation regions were collected, studied and analyzed. Namely, Belgorod Region, Komi Republic, Rostov Region, Chechen Republic, Bashkortostan Republic, Khanty-Mansi Autonomous Region, the Novosibirsk Region are presented in Table 1.

Table 1. Analysis of the regional to federal standards ratio for maintenance, repair and capital repair of regional and intermunicipal roads in prices of 2019.

No.	Subject of the Russian Federation	Ratio of regional to federal standard (coefficient 1.00)		
		maintenance	repair	capital repair
1	Belgorod Region (Central Federal District)	0.59(1.02*)	0.46 (1.07)	0.58 (1.07)
2	Komi Republic (North Western Federal District)	0.20(1.35)	1.76 (1.33)	1.85 (1.37)
3	Rostov Region (Southern Federal District)	0.80(0.83)	0.96 (0.82)	1.01 (0.86)
4	Chechen Republic (North Caucasian Federal District)	0.71(0.76)	0.93 (0.89)	0.96 (0.92)
5	Bashkortostan Republic (Volga Federal District)	0.40(1.00)	0.88 (0.68)	0.86 (0.66)
6	Khanty-Mansi Autonomous Region (Ural Federal District)	1.01(0.98)	1.06 (0.97)	1.06 (1.03)
7	Novosibirsk Region (Siberian Federal District)	1.04(1.06)	0.99 (0.93)	1.03 (0.96)
8	Republic of Sakha (Yakutia) (Far Eastern Federal District)	0.58(1.58)	1.21 (1.73)	0.90 (1.73)

Explanation: * in brackets there are the values of the ratio of the federal district to the whole Russian Federation standard

The sample of the listed regions was made according to the principle of considering one of the Russian Federation region from each federal district. For each region of the Federation, in strict accordance with the established local rules, the values of regional and federal standards were calculated for the corresponding federal district to which the studied region belongs. Estimated standards for constituent entities and federal districts are reduced to 2019 prices based on the approved consumer price deflator indices to the standards for the maintenance and deflator indices of investments in fixed assets to the repair and capital repair, in accordance with current rules are presented in Table 2.

Table 2. Calculation of standard and deviation. Capital repair (i = 1).

No.	Subjects of the Russian Federation	Standard		Deviation
		Regional (X_i)	Federal District (X)	X_i - X
1	Belgorod Region (Central Federal District)	7549	13991	-6442
2	Komi Republic (North Western Federal District)	24250	17912	6338
3	Rostov Region (Southern Federal District)	13230	11285	1945
4	Chechen Republic (North Caucasian Federal District)	12591	12063	528
5	Bashkortostan Republic (Volga Federal District)	11302	8695	2607
6	Khanty-Mansi Autonomous Region (Ural Federal District)	13866	13434	432
7	Novosibirsk Region (Siberian Federal District)	13441	12510	931
8	Republic of Sakha (Yakutia) (Far Eastern Federal District)	11805	22658	-10853

The results obtained show that within the federal districts there is a significant differentiation in the cost of standards. Especially this aspect is typical of the northern regions, such as Komi Republic, Republic of Sakha (Yakutia).

4 Assessment of the tightness of the relationship between regional and federal standards

The coefficient of variation refers to a measure of the relative spread of prices offered. It shows how much the average price range is from the average price value. According to the Federal Law (FL) No. 44 the coefficient of variation is an indicator of the spread for the selected price proposals of the tender, which is used when using the method of comparable market prices. This indicator can take the following values:

- less than 10% - in this case, the difference in prices is considered insignificant;
- from 10% to 20% - the spread is considered average;
- from 20% to 33% - the difference is recognized as significant, but acceptable;
- over 33% - the data is heterogeneous, when calculating the contract price it is not allowed to use data with a coefficient of variation over 33%.

According to the Federal Law No. 44 the permissible coefficient of variation is up to 33%. Calculations of the magnitude of the variance and the coefficient of variation in comparing the standards between regional and federal values for the overhaul of highways are presented in Table 3.

Table 3. Calculation of variance and coefficient of variation. Capital repair (i = 1).

No.	Subjects of the Russian Federation	Variance	Coefficient of variation		Interpretation of CV by FL No. 44
		σ	CV, units	CV, %	
1	Belgorod Region (Central Federal District)	6442	0.85	85.3%	The difference is not valid
2	Komi Republic (North Western Federal District)	6338	0.26	26.1%	The difference is significant, but acceptable
3	Rostov Region (Southern Federal District)	1945	0.15	14.7%	Average difference
4	Chechen Republic (North Caucasian Federal District)	528	0.04	4.2%	The difference is insignificant
5	Bashkortostan Republic (Volga Federal District)	2607	0.23	23.1%	The difference is significant, but acceptable
6	Khanty-Mansi Autonomous Region (Ural Federal District)	432	0.03	3.1%	The difference is insignificant
7	Novosibirsk Region (Siberian Federal District)	931	0.07	6.9%	The difference is insignificant
8	Republic of Sakha (Yakutia) (Far Eastern Federal District)	10853	0.92	91.9%	The difference is not valid

As a result, high values of the coefficient of variation were obtained in two regions - Republic of Sakha (Yakutia) and Belgorod Region. Determined heterogeneity of values, means that the standards used in budget allocations calculating for roads capital repair in these regions do not correspond to the federal level and require a reasonable revaluation. The results of determining of the regional and federal standards dispersion for the roads repair are presented in Table 4.

Table 4. Calculation of variance and coefficient of variation. Repair (i = 1).

No.	Subjects of the Russian Federation	Variance	Coefficient of variation		Interpretation of CV by FL No. 44
		σ	CV, units	CV, %	
1	Belgorod Region (Central Federal District)	3134	1.31	131.3 %	The difference is not valid
2	Komi Republic (North Western Federal District)	2203	0.24	24.2%	The difference is significant, but acceptable
3	Rostov Region (Southern Federal District)	730	0.15	14.7%	Average difference
4	Chechen Republic	202	0.04	4.2%	The difference

	(North Caucasian Federal District)				is insignificant
5	Bashkortostan Republic (Volga Federal District)	1054	0.23	23.1%	The difference is significant, but acceptable
6	Khanty-Mansi Autonomous Region (Ural Federal District)	441	0.08	8.1%	The difference is insignificant
7	Novosibirsk Region (Siberian Federal District)	356	0.07	6.9%	The difference is insignificant
8	Republic of Sakha (Yakutia) (Far Eastern Federal District)	2670	0.43	42.6%	The difference is not valid

An unacceptable difference in repair standards is also observed in two regions - Republic of Sakha (Yakutia) and Belgorod Region.

In this case, we should pay attention to the fact that the normative acts with the values of the standards were approved back in 2009 and 2012. Thus, the mechanism of applying the index-deflator of investments in fixed assets has shown its insolvency. For a period of more than 5 years it does not allow increasing the value of the standard to the level of real cash costs. Consequently, it is determined the need to review and update the standards values no later than 5 years.

Calculations of the magnitude of the variance and the coefficient of variation in between regional and federal values for road maintenance are presented in Table 5.

Table 5. Calculation of variance and coefficient of variation. Maintenance (i = 1).

No.	Subjects of the Russian Federation	Variance	Coefficient of variation		Interpretation of CV by FL No.44
		σ	CV, units	CV, %	
1	Belgorod Region (Central Federal District)	391	0.74	73.5%	The difference is not valid
2	Komi Republic (North Western Federal District)	1045	5.74	573.6%	The difference is not valid
3	Rostov Region (Southern Federal District)	24	0.03	3.3%	The difference is insignificant
4	Chechen Republic (North Caucasian Federal District)	50	0.08	7.7%	The difference is insignificant
5	Bashkortostan Republic (Volga Federal District)	537	1.46	146.1%	The difference is not valid
6	Khanty-Mansi Autonomous Region (Ural Federal District)	27	0.03	3.0%	The difference is insignificant
7	Novosibirsk Region (Siberian Federal District)	18	0.02	1.9%	The difference is insignificant
8	Republic of Sakha (Yakutia) (Far Eastern Federal District)	905	1.72	171.8%	The difference is not valid

In this case, the northern regions of the Komi Republic and Republic of Sakha (Yakutia) were determined for the standards for roads maintenance with significant heterogeneity and an unacceptable difference. This can be explained by the influence of the natural and climatic factor on increasing financial costs, which led to a deliberate understatement of standards in the regions due to the limited budget funds.

5 Conclusion

The results of the analysis of the data obtained in the course of the preliminary survey study of Russian rule-making practice showed that such regions of the Federation as Chechen Republic, Khanty-Mansi Autonomous Region, Novosibirsk Region, which are distinguished by their economic potential, political authority and leading position in their macro-territories, have shown the best results in tightness to federal standards. The value of the coefficient of variation was less than 10% and is stable for all types of work.

Insignificant deviation indicates the establishment of regional standards practically at the federal level. The results obtained are important for the regional executive power and can serve as an additional justification for the determination and establishment of target tariff policy. That is, when developing regional norms and rules, calculate and approve such sizes of basic standards that could be close to those established for federal highways.

References

1. V.N. Bogomolov, Report on the results of the control event "Checking the use of budgetary allocations aimed at the maintenance of public highways of federal, regional and intermunicipal importance in 2018 and the past period of 2019" (with the control and accounting bodies of the constituent entities of the Russian Federation), *Accounts Chamber of the Russian Federation*, 2 (2020)
2. P.V. Gulyaev, *Gornyi Zhurnal* 12, 23-26 (2018)
3. V.V. Nikiforova, E.R. Romanova, E.E. Grigorieva, *Gornyi Zhurnal* 3, 41-46 (2018)
4. E.E. Grigoryeva, *IOP Conference Series: Earth and Environmental Science* 302(1), 012138 (2019)
5. T.P. Egorova, A.M. Delakhova, *Proceedings of the International Conference "Aviamechanical engineering and transport" (AVENT 2018) Series: Advances in Engineering Research* 158, 123-129 (2018)
6. V.N. Leksin, B.N. Porfiryev, *Economy of Region* 3, 641-657 (2017)
7. V.N. Borisov, O.V. Pochukaeva, *Studies on Russian Economic Development* 27(2), 159-165 (2016)
8. T.P. Egorova, A.M. Delakhova, *Smart Innovation, Systems and Technologies* 138, 272-282 (2020)
9. T.P. Egorova, A.M. Delakhova, *IOP Conference Series: Materials Science and Engineering* 632(1), 012029 (2019)
10. D.V. Filippov, *IOP Conference Series: Materials Science and Engineering* 753(5), 052048 (2020)
11. S.V. Fedotov, T.F. Matyushenko, *Roads and bridges* 2(18), 11-20 (2007)
12. V.V. Gasilov, M.A. Preobrazhensky, S.S. Zamchalova, M.A. Shibaeva, *Science and technology in the road industry* 2, 10-12 (2002)
13. A.S. Eremeeva, *News of higher educational institutions. Series: Economics, Finance and production management* 2(12), 3-9 (2012)
14. E.A. Latkina, T.F. Matyushenko, *Roads and bridges* 2(38), 1 (2017)
15. I.A. Kurbatov, *Bulletin of the Moscow automobile and road state technical University (MADI)* 3(22), 60-63 (2010)
16. V.K. Apelin, *Roads and bridges* 18(2), 20-29 (2007)

Memorial narratives and symbolic images as a resource for the development of Northern urban space

Irena Khokholova[1*], *Natalia* Danilova[2], *Alyona* Tomaska[2], and *Kyunney* Pestereva[1]

[1] M. K. Ammosov North-Eastern Federal University, 58, Belinskogo str., Yakutsk, 677007, Russia
[2]Institute of Humanities and Problems of the Peoples of the North SB RAS, 1, Petrovskogo str., Yakutsk, 677027, Russia

Abstract. The article is devoted to the identification of symbolic markers and perception of the cognitive model of the space of the northern cities of the Republic of Sakha (Yakutia). The definition of "northern cities" is constructed based on the results of a study to identify various aspects of identity (national, territorial, ethnocultural) and is used by the authors when defining all cities of Yakutia. The main tools for constructing and regulating the cognitive map of a city are city monuments associated with historical memory and acting as components of state policy and a nation-building tool. The main research methods were historical and cognitive analysis and the method of questionnaire survey. The novelty of the article is a cross-cultural study of the perception of urban monuments as symbolic markers of urban space. The authors observe a difference in the perception of the symbolism of urban space between generational groups. It is concluded that each city, depending on its geographical and economic location, local characteristics, and national identity, has its own specific set of symbolic markers in the cognitive map of the city. The findings will contribute to the systematic study of the value attitudes of residents of modern Russian cities. Multilevel models of symbolic images associated with both national and territorial identities are being built.

1 Introduction

The city in its potential provides a dynamic course of economic, political, socio-cultural, communication, microevolutionary and other processes, acting as a matrix for the life arrangement and life support of citizens. Various socio-economic, political, and cultural factors influence the formation of the city's image. An important factor can be considered the potential contained in the cultural and symbolic sphere of urban space, which is demonstrated through various "city texts". According to D.N. Zamyatin and others, any city offers, as a rule, a multitude of symbolic images of different genesis, content, and structure, formed both by social and professional groups, and by individuals in the process of their purposeful activity [1]. These images can be presented as iconic symbolic markers of urban space.

2 Materials and methods

The theoretical and methodological basis was the work of famous Russian scientists representing the Moscow-Tartus semiotic school [2, 3], and the concepts of modern cross-cultural research were used. In particular, the concept of historical memory (place of memory) by Pierre Nora, conceptual provisions aimed at studying the socio-cultural life of the city and involving citizens in the process of modeling symbolic urban space, urban identity and territory [4, 5, 6]. To develop the concept of this article, theoretical articles by D.N. and N.Yu. Zamyatin and I.I. Mitin allows a methodological basis built for the presentation of the city as a text and the city as a symbolic space [7].

According to Yu.M. Lotman, the city is a complex semiotic mechanism, which is a "cauldron" of texts and codes, diverse and heterogeneous, belonging to different languages and levels [2]. V.N. Toporov includes in the concept "city text" all the information that streets, squares, gardens, monuments, buildings, townspeople, etc., openly or implicitly give us and "can be understood as a heterogeneous text to which a certain meaning is attributed and on the basis of which the system of signs realized in the text can be reconstructed" [3].

The city becomes a "place of memory", a focus of significant events, a repository and reproduction of historical memory. Historical memory is supported primarily by the cultural heritage embodied in material objects that become "places of memory". "Place of memory" embodies the unity of the spiritual and the material, which, over time, was preserved and passed from generation to generation through a remembering culture, becomes a symbolic element of the cultural and historical heritage [8]. Thus, the memory of the city is considered as a mechanism for preserving, broadcasting and reproducing the past socio-cultural experience of urban life for the construction of being in the present [9]. M. Castells argued that «there can be no spatial theory

* Corresponding author: iskhokholova@mail.ru

that would not be an integral part of the general social theory» [10]. He was supported by R. Ledrut and noted: «the city does not matter in itself, outside of the practice in which people participate at a certain time and in a certain place» [11].

Group self-identification is carried out not only with the help of official political, but also recognizable symbols that are significant for the community. These symbols are monuments, architectural structures, art objects, etc. Opportunities for interaction between the elite and society are often seen through the attitude of the authorities towards them.

The collection of primary sociological information on the RFBR and EISR project "The Symbolic Space of Northern Cities of the Republic of Sakha (Yakutia) in the Context of Socio-Political Processes" was carried out by a questionnaire survey of the urban population in 13 cities of the Republic of Sakha (Yakutia) (n = 919). A questionnaire survey on 2 thematic blocks: the role of the modern symbolic space of the cities of the Republic of Sakha (Yakutia) on the formation of political and national identity was carried out in September-December 2020. In accordance with the established sample, the questionnaire survey covered the population aged 18 years and older. The ethnic composition of the respondents is as follows: Sakha - 52.3%, Russians - 40.6%, indigenous peoples of the North - 2.3, other nationalities - 1.8, mestizos - 0.2%, did not indicate nationality - 2.7%.

3 Results

According to 2020, Yakutia includes 54 urban settlements, including: 13 cities, among which there are 4 cities of republican subordination, of which 1 city of republican significance - Yakutsk, the capital of the Republic. 3 cities are of republican subordination within the districts (Mirny, Nyurba, Pokrovsk); 9 cities of district (ulus) subordination; 41 urban-type settlements. The cities of Yakutia can be conditionally divided into the following districts according to several distinctive features into the oldest cities (Yakutsk, Vilyuisk, Verkhoyansk, Srednekolymsk, Olekminsk) and industrial cities (Neryungri, Aldan, Tommot, Lensk, Mirny, Udachny, Pokrovsk, Nyurba), northern cities (Verkhoyansk and Srednekolymsk). All northern cities have their own rich history, cultural heritage, in addition, they are symbolic centers of industrial development of the Permafrost territory. For example, Mirny is considered the "diamond capital", Neryungri is the "coal capital", Aldan is the "golden" province of Russia, and the oldest city Verkhoyansk, founded in 1638 as the Verkhoyansk prison, received the status of a city in 1822 and is the northernmost city and the center of the Pole of Cold. Northern cities in their potential provide a dynamic course of economic, political, socio-cultural, communication, microevolutionary and other processes, acting as a matrix for the life arrangement and life support of citizens in the harsh climatic conditions of the North.

In the context of modernization processes and globalization, the preservation and study of the ethnic specifics of the carriers of the "living traditional culture" is becoming one of the important factors in the implementation of socio-cultural policy. Today, there is a gradual understanding of cultural heritage not only as a system of spiritual values, but also as an image tool that can influence the development of the region. According to the results of questionnaire and spontaneous surveys, the presentation of the cultural heritage of the indigenous peoples of the North (Yakuts, Evens, Evenks, Yukagirs, Dolgans) is considered a promising direction in the republic and the formation of a new positive assessment scale for the perception of the image of the territory as a whole (98.0%) and the formation of national identity (86.0%).

It is known that cultural and symbolic capital is mainly concentrated in rural settlements. Nevertheless, in recent decades, in many Russian cities, including in the cities of Yakutia, significant progress has been made in organizing the urban landscape using the ethno-cultural potential, which shows that the urban population is quite adaptive to the political and socio-cultural changes that have taken place. Thus, the revival of national culture in the 90s of the twentieth century gave an impetus to the formation of a fundamentally new attitude to cultural heritage as a source of social, cultural and economic development of the region. In this context, special attention should be paid to symbolic images that construct and form the modern cultural environment (development resource) in the urban space and represent ethno-cultural brand codes: for example, in a questionnaire survey, respondents indicate such monuments as "Bull of Winter" - a symbol of Cold, "Peak of Cold" (Verkhoyansk), "Olonkhosut" (Mirny), "Kayury" (Neryungri), etc.

3.1 Symbolic images of the "North"

At the beginning of the XXI century, when the Russian regions made a qualitative leap in the implementation and development of socio-cultural projects, and the projects themselves became an effective tool for the development of territories, new cultural identity codes are being constructed. As a result of the actively pursued image policy in the republic to promote a positive image of the Pole of Cold, the geocultural concepts "cold", "north", "winter" become a brand tool and as metaphorical images experience a "second birth". The cold climate of the region began to be perceived as a source of additional benefits and opportunities for humanity and a healthy lifestyle, and the old stereotypical ideas that developed in the era of tsarist power, as about "the land of eternal winter and cold", "prison without bars", and in Soviet times "the north - as a raw material appendage" and "bear corner", began to leave the general perception of the Yakut region. Thus, the historical and cultural landscape as a panorama forms a "permanent space" and serves as an ideal position for seeing the world [12].

The promotion of the theme of Eternal cold, the positive brand of "Winter" on the world market contributed to the emergence of a new festival "Winter begins with Yakutia", which brings together all Santa Claus, Ded Moroz, etc., to light the first Christmas tree on December 1 on the main square of Yakutsk. The Lord of winter and cold Chyshaan, developed by the creative intelligentsia and became the modern embodiment of the mythical image of the Bull of Winter, has become a national winter brand in Russia.

In the socio-cultural space of Yakutsk city, a metaphorical image of the center of the Arctic space is gradually introduced. So, in Yakutsk, the Arctic State Institute of Culture and Art and the Arctic Research Center appeared, the Yakutsk State Agricultural Academy was renamed the Arctic State Agrotechnological University, the construction of the Arctic Center of Epos and Culture is planned, etc.

The image of the North/Cold received a second wind in the rebranding system "Yakutia – the country of permafrost", in which natural and climatic conditions become one of the main tools for constructing figurative and geographical models. Today, the cognitive perception of Verkhoyansk city is associated with the Cold Pole, and Srednekolymsk is represented as a topos, where the Arctic begins. In this context, the image of Yakutia for the development of extreme, ethnic and event tourism is actively promoted.

It would seem that the symbolic capital constructed within the framework of the new image and cultural policy could lay the foundation for the formation of the northern identity among the population of Yakutia, however, the results of the conducted sociological research showed a different result. So, among the indigenous population, regardless of nationality, "northerners" consider themselves only residents of the northern regions and the alien population, for whom the whole Yakutia is associated with the image of the North and is perceived as the Northern Region. 80.0% of the respondents of the Yakut nationality in the collective image of "their land" intertwine the image of the southern ancestral homeland, as the metaphor of the South as a "figure of memory" is translated throughout the cultural model of the Sakha people.

Thus, the concept of cultural heritage in different versions is considered as a product of the development and transformation of the urban landscape and a factor in the formation of national and territorial identity.

An important factor in the development of urban space can be considered the potential contained in memorial narratives that represent various significant events and personalities, urban mythopoetics, and ethno-cultural representations. In this regard, it is advisable to turn to the study of the objects of commemoration of urban space. The main attention was paid to the monuments, since it is through them that the nation-building landmarks (historical and cultural memory, identity, value landmarks) are reflected.

3.2 Memorial narratives of urban space

Information about the past and significant events of the country is broadcast to the population through the monuments. A survey in the cities of Yakutia showed that 63% of the towns inhabitants are interested in the country's politics, and only 54.4% are interested in the history of the republic, which shows that the level of interest is not high enough. At the same time, 67% of the respondents know in honor of which event or historical figure the monument was erected. In one question, the respondents were asked to name prominent figures of importance for Yakutia, 93% of the respondents named the figures of the Soviet era, only 6.6% of the answers mentioned modern politicians of Yakutia. 49% of the inhabitants see the Lenin monument as a symbolic center of the city, 23% - monuments to Yakut political representatives of the Soviet era, 15% - monuments to Cossack pioneers, Victory monuments (Memorials of Glory) - 7.6%. In the industrial cities of Yakutia, the concept "symbolic city space" is associated with the development of industry - the "Peace" memorial, etc. As the survey showed, the common history that the former Soviet republics went through is associated with key events in the history of the Soviet era (Revolution, Great Patriotic War, and others) and are factors in the formation of collective historical memory. Depending on the perception of the monument by the town inhabitants, it acquires its symbolic function, mythologeme and becomes an identifying marker of this locus. Meanwhile, not all monuments become iconic points. In Yakutsk, the following monuments were named as symbolic markers of the urban space by the majority of the surveyed citizens. The Monument to V.I. Lenin - the leader of the proletariat - 98%. The Memorial complex on Victory Square (1970-2005) - 95%. The Monument to the great son of the Yakut people P.A. Oyunsky is 80%. The Monument to the state and public figure, one of the initiators of the Yakut autonomy M.K. Ammosov is 80%. The Monument to the pioneer S.I. Dezhnev and his family is 78%; the Monument to the Janitor (70%); the Monument to the founder of the city P.I. Beketov (68%); the Monument to Mother (65%); the Monument to the founder of Yakut literature A.E. Kulakovsky (50%); the Monument to Academician V.P. Larionov (45%); the Monument to educator I. Veniaminov (35%);

It should be noted that not a single city has memorial complexes dedicated to legendary heroes or folklore motives of the indigenous peoples of Yakutia. Nevertheless, individual ethnic motives are included in the general compositional structure of the monuments (hitching post-serge, deer, choron, sledges, etc.). According to the survey, 70% of respondents would like to see monuments with a pronounced ethnic orientation in their cities (the first ancestor of Ellei Botur, historical personality Tygyn Darkhan, sacred horse, sacred deer, Ysyakh motives, epic heroes like Nurgun Botur, Tuyaaryma Kuo, Yuren Khosun, etc.).

As an integral part of the brand and image of the urban space, 35% of respondents indicated monuments dedicated to historically significant events related to the industrial development of the territory (monuments dedicated to miners, geologists in the cities Neryungri, Mirny, Aldan, etc.). 35% noted iconic personalities of ulus / district, "geniuses of the area" (a monument to N.G. Chernyshevsky in Vilyuisk, L. Petukhova in Nyurba, brothers Ksenofontov in Pokrovsk, etc.). And 40% marked local geocultural images (a monument to the Bull of Winter and the sign of the Pole of Cold in Verkhoyansk, for a mammoth in Srednekolymsk, for a diamond / brilliant in Mirny, etc.).

At the same time, the respondents indicated that monuments dedicated to residents, ordinary workers (a monument to a musher in Neryungri, an art object Dvornik in Yakutsk, etc.) and folk heroes (a monument to sniper Kulbertinov in Yakutsk, journalist N. Protopopova, art object "Doroobo" in Vilyuisk, etc.). So, for example, in Verkhoyansk, a sculptural monument dedicated to famous olonkho performer Daria Tomskaya-Chaika, the keeper of the local peculiarities of Verkhoyansk folklore, according to the majority of respondents (75%) on the cognitive map of the city, began to occupy higher positions (although it was installed relatively recently in 2019) than the previous brand code of the city as the capital of the Pole of Cold, the mythological image of the Bull of Winter (25% of the respondents).

If only historical memory is associated with monuments and memorials, then art objects and murals are a kind of author's representation of a modern vision of cultural heritage. Although street painting appeared all over the world in the 19th century, it has become popular in Russian cities, including Yakut ones, in the last 3-5 years. In Yakutia, not only in cities, but also in other settlements, they began to decorate the walls of buildings, which are mainly associated with both historical and cultural memory and local geocultural images. So, in Vilyuisk, a mural dedicated to Kate Marsden, the English sister of mercy, was realized; in Nyurba - the legendary founder of the Nyurba clans - Nyrbakaan; images of the Arctic space and nomadic life - in Srednekolymsk; to the symbols of Yakut culture - a girl in traditional clothes and jewelry; to the symbolic elements of the Ysyakh holiday - a choroon, a sacred horse, a hitching post, etc. in Yakutsk.

4 Conclusion

The symbolic space of cities in the northernmost region - the Republic of Sakha (Yakutia), which reveals different layers, different cultural levels, "pre-Soviet", "Soviet" and "global" - transnational layers, generally supports the image of a multiethnic region. The city-forming branch of the industrial northern region is expressed in the symbolic space of relatively young industrial cities that arose in the 20th century in connection with the development of resource components, economic and political development of the country.

Memorial complexes and monuments are cultural and symbolic objects that form, among other things, political identity, reflecting today the transformation of ideas about the region and serving as a resource for branding the territory. The symbolization of urban space is an integral part of the appropriation by the inhabitants of towns of certain elements that have a symbolic status. There is a difference in the perception of the symbolism of urban space between generational groups: youth (born in the post-Soviet era), middle and older generation (generation of the Soviet era), which is characterized by ignorance or forgetting of the country's historical events.

The significance of a monument as a tool and a symbolic marker of space, first, depends on the political context and its introduction into the collective memory through a constant "reminder" of its importance and significance in the socio-cultural and everyday life of the towns' inhabitants. In this connection, the urgent problem is the task of preserving the accumulated experience and values, places of memory, relics of the past, as well as ideas about the past that are stored in the memory of its inhabitants and the transfer of this knowledge to the younger generation.

Acknowledgments

The reported study was funded by RFBR and EISR according to research project No. 20-011-31324 "Symbolic space of the northern cities of the Republic of Sakha (Yakutia) in the context of socio-political processes" and also with the support of the Ministry of Education and Science of the Russian Federation within the framework of the Program of Fundamental and Applied Scientific Research of the Russian Academy of Sciences "Ethno-cultural diversity of Russian society and strengthening of the All-Russian identity" 2020-2022.

References

1. D.N. Zamyatin, *Modeling images of a historical and cultural territory: methodological and theoretical approaches.* (Moscow: Institute of Heritage, 2008)
2. Yu.M. Lotman, *Semiosphere. Culture and explosion. Inside thinking worlds. Articles. Research. Notes* (Sant-Petersburg: Art-SPB, 2000)
3. V.N. Toporov, *Petersburg and the "Petersburg text of Russian literature". Myth. Ritual. Symbol. Image. Research in the field of mythopoetic.* (Moscow: Progress-Culture, 1995)
4. V.L. Glazychev, *Urban Studies* (Moscow: Europe, 2008) Retrieved from: http://www.glazychev.ru/books/urbanistika/urbanistika.htm
5. D.V. Vizgalov, Branding of the city. (Moscow: Institute for Urban Economics Foundation, 2011)
6. A. Filko, *Sociodynamics*, **7**, 51 – 64, (2015)
7. N.Yu. Zamyatina, D.N. Zamyatin, I.I. Mitin, *Modeling images of a historical and cultural territory: methodological and theoretical approaches.* (Moscow: Institute of Heritage, 2008)
8. P. Nora, *France - memory.* (Sant-Petersburg: Publishing House of St. Petersburg University, 1999)
9. I.Yu. Solomina, *Social memory and culture: archetype, monument, oblivion* (Samara: Samara State Regional Academy, 2013)
10. M. Castells, *La question urbaine.* (Paris: François Maspero, 1972)

11. R. Ledrut, *Les images de la villes.* (Paris: Editions Anthropos, 1973)
12. I. Svirida, Slavic world. (Moscow: Progress-Tradition, 2007)

Transport logistics and interaction of participants in the russian transport market

Vladimir P. Smirnov [1,*]

[1] Russian customs Academy, Vladivostok branch, 16v, Strelkova str., Vladivostok, Russia

Abstract. Logistics processes are becoming more widespread in the Russian economy. Transport logistics is an integral part of logistics. Its functions are to select the optimal routes for the movement of economic goods in the interests of all participants in these processes. The purpose of this work is to substantiate the methods of solving the main problems of transport logistics in Russian conditions. The study uses a systematic approach. Based on it, algorithms for solving transport logistics problems in the Russian economy are refined, the choice of transport mode, carrier and other logistics partners is ensured, rational delivery routes are determined, the technological unity of transport and warehouse processes is established, and the optimization of transport process parameters is achieved. The emerging flow processes must be interconnected through the integration of transport, production, supply, sales, consumption, service, financing and information. In the associated flow processes, the economic interests of all participants in the transport market (the sender, who forms the material flows, the recipient and the carrier) will be rationally coordinated with the economic interests of other participants in the logistics process. This will help overcome technical, technological, informational, economic, financial, international, environmental, and labor problems. In this case, transaction costs are reduced by all participants in logistics chains, the risks of doing business are reduced, and a synergistic effect is achieved.

Keywords: tasks of transport logistics, a systematic approach

1 Introduction

Product is not fully ready if it is not delivered to the consumer who will use it. There are two types of gaps between producers and consumers: in space and in time. Accordingly, supply logistics, in-house logistics, and distribution logistics should be complemented by transport logistics and warehouse logistics.

The concept of transport logistics originated in the second half of the twentieth century. This term was applied in 1977 by Mundy et al. [9], in 1978 by Andrews et al. [2].

The term "transport logistics" is also widely used in the twenty-first century [5, 8, 9, 15]. However, some authors believe that definitions of transport logistics are not always accurate [3].

The transport function has undergone structural changes, which are reflected in the development of transport logistics [4]. A well-developed transport system provides higher efficiency, lower operating costs, and higher quality of service for logistics systems [14]. Lai et al. [8] investigated the performance indicators of the supply chain in the context of transport logistics and built a measurement tool for evaluating the efficiency and effectiveness of the work of shippers in transport logistics, service providers and consignees.

The consideration of the main components of the logistics system, such as warehousing, inventory, packaging and information flows in addition to transport, their systematic planning and at the same time respecting the limits of the associated costs and service levels is the main task of modern logistics [11].

The heads of logistics companies have learned from their practical activities their own, often unique, understanding of transport logistics.

They give transport logistics the following definitions:

- transport logistics is the whole process of transport operations, from the raw material and up to when it is turned into a product;
- transport logistics is a subtlety of cargo movement;
- transport logistics is the door-to-door cargo transportation, as well as the processing of documents and ensuring optimal time, market and price;
- transport logistics is preparation, sorting, packaging, distribution, transportation, storage, etc.;
- transport logistics is the organization and management of all supply chains from raw materials to delivery to the place of consumption;
- transport logistics is the process of ensuring the flow of goods from the time of loading to the place of delivery;

* Corresponding author: smirnov.vladimir@vfrta.ru

• transport logistics is a process in which a good, quick and thoughtful assessment of the current situation and a solution is made that provides the best possible result;

• transport logistics is a process that effectively plans a logistics chain from the initial service or point of sale to the end user.

It is necessary to bring together the positions of scientists and practitioners on the issue of the content of transport logistics.

2 Problem Statement

Logistics processes are becoming more widespread in the Russian economy. Russian specialists and practitioners' understanding of logistics and its components coincides with the ideas of their foreign colleagues.

However, in practice, Russian entrepreneurs often make mistakes when solving emerging problems. It is necessary to develop algorithms for solving logistics problems in the Russian economy, including transport logistics.

3 Research Questions

Transport logistics is an integral part of logistics. Its functions are to select the optimal routes for the movement of economic goods in the interests of all participants in these processes. The problems of transport logistics are:

• to reveal the peculiarities of transport logistics;

• to determine the logistics characteristics of different types of transport;

• to understand the logistics organization of transportation;

• to get acquainted with the solutions to the problems of choice in transport logistics.

4 Purpose of the Study

The purpose of this work is to substantiate the methods of solving the main problems of transport logistics in Russian conditions.

To do this, it is necessary to study the interaction of participants in the Russian transport market in solving these tasks, which allows overcoming technical, technological, informational, economic, financial, international, environmental, and labor problems, ensuring that transaction costs are reduced by all participants in logistics chains, reducing the risks of doing business, and obtaining a synergistic effect.

5 Research Methods

To achieve the goal of transport logistics (delivery of goods of the required quality and quantity at a specified time in the appropriate place with optimal costs) necessary to provide a pair of flow processes was based on the integration of transport, production, supply, sale, consumption, service, financing and information.

Transport logistics is based on rational coordination of the economic interests of the sender, who forms material flows, the recipient and the carrier with the economic interests of other participants in the logistics process. A special feature of transport logistics is the cooperative use of transport by all participants in the process of promoting material flows. To do this, the study uses an integrative approach.

6 Findings

Transport is a set of vehicles, communication routes, as well as structures and devices on these routes that jointly ensure the movement of goods and people from one point to another.

The following definition of transport logistics best reflects the application of an integrative approach: "Transport logistics is a crucial part in the supply chain that in its essence organizes, manages, optimizes and ultimately performs physical distribution of goods and information through the whole upstream and downstream chain in an efficient and effective manner. Its activities and functions go beyond the traditional transport function in so that they also include integration with other supply chain functions such as warehousing, accounting, marketing or customer relations, while taking into account the organizational, financial, commercial and operational aspects of the supply chain as a whole" [13].

The main elements of transport logistics are presented in table 1.

Table 1. Basic elements of transport logistics

The main elements of transport logistics	Specifications
Goods	Products accepted for transportation, as well as various property of individuals or legal entities
A path	The medium through which a vehicle moves to perform its function
Traction means	It means that result in the movement of rolling stock or goods
Rolling stock	Has its own characteristics depending on the type of transport
Containers and packaging	Ensure the safety of cargo, reduce time and increase convenience when performing cargo processing operations
Participants in the logistics process	The sender, the material flow generator, the recipient, and the carrier

From the point of view of the integrative approach, it seems erroneous to include Boldyreva et al. only three elements – transport links with suppliers and consumers, shipped goods – in the composition of the key elements of transport logistics [2].

In the transport logistics of Russia, as in the logistics of other emerging markets, there are diverse problems.

Financial problems are the following: lack of investments and lack of own funds for reconstruction of roads, renewal of rolling stock, signaling system, communication, improvement of service quality.

Technical problems are the following: 1) non-loading of rolling stock, in particular insufficient use of load-carrying capacity or useful internal volume of a vehicle's body; 2) the quality of transport services; 3) operational incompatibility of transport networks, in particular their exhaustive bandwidth.

Technological problems are the following: 1) the use of vehicles, which have already several times worked out their technical resources and carry out significantly higher emissions of harmful substances; 2) low quality of internal roads, which increase fuel consumption due to poor quality roads; 3) out-of-date technologies that do not meet current standards; 4) the backwardness of the transport infrastructure; low technical and technological level and level of organization of transportation management.

Information problems are the following: 1) complexity in the construction of optimal routes of traffic (transportation); 2) insufficient information support of carriers; 3) lack of information about programs that allow automation of the transportation process.

Economic problems are the following: 1) a constant increase in the cost of all types of transportation, including oil prices, fuel, lubricants, light and dark petroleum products, energy resources; 2) different conditions for payment for the use of infrastructure by carriers.

International problems are the following: 1) not predictable risks associated with changing climatic conditions; 2) the lack of a single register of vehicles; 3) failures of submission of motor vehicles in agreed dates for loading.

Environmental problems are the following: 1) pollution of the environment; 2) absence of additional charges for environmental pollution and noise.

Labor problems are the following: 1) insufficient number of skilled logistics personnel, including logistic operators; 2) insufficiently developed cooperation between commodity producers, which complicates the processes of integration and cooperation; 3) reducing the competitiveness of domestic carriers.

The logistics approach in the economy, including in transport, allows us to successfully solve emerging problems. First, the application of this approach eliminates a number of market failures, such as the tendency to establish monopoly control over the market, the uneven distribution of information in the market environment. Outsider companies overcome these dips after being included in the supply chain. Second, market actors involved in supply chains significantly reduce individual and overall transaction costs. Third, these subjects find themselves in the legal field and they have no interest in leaving from it.

Specialists justify the directions of solving the problems of transport logistics.

The directions of solving financial problems are as follows: 1) development of programs for the purchase of fixed assets and rolling stock in leasing; 2) development of targeted financing and investment programs, modernization, reconstruction and updating of rolling stock.

The directions of solving technical problems are as follows: 1) an effective renovation of the car park, capable of problems providing sufficient transport services of adequate quality and satisfaction; 2) optimization of the transportation process itself by collecting consolidated cargo from several senders, compiling optimal routes of routing, connecting a more efficient mode of transport on a certain route.

The directions of solving technological problems are as follows: 1) search for new products and technologies; 2) integration into the pan-European transport system; 3) increasing the level of transport security; 4) implementation of SAP software; 5) change of approaches to management of all economic processes; 6) to create a network of automated traffic management, with the reduction of reloading operations.

The directions of solving information problems are as follows: 1) the application of the latest computer technologies problems in transport and logistics for the processing of large amounts of information, the exchange of data in real time with minimal cost.

The directions of solving economic problems are as follows: 1) the return to the methodology for calculating the problems cost of transportation and taking into account not the market price per 1 km, but the individual profitability of each transportation; 2) comprehensive accounting of logistics costs; 3) refusal of cross-financing of passenger transportation.

The directions of solving international problems are as follows: 1) adaptation of the national legal framework to problems international conventions, agreements; 2) cooperation with foreign countries in the creation of transnational axes.

The directions of solving environmental problems are as follows: 1) reduction of atmospheric air and soil pollution; 2) reducing traffic noise.

The directions of solving labor problems are as follows: 1) improvement of transport legislation; 2) «erosion» of the borders with foreign countries, which will ensure the attractiveness of the transportation sector; 3) increase of transport operators, increase internal competition and promote tariff crash; 4) increase of wages of drivers.

The problem of choosing a carrier in logistics is solved by various methods (with concept of the primary connection between the choice of the carrier and the goods as the subject of transportation):

- the method of analyzing the costs and results of cargo delivery [6];
- the matrix method (on the criterion of minimum cost);
- the cost estimation method;
- the abstract carrier method;
- the method of accounting for technological parameters.

Many of these parameters relate more to the product than to the company's performance.

To ensure the technological unity of the transport and storage process and optimize the parameters of the transport process (increase the speed of transportation, reduce costs), it is necessary to significantly accelerate the introduction of information technologies and systems. However, experts note the insufficient implementation of information computer technologies and systems in logistics processes in the Russian economy [12].

7 Conclusion

The main tasks of transport logistics should be solved in a complex, with the mandatory application of an integrative approach. The logistics approach in the economy, including in transport, allows us to successfully solve emerging problems. The application of this approach eliminates a number of market failures, such as the tendency to establish monopoly control over the market, the uneven distribution of information in the market environment. Market actors involved in supply chains significantly reduce individual and overall transaction costs. These subjects find themselves in the legal field and they have no interest in leaving it. This will help overcome technical, technological, informational, economic, financial, international, environmental, and labor problems. In this case, individual and overall costs are reduced by all participants in logistics chains, the risks of doing business are reduced, and a synergistic effect is achieved. It is necessary to ensure that the understanding of transport logistics as an essential part of the supply chain, which inherently organizes, manages, optimizes and ultimately performs the physical distribution of goods and information throughout the upstream and downstream chain in an efficient and effective manner, is disseminated not only in the scientific field, but also among the managers of companies involved in logistics processes.

References

1. W.B. Andrews, R.A. Burnett, R.L. Engel. Transportation logistics for spent-fuel storage and disposal. Transact. of the Amer. Nuclear Society, **30**, 292–293 (1978)
2. L. Boldyreva, H. Zelinskaya, V. Krapkina, A. Komelina. Problems and solutions of transport logistics. Advances in Economics, Business and Management Research. 7th Int. Conf. on Model., Developm. and Strategic Manag. of the Econ. Syst. (MDSMES 2019), vol. 99 (2019), pp. 317–320.
3. K. Button. *Transport Economics* (Edward Edgar Publ., 2010)
4. L. Chatterjee, C.-M. Tsai. *Transportation Logistics in Global Value and Supply Chains* (Center for Transp. Stud. Boston Univ. 2002), p. 17. Retrieved from: http://www.bu.edu/transportation/WPSeries.html
5. I. Ehnert, L. Arndt, G. Mueller-Christ. A sustainable management framework for dilemma and boundaries in autonomous cooperating transport logistics processes. Int. of Environ. and Sustain. Developm., **5(4)**, 355–371 (2006)
6. J.H. Foggin. Improving motor carrier productivity with statistical process control technique. Transport. J., **24(1)**, 58–74 (1984)
7. K.-H. Lai, E.W.T. Ngai, T.C.E. Cheng. Measures for evaluating supply chain performance in transport logistics, Transportation Research. Part E: Logist. and Transport. Rev., **38(6)**, 439–456 (2002)
8. K.-H. Lai, E.W.T. Ngai, T.C.E. Cheng. An empirical study of supply chain performance in transport logistics. Int. J. of Product. Econ., **87(3)**, 321–331 (2004)
9. Y.H.V. Lun, C.-S. Lu, K.-H. Lai. Introduction to the special issue on transport logistics and physical distribution. Int. J. of Product. Econ., **122(1)**, 1–3 (2009)
10. R.A. Mundy, C.J. Langley, T.E. Gibson. Industry evaluation of a transportation/logistics curriculum. Transport. J., **17(1)**, 33–39 (1977)
11. A. Rushton, P. Croucher, P. Baker. *The Handbook of Logistics and Distribution Management: Understanding the Supply Chain* (Kogan Page Publ., 2014)
12. E.Y. Timofeeva, M.M. Pimonenko. International research projects in transport logistics. Transp. of the Rus. Fed., **1(20)**, 50–62 (2009)
13. D. Topolšek, K. Čižiūnienė, T.C. Ojsteršek. Defining transport logistics: a literature review and practitioner opinion based approach. Transp., **33(5)**, 1196–1203 (2018)
14. Y.-Y. Tseng, W.L. Yue, M.A.P. Taylor. The role of transportation in logistics chain. Proc. of the Eastern Asia Society for Transport. Stud. **5**, 1657–1672 (2005)
15. C.W.Y. Wong, K.-H. Lai, E.W.T. Ngai. The role of supplier operational adaptation on the performance of IT enabled transport logistics under environmental uncertainty. Int. J. of Product. Econ., **122(1)**, 47–55 (2009)

State support of JSC "Russian railways"

Olga I. Krushinskaia[1,*], *Natalia V.* Manchenko[1], *Daria A.* Dinets[1], and *Dmitriy S.* Artemev[2]

[1] Irkutsk State Transport University, 15, Chernyshevskogo str., Irkutsk, Russia
[2] Irkutsk National Research Technical University, 83, Lermontov str., Irkutsk, Russia

Abstract. The main sources of funds for railway investment projects in Russia are still state and regional budgets, funds from the National Welfare Fund, the net profit of Russian Railways, and pension savings from the Pension Fund of Russia. The reform of the system of the Ministry of Railways and the creation of JSC "Russian Railways", which began in 2016, has not yet released the state budget from financing its needs, and there has been no active use of private investment. Moreover, the monopoly began to seek to shift the financing of not only the modernization and construction of tracks, but even the maintenance of existing infra-structure on the shoulders of the state and shippers. From year to year, there is an increase in the total amount of state support for Russian Railways. Part of the received state support funds in the period under review is spent inefficiently by JSC "Russian Railways". The remaining budget funds that are not used by JSC "Russian Railways" additionally burden the economy, increasing inflation by various surcharges to tariffs, but the available budget funds are not selected in full, and billions of fines are paid for this. There are low rates of implementation of individual investment projects, overestimation of expenses for the purchase of equipment, an increase in the cost of construction, the cost of contracts for a number of objects exceeds the cost determined by state expertise, etc.

Keywords: state support, federal budget, regional budgets and off-budget funds

1 Introduction

The reform of the system of the Ministry of Railways and the creation of JSC "Russian Railways" were almost primarily justified by the prospects of attracting investment resources. It was assumed that the Corporation, unlike the Ministry, would actively use private investment and free the state budget from financing its needs. In December 2016, the government of the Russian Federation for this purpose increased the authorized capital of Russian Railways by 24.98 billion rubles; as recently as in April 2017 – by another 29.78 billion rubles.

However, it turned out somewhat differently: having taken a course to become an infrastructure company, the monopoly began to seek to shift the financing of not only the modernization and construction of tracks, but even the maintenance of existing infrastructure to the shoulders of the state and shippers.

In modern conditions, one of the factors hindering the country's economic development is the insufficient level of development of infrastructure sectors, including transport infrastructure.

Transport as a system-forming branch of the Russian economy cannot develop without modern infrastructure. It, in turn, is the most important condition for the effective use of the resource potential of both a particular region and the country as a whole. Modern infrastructure and effective management of the infrastructure complex contribute not only to increasing the productivity of the transport process and increasing the availability of transport services in accordance with social standards, but also to the growth of gross domestic product and the creation of new jobs, preserving the common economic space and transport mobility of the population. In a global economy, building and improving supply chains improves the competitiveness of the transport system on the world market and promotes the growth of exports of transport services.

The development of transport infrastructure is determined by the influence of a number of factors that are usually considered as infrastructure-forming [5]. At the same time, the creation and smooth operation of infrastructure facilities requires significant capital investments, taking into account the significant capital intensity and inertia of the industry, while most infrastructure projects are often unique and practically unprofitable.

According to the spatial development strategy, the implementation of infrastructure projects is de-signed to stimulate economic growth, primarily by removing structural restrictions, which will create a transport framework for economic development [9].

2 Problem Statement

According to various estimates, the additional need for infrastructure investment in the Russian Federation is at

* Corresponding author: tulasi1970@gmail.com

least 3.5 billion rubles per year, and for accelerated economic development – more than 7 trillion rubles per year. However, such funds are unlikely to be attracted in the coming years [3].

Today, the share of budget financing in the field of infrastructure construction in our country is more than 50 percent (Federal and Regional budgets). The share of private investment in the infrastructure project market remains low (from 3 to 5 percent). For comparison, it is from 30 to 50 percent abroad [2].

When determining the priorities of the investment strategy of JSC "Russian Railways", one of the goals is to implement a set of measures aimed at increasing the capacity of the infrastructure.

The new version of the Federal target program "development of the transport system" provides for financing activities from 2018 to 2021 in the amount of 7.75 trillion rubles [4].

The fact that the state has to invest significant funds in the Corporation to implement its investment projects for infrastructure development, we suggest analyzing the volumes and directions of state support for the period 2017–2019 (table 1, figures 1, 2, 3).

Table 1. State support from the Federal budget, regional budgets and extra-budgetary funds for the period of 2017–2019, billion rubles (compiled by the authors based on Social responsibility of Russian Railways, 2017, 2018, 2019).

Directions of state support	Period, year		
	2017	2018	2019
	Amount, RUB bln		
Direct support	37.8	43.2	42.8
Subsidies to compensate for losses in revenue resulting from state regulation of tariffs for services for the use of public railway transport infrastructure provided during the transportation of passengers in suburban traffic	35.4	36.8	37.1
Subsidies to finance part of the costs associated with the registration of intellectual property objects on foreign markets	0.001	0.004	-
Subsidy for reimbursement of expenses related to the maintenance of infrastructure for transportation on the Small ring of the Moscow railway	1.3	5.9	4.9[1]
Subsidies for the purpose of compensation of the lost income arising as a result of state regulation of tariffs for transportation of passengers by rail in suburban traffic in the territory of Krasnodar territory	0.7	0.07	0.07
Funding received from the social insurance Fund for planned preventive measures for injuries	0.3	0.3	0.4
Subsidies provided in connection with the regulation of electricity and heat tariffs	0.1	0.1	0.2
Subsidies for reimbursement of expenses for work on the liberation of territories that fall within the zone of reconstruction and construction of objects of the Small ring of the Moscow railway	0.05	0.02	-
Subsidies received for the maintenance of low-intensity sections of the railway	0.02	0.02	0.02
Cost recovery for reconstruction of public railway transport infrastructure	-	0.04	0.04
Subsidy for reimbursement of costs for carrying out design and survey work to assess the economic efficiency of transport infrastructure development in the Samara region for the organization of high-speed rail services	-	-	0.03
State support for shippers (consignees) through the company's budget	-	1.7	0.004
Subsidies for compensation of losses in income resulting from the establishment of preferential tariffs for grain transportation	-	1.7	0.004
Implementation of state projects through the budget of JSC "Russian Railways"	60.6	57.9	59.2[2]
Development of the Moscow transport hub (development of the MTU railway infrastructure):	26.8	23.1	47.3
• including from the Federal budget	17.4	12.1	7.02
• including from the Moscow budget (in 2019, including from the Federal budget through an inter-budget transfer from the Moscow budget)	9.4	11.0	40.2
Construction of the railway line Prokhorovka-zhuravka-Chertkovo-Bataysk	16.2	-	-
Speed and high-speed traffic	0.09	0.09	-
Comprehensive reconstruction of the Gorky-Kotelnikovo-Tikhoretskaya-Krymskaya Metro section with a bypass of the Krasnodar junction	17.6	31.9	-
Development of railway infrastructure on the Mezhdurechensk-Taishet section (Construction of the railway section and strengthening of power supply devices)	-	2.8	11.98
Compensation for losses related to the need to reconstruct the existing intersection of the Perm-Yekaterinburg highway with railway tracks on the 1446th kilometer of the Baharevka-Ferma stage	0.01	-	-
Compensation for losses related to the construction of pedestrian crossings over railway tracks in the Saransk city district	0.01	-	-
Implementation of state projects at the expense of the National Welfare Fund	-	20.0	19.5
Financing of the program of modernization of the railway infrastructure of the Baikal-Amur and Trans-Siberian railway lines with the development of throughput and transportation capacities from the NWF (Development of throughput and carrying capacity)	-	20.0	19.5
Transfer of real estate from state ownership to the authorized capital of JSC "Russian Railways"	-	0.8	0.3
Total government support	98.4	123.6	121.8
• including the Federal budget	96.0	97.2	96.64
• National Welfare Fund (NWF)	-	20	19.5
• regional budgets and extra-budgetary funds	2.4	6.4	5.66

[1] Refund of the unused balance of the subsidy (grant) in the amount of 0.3 billion rubles was made in 2020

[2] the Agreement is a long-term one with a deadline for commissioning facilities in 2022, and the remaining budget investments in accordance with the agreement are planned for 2020–2021.

https://doi.org/10.1051/shsconf/202111200045

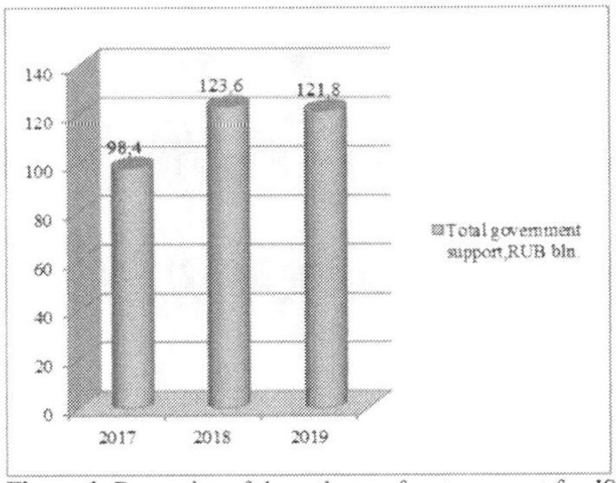

Figure 1. Dynamics of the volume of state support for JSC "Russian Railways" for the period of 2017–2019, billion rubles

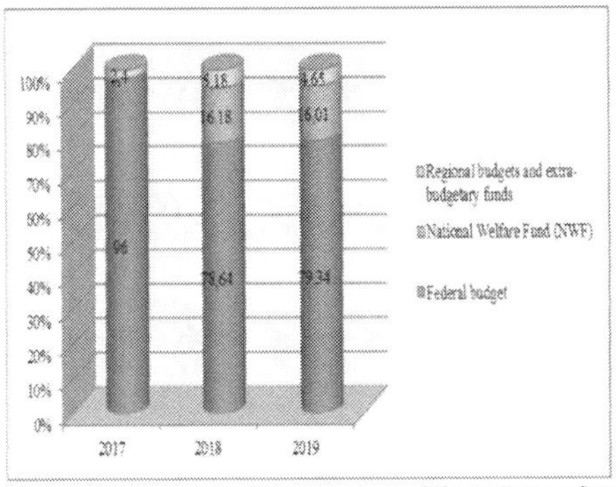

Figure 2. Dynamics of the structure of the volume of state support of JSC "Russian Railways" for the period of 2017–2019, %

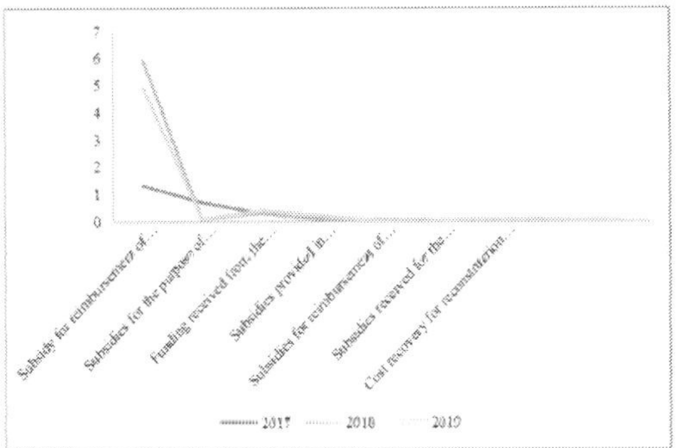

Figure 3. Dynamics of the volume of state support to JSC "Russian Railways" from regional budgets and extra-budgetary funds for the period of 2017–2019, billion rubles.

3 Research Questions

Data from table 1, figures 1 and 2 show that over the three-year period under review, the total amount of state support for Russian Railways increased by 23.4 billion rubles, or 23.8 %. From year to year, there are changes in the structure of state support. Since 2018, it has been implemented using funds from the National Welfare Fund (NWF). In 2019, compared to 2017, state support from regional budgets and extra-budgetary funds almost doubled, and state support from the Federal budget decreased by 2.21 %, respectively.

The government of the Russian Federation, on a long-term basis until 2030, is implementing state regulation of tariffs for the services of JSC "Russian Railways" for the use of public railway transport infrastructure in suburban passenger transport. Compensation for losses in the income of the infrastructure owner is carried out in the form of subsidies from the Federal budget. From year to year, there is an increase in the number of subsidies. Thus, in 2019, compared to 2017, its volume increased by 1.7 billion rubles, or 4.8 %.

In 2019, the implementation of state projects for the development of railway transport infrastructure continued, the financing mechanism of which was carried out through the budget of JSC "Russian Railways" through a contribution to the authorized capital. In addition, in 2018 and 2019, the National Welfare Fund of Russian Railways received 20 and 19.5 billion rubles, respectively, as a contribution to the authorized capital due to the issue of preferred shares, to finance the program of modernization of the railway infrastructure of the Baikal-Amur and Trans-Siberian railway lines with the development of throughput and transportation capacities. Real estate was transferred from state ownership in the form of a contribution to the company's authorized capital in the amount of 0.8 and 0.3 billion rubles, respectively, over the years.

4 Purpose of the Study

Direct financing of projects puts a significant burden on the state budget and does not provide an acceptable ratio of cost and quality of project execution [10].

The growth of state support for Russian Railways from regional budgets and extra-budgetary funds creates an additional burden on these budgets.

5 Research Methods

The accounts chamber of the Russian Federation, in accordance with the legislation, considers the results of the audit of budget execution and budget reporting on the execution of the Federal budget by the chief managers of budget funds. The main results are published on the website of the control body as the Board approves the relevant conclusions.

The Board of the accounting chamber approves the conclusion on the execution of the Federal budget in the Federal Agency for railway transport.

The objects of monitoring and control were the Federal Agency for railway transport (Roszheldor), JSC "Russian Railways".

6 Findings

In 2017, to continue the implementation of investment projects, JSC "Russian Railways" was provided with budget investments in the form of a contribution to the authorized capital in the total amount of 60.6 billion rubles. Projects are implemented within the framework of the investment program approved by the company's Board of Directors [1].

The analysis of the accounts chamber showed that the processes of planning budget allocations for making contributions to the authorized capital of Russian Railways and forming an investment program are not interrelated. The amount of contributions to the authorized capital in 2017 and the resulting balances at the beginning of the year exceeded the amount of Federal budget funding provided for in the investment program by 46.6 billion rubles.

When planning budget allocations for making contributions to the authorized capital, Roszheldor does not conduct a proper analysis of the possibility of using budget funds, as well as the pace of their actual development. This led to the formation of significant balances at the year end. Thus, when making changes to the Budget for 2017, Roszheldor's proposals for additional financing of the project "Complex reconstruction of the Gorky-Kotelnikovo-Tikhoretskaya-Krymskaya section with a bypass of the Krasnodar railway junction" for 7 billion rubles were taken into account. As a justification for the need for additional project financing in 2017 Roszheldor indicated the need to maintain the pace of work in order to complete its implementation on time. At the same time, at the end of 2017, the undeveloped balance of Federal budget funds allocated for the project implementation amounted to 9.8 billion rubles.

Roszheldor does not take into account the parameters provided for in the investment project passports when setting the terms of commissioning of a number of objects in contracts with Russian Railways. Thus, in accordance with the agreement of April 27, 2017, the deadline for commissioning individual objects within the framework of the project for the development of the Moscow transport hub is set for 2025. At the same time, according to the passport, the project implementation period and reaching its design capacity is 2020.

A low level of planning Roszheldora of volumes of the budgetary appropriations directed on financing of investment program of JSC "Russian Railways" annually, entails the failure of JSC "RZD" of the obligations accepted under contracts of transfer of shares in the expense of budget investments and, as a consequence, the non-implementation rates before the end of 2017, provided public investment.

According to the results of 2017, budget allocations in the amount of 3.7 billion rubles provided under the share transfer agreement dated April 27, 2017 were not disbursed. As a result, under the terms of the share transfer agreement dated April 27, 2017, the estimated amount of the fine is 5 % of the amount of the transferred contributions to the authorized capital of JSC "Russian Railways" (29.8 billion rubles) or 1.5 billion rubles.

One of the reasons for the formation of budget balances is the slow pace of implementation of individual investment projects. Thus, the project "Integrated development of the Mezhdurechensk-Taishet section" should be completed in 2019 to develop a promising cargo flow towards the ports of the Far East. The project has been funded from the Federal budget since 2013. Due to non-fulfillment of project implementation obligations, at the beginning of 2017, the remaining unused budget funds in the amount of 20.5 billion rubles were formed. These funds were transferred back in 2013, 2014 and 2015. In 2017 JSC "Russian Railways" accepted works on this investment project only for 3.3 billion rubles.

In accordance with the decisions taken by the Government, Russian Railways has been granted the right to use in 2017 the funds of unused contribution balances in the amount of 15 billion rubles for the implementation of the project for the construction of the Prokhorovka-Zhuravka-Chertkovo-Bataysk railway line and other projects. According to the explanation of Russian Railways, non-use of budget funds is due to non-fulfillment of its obligations by the contractor [1].

In general, Federal budget funds in the amount of 80.5 billion rubles were used to finance investment projects of JSC "Russian Railways" in 2017, which is 69 % of the total amount of contributions to the authorized capital and the balances of previous years at the beginning of the year. The amount of completed works is 68.9 billion rubles.

The analysis of the accounts chamber showed that the expenses of JSC "Russian Railways" for the purchase of equipment for a number of construction and reconstruction projects as part of the investment program projects in 2017, financed from the Federal budget, exceeded the estimated cost of equipment approved by the state expertise (taking into account the corresponding deflator set by the forecast of socio-economic development of the Russian Federation).

In particular, JSC "Russian Railways" signed contracts for the supply of equipment in 2017 for the object Complex reconstruction of the Kotelnikovo-Tikhoretskaya-Korenovsk-Timashevskaya-Krymskaya section with a bypass of the Krasnodar junction of the North Caucasus railway. Construction of the second track on the Forgotten-Polivyansky section for a total amount of 233.6 million rubles, which is 27 million rubles higher than the cost of equipment approved by the positive conclusion of the state expert examination of October 10, 2016, taking into account the deflator for 2017 (206.6 million rubles).

Based on the results of the audit of the use of NWF funds for the modernization of the railway infrastructure of the BAM and Trans-Siberian railway, in 2017 the parameters of the passport of the specified infrastructure project were adjusted. The start of the operation phase has been postponed from 2018 to 2020.

In the framework of adjusting the parameters of the passport of the project of JSC "Russian Railways" was a proposal of the accounting chamber about the direction of the project's financing income in the amount of 7.9 billion rubles received as interest on balances of funds of the NWF. In the new version of the project passport, Federal budget funding has been reduced by the specified amount.

In July 2017 JSC "Russian Railways" made additional advances to contractors at the expense of the National Welfare Fund in the amount of 3.7 billion rubles, which significantly exceeded their needs for 2017. As a result, the unused balance of NWF funds from contractors amounted to 3.2 billion rubles (86 % of the amount of additional advance payments). At the same time, in accordance with the procedure for settlements under contracts with contractors approved by the company's order, advance payments under contracts for construction and installation work are set off in full annually no later than December 31.

When calculating the initial (maximum) prices of construction and installation contracts in 2017, Russian Railways used deflators that did not correspond to the forecast of socio-economic development of the Russian Federation. This led to an increase in the cost of construction in the amount of 45.7 million rubles and, accordingly, expenses at the expense of the National Welfare Fund. In particular, according to the results of the state examination, the estimated cost of work on the object "Reconstruction of the locomotive maintenance point at the Ussuriysk station" in the prices of the 1st quarter of 2016 amounted to 405 million rubles. As a result of applying the deflator index that does not correspond to the forecast when calculating the initial (maximum) price of the contract, the price of the contract for work on the specified object amounted to 521.1 million rubles or exceeded the initial (maximum) price calculated taking into account the deflator set by the forecast by 17 million rubles.

Vadim Mikhailov, first Deputy General Director of Russian Railways, noted that in some cases, Russian Railways, without waiting for contributions from the Federal budget, finances investment projects at its own expense, so as not to interrupt the construction process, with subsequent reimbursement from the Federal budget. This allows for a number of investment projects to receive advanced execution.

The Board took the following action: on the results of monitoring of the investment program of JSC "Russian Railways" to send newsletters Roszheldora and JSC "Russian Railways"; the results of inspection of usage of funds of the NWF on the modernization of BAM and Transsib to send representation to the JSC "Russian Railways". They send reports on the results of events to the chambers of the Federal Assembly.

Long-term development program of JSC "Russian Railways" until 2025 envisages the development of public-private partnership in railway transport (including concessions) for the construction of new lines, involving public resources, investment infrastructure owner interested cargo owners and other "working arrangements" and an additional four trillion private investment [6].

Public-private partnership is a promising and long-term form of cooperation between the state and private business. It allows attracting technologies and management experience accumulated by businesses, stimulating investment activity of private capital, as well as improving the efficiency of public investment, including in infrastructure projects on railway transport. The variety of PPP forms and models used allows mutually beneficial distribution of risks between project participants. The use of public-private partnership mechanisms helps to increase the competitiveness of infrastructure projects in the investment resources market. Therefore, new legislative initiatives are needed to encourage private investment in infrastructure construction, including in railway transport.

Given the importance and role of transport infrastructure, discussions continue today about which mechanisms to attract private investment are most effective. The creation and modernization of transport infrastructure is, on the one hand, the task and responsibility of the state, since it has all the characteristics of a public good. On the other hand, its development is impossible without the participation of business, especially since there is a gap between the actual and necessary amounts of budget financing for infrastructure development.

Many European countries faced the "infrastructure gap" in the 1990s. In the context of budget constraints, this required more active involvement of private capital in the infrastructure sector, which was done and significantly accelerated the creation of new and modernization of existing infrastructure facilities.

Public-private partnership (PPP) can act as a mechanism for supporting infrastructure investments by the state, being a special scheme for implementing an investment project on mutually beneficial terms for both business and the state, it allows attracting "long money" from private investors to solve certain tasks that have, among other things, a social effect. In world practice, the most flexible and effective form of attracting infrastructure investment is a concession. The first concessions in Russia appeared in 2006, and today they account for almost 80 % of all projects that have passed the stage of commercial closure. Under concession agreements, infrastructure bonds are issued, which, in comparison with corporate bonds, have a number of advantages, including state guarantees and risk insurance. Concession and public-private partnership are the main forms of implementation of infrastructure projects. Their main difference is in the approach to the transfer of ownership of the object and the possibility of pledge of such rights.

This format can provide a multiplier effect in the economy as a whole and more actively accumulate investment resources for infrastructure construction.

7 Conclusion

1. JSC "Russian Railways", implementing its investment program, inefficiently spends the allocated funds of the Federal budget.

2. Undeveloped budget balances of JSC "Russian Railways" additionally burden the economy, increasing inflation by various surcharges to tariffs, but the available budget funds are not selected in full.

3. Having Achieved the next state support, the monopoly, it turns out, is not able to manage this money in a businesslike way and is obliged to pay billions in fines. Thus, at the end of 2017, 3.7 billion rubles were not spent. As a result, under the terms of the share

transfer agreement dated April 27, 2017, the estimated amount of the fine is 5 %, or 1.5 billion rubles.

4. The processes of planning budget allocations for making contributions to the authorized capital of JSC "Russian Railways" and forming the investment program are not interrelated. When planning budget allocations for making contributions to the authorized capital, Roszheldor does not analyze the possibilities of their use and the pace of their actual development. As a result, the remaining funds are accumulated. As a result, the amount of contributions to the authorized capital in 2017 and the generated balances at the beginning of the year exceeded the amount of funding from the Federal budget provided for in the investment program by 46.6 billion rubles. However, this approach applies not only to contributions to the authorized capital. When agreeing on the terms of implementation of Russian Railways projects, the regulator does not check the passports of these projects, increasing them, as happened with a number of objects of the investment program of the Moscow transport hub. As a result, their implementation period increased by five years.

5. One of the reasons for underutilization of budget funds is the low pace of implementation of individual investment projects. For example, the project "Integrated development of the Mezhdurechensk-Taishet section", which is necessary for developing a promising cargo flow towards the ports of the Far East, should be completed in 2019. The state budget has been funding it since 2013. At the same time, the plans were not implemented in the proper amount, and by the beginning of 2017, the balance of unused budget funds reached 20.5 billion rubles.

6. Overstatement of expenses for the purchase of equipment financed from the Federal budget, and an increase in the cost of construction and, accordingly, expenses at the expense of the National Welfare Fund were identified.

7. In general, Federal budget funds in the amount of 80.5 billion rubles were used to finance the monopoly's investment projects in 2017, which is 69 % of the total amount of contributions to the authorized capital and the balances of previous years by the beginning of last year. At the same time, the monopoly accepted completed works for only 68.9 billion rubles. In other words, 31 % of the funds were not used, and the budget spent 11.6 billion rubles more than the amount of work accepted.

8. The Monopoly did not give a clear confirmation of what and how the 20–26 billion rubles received from this additional 2 % of the tariff were spent.

9. JSC "Russian Railways", implementing the project "Development of public railway infrastructure on the Mezhdurechensk-Taishet section in 2016–2017 and the expired period of 2018", allowed inefficient use of budget funds and organized work at a low level. The project is designed to increase the capacity of the highway – this is necessary for the economic development of Khakassia, Kuzbass, the South of the Krasnoyarsk territory and the development of the Tyva coal deposits. It has been implemented since 2010, but the state-owned company cannot complete it in any way. According to the accounts chamber, it has received interest on funds allocated from the budget, finances work carried out without permission and at an inflated cost and makes unjustified advances to contractors. The total cost of the project is 45.6 billion rubles, of which 35.7 billion rubles are from the Federal budget, and 9.9 billion rubles are from Russian Railways. The state represented by the Federal Agency for railway transport even in 2015, fully met all their financial obligations. However, due to the low level of work organization, including the lack of project documentation that has passed state expertise in accordance with the established procedure, the goals of the project Integrated development of the Mezhdurechensk-Taishet section may not be achieved within the established time frame [7]. Construction deadlines are not being met for 13 of the 22 projects, they were extended from 2016 to 2020. In December 2018, the technical readiness of 8 objects was less than 50 %, and work on three objects did not even begin. The audit also revealed the risks of inefficient use of budget investments. Thus, the cost of contracts for a number of objects exceeds the cost determined by the state expertise. In addition, when forming the initial contract price, the company used deflator indices that do not correspond to the indicators of the socio-economic development forecast. As a result, the cost of contracts for the construction of a number of objects is overstated by 89.8 million rubles. This, according to the accounts chamber, entails risks of inefficient use of Federal budget funds for the specified amount.

Meanwhile, Russian Railways received income from placing budget investments in the accounts of credit organizations. So, on May 30, 2014, the state-owned company placed a Deposit in one of the banks in the amount of 12.5 billion rubles. Interest income – 2.4 billion rubles JSC "Russian Railways" spent on its business activities.

The passport of the infrastructure project was repeatedly re-approved: the values of quality indicators were reduced, one investment object worth 110 million rubles was excluded, but the total cost of the project was not reduced.

Moreover, the company financed construction on a number of sites that was carried out without permission. The audit showed that the financing of construction and reconstruction works in the period under review in the absence of design and estimate documents that have passed state expertise in accordance with the established procedure, entails risks of additional Federal budget expenditures. The cost of contracts for the construction of a number of objects exceeded the estimate made by the state expertise, for a total of 478.2 million rubles. A similar excess of hundreds of millions of rubles was allowed for the supply of equipment. In addition, JSC "Russian Railways" allowed the expenditure of funds for the author's supervision of a number of objects, while these costs were excluded from the estimates during the state examination, which again led to inefficient use of Federal budget funds.

Unjustified advance payments to contractors under contract agreements, which were practically not carried out in 2018, resulted in an increase in accounts receivable by 1.1 billion rubles, or 2.3 times. As of the date of completion of the audit, the volume of accounts receivable amounted to 1.9 billion rubles, 84 % of which are budget investments (1.6 billion rubles).

11. The Volume of unfinished investments of JSC "Russian Railways" in 2017 increased by 13 billion rubles and amounted to 43.7 billion rubles. The amount of income of JSC "RZD" in the form of percent from placing of means FNB on the account in VTB Bank for 2015–2017 amounted to 7.9 billion rubles.

As of January 1, 2018, total expenditures (since 2013) for the implementation of the Infrastructure project "Modernization of the BAM and Trans-Siberian railway infrastructure" amounted to 140.6 billion rubles, or 25 % of the total cost. In total, in 2017, construction of 16 planned facilities worth 4.9 billion rubles was completed. At the same time, permits for commissioning of the specified objects are not issued, and the objects have not been put into operation. In addition, work has not been completed on 6 objects, the construction of which should have been completed in 2017. The construction readiness of these objects ranges from 60 to 90 % [8].

Thus, all of the above indicates that JSC "Russian Railways" in terms of using state support funds needs increased control, it is necessary to carefully monitor the determination of the cost of work and equipment and the effectiveness of the organization of work, as well as to stop the practice of making capital investments without positive conclusions of the state expertise.

12. Infrastructure investment is considered one of the most effective tools for stimulating economic growth. Of course, the real dynamics of infrastructure investment will be determined primarily by the activity of the state. There is no doubt that the participation of the state is extremely important here, because it acts primarily as a subject of financial and legal guarantees, which will help maintain the necessary level of confidence in long-term investments, which are investments in the creation and modernization of infrastructure.

References

1. V.N. Bogomolov *Roszheldor does not conduct a proper analysis of the possibilities of using the budget funds of Russian Railways* (2018). Retrieved from: https://ach.gov.ru/news/roszheldor-ne-provodit-dolzhnyj-analiz-vozmozhnostej-ispolzovaniya-byudzhetnyh-sredstv-rzhd-33001
2. I.V. Bulynin. *International experience in the field of public-private partnership: analysis, assessment, use in Russian conditions* (2010). Retrieved from: https://cyberleninka.ru/article/n/mezhdunarodnyy-opyt-v-sfere-gosudarstvenno-chastnogo-partnerstva-analiz-otsenka-ispolzovanie-v-rossiyskih-usloviyah
3. Expert release Federal Law On PPP/MPP: Figures, Facts And Expert Assessment (2020). Retrieved from: https://investinfra.ru/novosti/nakdi-publikuet-ekspertnyj-reliz-federalnyj-zakon-o-gchp/mchp:-czifry-fakty-i-ekspertnye-oczenki.html
4. Federal target program "Development of the transport system of Russia (2010–2021)". Retrieved from https://fcp.economy.gov.ru/cgi-bin/cis/fcp.cgi/Fcp/ViewFcp/View/2010/264/
5. A.M. Kudryavtsev, L.N. Rudneva. *Transport infrastructure of the region: concept and factors of formation* (2013). Retrieved from https://cyberleninka.ru/article/n/transportnaya-infrastruktura-regiona-ponyatie-i-faktory-formirovaniya
6. Long-term development program of Russian Railways until 2025 (2019). Retrieved from: http://government.ru/docs/36094/
7. Report on the results of Audit of the use of the federal budget, aimed at the realization of investment projects with state participation "Development of railway infrastructure public on site Mezhdurechensk-Taishet" in 2016–2017 and the past period of 2018. Accounts Chamber Bull., **3** (2019). Retrieved from: https://ach.gov.ru/upload/iblock/66e/66e8599f958ff7a365b7561a8b01a6b8.pdf
8. Verification of the use of funds of the National Wealth Fund aimed at financing the infrastructure project "Modernization of the railway infrastructure of the Baikal-Amur and Trans-Siberian railways with the development of carrying and carrying capacity" in accordance with the order of the Government of the Russian Federation, dated 5 November 2013, no. 2044-r, for the period of the project. Report on the results of the control event. Accounts Chamber Bull. **6** (2018). Retrieved from: https://ach.gov.ru/upload/iblock/ffe/ffebcce7e0b4b6a9f9af5799a980898b.pdf
9. Social responsibility of Russian Railways. Retrieved from: https://company.rzd.ru/ru/9386
10. The planning of contributions to the authorized capital of Russian Railways was carried out by Roszheldor without taking into account the needs of the joint-stock company (2017). Retrieved from: https://ach.gov.ru/news/planirovanie-vznosov-v-ustavnyj-kapital-oao-rzhd-osushestvlyalos-roszheldorom-bez-ucheta-potrebnoste-30423

Digital transformation of education in the Arctic: case of North-Eastern Federal University

Petr Ivanov[1*], *Vasiliy* Savvinov[1], and *Raisa* Ivanova[2]

[1]North-Eastern Federal University (NEFU), 58 Belinsky str. Yakutsk, 677000, Republic of Sakha (Yakutia), Russia
[2]Mirny Polytechnic Institute (branch) of NEFU, 5/1 Tikhonova str. Mirny, 678175, Republic of Sakha (Yakutia), Russia

Abstract. The article presents the case of developing a program for the digital transformation of higher education in the northern territories of the Russian Federation through the example of the North-Eastern Federal University. It describes the principles and mechanisms of digitalization of basic processes in conditions of limited resources and remoteness from the leading scientific, educational and cultural centres. The authors suggest the development of a digital educational ecosystem to provide high quality and accessible education in remote areas.

1 Introduction

The process of digitalization of basic processes in universities is becoming one of the priority areas for the development of higher education all over the world. In Russia, the implementation of national projects "Digital Economy", "Education", "Science and University" and Strategies for scientific and technological development involves the transition to digital technologies, testing of digital university models and introduction of data-based management mechanisms.

At the same time, the studies of digitalization processes in leading Russian universities show that the gap between the digital future and the existing digital competencies of the education system is a significant problem [1]. Modern researchers indicate the following major challenges of education digitalization: lack of experience in setting priorities, insufficient personnel with the necessary competencies, limited resources, and decision-making lagging behind the requirements of modern time [2].

The pandemic of the new coronavirus infection revealed the growing digital divide in society, threat to the quality of education due to inequality in access to digital resources [3], [4]. The analysis of management decisions of universities during the pandemic showed "excessive rigidity and inflexibility of the educational process of universities [5]. An expert assessment revealed the need of a significant transformation of higher education management [1], [5], "necessary renewal and sufficient improvement to ensure stability" [6]. One of such solutions in many universities is the development of programs for digital transformation of universities.

Nowadays the automation of basic processes using end-to-end technologies, the development of online courses, implementation of blended forms of education and individual educational tracks, introduction of data-based management practices, and information security are the priority tasks of the North-Eastern Federal University (NEFU).

2 Results and Discussion

Currently, a program for digital transformation of the university is being developed, taking into account the potential and development zones. Digitalization of the main processes is a key principle for the development of NEFU for the medium term, along with such principles as social responsibility, priority of human interests, integration of education, science, applied research and production and concentration of resources for results.

2.1. NEFU program for digital transformation of the university

When developing the program, we took into account the influence of the heterogeneity of the environment and the territorial context of the development of education.
The processes of digitalization of basic processes are formulated taking into account the main constraints and problems of the university development:

- limited geography of reception due to territorial remoteness, extreme natural and climatic conditions, underdeveloped transport infrastructure of the macroregion, high cost of transfer;
- remoteness from the leading scientific, educational and cultural centres of the country, imperfection of the social and transport infrastructure of the macroregion, a complex logistic scheme that limit the geography of attracting talented students and world-class researchers. The number of students from the member states of the Russian Federation entering the higher education programs is half the number of foreign students (253 and

* Corresponding author: Petr-inf@mail.ru

529 in 2019, respectively). These circumstances also explain the small proportion of students with bachelor's and specialist degrees from other universities who apply for master's degree programs of NEFU;

- Non-personalization, mass character of education of bachelor's degree programs;

- Forced transfer of education to a distance learning format due to the COVID-19 pandemic.

Accordingly, the digital transformation of the university is associated with the following challenges:

- The growing complexity of socio-technical, socio-economic, political and cultural systems, the acute demand for new skills and knowledge;

- Active development of information and communication technologies that provide the ability to process "big data", the automation of cognitive processes in artificial intelligence systems, new ways of individual and collective learning;

- Multiple increase in information flow, intensive development of digital technologies, arrival of the "digital" generation;

- The development of network communities, the growth of the importance of networks, storage of information on the network and its availability from anywhere in the world;

- Growth of scientific and educational collaboration, expansion of opportunities for cooperation with domestic and foreign educational and scientific centres;

- A request from employers for the development of metasubject competencies for the environment with increasing uncertainty;

- Growing demand for innovations aimed at solving the problems of intensive socio-economic development of the macroregion;

- The implementation of large-scale national projects "Science", "Education", "Digital Economy", "Healthcare", etc.

The digital transformation of the university is considered by us as the formation of a digital educational ecosystem that ensures the quality and accessibility of education, integration into the international scientific community, and the satisfaction of the information needs of all stakeholders of the university's life.

The program is aimed at to improve the activities of the university, taking into account modern technologies and strategies based on the experience of leading domestic and foreign educational centres. The digital transformation of NEFU affects all the main processes of the university:

- Digitalization of the educational process;
- Digitalization of research activities;
- Digitalization of university management processes;
- Institutional changes;
- Upgrading IT infrastructure.

2.2 Directions of NEFU development program

When developing the program, we took into account the three basic directions of digitalization that most universities are implementing: the development of distance education, the improvement of information infrastructure, digitalization and optimization of business processes [1], [2], [7], [8]. A.V. Arakelov and M.A. Kodzheshau distinguish such tasks of the digital transformation of the university as the automation of all types of university activities; creation of a unified information and educational space; organization of electronic document management; creation of tools for operational management; creation of a holistic person-centered electronic information and educational environment [9, p.14].

Each of the directions of NEFU development program provides the activities aimed at digitalizing basic processes, introducing data-based practices and ensuring information security:

1. Building flexible educational tracks through the use of blended learning formats and the introduction of digital, including end-to-end, technologies.

2. Expansion of the range of programs of additional professional education, including in the distance form, ensuring scalability and availability of supplementary vocational education for the population of the macroregion.

3. Creation of a system of digital academic mobility of students.

4. Digitalization of the scientific and innovation process, the introduction of digital technologies in the organization of research activities. The transition to digital technologies, robotic systems, new materials and design methods, the creation of systems for processing large amounts of data, artificial intelligence.

5. Expansion of interaction with leading centers and collaborations in the field of digital technologies.

6. Upgrading of employees skills in the field of digital technologies.

7. Implementation of the concept of a "smart campus" in accordance with the goals of sustainable development, the creation of a modern and technological campus based on advanced facility management technologies.

8. Digitalization of university management processes: the development of the digital ecosystem of the university, the integration of digital services into a single digital ecosystem of the university, created on the principle of a "single window", the transition to a digital university management platform using data-based management practices [10].

2.3 Implementation of NEFU Digital Transformation Program

The digitalization process begins with updating the content of existing and forming new professional educational programs aimed to ensure the scientific and technological development of the Far East, sustainable development of the northern territories. They are implemented, inter alia, in the form of network interaction with educational, scientific organizations and enterprises of the real sector of economy. Today, NEFU provides 472 educational programs, including 83 in the field of digital economy and information and communication technologies, which is 17.6% of the total number of programs. In the 2019-2020 academic year a new international master's program "Convergence" started, which trains specialists in the field of modern digital technologies. The discipline "Introduction to end-to-end digital technologies" was added into the major educational programs of higher education of the bachelor's and specialist degrees. In order to expand the educational tracks of students, optional disciplines were introduced in the areas of end-to-end technologies of the digital economy (wireless communication, blockchain, artificial intelligence).

The 2020 pandemic has stimulated a completely total shift to distance learning technologies – both for teachers and students. A similar situation has developed in all universities of the country and the globe: "the rapid transition of one of the most massive systems of higher education in the world to a distance format - 450 universities in two weeks" [5], "the evolutionary period of transformation of education, which would have lasted for more than one decade" [6], "the pandemic has accelerated the speed of digitalization" [7], "the use of distance education using digital tools in higher education has increased particularly during the COVID-19 pandemic" [11].

It should be noted that before the transition to distance learning, the university had accumulated sufficient experience and technological groundwork for organizing distance learning in emergency situations. Thus, in 2018-2020 active work was carried out on the development of e-learning and distance learning systems of major educational programs. LMS Moodle (http://yagu.s-vfu.ru) is used as the main platform for the implementation of blended learning. During the pandemic, by mid-April 2020, an indicator of 100% was reached in terms of the fullness of distance learning environment with all the educational and methodological materials. At the moment, all 22179 subjects delivered in 154 major educational programs of NEFU are provided by the educational and methodological materials in electronic form and placed in the distance learning system.

The next direction is the construction of flexible educational tracks through the use of blended learning formats and the introduction of digital, including end-to-end technologies. NEFU takes part in a project to test educational models of blended learning using online courses as part of the priority project "Modern digital educational environment of the Russian Federation." This platform hosts 15 online courses in which students of NEFU and other universities are trained with confirmation of the learning outcomes.

Interaction with leading centres and collaborations in the field of digital technologies is expanding:

- NEFU is one of the co-founders of the University Consortium of Big Data Researchers together with Tomsk and Moscow State Universities, Northern (Arctic) and Ural Federal Universities, Higher School of Economics to conduct joint scientific and applied research and solve socially significant problems using collection and data analysis from social networks, university learning management systems (LMS), open platforms;

- NEFU participates in the network project of the NTI University 20.35 to promote artificial intelligence technologies and the development of relevant competencies among university staff and students. 100 educational organizations of the country are parties to the agreement;

- NEFU signed a Memorandum of Intent with the Agency for Strategic Initiatives to promote new projects, the National Technology Initiative Platform, the National Technology Initiative University 20.35. Within the framework of this memorandum, in 2019-2020, events have already been held, such as NEFU Design and Educational Intensive according to the University model 20.35, within the framework of which the University 20.35 services were tested for organizing the work of student project groups, building individual educational tracks, fixing a digital footprint and an increase in competencies.

- on the basis of NEFU, an educational program for advanced training "Data-based Management – Chief Data Officer in the authorities" was carried out, organized jointly with the Centre for Project Practices CDO of Tomsk State University [12].

Within the framework of the federal project for free of charge training in computer literacy of pensioners under the program of advanced training courses "Computer literacy of the population", together with the Union of Pensioners of Russia, 409 citizens of retirement age were trained on the basis of the Institute of Continuous Professional Education. As part of the implementation of the national projects "Digital Economy of the Russian Federation", "Demography", the College of Infrastructure Technologies has implemented a software and vocational education program "Skills of the Wise"; 216 residents of Yakutia of pre-retirement age have been trained.

Targeted work on the development of the electronic informational and educational system allowed creating a basis for the further development of the digital ecosystem of the university, capable to solve the problems of improving and developing the scientific and educational activities of the university. At the moment, NEFU operates an electronic informational and educational system, developed in accordance with the requirements of the Federal State Educational Standard for the conditions of education and taking into account the experience of Russian universities [1], [6]. The electronic informational and educational system of NEFU provides users with the following information services:

- Access to full curriculum with the work plans of subjects (modules) and internships;

- Access to electronic library systems and educational resources indicated in the work plans of the subjects;

- Fixing the course of the educational process, the midterm assessment results and the academic progress in general;

- Lectures and seminars, assessment of academic outcome, the implementation of which is provided with the use of e-learning, distance learning technologies;

- The creation of an electronic portfolio, including the storage of course papers, diploma thesis, scientific works, reviews and assessment of these works (under the tutor's supervision);

- Synchronous and (or) asynchronous interaction between the students, teaching and supporting staff via the Internet.

3 Conclusion

Thus, the electronic informational and educational system of NEFU implements the tasks set in the practice of higher education: equal access to information resources; implementation of curricula and plans; control of the educational process, ensuring synchronous and asynchronous interaction of participants of the educational process.

In general, the following stages of the program implementation can be distinguished:

Program launch stage (September-December 2020).

At this stage, the following steps were taken:

- Internal examination of the program at strategic sessions and expert discussions, its adjustment, taking into account the proposals and recommendations made;

- External expertise by the main stakeholders;

- External expertise at sessions of the Moscow School of Management SKOLKOVO;

- Financial evaluation of the program;

- Approval of the digital transformation program at the Academic Council of the University.

The stage of digital transformation of the main mechanisms (2021-2024)

In this context, all the main areas of activity of NEFU are being digitalized: educational, scientific, innovative. Infrastructural transformations are carrying out. Organization of work of interdisciplinary project teams. Interim assessment of program implementation and adjustment.

The stage of assessment and analysis (2025) is the final assessment of the achievement of the goal and solution of the program objectives and the implementation of the main indicators.

Thus, we can conclude that on the one hand the development of digitalization is explained by a rapidly changing world, growing heterogeneity of the environment, uncertainty of social and political processes. On the other hand such factors as limited resources, poor transport accessibility, remoteness from leading research and educational centres also promote the process of digital transformation of universities. This process consists, first of all, in the improvement of the quality and accessibility of education and increase in the efficiency of scientific research through the individualization of educational tracks.

References

1. A. Mikheev, Y. Serkina, A. Vasaev, *Current trends in the digital transformation of higher education institutions in Russia*, Ed. and Inf. Tech. (to be published)
2. E.V. Galazhinsky, *TSU: a model of digital transformation of the university*, EdCrunch Tomsk. Materials of the international conference on new educational technologies, Tomsk, 8-13 (2019)
3. S.K. Bekova, E.A. Terentev, N.G. Maloshonok, *Voprosy Obrazovaniya*. **1**, 1-15 (2021)
4. G. Coco, R. Lagravinese, G. Resce, *J. of Ec. Inequal.* **18**, 469 (2020)
5. D.A. Sergeev, V.N. Zimin, *Trends in the development of science and education*, **53-1**, 34-38 (2019)
6. E.A. Antyukhova, P.I. Kasatkin, *World Economy and international Relations*, **65**, **2**, 125-133 (2021)
7. A. Webb, R. W. McQuaid, C.W.R. Webster. *World Journal of Science Technology and Sustainable Development*, **18**, 1-19 (2021)
8. O. Tavstukha, O. Osivanova, T. Chelpachenko, A. Yudina, O. Ovchinnikov, L. Illarionova, Z. Gereev, N. Aleksandrova, *ropositos y Representaciones*, **9**, e1133 (2021)
9. A.V. Arakelov, M.A. Kodzheshau, *Distance educational technologies. Proceedings of the V International Scientific and Practical Conference*, Yalta, 9-15 (2020)
10. P.P. Ivanov, R.P. Ivanova, *Smart Innovation, Systems and Technologies*, **172**, 753-760 (2020)
11. U. Langegard, K. Kiani, S. Nielsen, PA. Svensson, *BMC Nursing*, **20 (1)**, 23 (2021)
12. V.M. Savvinov, P.P. Ivanov, *Year of digitalization at North-Eastern Federal University: concept, solutions, best practices* (Yakutsk, NEFU Publishing House, 2020)

Improving the quality of the railway transportation process

Viktor V. Kondratiev[1,*], *Viktoriya A.* Olentsevich[2], *Boris O.* Kuznetsov[2], and *Antonina I.* Karlina[1]

[1] Irkutsk National Research Technical University, 83, Lermontov Str., Irkutsk, Russia
[2] Irkutsk State Transport University, 15, Chernyshevsky Str., Irkutsk, Russia

Abstract. The safety of trains is the basis of the efficient operation of the railway transport system, the optimal organization of passengers and cargo transportation. All organizational and technical activities in railway transport must meet the requirements of safe and uninterrupted train traffic. Traffic safety is ensured by the maintenance of all facilities, railway tracks, rolling stock, equipment and mechanisms, technical devices and communications. In order to ensure a given level of train safety, it is essential that the technical facilities and personnel have the appropriate safety level. Comprehensive management of reliability, risks, cost of life cycle in railways is a technology that improves the quality of services. This technology is characterized by several components, namely, factors that affect the reliability and safety of railway facilities, as well as the means of achieving reliability and safety; risk and complete safety; cost of life cycle. As part of the security provision, the Russian Railways regularly upgrades its fixed assets. These are projects related to safety, which are based on life safety aimed at ensuring the technological sustainability of the production process, improving traffic safety and maintaining the existing throughput capacity of transport infrastructure. The assessment of the influence of technical failures and defects on the performance and economic criteria of rail transport will enable objective management decisions to be made in the area of fixed assets renewal in order to improve traffic safety and to introduce measures to reduce the share of non-productive expenditures of structural units.

Keywords: transport process, quality of transport services, technical equipment, transport infrastructure, non-productive expenditures, traffic safety.

1 Introduction

Competitiveness of transport services in the railway sector largely reflects the quality of transport services, which in turn is formed by the implementation of the following key indicators of the industry: train speed, cargo delivery time, cargo safety, the degree of satisfaction with the transport of goods of various volumes, the timeliness of technological operations, the regularity and rhythm of keeping the cargo moving according to the established plan of delivery of manufactured products, the complexity of transport services, the level of transport and logical services provided, etc. The values of the presented performance indicators of the railway transportation system (RTS), which fully form the quality of transport services, largely depend on the level of performance indicators of operational work presented in Figure 1. Operational work can be represented as a set of the following main indicators: technical reliability of transport infrastructure and technical equipment, transportation process organization safety, the degree of efficiency of using rolling stock in terms of time, the degree of efficiency of using rolling stock in terms of power, the degree of efficiency of using rolling stock in terms of load-carrying capacity. The same category includes non-productive rolling stock performance, as well as quality indicators of its use [2, 8, 14].

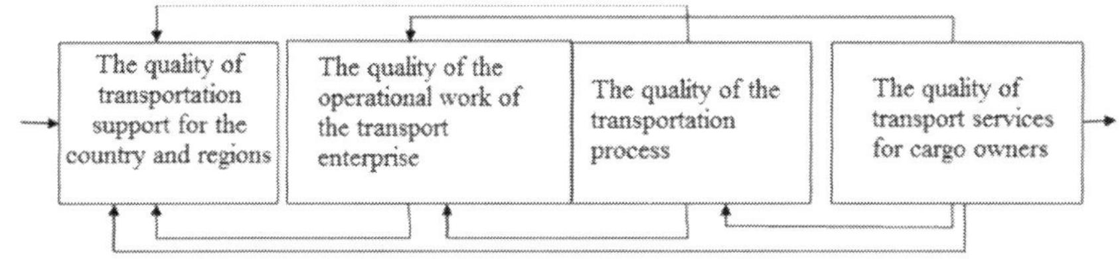

Figure 1. The scheme of transport service quality indicators of the railway transport system

* Corresponding author: kvv@istu.edu

The technical condition and reliability of the transport infrastructure and equipment have a significant impact on the performance of the operational targets of all structural units of the railway transportation system. Maintaining a high level of condition, health and reliability of the infrastructure is the main task of the Russian Railways company, as it is with these fixed assets that the entire amount of transport activities is implemented. This policy makes it possible to increase the interest and responsibility of the company and its structural units for the proper fulfillment of the obligations undertaken to carry out transportation by using infrastructure facilities in accordance with the technological and technical purpose; to ensure an acceptable level of wear and tear; to timely repair, maintain and replace individual elements; to carry out updates and modernization in accordance with up-to-date requirements for the quality of transportation of goods and passengers; to ensure the safety of train traffic, excluding or minimizing all possible types of risks [4, 9, 19].

To achieve the set objectives, within the territorial branches of Russian Railways OAO there is an infrastructure directorate, whose activities allow combining a number of separate types of railway transportation operations into a single system: track facilities, automation and telemechanics facilities, and operational degree of detail of the rolling stock. This infrastructure system operates within the existing structure of Russian Railways OAO Holding and is a large subdivision of the company, which defines its technical equipment performance indicators for each facility of the directorate (table 1).

Table 1. Indicators of performance evaluation of technical equipment of the infrastructure directorate

Infrastructure directorate service	Evaluation criteria	Indicators
Automation and telemechanics service	A volumetric indicator of the automation and telemechanics service operation	Number of failures (N_{fail})
	Natural indicators for evaluating the operation of technical equipment for the automation and telemechanics service	Availability factor (K_a), Failure intensity (λ_{fail}), Recovery time (Tr)
	An indicator of the automation and telemechanics service operation	Score evaluation of the operation (B)
	Economic criteria for evaluating the automation and telemechanics service operation	Economic consequences of technical equipment failures (EC_{fail})
Track service	Volumetric indicator of the track service operation	Number of failures (N_{fail})
	Natural indicators for evaluating the operation of technical equipment for the track service	Failure rate λ_{fail} and time to eliminate the consequences of failures
	A track service quality indicator	Score evaluation of the track state (B)
	Economic criteria for evaluating the track service operation	Economic consequences of technical equipment failures (EC_{fail})
Rolling stock service	Volumetric indicator of the rolling stock service operation	Number of failures (N_{fail})
	Natural performance indicators for evaluating the operation of the rolling stock service technical equipment	Failure rate λ_{fail} and time to eliminate the consequences of failures
	An indicator of the rolling stock operation quality	Corresponding to the target indicators in terms of technical equipment failures
	Economic criteria for evaluating the rolling stock operation	Economic consequences of technical equipment failures (EC_{fail})

Analysis of Table 1 shows that it is difficult to single out one main indicator, allowing one to significantly evaluate the activities of all the structural units of the infrastructure system of railway transportation, since its production, commercial, financial and economic activities have a complex structure. However, it should be noted that the main tasks of the railway transportation system to meet the indicators that characterize the fail-safety, rhythm and safety of the services provided by the infrastructure, namely, the number of events that caused train safety violations, the number and intensity of failures of technical equipment of the infrastructure. Still, the specific place of this system is held by indicators that characterize the safety of the transportation process, fail-safety, the regularity and rhythm of the maintenance of the infrastructure, the number and intensity of technical equipment failures. In connection with these factors, it is advisable to conduct a study in the railway transportation system by the example of the operation of the automation and telemechanics sector. Ensuring the reliability of the operation of signaling, centralization and blocking devices largely leads to the maintenance of high levels of train safety, smoothness and timeliness of production of technological operations. The implementation of the full schedule of trains and daily schedules of railway stations ensures the optimal operation of all the structural units of the railway transportation system. It also affects most of the industry's financial and economic performance, such as: productivity, non-productive expenditures of transport companies and service users, operating costs, freight costs, as "technical equipment failure" indicators and the quality of production and technical processes are in direct relationship and affect the performance and functioning indicators of all transport enterprises of the railway transportation system [3, 5, 6].

According to the maintenance procedures of transport infrastructure and equipment established in the railway transportation system, the qualitative characteristics of technical maintenance of devices of automation and telemechanics are reflected in the scoring evaluation of their condition and performance, taking into account the reliability of their operation. The technical means that provide the transportation process are characterized by reliability indicators, which in turn should have the properties of fail-safety, durability, repairability.

In accordance with the goals and objectives set by Russian Railways OAO, all the structural subdivisions included in the sector of automation and telemechanics are required to provide functions aimed at:

- involving the maintenance of means of railway automation and telemechanics in a technically sound state within the established boundaries of production activities;
- preventing and eliminating violations of their normal operation in accordance with the regulatory legal acts of the Russian Federation, the regulations of Russian Railways OAO (RZD OAO) and the railway;
- ensuring the safety of train traffic and performing shunting operations at the objects of railway transportation system;
- implementation of measures aimed at improving the reliability and continuity of operation of the means of large-scale railway automation and telemechanics, efficiency of their functioning and economy of their operation.

2 Problem Statement

Organization traffic safety is ensured by the maintenance of all objects, railways, rolling stock, equipment and mechanisms, technical devices and communications.

3 Research Questions

Methods of organization traffic safety is ensured by the maintenance of all objects, railways, rolling stock, equipment and mechanisms, technical devices and communications.

Economic efficiency of maintenance of facilities, railways, rolling stock, equipment and mechanisms, technical devices and communications.

1. Research of failures of technical means of the railway transportation system. As practice shows, it is the failures of the infrastructure system elements and the technical means responsible for ensuring the transportation process that have the maximum impact on the safety of trains in the railway station, which is the main indicator of the quality of the transportation process, in particular, the indicator of the quality of transportation services [1, 13, 16].

As a result of the failure of infrastructure system and technical equipment elements, malfunctions in their operation, impairment of their performance, there are non-scheduled detention of cargo and passenger trains, leading to an increase in the cargo delivery, as well as an increase in operating costs and an growth of the cost of the transportation process, since additional labor and material resources will be used to eliminate them. There is also a possibility of an increase in the risk of loss of existing users of the services of the railway transportation system and a decrease in the likelihood of attracting new ones. This is a result of their transition to road transport as the most attractive one for the organization of the transportation process, which will lead to the loss of potential revenues of Russian Railways OAO, a decrease in its level of competitiveness.

From an economic point of view, the failure of an element of the infrastructure system and technical equipment is non-productive losses of the railway transportation system, arising from the inconsistency of the regulatory parameters of the transportation process to the actually established ones. These economic losses resulting from failure are divided into two categories: direct and indirect losses. Figure 2 shows the constituents of the costs.

The economic damage of the railway transportation system operation caused by the unproductive loss of transportation process time is primarily due to an increase in the share of the costs in terms of train-hours of non-scheduled detention of cargo, suburban and passenger trains. The amount of these expenditures is determined by the method of expenditure rates [15]. The cost estimate of the expenditures associated with delays of trains of corresponding categories due to the equipment failure is defined as a product of the delay time of the transportation process and the increased expenditure rate:

$$\Sigma E = \Delta Nt_{car} * e_{car}^{Nt} + \Delta Nt_{pass} * e_{pass}^{Nt} + \Delta Nt_{sub} * e_{sub}^{Nt}, \quad (1)$$

where ΔNt_{car} is the train-hours of delays of cargo trains caused by failures of infrastructure system and technical equipment elements, failures in their operation, impairment of their serviceability in the relevant subdivision of Russian Railways OAO;

ΔNt_{pass} is the train-hours of delays of passenger trains caused by failures of infrastructure system and technical equipment elements, failures in their operation, impairment of their serviceability in the relevant subdivision of Russian Railways OAO;

ΔNt_{sub} is the train-hours of delays of suburban electric trains caused by failures of infrastructure system and technical equipment elements, failures in their operation, impairment of their serviceability in the relevant subdivision of Russian Railways OAO;

e_{car}^{Nt}, e_{pass}^{Nt}, e_{sub}^{Nt} are the increased expenditure rate of a train-hour of the relevant category train detention, rubles.

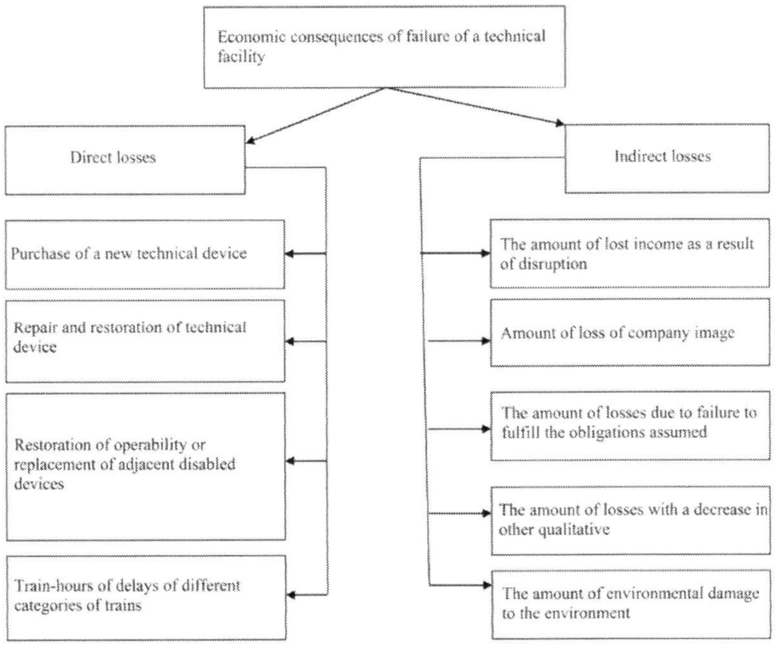

Figure 2. Economic losses resulting from the failure occurrence

Table 2 shows the calculation of the economic damage caused by the delay of trains of all categories of traffic, due to failures of infrastructure system and technical equipment elements, failures in their operation, impairment of their serviceability for 2019 in the Irkutsk division of signaling, centralization and blocking (an object of study).

Table 2. Economic damage caused by train delays in all categories of traffic due to failures of infrastructure system and technical equipment elements

The name of the meter	Expendable rate, rubles	The meter, train-hours	Non-productive expenditures, rubles
A type of activity – cargo transportation			
Train-hours of a cargo train (electric traction)	1884.05	256	482316.8
A type of activity – passenger transport			
Train-hours of a passenger train in the long-distance traffic	368.06	56	20611.36
Train-hours of a passenger train in the suburban traffic	325.59	12	3907.08
Total economic damage	–	324	506835.24

2. According to the established technology of operation of the railway transportation system, the malfunction of infrastructure elements and technical equipment is taken into account using indicators such as events, failures and delays of trains, and the same indicators are classified as failures of the first, second and third groups. Events are understood to be malfunctions of technical equipment, which have caused serious violations of train traffic safety. Failure of technical equipment is understood to be a malfunction, which is an impairment of the object's serviceability. It is established that the period of elimination of the event should not exceed 12 hours, the period of elimination of the failure – 2.5 hours and the period of elimination of the train delay – 0.33 hours.

Taking into account the time factor and the severity of the consequences, in accordance with the "Regulations on the procedure for official investigation and registration of traffic accidents and other related to violation of traffic safety rules and operation of railway transport" [7, 18], depending on the duration of the break in the transportation production process, three categories of failures are distinguished:

- the first category includes failures that resulted in a delay of 6 minutes or more for a passenger or suburban train; for a delay of a cargo train at the station-to-station block (station) for an hour or more or led to cases of traffic safety violations in train service or shunting operation;
- the second category includes failures that led to a train delay at the station-to-station block (station) lasting from 6 minutes to 1 hour, or when the impact has led to a deterioration in performance, excluding delays of trains belonging to the first category;
- the third category includes failures that do not have consequences related to failures of the first and second categories.

A generalization of information about the failures that occurred during this year in comparison with the previous one and with an indication of the established reasons is provided in the "Analysis of the operation of the technical means of the division". The analysis showed that the maximum number of failures – 82 – was made in 2018. The dynamics of quality indicators is presented in Table 3.

Table 3. Dynamics of quality indicators

Indicator	Period of time		
	2017	2018	2019
Failures, pc.	54	82	70
Number of points, points	5.2	3.34	3.92

4 Economic assessment of the possible consequences of a failure

The economic assessment of the possible consequences of a failure of the infrastructure system and technical equipment, their malfunctions, disruptions in their performance predetermines the amount of capital expenditures and the priority of their investment. Table 4, according to the category of the structural subdivision, presents an algorithm for evaluating its operation, depending on the amount of possible economic losses.

Table 4. Evaluation of the operation of the studied object depending on the magnitude of possible economic losses

A category of the studied object	Reliability of an infrastructure element or a technical equipment	The flow of technical failures at a division, 1/h*10^{-n}	Possible economic losses (per month), thousands of rubles
I	High degree	$10^{-5} \leq \lambda < 10^{-3}$	50
II	Medium degree	$10^{-3} \leq \lambda < 10^{-1}$	50 – 4 000
III	Low degree	$\lambda \geq 10^{-1}$	> 4 000

Ensuring a sufficiently high degree of reliability in the operation of infrastructure system and technical equipment elements minimizes the amount of possible economic damage in the organization of the transportation process. This strategy is the most preferable to the effective organization of the operation of the studied object [10, 12, 17].

It is most appropriate to invest in the renovation of infrastructure system and technical equipment elements for the third category in terms of a studied object that has minimal reliability, and the economic damage to the performance indicators of the railway transportation system is significant and, along with that, may affect the continuity and safety of operation, lead to an increase in non-productive costs and a decrease in competitiveness.

Based on the data of Table 2 and 3, we calculated the magnitude of the intensity of failures of the infrastructure system and technical equipment elements (λ) in terms of the studied object based on the average time of between failures (T_o), by formulas (2, 3):

$$\lambda = \frac{1}{T_0}, \quad (2)$$

$$T_0 = \frac{365*24-\Delta Nt}{n} \quad (3)$$

where ΔNt is train-hours of non-scheduled delays of trains of different categories (table 2);

n is the number of failures of infrastructure system elements and technical equipment that caused train delays (table 3):

$$T_0 = \frac{365*24-324}{70} = 120{,}5 \text{ hours.}$$

Failure intensity in general in terms of the studied object:

$$\lambda = \frac{1}{120{,}5} = 8{,}29 * 10^{-3}$$

5 Research Methods

In accordance with the evaluation parameters, the studied object falls to the first category, characterized by a low degree of reliability of the infrastructure system and technical equipment elements, as well as the high amount of economic damage arising from possible failures. The economic evaluation of the consequences of the failure of infrastructure system and technical equipment elements determines the size of capital contributions, priorities and the direction of their investment. In order to establish a priority range of investments in terms of the studied object, it is necessary to determine the proportion of economic damage (γ_{ed}), which will occur at the railway station and on the section of the railway track, in the overall amount of economic consequences of failures of infrastructure system and technical equipment elements. The priority range of investments in the renovation of elements of infrastructure system and technical equipment is presented in Table 5.

Table 5. The necessity to update the elements of the infrastructure system and technical equipment

The share of economic damage to the railway station and the section of the railway line of the studied object of category III, %	The investment order
$\gamma_{ed} > 60$	First order of investment
$40 < \gamma_{ed} \leq 60$	Second order of investment
$\gamma_{ed} \leq 40$	Third order of investment

According to Table 6, it is clear that if the proportion of economic damage to the railway station and the section of the category III railway line is more than 60 per cent of the total economic impact, the possibility of prior investment in the renovation of key elements of the infrastructure system and technical equipment at these railway stations and sections should be considered [11]. The algorithm of making managerial decisions to determine the order of updating the technical base of the subdivisions of automation and telemechanics facilities is presented in Figure 3.

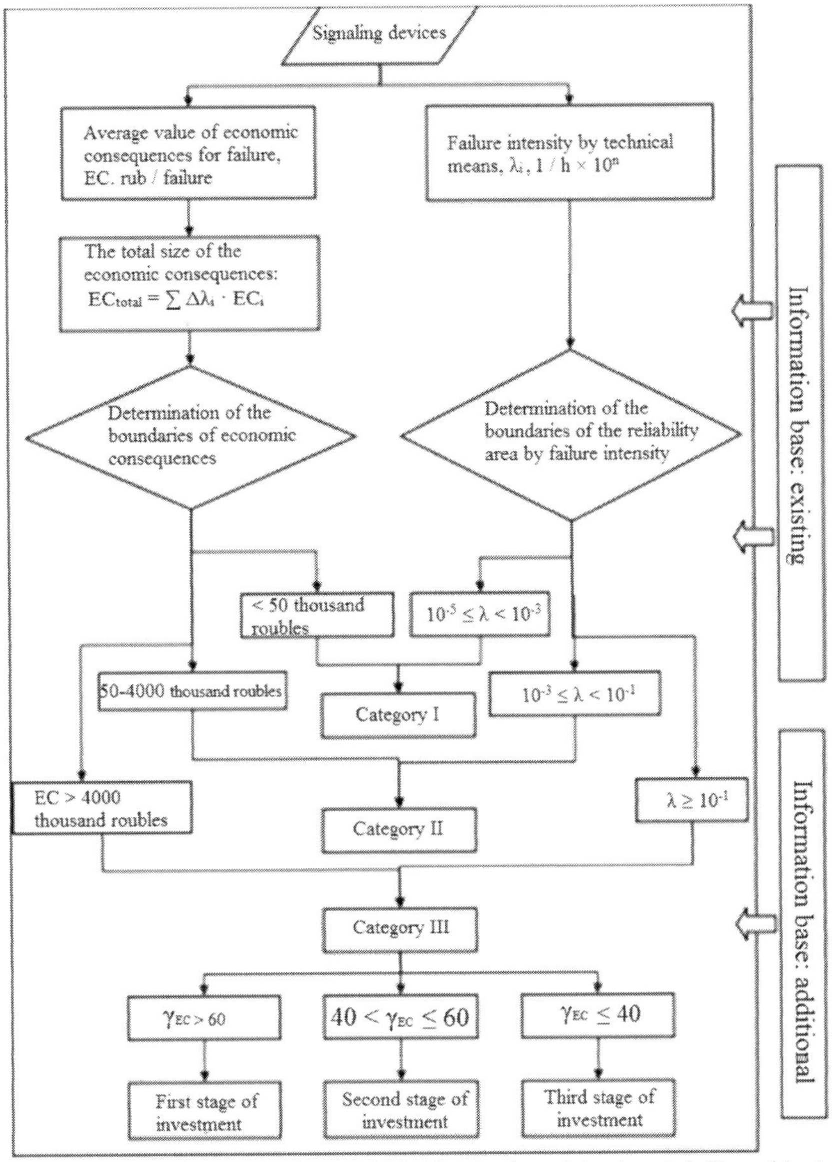

Figure 3. Algorithm for the order of managerial decision aimed at updating the technical base of the studied object]

6 Findings

Comprehensive management of reliability, risks, cost of life cycle in railways is a technology that improves the quality of services. This technology is characterized by several components, namely, factors that affect the reliability and safety of railway facilities, as well as the means of achieving reliability and safety; risk and complete safety; cost of life cycle. As part of the security provision, the Russian Railways regularly upgrades its fixed assets. These are projects related to safety, which are based on life safety aimed at ensuring the technological sustainability of the production process, improving traffic safety and maintaining the existing throughput capacity of transport infrastructure. The assessment of the influence of technical failures and defects on the performance and economic criteria of rail transport will enable objective management decisions to be made in the area of fixed assets renewal in order to improve traffic safety and to introduce measures to reduce the share of non-productive expenditures of structural units.

7 Conclusion

In the course of the completed study, the task to economically assess the effectiveness of quality management by the example of the subdivision of the infrastructure block of railway transport – the economy of automation and telemechanics – has been set and solved. The initial data for calculation are data on technical equipment failures. Taking into account economic losses in a structured form will allow one to demonstrate the real production situation in a particular subdivision or at the appropriate section of the signaling, centralization and blocking division. Assessment of the divisions of signaling, centralization and blocking, based on the principle of proportionality of economic consequences and parameters of reliability of technical means, will make it possible to determine the order of investment of the renovation of the company's fixed assets, affecting traffic safety. Possible economic damage, taking into account the parameters of the reliability of technical means, can be forecasted on any transport infrastructure object in the railway transportation system.

References

1. G.V. Bubnova. *Freight marketing management models* (Marshrut Publ., Moscow, 2003)
2. T.A. Bulokhova. More on the problem of improving the quality of transport services. *Modern approaches to transport management and logistics. A coll. of mat. of the All-Rus. Sci. and Pract. Conf.* (Moscow State Transp. Univ. Publ., Moscow, 2016)
3. T.A. Bulokhova. Modeling assessment of the quality of public transport services. Modern technol. Syst. Anal. Model., **1(65)**, 128–136 (2020)
4. T.A. Bulokhova, A.P. Bulokhov. Improving methodical approaches to assessing the quality of passenger transportation. *Transport Infrastructure of the Siberian Region. Mat. of the 6th Int. Sci. and Pract. Conf., dedicated to the 40th anniversary of the found. of the Irkutsk State Transp. Univ.*, vol. 2 (IrGUPS Publ., Irkutsk, 2015), p. 62
5. S.S. Gromyshova, N.P. Astashkov, V.A. Olentsevich, O.V. Lobanov. Assessment of the level of safety of complex transport systems in order to increase their competitiveness in the transport service market. Modern technol. Syst. Anal. Model., **2(62)**, 250–259 (2019)
6. V.A. Makeev, E.E. Oksenyuk. Transport quality system: sequence of construction and evaluation of parameters. The Herald of RSTU, **2** (2013)
7. O.S. Maysak. SWOT analysis: object, factors, strategies. The problem of finding links between factors. Casp. Magaz.: manag. and high technol., **1** (2013)
8. V.A. Olentsevich, Y.I. Belogolov, G.N. Kramynina. Set of organizational, technical and reconstructive measures aimed at improvement of section performance indicators based on the study of systemic relations and regularities of functioning of railway transport system. IOP Conf. Ser. Mat. Sci. and Eng., **832**, 012038 (2020)
9. *Railway transport in Russia: challenges until 2025* (Institute of Natural Monopolies, 2019)
10. A.Yu. Ryazanov. Logical and mathematical representation of the work of transport and economic systems. *Reliability and quality. Proc. of the Int. Symp., in 2 volumes* (PSU Inform. Publ. Center, Penza, 2007), 406 p
11. G.V. Savitskaya. *Analysis of the company's business activities*, 2nd ed., update. and rev. (Ekoperspectiva SEP, Minsk, 1998)
12. M.O. Severova. Issues of assessing the efficiency of enterprises of railway transport infrastructure. Sci. probl. of transp. in Siber. and the Far East: an acad. J., **1**, 24–26 (2011)
13. M.O. Severova, Yu.A. Shkudnova. On the issue of assessing the effectiveness of railway infrastructure. The Herald of Siber. State Transp. Univ., **26**, 112–119 (2012)
14. L.V. Shkurin, S.V. Brazhnikova. *Competitiveness of the transport company: economic assessment of the efficiency and quality of production activities of the structural units of railway transport*, Monograph (Emperor Nicholas II Moscow State Transp. Univ. Publ., Moscow, 2016)
15. N.P. Tereshina, V.G. Galaburda, V.A. Tokarev et al. *Economics of rail transport*, Textbook (Ed. and Methodol. Center for Ed. in Railway Transp., Moscow, 2011).
16. *The development strategy of the Federal Passenger Company until 2030.* Report. Mat. Accord. to the techn. enquiry (2018)
17. V.D. Vereskun. *Organizational and technological reliability and efficiency of functioning of railway production facilities* (SSTU Publ., Novosibirsk, 2010)
18. A.A. Vovk. *Assessment of the efficiency of transport production and reserves of its growth*, Monograph (Kroma Publ., Moscow, 2000)
19. V.I. Yakunin. Infrastructure cannot be saved on if we want to create points of economic growth Gudok, **127**, 3 (2010)

The concept of sustainable development as the foundation for changing education in the North

Vasiliy Savvinov[1*]

[1]North-Eastern Federal University (NEFU), 58 Belinsky str, Yakutsk, Republic of Sakha (Yakutia), Russia

Abstract. The article reveals the experience of the implementation of the Sustainable Development Goals in Russian universities based on the case study of North-Eastern Federal University (NEFU). The article presents a comparative analysis of strategic programs to manage the development of universities in the North of Russia and the northern countries of Europe and America in the context of global changes and growing uncertainty of the environment. It shows NEFU's groundwork for the implementation of the sustainable development model of the northern territories and justifies the key principles and the directions of change in the academic and innovative activities of the university related to the implementation of the UN Sustainable Development Goals.

1 Introduction

During the intensive development of the northern territories, primarily hydrocarbon production and the Northern Sea Route development, the issues of anthropogenic impact on northern ecosystems, ensuring the protection of population and the northern territories from natural and non-natural emergencies, and protecting indigenous peoples' living environment and traditional lifestyle are of particular relevance.

The studies of current modernization of local and foreign universities' academic activities show that sustainable development concept is becoming one of the most popular methodological foundations for the management of education development. It is also mentioned that "universities have an integral role in the development of communities and they possess a social responsibility to be agents of change in relation to society's socio-economic, political, and environmental issues" [1].

The article discusses main methodological approaches to the formulation of the development priorities of the North-Eastern Federal University (NEFU), reveals the principles, priorities, key focuses, and management mechanisms that ensure NEFU's dynamic development based on the implementation of sustainable development concept.

2 Results and Discussion

Currently, a number of factors have a limiting influence on the development of the northern territories and the arrangement of human life in the North:

- extreme climate and the intensity of climate change;
- strong disconnection in the vast territory of isolated settlements, their low transport accessibility and connectivity in the North and Far East;
- population decline, skilled workforce outflow;
- compared to the Russian average, low living standards limiting new workforce influx and, on the contrary, generating population outflow from the macroregion;
- main production orientation on the development of extractive industries and low availability of infrastructure;
- general instability of the socio-economic situation makes the North of Russia and the Far East geopolitically vulnerable [2, p. 29-30].

Due to the growing human impact on northern ecosystems, global climate change, the intensive development of natural resources, and the increasing military and economic presence of foreign states in the macroregion, the issue of ensuring the sustainability of the northern territories becomes acute.

NEFU's unique geopolitical position determines the priorities of its long-term development – the implementation of projects for the advanced development of the Far East and the North of Russia. NEFU's territory in accordance with the Spatial Development Strategy of the Russian Federation for the period up to 2025 covers two major priority geostrategic macroregions of the country – the Far East and the Arctic Zone of the Russian Federation.

The features of this stage of the university's development are determined, on the one hand, by the need to train competitive staff for the reclamation of the macroregion's significant natural resource potential, the modernization of transport and energy infrastructure, and the formation of highly efficient non-resource industries. On the other hand they are determined by the aims of

* Corresponding author: dagasia@mail.ru

youth retention. The living standards of the population of remote and secluded localities are not comparable to the national average. Continuing decline in population, general instability of the socio-economic situation make this part of Russia geopolitically vulnerable. Vast territory alongside low population density, as well as the availability of ample mineral resources that can become the basis for the implementation of highly profitable projects, create systemic tension, especially from the point of view of international competition for untapped space [3; 3].

North-Eastern Federal University bases its' activities on *sustainable development ideology* proclaimed by the United Nations reflected in the 17 Sustainable Development Goals [4]. The choice of the mission, the formulation of the goals and key principles of the university's development are based on the comparative analysis of international and Russian higher education practices, the assessment of the environmental factors' impact, and team potential.

During the elaboration of the North-Eastern Federal University's development program for 2021-2030, emerging global trends in higher education development were taken into account. The analysis shows that the implementation of sustainable development principles and goals in universities' activities is becoming one of the most pronounced global trends in higher education development. University community started implementing sustainable development principles almost immediately after the publication of the Brundtland Report. As it is known, the first attempt to transition universities to these principles is the Talloires Declaration (1990) [5, p. 58]. Hilligje van't Land, Secretary General of the International Association of Universities (IAU) and Executive Director of the International Universities Bureau, notes that "sustainable development has been on the agenda of universities since the Brundtland Report [published by the World Commission on Environment and Development in 1987], so universities have been asked to embark on the whole agenda and ensure that the hundreds and thousands of graduates they produce have a good understanding of sustainable development and their role in society, especially when they take up leadership roles." [6]

The study of the experience of universities in northern countries and regions of Europe and America and the Far North of Russia, conducted in 2020-2021, shows that most of the northern universities focus on increasing their influence on the sustainable development of the northern territories. The mission of the University of Eastern Finland has the following statement "the high standard of research and education of our university builds a global future and strengthens the vitality" [7] of the region. Arctic University of Norway defines its' mission as "to protect the Arctic and its invaluable terrain. For this reason, the universities main teaching expertise lies in scientific fields such as polar environment, climate research, telemedicine, medical biology and fishery science." [8]

Russian universities' policy papers present a similar vision of strategic development prospects in the context of intensive natural and social changes. Thus, Pitirim Sorokin Syktyvkar State University aims to promote "sustainable innovative and advanced socio-economic development of the Komi Republic and the European North of Russia based on the internationalization and integration of education, science, and production." [9, p. 13] Vitus Bering Kamchatka State University's mission is "to promote sustainable socio-economic development of the Kamchatka Territory and Russia as a whole through the formation of human and intellectual capital." [10, p. 3]

Currently, there is a significant amount of research on the implementation of the sustainable development concept in higher education organizations. In particular, O.V. Saginova, Y.L. Saginov, and A.I. Grishin distinguish the following interrelated concepts of a sustainable university based on comparative analysis:

1. The concept of a green university implementing the principles of environmental safety and protection.
2. The concept of a smart university implementing modern technologies for managing main processes.
3. The concept of a healthy lifestyle aimed at creating conditions for harmonious and comprehensive development, healthy eating, sports, etc.
4. The concept of managing social and cultural diversity focused on the education of national, racial, and religious tolerance, on the interaction and mutual enrichment of cultures.
5. The concept of historical and cultural heritage management based on the use of historical and cultural values of the country, region, city, industry, university as intangible assets of development [11, p. 217].

Since 2017, North-Eastern Federal University has been publishing reports on the achievement of the Sustainable Development Goals on its' official website [12]. Their analysis shows that the university has groundwork for the implementation of a university model – a platform for sustainable development of the North:

- in 2019 and 2020, NEFU was ranked within the top 400 worldwide and for the second year was 15th in Russia in the *University Impact Rankings* which assesses the universities of the world by the level of their impact on the sustainable development of society.

The university implements *educational programs focused on training staff considering the values and goals of the sustainable development of the northern territories* (geographical foundations of the sustainable development of recreation and tourism; the sustainable development of the Arctic; environmental protection and rational use of natural resources; industrial ecology; permafrost forestry; rational use of land in northern and Arctic territories, etc.)

Since 2014, NEFU has been participating in the implementation of the joint program for training experts that play leading roles in ecology, culture, and sustainable development in the regions of the Far East and the Arctic - RJE3 [13]. The program is implemented jointly with University of Hokkaido (Japan) and Far Eastern universities in such areas as environmental assessment, cultural diversity, soil and production, regional resources development, and disaster management.

Research is also based on the Sustainable Development Goals. Since the first stages of NEFU's development program realization, longitudinal studies of the peoples' of the North languages dynamics, the adaptation of the indigenous peoples' of the North culture to global natural and social changes have been organized. Major ethnographic expeditions are conducted in the places of compact residence of the indigenous minorities of the North. Since 2011, an Arctic multilingual portal is being developed – arctic-megapedia.ru, which presents the indigenous minorities of the Far East: Yukagirs, Evens, Evenks, Dolgans, Chukchi, Nanai, Negitals, Udege, Koryaks, etc.

There are annual scientific expeditions for studying the dynamic of changes in northern ecosystems influenced by anthropogenic processes, the evolution of permafrost landscapes influenced by globalization. There is also an operating system of educational and research polygons and stations for monitoring natural processes. NEFU participates in the international scientific project COPERA — C budget of ecosystems and cities and villages on permafrost in eastern Russian Arctic [14]. The project involves the University of Hokkaido, Japan Agency for Marine-Earth Science and Technology, and the University of Alaska Fairbanks. The aim is to study the "carbon dioxide budget" in the Arctic zone of Yakutia by combining the results of monitoring permafrost and carbon dioxide emissions into the atmosphere with the dynamics of socio-economic indicators.

A comprehensive assessment of North-Eastern Federal University's role and potential in regional development according to the OECD methodology showed that a significant direction of the university's development is the preservation and development of the languages and culture of the peoples of the North. It includes not only academic, but also outreach activities in history and culture, multilingual multicultural education, social engineering, and cultural policy.

NEFU is one of the organizers of the *Northern Sustainable Development Forum* – an annual international meeting of experts to discuss problems and prospects for the implementation of sustainable development goals and values in northern countries and regions, directions of international and interregional cooperation for the balanced development of the North.

NEFU implements outreach and social initiatives aimed to promote sustainable development values and goals ("Waste to Income" project, which is a platform for the exchange of experience for schoolchildren, students, postgraduates, young researchers and various specialists on the issues of waste disposal and recycling; "Open. Lectures" project, summer lectures in the park of Yakutsk; "Our University_Lectures" series of popular science lectures; volunteer clean-ups and environmental task forces, etc.) [15]

As it is mentioned at the beginning, the strategic goal of NEFU's development was concretized considering sustainable development ideology. The university's mission is to form the new generation of professionals who realize the values and goals of sustainable development of the North and the Far East, influencing the solution of global issues.

The university's development program forms the model of the sustainable development of the northern territories from three components of sustainability:

- **Sustainable society**, as a combination of different social groups with a variety of values and traditions, interests and needs, history and culture, living together in extreme climate of the North;
- **Sustainable environment**, as a natural space of conflict-free, non-destructive coexistence of animate and inanimate nature representatives;
- **Sustainable economy of the North**, as a condition to create a living space for a human in extreme conditions, the basis of his material well-being, comfort of living, and high living standards.

These areas are independent areas of knowledge, but at the intersection of these areas, the spheres of interaction "human and nature", "nature and economy", "economy and human" arise. The imbalance in any of these subsystems disrupts the stability of the system as a whole.

NEFU's development program is based on the following **academic areas** corresponding to the priority development areas of science and technology of the Russian Federation, meeting the objectives of socio-economic development of northern territories and the Far East, which comply with the UN Sustainable Development Goals and the objectives of national projects:

1. Sustainable economy: economic development and well-being;
2. High living standards;
3. Sustainable society: social justice;
4. Favourable environment;
5. Sustainable nature: environmental well-being;
6. Lean environmental management;
7. Digital technologies.

The comparative analysis of the priority areas of the universities of the Far North of Russia and the northern countries of Europe and America shows the relevance of the selected priorities to the global research agenda and the conceptual foundations of higher education development in increasing uncertainty in the environment.

The transition to a new development model is based on the following **key principles of the development of the university**:

1. *Social responsibility* – the university's responsibility for the results of the impact of its decisions and activities on society and nature, considering ethical standards and possible social, economic, and environmental consequences.
2. *The priority of human interests* – creating a comfortable space for the full realization of students and employees' potential.
3. *The integration of education, science, applied research and production.*
4. *Digital transformation of the university.*
5. *The concentration of resources on the result* – focusing resources on promising areas of development.

3 Conclusion

Due to the growing human impact on northern ecosystems, global climate changes, and the increasing military and economic presence of foreign states in the macroregion, the issue of ensuring the northern territories' sustainability becomes pressing. In these circumstances, universities base their activities on sustainable development values and goals.

A brief analysis of the activities of universities in the North of Russia and northern countries of Europe and America, the results of modelling NEFU's development program show the demand for new strategic initiatives to respond flexibly to the growing unpredictability of global processes. The comparative analysis allows concluding that the necessary institutional and infrastructural conditions have been created for the further progressive development of the North-Eastern Federal University and the implementation of large-scale research, educational, and innovative projects aimed at the sustainable development of the Far East and the North of Russia.

References

1. M. Mbah, A. Johnson, F. Chipindi, *IJED*, **83**, 102355 (2021)
2. V.M. Savvinov, *University is an intellectual, innovative and spiritual platform for sustainable development of a macroregion:* monograph, (M.K. Ammosov North-Eastern Federal University, Yakutsk, 2021)
3. Order of the Government of the Russian Federation No. 2464-r *The national program for the socio-economic development of the Far East for the period up to 2024 and for the long term until 2035,* Retrieved from: https://www.garant.ru/products/ipo/prime/doc/74587526/
4. Decree of the President of the Russian Federation dated March No. 164 *The fundamentals of the state policy of the Russian Federation in the Arctic for the period up to 2035,* Retrieved from: http://www.garant.ru/products/ipo/prime/doc/73606526/
5. *Sustainable Development Goals.* Retrieved from: https://www.un.org/sustainabledevelopment/ru/sustainable-development-goals/
6. Kuznetsov V.V., Lukina A.V., Malova D.V., *Bulletin of the Plekhanov Russian university of Economics,* 1 (91), 56-64 (2017)
7. Brendan O'Malley, *Universities are key to achieving sustainable development,* Retrieved from: https://www.universityworldnews.com/post.php?story=20200711094917938
8. *Seizing a sustainable future. Strategy 2030.* Retrieved from: https://www.uef.fi/en/strategy-2030.
9. UiT The Arctic University of Norway. Retrieved from: https://www.timeshighereducation.com/world-university-rankings/uit-arctic-university-norway
10. *Development program of the federal state budgetary educational institution of higher education "Pitirim Sorokin Syktyvkar State University" for the period 2017-2021.* Retrieved from: https://www.syktsu.ru/sveden/common/progamma_strategii 3% 20% D0% B8% D1% 8E% D0% BB% D1% 8F% 202017.pdf
11. *Report on self-examination of the university* Retrieved from: http://www.kamgu.ru/sveden/files/Otchet_o_samoobsledovanii_vuza_na_04/01/2020.pdf
12. Saginova O.V., Saginov Yu.L., Grishin A.I., *Sustainable development of the university.* Retrieved from: https://cyberleninka.ru/article/n/ustoychivoe-razvitie-universiteta/viewer.
13. *Report on Achieving the Sustainable Development Goals, 2019, North-Eastern Federal University.* Retrieved from: https://www.s-vfu.ru/universitet/rukovodstvo-i-struktura/strukturnye-podrazdeleniya/DSR/Docs/%D0%A1%D0%92%D0%A4%D0%A3_%D0% A6% D0% A3% D0% A0.pdf
14. *The program of training experts for leadership in the field of ecology, culture and sustainable development in the regions of the Far East and the*

Arctic, RJE3. Retrieved from: https://www.s-vfu.ru/universitet/rukovodstvo-i-struktura/strukturnye-podrazdeleniya/ums/RJE3/

15. Gavrilieva T.N., Sugimoto A., Fuji M., etc. *Higher education in Russia,* 27 (7), 52-65 (2018)
16. Resolution of the State Assembly (Il Tumen) of the Republic of Sakha (Yakutia) No. 46-VI. *The strategy of socio-economic development of the Republic of Sakha (Yakutia) until 2032 with the definition of a target vision until 2050,* Retrieved from: http://docs.cntd.ru/document/550299670

Economic efficiency of the introduction of the automated locomotive reliability management system

Viktoriya A. Olentsevich[1,*] *Viktor V.* Kondratiev[2], *Boris O.* Kuznetsov[1], and *Antonina I.* Karlina[2]

[1] Irkutsk State Transport University, 15, Chernyshevsky Str., Irkutsk, Russia,
[2] Irkutsk National Research Technical University, 83, Lermontov Str., Irkutsk, Russia

Abstract. Railway transport has its characteristic features: a specific investment structure, its relatively high capital intensity and duration of implementation of innovative projects, technological continuity of the production process, and much more. The specified features are also appropriate to service locomotive depots and actively influence the development of methodological approaches to assessing the economic efficiency of innovative projects. Of course, the main task of innovative projects is not to increase their number, but to increase the long-term efficiency of the transport company. To achieve this goal, according to the authors, it is advisable to use the balanced scorecard (BSC), which is an element of the company's strategic management, but it must be adapted to innovative projects. The use of the balanced scorecard in the service locomotive depot of the railway transport system will allow the industry to reach a new level of functioning and make it possible to assess the economic efficiency of innovative projects as objectively and accurately as possible. The proposed methodology for assessing the effectiveness of an innovative project makes it possible, on the one hand, to economically justify the project, on the other hand, to monitor production efficiency both for the purposes of the project itself and to stimulate the innovative activity of the enterprise. With that, the result of an innovative project is considered not as static, but as a set of indicators for its implementation, which has specific dynamic, economic, financial, technical, technological and other characteristics.

Keywords: railway transport system, service locomotive depot, target map, planning horizon, locomotive yardtime, production efficiency monitoring.

1 Introduction

One of the key tasks stipulated by the Strategy for the Development of Railway Transport of the Russian Federation until 2030 is the effective use of the transit potential of Russian railways with reference to the development of sections of international transport corridors passing through the territory of Russia. Its effectiveness of the formation is ensured on the basis of the principle of the proper quality of transportation, including the timeliness of delivery and safety of goods [3, 10].

Ensuring the competitiveness of enterprises of the railway transport system is impossible without carrying out work on their technical re-equipment, the use of modern technologies, the development of new products that are in demand by consumers of transport and related services. It should be borne in mind that these measures, which should be carried out within the entire transport complex, are not effective without a comprehensive solution of problems aimed at improving the management system, labor organization, the introduction of advanced practices in the scientific organization of labor and modernization, equipment maintenance [2, 6, 11].

The locomotive complex, as one of the key links in the production unit of Russian Railways OAO, significantly affects the overall efficiency of the Company and cannot be kept out of the ongoing transformations. The main functions of repair enterprises and sectors in the railway transport system are to perform the following types of work: ensuring the technical functional state of equipment (and other fixed assets); reduction of unproductive yardtime of rolling stock; reduction of repair and maintenance costs while maintaining the repair quality; scientifically substantiated specification of labor costs for individual operations to perform repair and restoration work; automation of production processes, etc. [8].

Analysis of the world experience of operating the locomotive fleet shows that the most successful relationship model is that one, in which the manufacturer of this equipment bears the main responsibility for the technical condition throughout the entire life cycle of the locomotive to its owner. When introducing this procedure, rolling stock manufacturers have additional incentives: to invest in improving the locomotive design; increasing the level of its reliability and uninterrupted operation; supply of the necessary modern technological

* Corresponding author: olencevich_va@mail.ru

equipment; automation of the production cycle; staff motivation [7, 13, 15].

The use of BSC in a service locomotive depot will allow the enterprise to reach a new level of functioning and will make it possible to assess the economic efficiency of innovative projects as objectively and accurately as possible.

2 Problem Statement

Using a balanced scorecard (BSC) in estimating economic efficiency of the introduction of the automated locomotive reliability management system

3 Research Questions

The BSC target map is applicable to the service locomotive depot

BSC works on the basis of measuring and evaluating performance according to a set of optimally selected indicators that measure the goals of an innovative project and reflect all its aspects, both financial and non-financial. Therefore, one of the most important measures in the preparation of the BSC application is the selection of key performance indicators (Key Performance Indicators, KPIs), since the use of the selected indicators will make it possible to make the goals of the innovation project more specific and measurable [1, 5, 12].

The number of areas considered in the BSC may vary, and therefore it is advisable to use this system in relation to the service locomotive depot in three main areas: finance, internal business processes and personnel management.

In these areas it is necessary to clearly formulate the ultimate goals that are pursued by the service locomotive depot as part of an innovative project. For example, in the direction of "finance" it is advisable to choose the profitability of an innovative project as the main goal, since it is the achievement of this goal that will lead to the successful implementation of the project as a whole. In the direction of "business processes" it is necessary to improve, first of all, critical internal production processes. Thus, the following goals can be formed in this direction: improving the quality of technical inspection and current repair of locomotives, as well as reducing the yardtime of locomotives for these types of repairs.

In the area of "personnel and systems", the perspective of personnel training and increasing the level of automation are designed to provide long-term growth and improvement.

Based on the above, we will draw up a map of BSC goals applicable to a service locomotive depot (Fig. 1).

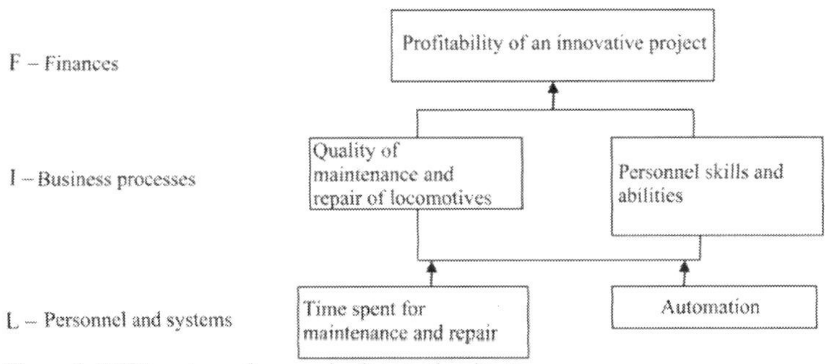

Figure 1. BSC target map for a service locomotive depot

After setting the BSC goals, it is necessary to establish a set of indicators, with the help of which we will evaluate the effectiveness of achieving each of the goals and the entire innovative project, table 1.

Table 1. Key Performance Indicators (KPI)

Direction	Goals	KPI Indicators
F	F1. Profitability of an innovative project	Net present value, NPV
		Discounted project payback period, DPP
I	I1. Reducing the time for maintenance and repair of locomotives	Yardtime of locomotives for unscheduled repairs
	I2. Improving the quality of maintenance and repair of locomotives	Locomotive technical availability coefficient
L	L1. Automation	Automation level

The proposed methodology for assessing the effectiveness of an innovative project makes it possible, on the one hand, to economically justify the project, on the other hand, to monitor production efficiency both for the purposes of the project itself and to stimulate the innovative activity of the enterprise. With that, the result of an innovative project is considered not as static, but as a set of indicators for its implementation, which has specific dynamic, economic, financial, technical, technological and other characteristics [9, 14, 16].

Assessment of the economic efficiency of the introduction of an automated system for locomotive reliability management based on the proposed method.

Based on the proposed methodology, using the balanced scorecard system (BSC) developed by the authors in relation to the service locomotive depot, we will assess the economic efficiency of the introduction of the automated locomotive reliability management system (ALRMS) in the railway transport system.

The first indicator to be assessed within the BSC is net present value (NPV). To calculate it, you need a cash flow (CF) determined by the formula (1):

$$CF = CI - CO, \quad (1)$$

where CI (Cash Inflows) is the cash inflow;

CO (Cash Outflows) is the cash outflow.

As previously noted, the planning horizon is 5 years. In this case the discount rate (r) equal to 10% is taken to determine the discounted cash flows. Calculations are given in table 2.

Table 2. Determination of net present value (NPV)

Period of time	Initial costs, IC	Cash Inflow, CI	Cash Outflow, CO	Cash flow, CF	Discounted cash flow
0	3.494	-	-	-	-
1	-	3.274	5.558	-2.284	-2.077
2	-	3.616	3.835	-0.218	-0.180
3	-	4.345	1.976	2.369	1.780
4	-	4.172	1.219	2.953	2.017
5	-	4.236	0.384	3.852	2.392
NPV					0.437

NPV = (-2.077 – 0.18 + 1.78 + 2.017 + 2.392) – 3.494 = = 0.437 million rubles.

The net discounted income from the implementation of ALRMS is greater than zero and is equal to 0.437 million rubles, which indicates the attractiveness of the investment project, while the project provides coverage of future costs.

The next indicator that needs to be assessed in the framework of the BSC is the discounted payback period of the project (Table 3).

Table 3. Determining the discounted payback period (DPP)

Period of time	Initial costs, IC	Cash flow, CF	Discounted cash flow	Discounted cash flow running total
0	3.494	-	-	-
1	-	-2.284	-2.077	-2.077
2	-	-0.218	-0.180	-2.257
3	-	2.369	1.780	-0.477
4	-	2.953	2.017	1.540
5	-	3.852	2.392	3.931

According to the calculations (Table 3), the amount of cash receipts from the implementation of ALRMS will pay back the initial project costs in the fifth year, the payback period of the project is 5 years.

So, in the direction of "finance" using the key performance indicators of the project, an assessment is given that indicates the prospects of the project and the goal of "profitability of the project" has been achieved.

The next step is to assess the KPI in the direction of "business processes". The first indicator that needs to be assessed is the yardtime of locomotives for emergency repairs. The analysis showed a decrease in yardtime due to the elimination of violations of operating modes and due to the identification of pre-failure conditions during emergency repairs.

4 Purpose of the Study

The next indicator to be assessed is the coefficient of technical availability of the locomotive, calculated by the formula (2):

$$\text{CTA} = \frac{t_{wc}}{t_{wc}+t_{ic}+t_{sch}}, \quad (2)$$

where t_{wc} is the total time of the locomotives staying in a working condition in the considered period of operation, hours;

t_{ic} is the total time spent by locomotives in an inoperative condition for emergency types of repairs in the considered period of operation, hours;

t_{sch} is the total time of the locomotives staying in an inoperative condition due to their scheduled maintenance and repairs in the considered period of operation, hours;

Analysis of the values of the coefficient of the technical availability of a locomotive by elements showed that a decrease in the locomotive yardtime for emergency repairs should lead to an increase in its value. It should be noted that predicting changes in this indicator is a very difficult task, since it is necessary to take into account many factors that affect both the operating time of locomotives and their yardtime for planned and emergency types of repair, therefore the authors build their predictions on the experience of using ALRMS in the service locomotive depot of the railway station Zima of the East Siberian railway. So, during the use of ALRMS in this depot, it was possible to increase the technical availability coefficient of the locomotive by 1.02 %. Therefore, for further research, we will take this value as a basis. Predicted values of the indicator of the locomotive's technical availability coefficient are presented in Table 4.

Table 4. Predicted values of the indicator of the locomotive's technical availability coefficient in the service locomotive depot

Indicator	Year 2020	Year 2021	Year 2022	Year 2023	Year 2024
Locomotive technical availability coefficient	0.914	0.924	0.933	0.942	0.954

5 Research Methods

So, the indicators to be assessed within the framework of the BSC system in the direction of "business processes" have been achieved, which in turn leads to the achievement of the set goals:

- reducing the yardtime of locomotives in the process of maintenance and current repair;
- improving the quality of maintenance and current repair of locomotives.

The last and most important indicator when assessing the implementation of an ALRMS project using a balanced scorecard is the level of automation. The main goal of automation is to improve quality, while the automated process is more stable than the process performed manually. The automation level is assessed by the authors using a questionnaire survey of experts according to the point system. Representatives of the service locomotive depot and service users will act as experts. The indicators will be assessed by experts based on the following adopted assessment system [4]:

- 0 – the process is poorly organized;
- 1 – the process is satisfactorily organized;
- 2 – the process is well organized.

Data from a questionnaire survey of experts on the level of automation of ALRMS are presented in Table 5.

Table 5. Results of the questionnaire survey of experts

Direction of assessment	Expert 1	Expert 2	Expert 3	Expert 4	Expert 5	Expert 6	Expert 7	Expert 8	Expert 9	Expert 10
Consistency of the ALRMS actions	2	1	2	1	2	1	2	1	2	1
The level of ALRMS integration into the production process	1	2	1	2	2	1	2	1	2	2
The ALRMS integration level for coordination with existing software platforms	2	2	2	1	1	2	2	1	2	1
Independence of the ALRMS execution	1	1	2	2	2	2	1	2	2	2
Coordination of ALRMS with the economic objectives of the enterprise	1	2	1	2	1	2	1	1	2	2
Using the functionality of the ALRMS software	2	2	2	1	2	1	2	2	2	1

Further, based on the data in Table 5, we estimate the level of automation using the formula (3):

$$A = \frac{\sum a_i}{N \cdot a_{max}}, \quad (3)$$

where A is the level of automation;

$\sum a_i$ is the number of points for which the automation was assessed by a specific expert;

N is the number of experts participating in the questionnaire survey, persons;

a_{max} is the maximum score according to the questionnaire survey results.

The results of the questionnaire survey conducted are presented in Table 6.

Thus, according to (3), the level of automation can be assessed as high.

- A = 0.2 – 0.35 is the low level of automation;
- A = 0.36 – 0.65 is the average level of automation;
- A = 0.66 – 1 is high level of automation.

Further, it is necessary to determine the degree of consistency of experts' opinions through the coefficient of concordance, using well-known methods of mathematical statistics, since if the parameters recommended by statistics are not met by the degree of consistency of opinions, the results of this survey are not taken into account in the future [16]. Calculations of the coefficient of concordance (Table 7), the value of which was 0.635, indicate a fairly high degree of agreement among experts.

Table 6. Assessment of the level of automation after the implementation of ALRMS

Direction of assessment	Number of experts, persons	Total points	Maximum score on the questionnaire	Automation level
Consistency of the ALRMS actions	10	15	2	0.75
The level of ALRMS integration into the production process		16	2	0.80
The ALRMS integration level for coordination with existing software platforms		16	2	0.80
Independence of the ALRMS execution		17	2	0.85
Coordination of ALRMS with the economic objectives of the enterprise		15	2	0.75
Using the functionality of the ALRMS software		17	2	0.85
Average score				0.80

Table 7. Results of the questionnaire survey of experts

Direction of assessment	Expert 1	Expert 2	Expert 3	Expert 4	Expert 5	Expert 6	Expert 7	Expert 8	Expert 9	Expert 10	Total	Sum of squares of differences
Consistency of the ALRMS actions	4	5	6	4	4	4	4	2	4	4	41	30.25
The level of ALRMS integration into the production process	3	3	3	2	3	3	3	1	3	2	26	90.25
The ALRMS integration level for coordination with existing software platforms	2	2	4	3	2	2	2	3	2	1	23	156.25
Independence of the ALRMS execution	1	1	1	1	1	5	1	4	1	3	19	272.25
Coordination of ALRMS with the economic objectives of the enterprise	5	4	5	5	6	4	5	5	5	5	49	182.25
Using the functionality of the ALRMS software	6	6	2	6	5	6	6	6	6	6	5	380.25
Total	21	21	21	21	21	21	21	21	21	21	213	1111.5

6 Findings

The main task of innovative projects is not to increase their number, but to increase the long-term efficiency of the transport company. To achieve this goal, according to the authors, it is advisable to use the balanced scorecard (BSC), which is an element of the company's strategic management, but it must be adapted to innovative projects. The use of the balanced scorecard in the service locomotive depot of the railway transport system will allow the industry to reach a new level of functioning and make it possible to assess the economic efficiency of innovative projects as objectively and accurately as possible. The proposed methodology for assessing the effectiveness of an innovative project makes it possible, on the one hand, to economically justify the project, on the other hand, to monitor production efficiency both for the purposes of the project itself and to stimulate the innovative activity of the enterprise. With that, the result of an innovative project is considered not as static, but as a set of indicators for its implementation, which has specific dynamic, economic, financial, technical, technological and other characteristics

7 Conclusion

The difficult economic conditions in which both the state in general and the railway industry in particular find themselves, the slowdown in economic growth of the industry, and a decrease in traffic volumes require a competent implementation of the policy of effective cost and financial resource management.

The goal of innovative projects implemented in the railway transport system in order to increase the level of its competitiveness in the domestic and foreign transport market is not to increase their number, but primarily to increase the long-term efficiency and smooth operation of the transport company. To achieve this goal, according to the authors, it is advisable to use the balanced scorecard (BSC), which is an element of the company's strategic management, but it must be adapted to innovative projects. On the basis of the proposed methodology, using the balanced scorecard system (BSC) developed by the authors in relation to the service locomotive depot, the economic efficiency of the introduction of the automated locomotive reliability management system (ALRMS) in the railway transport system has been estimated. With regard to the service locomotive depot, it is advisable to use this system with three main indicators: finance, internal business processes and personnel management. All the goals set within the BSC have been achieved, which indicates the effectiveness and success of the ALRMS project being implemented.

References

1. L.V. Eichler, M.A. Zotkina. Improvement of cost management methodology for a freight road transport enterprise in modern conditions. *Economiczne nauki. Przemysl: Nauka i studia. Materialy VII Miedzynarodowej naukowi-praktycznej konferecji "Dynamika naukowych badan*, vol. 3 (2011), pp. 80–84
2. N.N. Grigorieva, V.A. Olentsevich, N.P. Astashkov. Formation of new principles and models of the industry structural units in the context of the introduction of digital technologies. Modern technol. Syst. Anal. Model, **65(1)**, 156–165 (2020). DOI: 10.26731/1813-9108.2020.1(65).156-165.
3. *The Development Strategy of Russian Railways Holding for the period until 2030* (Rus. Railways OAO, Moscow, 2013)
4. S.G. Kamshilov. Determination of the level of automation of business processes at an industrial enterprise. Bull. of the Chelyabinsk State Univ., **1**, 56–61 (2015)
5. R. Kaplan, D. Norton. *Balanced Scorecard* (Biznes Olymp Publ., Moscow, 2002)
6. N.V. Krol, N.N. Grigoryeva, V.A. Olentsevich, R.Yu. Upyr. Ways to improve the economic mechanism in the development of the customer base of freight traffic. Modern technol. Syst. Anal. Model., **4(64)**, 155–161 (2019)
7. Maintenance of rolling stock in European countries. World Railways, **4**, 50–53 (2009)
8. N.L. Mikhalchuk. The relevance of the development of a locomotive repair complex in the transition to full maintenance service. *Prospects for the development of locomotive service. Mat. of the First Int. sci. and pract. conf.* (TMKh-Service, Moscow, 2014)
9. D. Norton. Implementation of BSC: advice from the author of the concept. Finan. Direct., **4** (2006)
10. *Official site Russian Railways OAO*. Retrieved from: www.rzd.ru
11. V.A. Olentsevich, N.P. Astashkov. Methods of attracting customers to the services of the transport and logistics business unit in order to increase the competitiveness of the Russian Railways holding company. Transp. Infrastruct. of the Siber. Reg., **1**, 4–8 (2019)
12. M. Porter. *Competitive strategy: methods for analyzing industries and competitors*. Trans. from English (Alpina Business Books, Moscow, 2007)
13. N.I. Pyzhyanov. On the way to high-quality service maintenance of the locomotive fleet (domestic and foreign experience). Online magaz. "Sci. Stud.", **8(3)** (2016)
14. *Prospects for the development of locomotive service. Mat. of the First Int. sci. and pract. Conf.* (TMKh-Service OOO, Moscow, 2014)
15. G.V. Silicheva, N.I. Pyzhyanov. Problems and prospects of development of the locomotive repair complex of "Russian Railways" OAO. Modern technol. Syst. Anal.. Model., **2(46)**, 143–147 (2015)
16. M.A. Zotkina. Formation of cost management costs based on a balanced scorecard. In: *Actual problems of economics and management of Russian enterprises*. The collection of scientific papers, vol. 4 (pp. 43–48) (SibADI Publ., Omsk, 2009)

Use of a digital platform in the selection of socially significant projects (through the example of the lisp in the Yakutia)

L Ushnitskaya*, *A* Feofanova

North-Eastern Federal University, 58, Belinsky ave., Yakutsk, 677000, Russia

Abstract. The relevance of the research issue is explained by the global transition to digital format. Under the conditions of digitalization of all spheres of life, it turned out to be necessary to conduct a digital dialogue between state executive authorities and population. The Republic of Sakha (Yakutia), being the leading region of the implementation and development of IT technologies in the Far East, has introduced the use of a digital platform to resolve local matters. The paper discusses the experience of the implementation of the Local Initiatives Support Program. The transformation of the program was studied, taking into account the specific features of the republic, the presence of the Arctic territories in it, as well as the use of a digital platform as a tool to improve the implementation of programs.

1 Introduction

Nowadays participatory budgeting is quite well developed and is presented by different initiatives depending on the needs of countries all over the world. One of these practices is the Local Initiatives Support Program (hereinafter referred to as LISP). LISP has been implemented in the Russian Federation since 2007. The variety of the implementation of LISP projects in the constituent entities of the Russian Federation is reasoned by not only to different management processes at all levels, but also by special factors in the territory of the implementation. As of the end of 2019, LISP was already being actively implemented in over 61 regions of the country. In the Republic of Sakha (Yakutia), LISP is one of the programs for solving socially important local problems.

During the implementation of this program the important social and institutional effects of a medium-term nature are achieved as a result of the large-scale implementation in the territory of the Republic of Sakha (Yakutia) of public infrastructure development projects based on local initiatives in addition to the direct results in the form of the development of municipal infrastructure. These effects are as follows: the increase in social activity of the population, the decrease in dependent rate, the growth of mutual trust between the population and the authorities, etc. Despite these results, the republic does not cease to improve the practice of participatory budgeting.

2 Materials and methods

LISP is one of the initiatives of participatory budgeting. It is based on the concept of direct involvement of citizens in the solution of local issues and, in particular, in the decision on the choice of priority areas for spending part of budget funds at the municipal, regional or national level, as well as subsequent control over the implementation of these decisions.

3 Results and discussion

The Republic of Sakha (Yakutia) started the implementation of the LISP in 2017. In 2017, only 5 pilot territories - municipalities, which included urban and rural settlements that were part of the municipal districts: Verkhnevilyuisky, Churapchinsky, Ust-Aldansky, Khangalassky settlements (districts) and the urban district "Zhatay" participated in the LISP. Since 2019, all 445 municipalities of the republic have been participating in the LISP.

The most significant factor for winning the tender selection of the LISP is the activity (number) of the population that was present at the meeting during the choice of a particular project. In total, 928 thousand people live on the territory of the republic of participating municipal organizations, 203 thousand people or 22% took part in preliminary discussions of projects for 2020, 86 thousand people or more than 9% of the total population took part in the final meetings. There is the increase in the dynamics of population activity over the years of the implementation of the LISP.

Table 1. The number of meetings in the municipalities of the Republic of Sakha (Yakutia) organized by the LISP for 2018-2020, thousand people

*Corresponding author: ulei50@mail.ru

 https://doi.org/10.1051/shsconf/202111200050

Years	Number of residents in municipalities	Participated in the preliminary identification of the problem	Took part in the final meetings
2018	586	165	82
2019	927	183	85
2020	928	203	86

The main result of the implementation of the participatory budgeting mechanism is the realization of the potential of citizens in the solution of specific socially significant tasks, which is confirmed by the activity and interest of the population, a change in the attitude of people to their own role in the development of territories, control by the population over the spending of program funds, the increase in the level of confidence in the actions of authorities in general.

Table 2. Information about the meetings and the projects for 2018-2020

	2018	**2019**	**2020**
Number of meetings	528	627	551
Number of discussed projects	1545	2225	1832
Number of projects submitted for tender	503	565	503
Number of winning projects	451	443	332

If we talk about the Arctic territory of the republic, then we can confidently say that the residents of the Arctic territories are as much interested in the improvement of the territory as the residents of the central regions. Throughout the entire time of the implementation of the LISP on the territory of the republic, from the municipalities of the Republic of Sakha (Yakutia), which are part of the Arctic zone of the Russian Federation, 286 applications were submitted to tender (18% of the total number of applications in the republic), including 94 in 2018 applications, 98 applications –in 2019, 94 applications - in 2020.

Due to the repeated appeals from the heads of municipalities to the regulatory legal act governing the procedure of tender selection and allocation of subsidy funds for the LISP, the particular changes were made regarding the creation of a separate window when conducting tender selection for municipalities of the Arctic regions. As practice shows, the municipalities of the Arctic regions can not compete on equal terms with the central regions, which is due to the low number of residents of settlements, the remoteness of territories from each other and special conditions for the timing and cost of the delivery of materials and high costs in the implementation of projects.

Earlier, by the Decree of the President of the Russian Federation V.V. Putin dated May 13, 2019 N 220 "On Amendments to the Decree of the President of the Russian Federation of May 2, 2014 N 296" on the Land Territories of the Arctic Zone of the Russian Federation additional 8 regions of Yakutia were included. Thus, all northern regions of the republic are now classified as arctic.

The additional window for arctic regions will help to increase the number of participants, the number of winners among the municipalities of the arctic regions and the elimination of competition with the rest of the regions. Thus, the projects of the municipalities of the Arctic regions will compete on equal terms between similar projects.

In 2018, citizens usually gave preference to the projects on the arrangement of sports grounds, the next most important projects were related to the organization of water supply and landscaping. In 2019, the share of projects aimed to improve water and heat supply facilities increased, followed by projects on the arrangement of sports grounds. In 2020, the first was the share of projects related to the improvement of places of residence and aimed at the improvement of territories (these are the places of public entertainment, landscaping of parks and squares, adjacent territories, arrangement of sidewalks, landscaping of burial sites, lighting projects, etc.). The second was the share of water and heat supply facilities. The activity of the population of the Arctic regions is presented in Table 2.

Table 3. The number of meetings of the municipalities of the Arctic regions of the Republic of Sakha (Yakutia) according organized by the LISP for 2018-2020

Years	The number of residents living in the municipalities of Arctic regions	The number of residents who took part in the preliminary identification of the problem	The number of residents who took part in the final meetings
2018	67 583	15 529	9 262
2019	67 674	26 006	10 166
2020	67 625	25 847	12 958

In general, we can note the increased interest in the project on the part of the population and municipalities and participation in in it. The achieved results, as well as the fairly high quality of documents prepared directly in municipal settlements, indicate the ability and readiness of the municipalities of the Republic of

Sakha (Yakutia) and the population to implement larger-scale projects related to the participation of the population.

The amount of subsidies from the State budget of the Republic of Sakha (Yakutia) for co-financing projects for the development of public infrastructure based on local initiatives is 500.0 million rubles annually. The figures show that co-financing for the project from all sources is increasing every year by an average of 27.5%. This indicator also shows the high interest of the population of the republic in the development of the place of residence and the improvement of public spaces. For each ruble of subsidies allocated from the state budget of the Republic of Sakha (Yakutia), an additional 37.2 kopecks of co-financing is attracted.

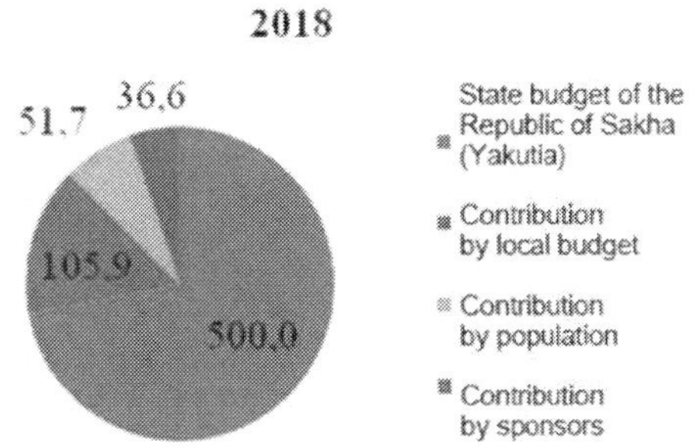

Figure 1. The ratio of state budget funds and additional co-financing for 2018, million rubles

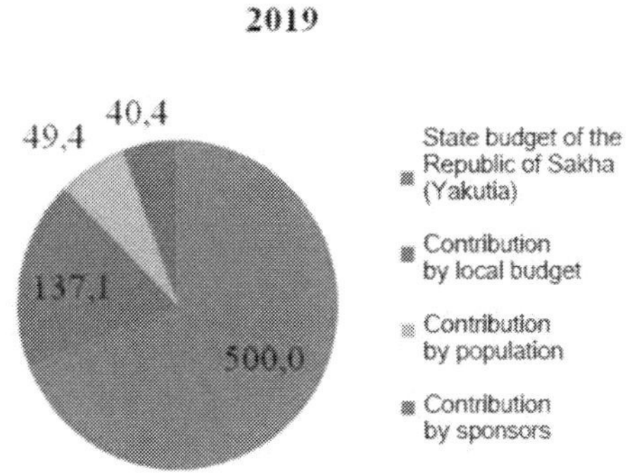

Figure 2. The ratio of state budget funds and additional co-financing for 2019, million rubles

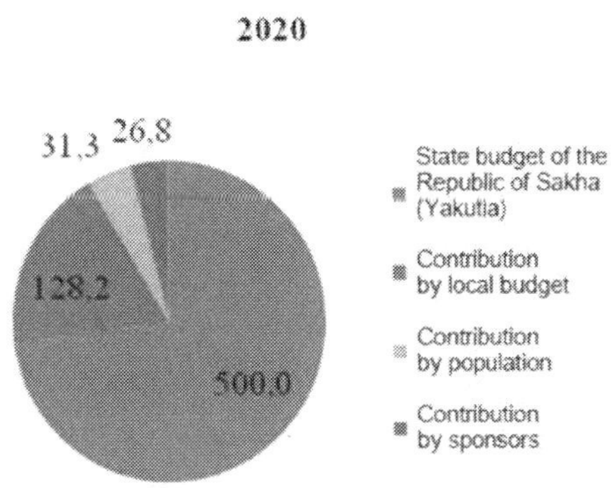

Figure 3. The ratio of state budget funds and additional co-financing for 2019, million rubles

Starting from 2020, the LISP became a part of the republican project "Hometowns and Village" (hereinafter referred to as HaV), which united all programs with the initiative of citizens. The project included republican programs implemented on the principles of the support of the initiatives of citizens, such as: the Program for Supporting Local Initiatives, the All-Republican Movement of Good Deeds "My Yakutia in the 21st Century", the Regional Project "Formation of a Comfortable Urban Environment" and Integrated Development of Rural Areas of the Republic of Sakha (Yakutia). The project was established in accordance with the instructions of the Head of the Republic of Sakha (Yakutia) A.S. Nikolaev.

The purpose of the HaV is to enhance the participation of the residents of the Republic of Sakha (Yakutia) in the solution of local problems, the availability and ease of the participation of the republic population in the selection of projects to create comfortable living conditions for them and to increase the social and civic activity of citizens.

The tasks of the HaV are:

- The increase in the income and employment of the population;
- The increase in the efficiency of spending budget funds;
- The intensification of the interaction of local self-government bodies of municipalities and the executive power and residents of the Republic of Sakha (Yakutia);
- The increase in the level of information transparency in the activities of local self-government bodies of municipalities and the executive power of the Republic of Sakha (Yakutia);
- The improvement of the quality of the solution of local issues by local self-government bodies of the Republic of Sakha (Yakutia);
- The intensification of control by the population over the exercise of the powers of local self-government bodies of municipalities and the executive power of the Republic of Sakha (Yakutia);
- The increase in public satisfaction with the quality of work of local self-government bodies of municipalities and the executive power of the Republic of Sakha (Yakutia).

For the optimization of work with the population within the framework of the HaV, the following activities are carried out:

- Simultaneous organization of meetings of the population at the 1st stage, at which projects will be selected;
- The introduction of an integrated information portal "OneClickYakutia", where any citizen of the republic can get full information about the implementation of a project on the territory of a particular municipality, as well as about the timing of voting and projects proposed for discussion;
- A general interdepartmental tender commission, which will carry out work on the consideration of projects based on the initiatives of citizens of the Republic of Sakha (Yakutia), chaired by the First Deputy Chairman of the Government of the Republic of Sakha (Yakutia).

The implementation of the HaV consists of 4 stages:

- Stage 1 is preparatory and it includes: training of the representatives of municipalities, organization of meetings with the public, distribution of information on the acceptance of tender documents;
- Stage 2 is tender selection of projects, it includes: the formation of tender applications, their submission on the site "OneClickYakutia", work with tender documents, preparation of materials for the meeting of the interdepartmental commission and the tender selection of projects;
- Stage 3 is the allocation of funding, it includes: the work of chief administrators of budgetary funds on the preparation of the necessary regulatory legal acts and work with the Ministry of Defense to conclude agreements for financing projects;
- Stage 4 is the implementation of projects by the winners of the tender selection; it includes the organization of tender procedures for the selection of contractors, implementation of projects and submission of reports on implementation.

The main achievement of the HaV implementation was the introduction of an integrated information digital portal "OneClickYakutia". The largest digital module, the LISP program, has been fully transferred to the digital platform "OneClickYakutia".

According to the LISP, the selection of initiative projects for 2021 was online using the digital platform "OneClickYakutia" and social networks. Today, 182 ideas have been proposed by initiative groups on the OneClickYakutia digital platform. About 34,926 residents of the republic took part in online meetings. On the LISP website, 457 reports of the final meetings were uploaded, a total of 1974 projects were discussed, and 477 applications were created for the participation in tender selection.

3 Conclusion

As a result of the introduction of an integrated information portal "OneClickYakutia", any citizen of the republic can receive full information about the implementation of a project in the territory of a particular municipality, as well as about the timing of voting and projects proposed for discussion. It also provides a one-time acceptance of tender application using the digital platform "OneClickYakutia".

We can state that both the online project selection stage and the final voting stage were tested and proven in the pilot mode on the digital LISP module. The municipalities, including those of the Arctic regions organized meetings and final voting on an integrated digital platform 'OneClickYakutia". Therefore it gave the greatest effect in the attraction of participants to the discussion on the choice of an initiative project. This innovation, in the form of digital voting, in the process of resolving issues of local importance, greatly facilitated and improved the implementation of initiative projects in the context of openness and accessibility of data for the population and the effective activity of the government.

This experience of the implementation of digital voting in the LISP within the framework of the HaV can be repeated and recommended to other programs implemented on the principles of the support of the initiatives of citizens, as well as at the level of other regions of the country.

References

1. Decree of the Head of the Republic of Sakha (Yakutia) No. 1542 *"On the project Hometowns and villages"*, (2020)
2. *Guidelines for the preparation and implementation of participatory budgeting in the Russian Federation of the Ministry of Finance of the Russian Federation*, (2020)
3. *Report of the Ministry of Finance of the Russian Federation on the best practices for the development of participatory budgeting in the constituent entities of the Russian Federation and municipalities* (2020)
4. *Integrated digital platform "OneClickYakutia"*, module of the Program for Supporting Local Initiatives of the Republic of Sakha (Yakutia) Retrieved from: https://ppmi.yakutia.click
5. Vagin V.V., Pominov K.V.a, et al. *50 questions about participatory budgeting: a tutorial* (Moscow: Filin, 2018)

International legal treaty as a basis for scientific cooperation in the Arctic regions

O Maksimova*, *and A* Armashova

Moscow University for the Humanities 5, Yunosti St., Moscow, 111395, Russia

Abstract. The paper analyzes international treaties related to the activities of states in the Arctic regions. These treaties constitute the legal basis for scientific cooperation. Due to the threat of climate change and global warming, the preservation of the Arctic ecosystem is becoming one of the urgent tasks for global scientific community. Russia, with its vast Arctic territories, can play a key role in joining the efforts of scientists from different countries. In the modern world, international cooperation is impossible without the established system of international legal treaties. The main instrument of international law for cooperation in the Arctic is the 1982 UN Convention on the Law Of the Sea. The success of scientific events in the Arctic depends on the results of the activities of the Arctic Council and on the effectiveness of activities, including research during the implementation of the Strategy for the Development of the Arctic Zone of the Russian Federation.

1 Introduction

Russia is a groundbreaker and has long been a leader in the field of polar research. National scientific research in the Arctic began in the 18th century, when the Great Northern Expedition (1733–1743) of Vitus Bering was organized [1].

During the Soviet period, there was qualitative breakthrough in the development of the Arctic. During this period regular scientific research began to be carried out and reliable knowledge was obtained about this inaccessible region of the Earth.

It is important to note that by the middle of the 20th century, international customs had developed in the Arctic, which regulated the development of this territory and the procedure for scientific research was established in Russian legislation.

Since 1982, a new stage in the development of the legal regime in the Arctic has begun. It is associated with the adoption of the UN Convention on the Law of the Sea. This document established a universal legal regime for maritime spaces, including the basic principles of the implementation of human activities in the Arctic maritime spaces. The Russian Federation confirmed this international treaty on February 6, 1997.

2 Results

The 1982 Convention on the Law of the Sea has significantly expanded the range of subjects of international law that may have scientific interests in the Arctic and the Arctic Ocean. Any state has the right to take part in scientific research of maritime spaces (for example, on the basis of Articles 238 and 254 of the Convention).

Nowadays in addition to the Arctic states, international organizations are actively involved in scientific research in the Arctic. For example, the Arctic Reference Observation Network, the International Arctic Science Committee (as an organization specially created in 1990 to enable scientists from all over the world to work together regardless of nationality), University of the Arctic, Forum of Arctic Research Operators, International Network for Ground Research and Monitoring in the Arctic, World Meteorological Organization, International Council for the Exploration of the Sea, Pacific Arctic Group, Arctic Council, Northern Forum, etc.

For the first time in the history, all states received an undeniable right to conduct marine scientific research, regardless of their geographical location (Article 238 "The right to conduct marine scientific research"). Moreover, the Convention established general principles for the conduct of marine scientific research. The important principle is that they are conducted exclusively for peaceful purposes. It is necessary to take into account that "marine scientific research activities do not create a legal basis for any claims on any part of the marine environment or its resources" (Article 241 "Non-recognition of marine scientific research as a legal basis for claims"). This provision is a significant limitation and radically changed the situation in the Arctic region.

The Convention has a special section 2 of part XIII devoted to international cooperation. According to the Convention, "states and competent international

*Corresponding author: omaksimova@yandex.ru

organizations cooperate by bilateral and multilateral agreements in the creation of favorable conditions for marine scientific research in the marine environment and in the combination of the efforts of scientists in the study of the essence of phenomena and processes occurring in the marine environment and the relationship between them (Article 243 "Creation of favorable conditions")".

The important rule is enshrined in Article 246 "Marine scientific research in the exclusive economic zone and the continental shelf". Paragraph 3 of this article establishes that "the coastal states, under normal circumstances, give their consent to the implementation of marine scientific research projects by other states or competent international organizations in their exclusive economic zone or on their continental shelf in accordance with this Convention exclusively for peaceful purposes and to enhance scientific knowledge about the marine environment for the benefit of all mankind. To this end, coastal states shall establish the rules and procedures in order to ensure that there is no undue delay or refusal of such consent. "

"One of the important issues in the field of the use of marine spaces of marine resources for Russia is the legal regulation of the Northern Sea Route, which mainly runs through the seas of the Arctic Ocean, in the Arctic zone, and is a route passing through the historical sea space of Russia" [2, 94]. It is necessary to note that for the regulation of the water area of the Arctic Ocean and its seas, the Convention provides the section 8 "Ice-covered areas".

There are a large number of international treaties in the world related to the Arctic (today there are more than a hundred). However there is still no general international treaty on the Arctic or any special rules, international legislative acts on the compliance with the environmental safety measures [3]. Therefore, the above mentioned 1982 UN Convention on the Law of the Sea remains the most important document.

The various aspects of the legal regime in the Arctic are regulated by a number of international regional agreements. These include: Arctic Environment Protection Strategy 1991, Kirkenes Declaration 1993, Nuuk Declaration on Environment and Development in the Arctic 1993, Tromsø Declaration 1993, Declaration on the Establishment of the Arctic Council 1996, Ilulissat Declaration of five states coastal to the Arctic Ocean 2008, Kiruna Declaration 2013, Iqaluit Declaration 2015 and others.

In addition to these documents, which are important for the entire Arctic region, a number of bilateral agreements between countries are distinguished, such as the Treaty between the Russian Federation and the Kingdom of Norway on the delimitation of maritime spaces and cooperation in the Barents Sea and the Arctic Ocean in 2011 and many other similar agreements.

Eight Arctic governments signed the Agreement on the Expansion of International Arctic Scientific Cooperation on May 11, 2017 at a meeting organized in the framework of the Arctic Council in Fairbanks, Alaska. This agreement entered into force on May 23, 2018 [4]. The purpose of the Agreement was to expand cooperation in scientific activities in order to increase efficiency and effectiveness in the development of scientific knowledge about the Arctic. The agreement facilitates the access of scientists from Arctic governments to the Arctic territories that are defined by each government, including the entry and exit of people, equipment and materials, access to research infrastructure, facilities and data. The agreement also encourages the parties to promote education, career development, training opportunities and activities related to traditional and local knowledge.

However, this Agreement is essentially a framework document, which implies the achievement of individual agreements within the framework of scientific cooperation in the Arctic if necessary. At the same time, the achieved special agreements should not contradict the norms of international law, as well as the laws, bylaws and procedures of the coastal states.

At the same time, the Agreement, encouraging the member states to make "maximum efforts" to facilitate the entry and exit of people with all the equipment necessary for research, does not imply the obligatory introduction of special documents which allow free border crossing. Article 15 of the Agreement on the resolution of any dispute through direct negotiations is also considered imperfect, since it does not offer a clear solution to the problem if the parties involved in conflict can not come to a compromise during the negotiations.

The Agreement covers the territories designated by the member states of the Arctic Council themselves. At the same time, it is emphasized that the mention of these territories does not mean the recognition of maritime rights or the delimitation of borders between states, which, in turn, does not lead to the resolution of territorial claims in this region.

Summing up the study of this Agreement, the following conclusions can be drawn. Firstly, this agreement testifies to the intensification of the authority of the Arctic Council, under the auspices of which it was prepared, since, despite geopolitical contradictions, it managed to join the efforts of interested states. Secondly, the Agreement brought together not only the member states of the Council, but also all interested countries, which is important for the development of Arctic science. Thirdly, we observe that there is a gradual movement towards the unification of the Arctic region. This indicates the possibility of the transformation of the Arctic region into a demilitarized and exclusively used for peaceful purposes territory by the analogy with the Antarctic Treaty of 1959.

Article 1 "Terms and definitions" of the above mentioned Agreement of May 11, 2017 defines scientific activity as: "the efforts to improve understanding of the Arctic through scientific research, monitoring and assessment. Such activities may include, but are not limited to, the following activities: planning and implementation of research projects and programs, expeditions, observations, monitoring initiatives, surveys, modeling and assessments; staff training; planning, organizing and conducting scientific seminars, symposia, conferences, workshops and meetings; collection, processing, analysis and exchange of scientific data, ideas, results, methods, experience, as well as traditional and local knowledge; development of sampling methodologies and protocols; preparation of publications; as well as the development, implementation and use of logistics support for research activities and research infrastructure".

This Agreement provides a fairly impressive list of scientific activities. However, not all of these activities are actively developing. International scientific expeditions seem to be a more complicated type of cooperation that requires serious organizational and material investments.

The study of international treaties and agreements on Arctic issues allows concluding that the implementation of international cooperation in the Arctic in general and scientific cooperation in particular, is associated with a system of international treaties, which present a multi-level hierarchical system.

1. The upper level, represented by the 1982 Convention on the Law of the Sea, defines the belonging of the Arctic Ocean to certain countries and opens up the possibility for non-Arctic countries to conduct various activities in open neutral waters. On the one hand, the inaccessible areas of the subpolar part of the ocean do not contribute to efficient economic activities. On the other hand, it is of undoubted interest for scientific research and environmental monitoring.

2. The middle level is represented by the most important document regulating scientific activities in the Arctic. This is an Agreement signed in 2017 by the governments of 8 Arctic states in order to intensify international Arctic scientific cooperation. The purpose of this document is "to intensify cooperation in the field of scientific activities in order to increase the effectiveness and efficiency of the development of scientific knowledge about the Arctic". However, the main achievement of the Agreement is the opening of access to research areas.

3. The lower, practical level is represented by direct agreements between states, regional governments, international organizations, as well as scientific organizations, which define the Arctic territories as territories for scientific research, or the subject of these scientific studies is the issues related to the Arctic.

This existing system of agreements provides the opportunity for scientific organizations from any country in the world to be involved in scientific cooperation, and even become the initiator of joint scientific projects. Scientific cooperation in the Arctic also face the same problems that are typical of cooperation in other areas in this region: "international cooperation in the Arctic is characterized by some uncertainty, approbation of conflicting principles and a lack of informed decisions on interaction with foreign partners" [5, 42].

The main directions of scientific research and problematic issues can be determined on the basis of documents and materials that are provided by participants in scientific events in the Arctic. Thus, from September 2019 to September 2020, the biggest-ever in the history Arctic expedition Mosaic was organized. The main purpose of which was to get a fundamental understanding of the Arctic as the epicenter of global warming and its impact on climate change. The scientists from 17 countries participated in the project, including Russia, Germany, France, USA and China. The research results were delivered to 80 research institutes. The mission led to unprecedented measurements of what happens in the dark Arctic months to the atmosphere, ocean and ecosystem and sea ice.

While the biggest part of this data remains to be analyzed, early results indicate that the atmosphere of unexplored wilderness with temperature layers is so high that the air at the top of the ship was sometimes 5°C warmer than the one closest to the ice. In the ocean, the researchers have found a flourishing ecosystem in dark waters that they believe will be relatively calm.

Scientists are confident that the Arctic ice is disappearing. The shrinking thickness of ice is accelerating global warming. As glaciers in Greenland and other Arctic glaciers lose ice, they are contributing to sea level rise, potentially exposing millions of people to flooding. Nearly all of the dramatic, high-profile impacts of climate change, from alarming coastal erosion to intense and frequent fires, are already happening in the Arctic at a rapid pace.

These expeditions will be included in climate models that will help scientists understand how life in the Arctic will change as it warms and how quickly the rest of the Earth heats up [6].

The most popular and successful type of international scientific cooperation in the Arctic is to plan, organize and conduct scientific seminars, symposia and conferences.

An illustrative and successful example of an international scientific conference on the problems of the Arctic is the Northern Forum on Sustainable Development, which was held in Yakutsk from the 24th to the 28th September, 2019. The representatives of eight circumpolar states (Russia, Canada, USA, Iceland, Finland, Norway, Sweden, Denmark) and the countries of Northeast Asia (China, South Korea and Japan) attended this event, as well as a delegation of the Ministry for the Development of the Far East and the Arctic of the Russian Federation, Far East agencies for attraction of investments and export support, Nenets and Khanty-Mansi Autonomous District, Kamchatka and Khabarovsk Territories.

The Northern Forum on Sustainable Development was held for the first time in Yakutsk in 2019. A total of 111 guests came to the forum, including 51 foreign guests from 14 countries including the circumpolar ones (Iceland, Canada, USA, Finland, Norway, Russia) and the countries of Northeast Asia (China, South Korea and Japan), and also from Germany, France, Great Britain, Lithuania, Switzerland.

The main organizers of the Northern Forum on Sustainable Development were the International Non-Governmental Organization of the Northern Regions "Northern Forum", the Government of the Republic of Sakha (Yakutia), North-Eastern Federal University named after M.K. Ammosov.

The co-organizers of the event were: Project Office for the Development of the Arctic (Moscow), Russian International Affairs Council (Moscow), St. Petersburg State University (St. Petersburg), Surgut State Pedagogical University, Siberian Federal University (Krasnoyarsk).

The following international organizations took part in the organization of the Northern Forum on Sustainable Development in 2019: International Arctic Science Committee (IASC), Arctic University, Institute of Circumpolar Reindeer Husbandry of Arctic University, International Reindeer Center, Association of "Reindeer Herders of the World", Education Committee of the International Association of permafrost studies, Association of Mayors of the World Winter Cities (Sapporo, Japan). The scientific program of this conference was rich in various events [7]. Scientists from Russia and China discussed cooperation in the Arctic at a round table at the Northern Forum on Sustainable Development and bilateral cooperation in the Arctic. The most interesting discussion at the round table was the discussion of the issues of institutional cooperation and the legal framework for future partnership".

The International UNESCO Chair of the North-Eastern Federal University presented an international project on digitalization of the linguistic and cultural heritage of the indigenous peoples of the North and the Arctic on September 27, at the project session of the Northern Forum on Sustainable Development.

During this event many scientific topics and developments related to the problems of the Arctic were covered. The Forum brought together an authoritative representation of scientists from academic institutions and universities. Thus, the Northern Forum on Sustainable Development, held in 2019, served as a platform to establish international scientific ties between Russian scientists and their foreign colleagues. Thus, the scientists from the University of Laval (Quebec, Canada), University of Northern British Columbia (Prince George, Canada), Oceanological University of China (Qingdao, China), Shanghai Institute of International Studies (Shanghai, China), Arctic University of Norway (Tromsø, Norway), Arctic Center of the University of Lapland (Rovaniemi, Finland) attended this forum. Moreover, it is necessary to note that "Russia is open to international cooperation in the Arctic; however, on the other hand, the national security of its northern borders is the most important priority of its policy" [8, 281].

Let us mention another major event as an example of international cooperation in the format of a scientific forum. The 5th International Arctic Forum "The Arctic - Territory of Dialogue" was on April 9-10, 2019 in St. Petersburg. More than 3,600 people from 52 states, including China, Norway, Finland, Sweden, USA, Denmark, Iceland, Canada and Japan attended this event.

The participants discussed the issues of the Northern Sea Route (NSR), environmental preservation, and rational use of natural resources, interaction in security provision, sustainable development of indigenous peoples of the North and the development of international cooperation. The heads of state in their speeches emphasized the significant role of the Arctic Council in the solution of these issues.

The proposal announced at the forum to create a scientific and educational center in the Arctic, which will ensure the development of fundamental science and solve practical problems of the development of the Arctic, also seems promising. The joint development of the educational standard for this center by the countries participating in the Arctic Council will allow training highly qualified Arctic specialists for international projects [9].

3 Discussion

When we talk about the achievements of Russian scientific expeditions that managed to reach different territories including the Pacific coast and begin the development of northeastern Russia, it is necessary to mention the contribution of the indigenous peoples of Siberia. Thus, the Yakuts and other peoples of Eastern Siberia in the 18th - 19th centuries rendered invaluable assistance in the transportation of goods for expeditions on Russian Arctic development, providing their horses for transportation. The details about this contribution of the Siberian peoples are described in the work of M.M. Fedorov "The legal status of the peoples of Eastern Siberia (the 17th-early 19th century)" [10].

In our opinion, the Russian model of scientific cooperation in the Arctic should take into account the priorities of state policy in the Arctic, the benefits of direct interaction with scientists from other countries, the possibilities of international organizations in which Russia and its constituent entities participate. The priority tasks of the "Strategy for the development of the Arctic zone of the Russian Federation and provision of national security for the period up to 2035" are defined the tasks of social development [11]. Consequently, the priority topics of scientific research in the Arctic should primarily include the issues of socio-economic development.

International scientific cooperation should be based on the principle of priority of the interests of the Arctic states. A timely decision is the creation of the Scientific Arctic Council under the Government of the Russian Federation, which determines the priorities and topics of scientific research in the region, including those carried out in cooperation with international scientists. It is necessary to coordinate the activities of more than 500 organizations in 50 regions conducting scientific research on Arctic topics [12].

The problem of the contradictory nature of the norms of international agreements and the norms of national legislation is a problem of legal science. Today Russia is developing modern legal regulation for a hard-to-reach territory, such as the Arctic. A problematic issue is the question of the definition of rights and their guarantees for legal entities in the Arctic zone. According to Article 2 of the International Covenant on Economic, Social and Cultural Rights "the rights reflected in this Covenant will be exercised without any discrimination, such as race, color, sex, language, religion, political or other beliefs, national or social origin, property status, birth or other circumstance". Thus, the provision of additional guarantees and expansion of the scope of rights to individual subjects of law may be in conflict with an international treaty.

Some proposals and solutions developed by the scientific community on the problems of legal regulation in the Arctic can be found in the final documents of major scientific conferences [13].

Conclusions

1) With the adoption of the 1982 convention, the scientific activity of scientists from different countries has increased. Especially the activity on the study of the Arctic by scientists from non-Arctic countries has increased, for example, China, Japan and South Korea.
2) There are certain scientific traditions and customs in the scientific study of the Arctic, but they are not sufficiently taken into account by modern international law, which often provokes the contradictions between participants in international relations.
3) The regulatory framework for international cooperation in the Arctic is not sufficiently developed and analyzed by national legal scholars and requires a more thorough study in order to develop an optimal model for the development of the Russian Arctic, including the development and implementation of priority areas of scientific research.
4) The clear priorities in the definition of the goals of scientific research and reduction of secondary topics will allow Russian scientific organizations to enter into contractual relations in the field of scientific cooperation in the Arctic, make them more effective, give an opportunity to abandon topics which are not useful or contradictory in terms of the national interests of Russia.
5) Scientific and scientific-practical conferences are one of the instruments for the development of contractual relations in the field of international scientific cooperation in the Arctic. In particular, it is the Northern Forum on Sustainable Development being held in Yakutsk. It should be assumed that the role of conferences and other formats of scientific interaction are not to define or approve the agenda of international scientific research, but to create an atmosphere of trust between scientists, assess the capabilities and style of work, develop a common scientific discourse and conceptual apparatus.
6) There is no doubt that the content of reports and scientific discussion should become the primary material for the formation of proposals on priority topics of scientific research, which should be evaluated and implemented at the state level.

References

1. Pasetskiy V.M. *Russian discoveries in the Arctic: in 2 parts.* (St. Petersburg: Admiralty, 2000)
2. Maksimova O.D. *Global Dialogue on Sustainable Development Goals: Legal Dimension (to the 75th Anniversary of the UN)*, p. 94-98 (Moscow, 2020)
3. The final public resolution on the socio-economic development of the Arctic zone of the Russian Federation for 2019, *IX International Forum "Arctic: Present and Future.* (SPb., 2019) Retrieved from: http://www.forumarctic.com/upload/conf2019/resolution/Rezolution-2019.pdf
4. *Agreement on the expansion of international Arctic scientific cooperation in 2017* Retrieved from: https://docs.cntd.ru/document/456059402
5. Chumakov D.S. *Bulletin of Moscow University, 25. International relations and world politics.* 2, 41-61, (2011)
6. *MOSAiC Multidisciplinary drifting Observatory for the Study of Arctic Climate Science Plan* Retrieved from: MOSAiC Expedition (mosaic-expedition.org)
7. Final report for 2019. *Northern Forum on Sustainable Development,* September 24-28, Retrieved from:: https://nsdf.ru/report/
8. Nikolaev A.N., Maksimova D.D., Kugunurov V.V. *University - Intellectual, Innovative and Spiritual Platform for Sustainable Development of the Macroregion.* (Yakutsk, 2021)
9. *Results of the International Arctic Forum "The Arctic - Territory of Dialogue"* Retrieved from: https://roscongress.org/upload/medialibrary/f14/itogi-v-mezhdunarodnogo-arkticheskogo-foruma_rus.pdf
10. Fedorov M.M. *Legal status of the peoples of Eastern Siberia (XVII - early XIX century).* (Yakutsk, 1978).
11. Decree of the President of the Russian Federation of October 26, 2 No. 645 *"On the Strategy for the Development of the Arctic Zone of the Russian Federation and Ensuring National Security for the Period until 2035"* (2020) Retrieved from: https://www.garant.ru/products/ipo/prime/doc_/74710556/
12. Trutnev Y. Meeting on the organization of research and development in the Arctic zone Retrieved from: http://government.ru/news/40437/
13. Final public resolution on the socio-economic development of the Arctic zone of the Russian Federation for 2020, *XI International Forum "Arctic: Present and Future"* Retrieved from: http://www.forumarctic.com/conf2020/

Factors of formation and development of the transport network in areas inhabited by indigenous small-numbered peoples of the North in Yakutia

Antonina Savvinova[1,*], *Viktoriya* Filippova[1,3], *Alexey* Borisov[2], *Galina* Gnatyuk[1] and *Alla* Fedorova[1]

[1]M.K. Ammosov North-Eastern Federal University, Institute of Natural Sciences, 58, Belinskogo str., Yakutsk, 677000 Russia
[2]M.K. Ammosov North-Eastern Federal University, Mining Institute, 58, Belinskogo str., Yakutsk, 677000 Russia
[3]Institute for Humanities Research and Indigenous Studies of the North, Siberian branch, Russian Academy of Sciences, 1 Petrovskogo str., Yakutsk, 677027, Russia

Abstract. The paper considers the natural, historical, economic and geographical factors of the formation and development of the transport network in the areas inhabited by Indigenous Small-Numbered Peoples of the North in Yakutia. A brief analysis of historical periods of transport network development in Yakutia was carried out. A retrospective statistical analysis of the indicators of the road network development and its access for the areas inhabited by Indigenous Small-Numbered Peoples of the North: the length and density of public roads, the coefficients for assessing the level of the road network and transport was chosen as the study method. The analysis was based on the Engel, Goltz, Uspensky coefficients, as well as the density indicator of the road network as key indicators. Taking into account all coefficients the ranking of areas inhabited by Indigenous Small-Numbered Peoples of the North showed that high level of transport security of the population is typical for the areas with industrial development and a well-developed road network – Neryungrinsky, Mirninsky and Aldansky. The average level is typical for most areas that have year-round regional roads and winter roads. The lowest level of transport security of the population is observed in the Allaikhovsky and Oleneksky districts.

1 Introduction

The development of transport communication, active construction of roads, improvement of transport infrastructure, especially in developed territories, almost erased the borders between the regions. Economic, cultural, social communications are established between them that have a positive economic and social effect: formation of a single economic space, intensification of trade and steady growth of interregional trade, development of the labor market, etc. There is insufficient scientific knowledge of the impact of the state of the transport network on the socio-economic situation of the territories of residence and traditional nature management of the Indigenous Small-Numbered Peoples of the North (ISNPN). In remote regions of the North with a poorly developed transport network and great distances, there are still problems in the development of transport communications. The current structure of the transport network in areas inhabited by Indigenous Small-Numbered Peoples of the North in Yakutia is associated with natural and climatic conditions, with the history of settlement and the type of economic use of the territory.

The places of residence of the ISNPN of Yakutia include the geographical zones of the tundra, forest tundra and the northeastern part of the forest zone. With the exception of small areas in the southwest, almost the entire territory is in the zone of continuous permafrost. The vast territory of the region represents a combination of both latitudinal-zonal and highland landscapes. The permafrost landscape map of Yakutia shows 143 different combinations of permafrost landscapes. Tundra landscapes occupy 10.3%, northern taiga – 26.4%, middle taiga – 27.5%, mountain deserts – 1.9%, mountain tundra – 9.4%, subalpine shrubs – 2.6%, mountain sparse wood – 20.3%, mountain taiga – 1.5% of the entire territory of Yakutia. In general, flat permafrost landscapes occupy about 64.2%, and mountain landscapes – about 35.4%, water bodies – 0.4% [1].

The studying areas are represented by all relief forms: mountains, plateaus, intermountain basins and lowlands. A significant area of the territories is occupied by plateaus and highlands – the relief is characterized by a combination of flat hills, intermountain depressions and low ridges being the habitats of commercial fur-bearing animal species for hunting, and is suitable for the development of taiga reindeer herding. The coastal territories of Yakutia are bordered by lowlands, which for a significant length retain a uniform flat-hilly relief with low absolute heights, with a weak inclination to the sea – they are the most favorable for the development of tundra reindeer herding.

All major rivers – Lena, Yana, Indigirka, Kolyma, Olenek, Anabar – are used for river transport during

* Corresponding author: sava_73@mail.ru

navigation, along which most of the cargo for the northern regions is imported. They are characterized by high spring flood, annually repeating small summer and autumn floods and long and low water season [2]. Rivers and lakes are abundant in fish and contribute to the development of one of the traditional economic activities of the ISNPN – fishing. The climate in most of the territory belongs to the Arctic and subarctic continental climatic zone, the main characteristics of which are dryness and significant fluctuations in both daily and seasonal temperatures. Thus, the natural conditions of the places of residence of the ISNPN in the Republic of Sakha (Yakutia) are characterized by extreme climatic conditions, the predominance of tundra, forest tundra and taiga characterized by wide biodiversity.

2 Factors of transport network formation in Yakutia

The development of the vast territory of Yakutia and related territories by indigenous peoples took place in different historical periods. Each people that inhabited the region developed their own system of transport link, which was determined not only by the presence of official roads, paths, but also by the way of life and economic activities of the population. If we analyze the history of the development of the transport network in Yakutia as a whole, then generally accepted periods are distinguished: pre-Soviet, Soviet and post-Soviet.

Before joining Russia Yakutia had no regular routes of communication. In the 17th century, waterways prevailed with the construction of strongholds from the Lena River with access to the Pacific coast. Pathbreakers walked along the river, animal transport also began from the rivers, and sledding was laid along the rivers in winter. The main cargo transportation was carried out by waterways. In 1738, the Irkutsk-Yakutsk tract was established in the system of ways of communication, which was laid along the left bank of the Lena River from Vitim to Yakutsk in completely uninhabited places. For several centuries the route from Irkutsk Region remained the main transport axis for the delivery of goods to the region. During the Second Kamchatka Expedition under the leadership of Captain-Commander V.I. Bering, there was a need for stable communication between Yakutsk and the Port of Okhotsk. From 1650 to 1720, the Yakut-Okhotsk route of water-land communication operated, which ran through the places of residence of the Evenks. Since 1845, the connection of the Ayan port with Yakutsk became regular and since 1852 the Yakutsk-Ayan tract, with a length of 1120 versts, becomes operational year-round [3].

In the Soviet period the rapid growth of the gold mining industry in the south of the republic began from the 1920s. At that time, the main population of the regions of Southern Yakutia were the Evenks engaged in traditional types of husbandry. Gold mining and the construction of the year-round Amur-Yakutsk highway (Bolshoy Never-Yakutsk) turned the southern part of the republic into one of the major centers of the country's mining industry. In 1932, the construction of the Yakutsk-Magadan highway began to ensure gold mining in the north-east of Yakutia. In parallel, the development of the Northern Sea Route was underway, which subsequently played a positive role in ensuring the northern import to ISNPN residences in the Arctic regions of Yakutia. Continuous air communication was established both between district centers within the republic center and with large cities of the USSR. In the 1950s diamond-bearing deposits were developed in Western Yakutia in the area of the Mirny city, and a transport hub was formed for the development of a new industry, roads to mining sites appeared. A large supply base was created in the city of Lensk, conveniently located on the main river highway of the Lena River. The diamond-producing region (Mirny, Aikhal, Udachny, Chernyshevsky) with a new transport supply base was connected by a highway stretching about 400 km.

The establishment of the South Yakutsk Territorial-Production Complex (TPC) led to the construction of the Baikal-Amur Mainline – Tynda – Berkakit railway line in the 1970s, which connected Yakutia with the Baikal-Amur and Trans-Siberian main lines.

Today, the transport network not only provides the need for the economy and the population in transportation, but also serves as a material basis for the socio-economic development of the region. The main volume of goods is transported by road transportation, which occupies 45% and by inland water transport in the short navigation period, the share of which in cargo turnover is 44%, railway – 10%, others – 1% [4].

The Republic is characterized by the weak development of the road network, the territorial isolation of individual roads, as well as the significant development of the departmental road network, which actually performs the functions of the public road network. The road network in the region has a length of over 30 thousand km, of which more than half are temporary (seasonal) roads – winter roads. To date, more than 300 settlements of the republic have not been provided with year-round access to the public network with paved roads. Only 16 out of 34 districts are provided with such a transport link. Air transport remains the main year-round passenger mode of transport and plays a huge role in the socio-economic and cultural development of the republic. Aviation accounts for 84% of passenger turnover. On July 27, 2019, the first passenger train in the history arrived at Nizhny Bestyakh station; it opened traffic on the Amur-Yakut Railway throughout its length. Infrastructure facilities and a railway station at the Nizhny Bestyakh station were built. This country's largest railway construction in the last 25 years is a key project for the creation of a support transport network in the region and the main strategic object of transport construction. Thus, the socio-economic development of the vast territory of Yakutia is largely determined by the state and effectiveness of the transport system, which plays an exceptional role in ensuring its life, creates an uninterrupted supply of strategically important sectors of the economy.

https://doi.org/10.1051/shsconf/202111200052

The settlement of the population of Yakutia is characterized by high dispersion and a large number of scattered rural settlements with low population. Transport accessibility of settlements plays an important role in the settlement pattern of the population in the areas of residence of the SIPN. The historically established settlement system, primarily the scarcity of people and the dispersion of settlements, low density, branching and poor quality of local roads, have largely determined the territorial organization of the population, the level of socio-economic development and the mobility of the population of the studied territories.

In the conditions of the industrialization policy pursued in the Soviet period, a new settlement framework was formed laying the basis for various settlements with different size and functions, which were dispersed throughout Yakutia, and the number of which was supported by state-stimulated migrations (northern allowances, Komsomol construction projects, etc.). The spatial location of these settlements was characterized by high fragmentation, which in many ways was determined by their monofunctionality (mining centers), as well as the actual underdevelopment of the transport infrastructure in the territory.

3 Current state of the transport network in the places of residence of ISNPN

According to the Order of the Government of May 8, 2009 No. 631-p, 21 districts with rural settlements (naslegs) in Yakutia are included in the list of places of traditional residence and traditional economic activity of Indigenous Small-Numbered Peoples of the North of the Russian Federation [5]. In total, according to the Law "On the List of Inaccessible and Remote Areas in the Republic of Sakha (Yakutia)" (as amended by the Law of the Republic of Sakha (Yakutia) No. 1249-V of 25.04.2017), 185 settlements and 163 production sites in Yakutia belong to hard-to-reach areas [6]. Areas inhabited by Indigenous Small-Numbered Peoples of the North are the most inaccessible territories of the Republic of Sakha (Yakutia). Among 18 districts of Indigenous Small-Numbered Peoples of the North (except for Mirninsky, Neryungrinsky and Ust-Maysky districts), 55 settlements are considered as hard-to-reach areas.

The length of public and non-public roads in 21 areas inhabited by Indigenous Small-Numbered Peoples of the North is 19668 km. The length of paved roads in 11 districts is less than 100 km, in 9 districts – from 100 to 1000 km, and only in the Mirninsky district the length of such roads exceeds 1000 km (1,037 km). In 4 districts, the share of paved roads is more than 50% – Aldansky, Mirninsky, Oymyakonsky and Tomponsky. But this does not mean that paved roads are available to all settlements where the Indigenous Small-Numbered Peoples of the North live. Paved roads mainly pass to district centers. Thus, most of the places of residence of the Indigenous Small-Numbered Peoples of the North do not have a year-round connection and are transport-isolated. Ground and freight transportation are available only through winter roads running along rivers and passages that operate from the end of December to mid-April.

If the density of the network of paved public roads in the Republic of Sakha (Yakutia) by the end of 2020 was 4.2 km per 1000 km^2 territory, then in 21 areas inhabited by Indigenous Small-Numbered Peoples of the North the average value is 0.8 km per 1000 km^2. The density of the road network is associated with the level of economic development of the districts. In 9 out of 21 districts the road density is from 2 to 10 km per 1000 km^2, and in 12 districts – from 10 to 15 km per 1000 km^2.

The average density of public and non-public roads per 100 km^2 of the ISNPN area is 1 km per 100 km^2. In 7 districts the density is less than 1 km per 100 km^2, in 12 districts – 1.0-1.5 km per 100 $km^2$and in only 2 districts – 1.5-1.6 km per 100 km^2.

The Engel, Goltz, Uspensky and Vasilevsky coefficients, as well as the density indicator of the road network, were used as key indicators to assess the level of provision of the studied territories with a road network and transportation. The Engel coefficient allows obtaining a generalized estimate of the region's security of the transport network – this is the ratio of the length of transport routes to the developed area and the population. However, the Engel coefficient may slightly distort the level of transport development of these territories, given that the population of different districts, regions and countries may vary significantly. Therefore, the Goltz coefficient was also calculated, which shows a more verified picture of the level of transport development showing the level of security of the transport network between settlements [7].

In order to fully assess the development of the road network of a particular district and its constituent municipalities it is not enough to use only the Engel and Goltz coefficients, despite the fact that they allow taking into account two important factors that especially affect the state of the studied object – the area and the number of settlements. Both indicators have one significant drawback: inability to assess the level of transportation of the territory under study. Other indicators are already coping with this task – Uspensky and Vasilevsky coefficients [8].

The Engel, Goltz, and Uspensky coefficients do not have normative values, and the higher the value of these coefficients, the more the region is provided with transport infrastructure. These indicators are designed to build rankings for the provision of territories with transport infrastructure [9, 10].

4 Conclusion

Considering that some of these coefficients show a slightly distorted perspective of the transport development in the territories, we believe that in the conditions of Yakutia it is most optimal to use the aggregate coefficient, which gives a more approximate scenario of the level of security of the studied territories by the transport network (Table 1).

Table 1. level of provision of the studied territories by the transport network

Region	Engel coefficient-1 (Ke-1)	Goltz coefficient-1 (Kg-1).	Uspensky coefficient-1 (Ku-1)	Sum of coefficients	Ranking
Mirninsky	0.013	47.370	0.042	47.425	1
Neryungrinsky	0.014	78.422	0.066	78.502	1
Aldansky	0.019	22.880	0.090	22.989	1
Tomponsky	0.038	9.265	0.197	9.500	2
Anabarsky	0.025	8.734	0.167	8.926	2
Olekminsky	0.038	8.448	0.230	8.716	2
Kobyaysky	0.041	7.893	0.265	8.199	2
Oymyakonsky	0.042	7.007	0.205	7.254	2
Ust-Yansky	0.059	6.451	0.416	6.926	2
Verkhnekolymsky	0.029	6.464	0.111	6.604	2
Ust-Maisky	0.048	6.244	0.288	6.580	2
Abyisky	0.047	6.227	0.305	6.579	2
Eveno-Bytantaysky	0.047	6.150	0.372	6.569	2
Bulunsky	0.047	5.577	0.484	6.108	2
Verkhoyansky	0.054	5.757	0.249	6.060	2
Srednekolymsky	0.039	5.472	0.290	5.801	2
Zhigansky	0.017	5.638	0.087	5.742	2
Momsky	0.060	4.760	0.736	5.556	2
Nizhnekolymsky	0.034	4.032	0.273	4.339	2
Oleneksky	0.032	3.610	0.402	4.044	3
Allaikhovsky	0.063	3.134	0.631	3.828	3

Our calculations of the total coefficient and ranking by ISNPN areas of residence in Yakutia showed the following results:

– high level of transport security of the population is typical for the areas of their residence with industrial development – Neryungrinsky, Mirninsky and Aldansky. The railway and the Lena federal highway pass through the territories of the first two districts, and the Vilyuy federal highway passes through the territory of the Mirninsky district;

– average level is typical for most areas that have year-round federal roads, such as the Kolyma road, republican roads: Kobyai, Amga, winter roads: Arctic, Yana, Verkhoyanye, Bulun, Indigir, Sebyan and intra-district winter roads pass through the territories of the districts;

– lowest level of transport security of the population is observed in Allaikhovsky and Oleneksky districts.

Thus, despite the changes towards improving transport accessibility in the republic, most territories, especially Arctic ones, are in the conditions of transport isolation. The development of transport infrastructure is necessary to provide the population with food and industrial goods year-round, stimulate trade, increase the availability of medical and other social services, which will improve the level and quality of life in the places of residence of the Indigenous Small-Numbered Peoples of the North.

Acknowledgements

The study was carried out on the basis of the grant of the Russian Science Foundation (project No. 21-17-00250).

References

1. A.N. Fedorov, N.F.Vasilyev, Y.I. Torgovkin, A.A. Shestakova et al. *Geosciences*, **8**, 465 (2018).
2. A.I. Borisov, G.A. Gnatyuk. *Moscow Economic Journal*, **5 (3)**, 63-75 (2018).
3. P.L. Kazaryan, *Management of the metropolis*, 6, 53-63 (2008).
4. *Results of the operation of the transport complex and road facilities of the Republic of the Sakha (Yakutia)* Retrieved from: https://mintrans.sakha.gov.ru/
5. *List of places of traditional residence and traditional economic activity of Indigenous Small-Numbered Peoples of the North of the Russian Federation according to the Government Order of May 8, 2009 No. 631-p.* Retrieved from: https://www.garant.ru/products/ipo/prime/doc/95535/
6. *On the list of inaccessible and remote areas in the Republic of Sakha (Yakutia)* Retrieved from: https://docs.cntd.ru/document/802010254
7. D.F. Dabiev, U.M. Dabieva *International Journal of Applied and Basic Research*, **11**, 283-284 (2015).

8. S.A. Kozhevnikov *Economic and social changes: facts, trends, forecast,* **12 (6)**, 91-109 (2019).
9. N.V. Volkova, N.V. Svistelnik. *World of Economics and Governance,* **18 (2),** 101-120 (2018).
10. T.I. Tokhirov. *Bulletin of Dagestan State University,* **33 (3)**, 79-83 (2018).

AUTHOR INDEX

AUTHOR INDEX